# Quantum Electronics for Atomic Physics and Telecommunication

# Quantum Electronics for Atomic Physics and Telecommunication

*Second Edition*

Warren Nagourney
*Seattle, Washington*

OXFORD
UNIVERSITY PRESS

**OXFORD**
UNIVERSITY PRESS

Great Clarendon Street, Oxford, OX2 6DP,
United Kingdom

Oxford University Press is a department of the University of Oxford.
It furthers the University's objective of excellence in research, scholarship,
and education by publishing worldwide. Oxford is a registered trade mark of
Oxford University Press in the UK and in certain other countries



First Edition published in 2010
Second Edition published in 2014

Impression: 1



Published in the United States of America by Oxford University Press
198 Madison Avenue, New York, NY 10016, United States of America

British Library Cataloguing in Publication Data
Data available

Library of Congress Control Number: 2013958255

ISBN 978–0–19–966548–8

*To my wife, Patricia.*

# Preface to second edition

This book is based upon a series of lectures which I gave for a graduate level class in quantum electronics in the University of Washington physics department. I assumed a working knowledge of intermediate electromagnetic theory, quantum mechanics and optics. I took a slightly different approach from most books and courses on the subject by deliberately slanting the material so that it would be more relevant to atomic physics experimentalists than, perhaps, to workers in the telecommunication industry.

As a result of my experimental atomic physics orientation, I have included topics not usually found in traditional texts on quantum electronics. An example of this is the application of nonlinear optics to the synthesis of coherent radiation in regions of the spectrum where lasers don't work very well or at all. Most books cover the theory of second-harmonic generation, sum frequency mixing and parametric processes, but avoid discussions of details of a practical frequency synthesis system. We therefore discuss such matters as the optimum cavity geometry for "enhancement" of nonlinear processes, "impedance matching" into the cavity, assessment and correction of cavity astigmatism and practical techniques for mode-matching into the cavity. These issues receive scant attention in most books on quantum electronics.

Another area not discussed in most texts is the frequency stabilization of lasers to cavities and of cavities to lasers. The former situation is encountered when one constructs a frequency standard or seeks to observe an atomic transition over long periods of time. The latter is a necessary condition for the use of a build-up cavity in nonlinear frequency synthesis. Both approaches require some knowledge of control system theory, and a review of this subject is provided. In addition, two important techniques for generating a "discriminant" for frequency locking are discussed.

The remainder of the book covers topics in a manner which is similar to that of other textbooks on quantum electronics. Thus, we provide a fairly conventional discussion of Gaussian beams, "standing-wave" cavities, continuous-wave laser theory, electro-optical and acousto-optical modulation and some nonlinear optics theory. Due to their ubiquity in every atomic physics laboratory, semiconductor diode lasers are discussed in considerable detail.

The following topics have been added in the second edition.

I have included a complete chapter on fiber lasers, which are rapidly becoming the laser of choice for many applications in atomic physics labs where ruggedness, simplicity and frequency stability are important. I have also added a section on the mode-locked fiber laser in the relevant chapter. To prepare the reader for the fiber laser material, an introduction to guided waves in optical fibers and waveguides is provided.

I have briefly described a recently developed technique which is very promising for generating an octave-spanning frequency comb. It uses parametric four-wave mixing in the "whispering gallery modes" of a microresonator. If it is successful, it has the potential to enormously simplify optical-frequency metrology and the realization of an atomic clock from an optical-frequency standard.

The remaining changes are small improvements to the first edition. For example, in order provide greater insight into the optical Bloch equations, I have included a section on the interesting connection between a two-level atom and a quantum-mechanical spin-1/2 system. I have further included some additional plots which will be pedagogically useful in teaching about the optical Bloch equation (Chapter 5). Finally, I have included a section on periodically poled optical waveguides for the enhancement of nonlinear optical processes. This technique should eventually supplant the use of build-up cavities for nonlinear enhancement and is simple, robust and potentially inexpensive.

The inclusion of *telecommunication* in the title to the second edition is mainly due to the addition of a detailed chapter on fiber optics to the new edition. It was felt that this chapter, together with the chapters on semiconductor lasers, provided sufficient material of interest to workers in telecommunication to warrant a title change.

## Acknowledgements

I would like to thank Dr. Eric Burt, Professor Steve Lamoreaux and Professor Ady Arie for reading portions of the manuscript and making helpful suggestions.

# Contents

# 1
# Gaussian beams

## 1.1 Introduction

Atomic physics calculations involving radiative interactions often model the radiation as a plane wave, which is appropriate when studying individual atoms but is a gross oversimplification when considering larger-scale interactions, such as laser gain and nonlinear optics. Studies of the latter sort require a more realistic description of the fields. The *Gaussian beam model* serves this purpose very well: it is simple, elegant and realistic, since it is an excellent approximation to the lowest transverse mode of a laser (and the lowest mode of an optical cavity).

A full description of Gaussian beams will be given, beginning with the use of the *complex beam parameter*, $q$, to completely describe such a beam and ending with the extremely useful $ABCD$ rule, which facilitates the study of Gaussian beams in the presence of optical elements such as lenses, spherical mirrors and crystals for laser gain or nonlinear optics. The approach will be fairly complete but with several slightly more "advanced" topics such as astigmatism and higher-order modes deferred until a later chapter.

## 1.2 The paraxial wave equation

In a region of space distant from currents and charges, it can be shown (using Maxwell's equations) that the six Cartesian components of the electric and magnetic fields satisfy the *scalar wave equation*

$$\nabla^2 u - \frac{1}{c^2}\frac{\partial^2 u}{\partial t^2} = 0, \tag{1.1}$$

where $u$ is any field component ($E_x, E_y, E_z, B_x$, etc.). If we assume that the wave is monochromatic with a time dependence given by $e^{i\omega t}$, then this equation becomes

$$\nabla^2 u + k^2 u = 0, \tag{1.2}$$

where $k$ is the magnitude of the *wave vector* of the wave. We assume at present that the propagation medium is homogeneous and isotropic and can be described by a *refractive index*, $n$, so that

$$k = \frac{\omega n}{c} = \frac{2\pi n}{\lambda}, \tag{1.3}$$

where $\lambda$ is the *vacuum* wavelength of the radiation and $c$ is the speed of light in a vacuum.

The *paraxial assumption* requires that the normals to the wavefronts make a small angle ($\ll 1$ radian) with some fixed axis, which we will take as the $z$-axis of a Cartesian coordinate system in all that follows. We will assume that the beam differs only slightly from a plane wave and can therefore be modeled as a plane wave which is *spatially modulated* by a slowly varying function of the coordinates, $\psi(x, y, z)$,

$$u(x, y, z) = \psi(x, y, z)e^{-ikz}. \tag{1.4}$$

In order for the paraxial condition to be satisfied, $\psi(x, y, z)$ must vary much more slowly along the $z$-axis than it does along the $x$- and $y$-axes. Thus, when substituting eqn 1.4 into the wave equation, $\partial^2\psi/\partial z^2$ can be ignored compared to the other second derivatives and we obtain,

$$\text{Paraxial wave equation:} \quad \nabla_t^2\psi - 2ik\frac{\partial\psi}{\partial z} = 0, \tag{1.5}$$

where $\nabla_t^2$ is the *transverse Laplacian* $(= \frac{\partial^2}{\partial x^2} + \frac{\partial^2}{\partial y^2})$.

## 1.3 Gaussian beam functions and the complex beam parameter, $q$

A particularly simple solution of the paraxial wave equation (eqn 1.5) can be obtained by substituting a *trial solution* of the form,

$$\psi(x, y, z) = \exp\left\{-i\left(P(z) + \frac{k}{2q(z)}r^2\right)\right\}, \quad r^2 = x^2 + y^2, \tag{1.6}$$

where the functions $P(z)$ and $q(z)$ are to be determined. We are looking for cylindrically symmetric solutions at present (hence, $\psi$ depends only upon $r$ and $z$); we will consider solutions with azimuthal dependence in the next chapter when we study higher-order modes. Comparing terms of equal power in $r$, one obtains:

$$\frac{dq(z)}{dz} = 1 \tag{1.7}$$

$$\text{and} \quad \frac{dP(z)}{dz} = -\frac{i}{q(z)}. \tag{1.8}$$

Equation 1.7 is trivially solved, yielding the very important $z$-dependence of $q$:

$$q(z_2) = q(z_1) + (z_2 - z_1). \tag{1.9}$$

where $z_1$ and $z_2$ are any two points on the $z$-axis. The *phase term*, $P(z)$, will be discussed later.

The *complex beam parameter*, $q$, can be written in terms of two real parameters, $R$ and $\omega$, using the following expression (which might appear somewhat arbitrary at first sight):

$$\frac{1}{q} = \frac{1}{R} - i\frac{\lambda}{n\pi\omega^2}. \tag{1.10}$$

The interpretation of the quantities $R$ and $\omega$ becomes clearer when eqn 1.10 is substituted into eqns 1.6 and 1.4,

$$u(x, y, z) = \exp\left\{-i\left(P(z) + kz + k\frac{r^2}{2R}\right) - \frac{r^2}{\omega^2}\right\}, \tag{1.11}$$

where we have used $k = 2\pi n/\lambda$. The field components, $u$, have a Gaussian radial dependence with $e^{-1}$ radius of $\omega$ and a *phase shift* given by the three bracketed terms in eqn 1.11. The interpretation of $R$ as the *wavefront radius* can be seen by considering the $r$-dependence of *surfaces of constant phase* (i.e., wavefronts):

$$z + \frac{r^2}{2R} = const, \tag{1.12}$$

where the term $P(z)$ has been left out since (as will be shown later) it varies with $z$ much more slowly than the $kz$ term. It should be readily evident that eqn 1.12 is an approximation to an equation for a sphere of radius $R$. The approximation is valid when $r \ll R$, which is equivalent to the paraxial assumption. Equation 1.12 contains the *sign convention* for wavefront and surface radii:

**Sign convention:** *When the source or center of curvature is to the left of a wavefront (propagating left to right) or spherical surface, the radius is positive; otherwise it is negative.*

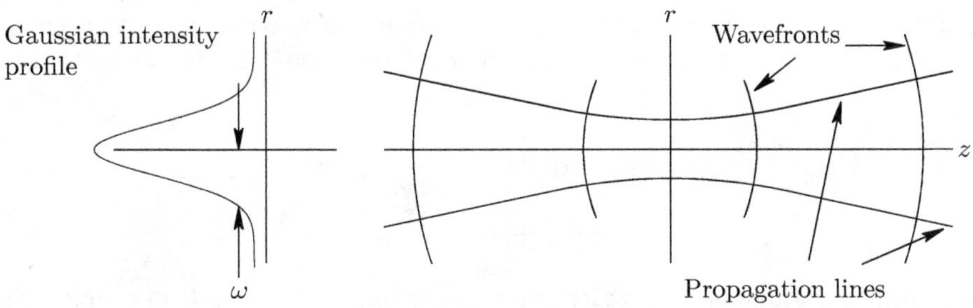


**Fig. 1.1** A Gaussian beam near a focus showing the $e^{-2r^2/\omega^2}$ radial intensity profile and the spherical wavefronts.

In summary, we have found one solution of the source-free wave equation in the paraxial approximation and shown that it has a Gaussian radial dependence (i.e., it is a *Gaussian beam*) and that all of the properties of the Gaussian beam are contained in a *complex beam parameter*, $q$, which depends upon the beam radius, $\omega$, and wavefront radius, $R$, and transforms very simply with $z$ as given by eqn 1.9.

## 1.4 Some Gaussian beam properties

Figure 1.1 illustrates the focusing property of a Gaussian beam: it does not come to a *focus* at a point but rather achieves a *minimum spot size* (called a *waist*, whose radius is $\omega_0$) when the wavefront becomes a plane (infinite radius). The corresponding value of $q$ at a waist is found by setting $R = \infty$ in eqn 1.10:

$$\text{At waist:} \quad q \equiv q_0 = i\frac{n\pi\omega_0^2}{\lambda}. \tag{1.13}$$

If $z$ is the distance from the waist, the other values of $q$ are (from eqn 1.9):

$$q(z) = q_0 + z = i\frac{n\pi\omega_0^2}{\lambda} + z. \tag{1.14}$$

From this, we can obtain the very useful results that, given an arbitrary $q(z)$,

$$\text{Distance to waist} = -Re\{q(z)\} \quad \text{and,} \tag{1.15}$$

$$\text{Radius of waist} = \sqrt{\frac{\lambda}{n\pi} Im\{q(z)\}}. \tag{1.16}$$

Using eqn 1.14 and eqn 1.10, we can obtain $\omega$ and $R$ as a function of the distance ($z$) from a waist:

$$\omega(z) = \omega_0 \left[1 + \left(\frac{\lambda z}{n\pi\omega_0^2}\right)^2\right]^{1/2} \tag{1.17}$$

$$R(z) = z\left[1 + \left(\frac{n\pi\omega_0^2}{\lambda z}\right)^2\right]. \tag{1.18}$$

Equation 1.17 clearly shows the hyperbolic shape of the beam contour and allows one to define the *Rayleigh length* as the distance from the waist to the place where the spot size increases by a factor of $\sqrt{2}$ and the *confocal parameter* as twice the Rayleigh length:

$$\text{Rayleigh length} \equiv z_R = \frac{n\pi\omega_0^2}{\lambda} \tag{1.19}$$

$$\text{Confocal parameter} \equiv b = \frac{2n\pi\omega_0^2}{\lambda}. \tag{1.20}$$

These definitions allow us to rewrite the $z$-dependence of $\omega$ and $R$ in a much simpler form:

$$\omega(z) = \omega_0 \left[1 + \left(\frac{z}{z_R}\right)^2\right]^{1/2} \tag{1.21}$$

$$R(z) = z + \frac{z_R^2}{z}. \tag{1.22}$$

The Rayleigh length and confocal parameter are illustrated in Fig. 1.2.

It is often useful to work with $1/q$ which we will *define* as $q_1$. We can then easily show that the distance to the waist, the Rayleigh length and the waist size can be written as:

$$\text{Distance to waist:} \quad z = -\frac{Re\{q_1\}}{|q_1|^2} \qquad \left(q_1 \equiv \frac{1}{q}\right) \tag{1.23}$$

$$\text{Rayleigh length:} \quad z_R = -\frac{Im\{q_1\}}{|q_1|^2} \tag{1.24}$$

$$\text{Waist size:} \quad \omega_0^2 = \left(\frac{\lambda}{n\pi}\right) z_R = -\left(\frac{\lambda}{n\pi}\right)\frac{Im\{q_1\}}{|q_1|^2} \tag{1.25}$$

$$\text{Spot size:} \quad \omega^2 = -\frac{\lambda}{n\pi Im\{q_1\}}. \tag{1.26}$$

Equations 1.21 and 1.22 allow us to determine $R$ and $\omega$ at a distance $z$ from the waist. Occasionally we would like to know the size of the waist if we are given $1/q$

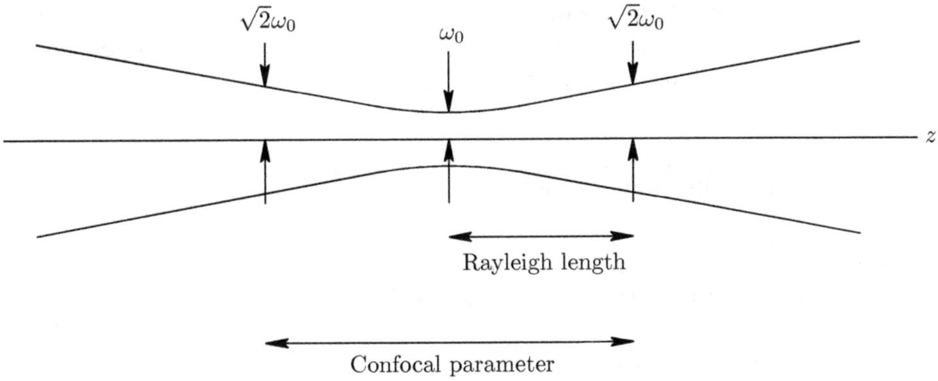


**Fig. 1.2** An illustration of the definitions of the Rayleigh length, $z_R$, and confocal parameter, $b$.

(actually, $R$ and $\omega$) at some place other than at the waist. Using eqn 1.25 and the definition of $1/q$, we obtain:

$$\omega_0 = \frac{\lambda R \omega}{\sqrt{(\pi n \omega^2)^2 + (\lambda R)^2}}. \tag{1.27}$$

## 1.5 The phase term: Gouy phase

The calculations up to this point have ignored the $P(z)$ term, justifying this by assuming that its $z$-dependence is much slower than that of the $kz$ term. Now that a solution for $q$ as a function of $z$ has been obtained, we can substitute it into eqn 1.8:

$$\frac{dP(z)}{dz} = -\frac{i}{q(z)} = -\frac{i}{z + i(n\pi\omega_0^2/\lambda)}. \tag{1.28}$$

Integrating this (from the waist, where $z = 0$) yields,

$$iP(z) = \ln[1 - i(\lambda z/n\pi\omega_0^2)] = \ln\sqrt{1 + \left(\frac{\lambda z}{n\pi\omega_0^2}\right)^2} - i\tan^{-1}\left(\frac{\lambda z}{n\pi\omega_0^2}\right). \tag{1.29}$$

The first term above gives the expected amplitude decrease with $z$ and the second term gives an additional *phase shift*, which is rapidly changing with $z$ when $|z| < z_R$. The complete expression for a field component of a Gaussian beam can now be written down (where an overall proportionality constant of 1 is assumed):

$$u = \frac{1}{\sqrt{1 + \left(\frac{z}{z_R}\right)^2}} \exp\left\{ i\tan^{-1}\left(\frac{z}{z_R}\right) - ik\left(z + \frac{r^2}{2R}\right) - \frac{r^2}{\omega^2} \right\}. \tag{1.30}$$

The additional phase shift is often called the *Gouy* phase, after the French physicist who discovered in 1890 that a converging light beam underwent a 180° phase shift as one traversed the focus along the direction of propagation. The total phase evolution of the Gaussian beam (on axis) can be written as the sum of two components: the *normal* phase shift and the *Gouy phase*:

$$\text{Phase shift of Gaussian beam} = i\phi = \overbrace{ikz}^{\text{normal}} \overbrace{- i\tan^{-1}\frac{z}{z_R}}^{\text{Gouy}}. \tag{1.31}$$

The Gouy phase changes occur largely within one Rayleigh length of the waist, justifying the assumption that the normal phase shift is a much more rapid function of $z$ (since it varies significantly over the distance of one wavelength and $z_R \gg \lambda$ in the paraxial approximation). The Gouy phase will play a large role in determining the resonant frequencies of an optical cavity, where it causes the location of the resonances to be determined by the transverse mode number in addition to that of the axial mode. The $z$-dependence of the Gouy phase is shown graphically in Fig. 1.3.

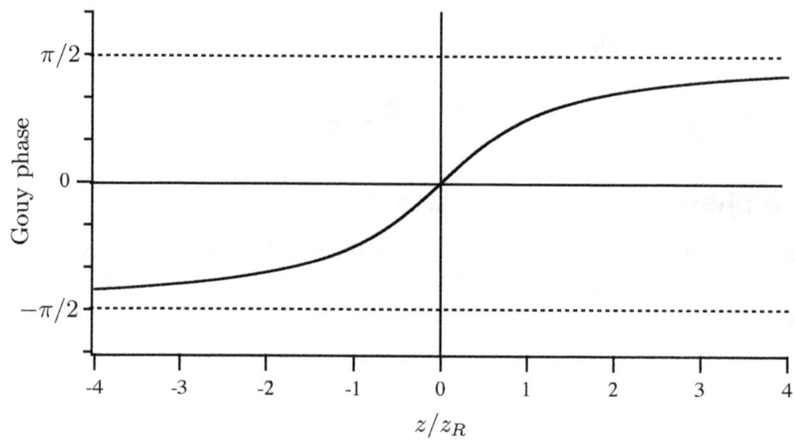


**Fig. 1.3** The $z$-dependence of the Gouy phase.

## 1.6 Simple transformation properties of the complex beam parameter

We have shown that the behavior of a Gaussian beam is determined solely by a single complex parameter, $q$, from which one can determine the beam's spot size and wavefront radius of curvature at all points along the beam's path. To complete our discussion of this beam model, we will determine how $q$ is transformed in the four most common situations encountered in the laboratory: free-space propagation, refraction by a *thin* lens, reflection by a *thin* spherical mirror and transmission through a slab of some transparent material with refraction index, $n$. Using these transformation rules, one can completely determine the behavior of a Gaussian beam as it progresses through most pieces of apparatus found in the laboratory. This approach of directly calculating the transformations of $q$ is very useful but tedious; in the next section we will describe a much simpler technique for accomplishing the same task.

**Free-space:** Free-space propagation from $z_1$ to $z_2$ is immediately obtained from eqn 1.9.

$$q(z_2) = q(z_1) + z_2 - z_1. \tag{1.32}$$

**Thin lens:** Assume that the spot size does not change. We are thus only interested in how the wavefront radius changes upon traversing a thin lens.

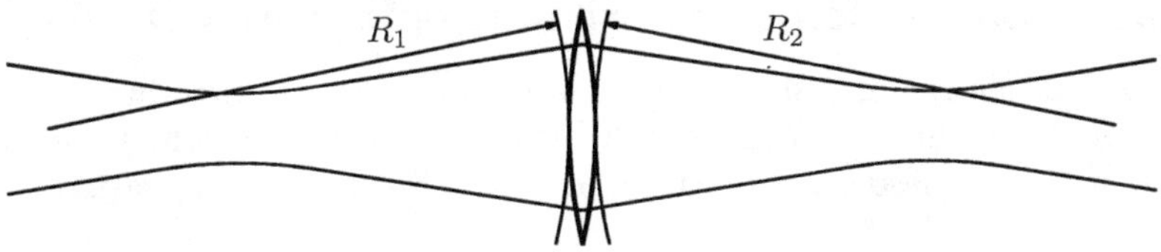


Since the wavefront radii obey the simple lens formula (where $i$ is the distance between the image and the lens, $o$ is the distance between the object and the lens and $f$ is the focal length),

$$\frac{1}{i} + \frac{1}{o} = \frac{1}{f} \longrightarrow \frac{1}{R_1} - \frac{1}{R_2} = \frac{1}{f} \tag{1.33}$$

(the sign change is due to our observance of the sign conventions for radii).

$$\text{Since } \omega_1 = \omega_2, \quad \frac{1}{q_1} - \frac{1}{q_2} = \frac{1}{f} \longrightarrow q_2 = \frac{q_1}{1 - q_1/f}. \tag{1.34}$$

**Thin spherical mirror:** A thin mirror is described by the same formula as the lens, with $f = R/2$, where $R$ is the radius of curvature of the mirror.

$$q_2 = \frac{q_1}{1 - 2q_2/R}. \tag{1.35}$$

**Slab of thickness, $d$, and index, $n$:** We assume that the beam is *normally incident* on the input and output faces of the slab (non-normal incidence introduces *astigmatism*, which will be treated in the next chapter). Three situations need to be considered: transition through the input face, translation by distance $d$ and transition through the output face. We assume that the slab is surrounded by a medium of unity refractive index. From the expression for $1/q$ (eqn 1.10), one can see that the imaginary term has a $1/n$ dependence. To demonstrate that the real part also has this dependence, one need only assume that the normals to the spherical wavefronts obey Snell's law. Thus, the transition $q_0 \rightarrow q_1$ across the input face becomes

$$q_1 = nq_0. \tag{1.36}$$

The internal traversal of the slab ($q_1 \rightarrow q_2$) satisfies eqn 1.9:

$$q_2 = q_1 + d. \tag{1.37}$$

Finally, the transition ($q_2 \to q_3$) across the output face satisfies

$$q_3 = \frac{1}{n} q_2. \tag{1.38}$$

Thus, the overall transition across the slab is described by

$$q_3 = q_0 + \frac{d}{n}. \tag{1.39}$$

## 1.7 Matrix formulation of paraxial ray optics: ABCD rule

We will make a slight digression to ray optics in order to introduce a very useful formalism for describing ray propagation using matrices. We will then discuss the remarkable utility of these same matrices in determining the propagation of Gaussian beams.

We will first consider a single paraxial ray in an axially symmetrical optical system. This ray is completely described by a two-element column vector containing the *height*, $y$, of the ray above the axis and the *slope*, $y'$, of the ray with respect to the $z$-axis:

$$\text{Ray vector:} \quad \begin{pmatrix} y \\ y' \end{pmatrix}. \tag{1.40}$$

When considering the four situations discussed earlier (free-space propagation, lenses, mirrors and slabs), the ray vector is subject to a *linear transformation* represented by a *ray matrix*, called the *ABCD matrix*:

$$\begin{pmatrix} y_2 \\ y_2' \end{pmatrix} = \begin{pmatrix} A & B \\ C & D \end{pmatrix} \begin{pmatrix} y_1 \\ y_1' \end{pmatrix}, \tag{1.41}$$

where the subscripts $_{1,2}$ are for the ray before and after the transformation, respectively. A description of our four transformations using ray matrices will now be obtained.

**Free-space propagation:** The ray matrix is simply obtained using the equation for a straight line (we assume the propagation distance is $d$):

$$y_2' = y_1', \quad y_2 = y_1 + y_1' d \longrightarrow \begin{pmatrix} A & B \\ C & D \end{pmatrix} = \begin{pmatrix} 1 & d \\ 0 & 1 \end{pmatrix}. \tag{1.42}$$

**Thin lens:** The calculation is similar to that for the $q$-transformation. Again, the ray height is unchanged ($y_2 = y_1$) and the slope change is obtained using the lens formula (eqn 1.33):

$$y_2 = y_1, \quad y_2' = -\frac{y_2}{i}, \quad y_1' = \frac{y_1}{o}. \tag{1.43}$$

From the lens formula, we obtain:

$$y_2' = y_1' - y_1/f \quad \longrightarrow \quad \begin{pmatrix} A & B \\ C & D \end{pmatrix} = \begin{pmatrix} 1 & 0 \\ -1/f & 1 \end{pmatrix}. \tag{1.44}$$

**Thin spherical mirror:** The ray matrix is the same as for the thin lens, replacing $f$ with $R/2$:

$$\begin{pmatrix} A & B \\ C & D \end{pmatrix} = \begin{pmatrix} 1 & 0 \\ -2/R & 1 \end{pmatrix}. \tag{1.45}$$

(In order to use this approach with a mirror, one must redefine the *positive* $z$-direction *after* a reflection.)

**Slab of thickness, $d$:** From Snell's law in the paraxial approximation, the change in y can be easily determined:

$$y_2 = y_1 + y_3' d = y_1 + y_1' \frac{d}{n}, \tag{1.46}$$

where we assumed that sines of the angles are equal to the slopes and used Snell's law at the input face. The quantity $y_3'$ is the slope of the ray *inside the slab.* The slope doesn't change when traversing the crystal since the exit angle of the ray is the same is its entry angle (for parallel crystal faces). Thus,

$$\begin{pmatrix} A & B \\ C & D \end{pmatrix} = \begin{pmatrix} 1 & d/n \\ 0 & 1 \end{pmatrix}. \tag{1.47}$$

A useful property of all of the above ABCD matrices is the fact that they are *unimodular*:

$$\det \begin{pmatrix} A & B \\ C & D \end{pmatrix} = AD - BC = 1. \tag{1.48}$$

This property also holds for *composite* systems composed of any number of the above elements since the determinant of a product of matrices is the product of the determinants of the individual matrices.

A remarkable connection between the ray matrices and $q$-transformations can be seen in Table 1.1. From the slightly odd way in which the $q$-transformations are presented, it should be clear that for these four cases the following rule holds:

$$q_2 = \frac{Aq_1 + B}{Cq_1 + D} \Longleftrightarrow \begin{pmatrix} A & B \\ C & D \end{pmatrix}. \tag{1.49}$$

Stated in words: *For a free-space propagation, thin lens, thin mirror or slab, knowledge of the ABCD matrix immediately allows one to write down the q-transformation using the expression on the left side of eqn 1.49.*

For this rule to be generally useful, we need to find its analogue (if it exists) for a composite system composed of varying numbers of all four of the elements discussed above. To do this, we will first consider a system composed of any two elements and investigate the composite $q$-transformation for such a system. We therefore have two transformations of the following form (labeling them by their $ABCD$ matrices):

$$q_0 \overset{ABCD}{\longrightarrow} q_1 \overset{A'B'C'D'}{\longrightarrow} q_2. \tag{1.50}$$

The $ABCD$ matrices for the two transformations are:

**Table 1.1** Relations between $q$-transformations and ray matrices.

| Case | $q$-transformation | Ray matrix |
|---|---|---|
| Free-space | $q_2 = q_1 + d = \frac{(1)q_1 + (d)}{(0)q_1 + (1)}$ | $\begin{pmatrix} 1 & d \\ 0 & 1 \end{pmatrix}$ |
| Thin lens | $q_2 = \frac{q_1}{1 - q_1/f} = \frac{(1)q_1 + (0)}{(-1/f)q_1 + (1)}$ | $\begin{pmatrix} 1 & 0 \\ -1/f & 1 \end{pmatrix}$ |
| Mirror | $q_2 = \frac{q_1}{1 - 2q_1/R} = \frac{(1)q_1 + (0)}{(-2/R)q_1 + (1)}$ | $\begin{pmatrix} 1 & 0 \\ -2/R & 1 \end{pmatrix}$ |
| Slab | $q_2 = q_1 + d/n = \frac{(1)q_1 + (d/n)}{(0)q_1 + (1)}$ | $\begin{pmatrix} 1 & d/n \\ 0 & 1 \end{pmatrix}$ |

$$q_0 \longrightarrow q_1 : \begin{pmatrix} A & B \\ C & D \end{pmatrix} \qquad q_1 \longrightarrow q_2 : \begin{pmatrix} A' & B' \\ C' & D' \end{pmatrix}. \tag{1.51}$$

Applying the rule twice:

$$q_2 = \frac{A'q_1 + B'}{C'q_1 + D'} = \frac{(A'A + B'C)q_0 + (A'B + B'D)}{(C'A + D'C)q_0 + (C'B + D'D)}. \tag{1.52}$$

The terms in brackets are just the components of *the matrix product of the two matrices.* Thus the rule is valid for composite systems provided that one uses the composite matrix formed by taking the matrix product of the individual matrices. It is obvious that an analogous rule holds for the ray optics case. We therefore obtain the $ABCD$ rule:

**ABCD rule:** *The overall q-transformation for a complex system composed of thin lenses, thin mirrors, free spaces and slabs can be obtained by determining the ABCD matrices of the individual components, multiplying the matrices together and applying eqn 1.49 to the composite matrix.* In addition to the four elements mentioned above, any element which causes a $q$-transformation given by the left side of eqn 1.49 can be included in the $ABCD$ rule.

As a practical matter, it should be noted that the component $ABCD$ matrices are written down in *the reverse order from left to right to the order in which the beam encounters the various components.* We will see several examples of this "rule" in the next few chapters.

## 1.8 Further reading

This chapter provided an essential introduction to Gaussian beams and their manipulation using the complex beam parameter and the $ABCD$ rule. Much of the formalism is based upon the pioneering work on the subject conducted by the researchers at Bell

Laboratories in the 1960s and is described in the classic paper by Kogelnik and Li (1966). Other books on the subject include the excellent textbook by Yariv (1989), the tome by Siegman (1986) and numerous other books on laser theory. Three excellent representative samples of the latter group are Verdeyen (1995), Milonni and Erberly (1988), and Svelto (2004). The background material in classical optics can be obtained from the excellent textbooks by Hecht (2002) and Jenkins and White (1957) and from the well-known treatise by Born and Wolf (1980).

## 1.9 Problems

(1.1) Calculate the ABCD matrix for a ray entering a spherical dielectric interface from a medium of refractive index $n_1$ to a medium of index $n_2$, with radius of curvature $R$ ($R > 0$ if center is to left of surface). Make the usual paraxial approximations.

(1.2) You are given an optical system described by an ABCD matrix in which both the input plane and the output plane are at beam waists. If the input waist is $\omega_1$, determine the conditions on the matrix elements and the output waist, $\omega_2$ (in terms of ABCD and $\omega_1$).

(1.3) Find the ABCD matrix of a "Gaussian" aperture whose field transmission $t(x, y)$ is given by

$$t(x, y) = \exp\left[-\frac{x^2 + y^2}{2\omega_a^2}\right].$$

(1.4) It is often necessary to accurately measure the spot size of a Gaussian beam. Below are two possible methods for doing this in a laboratory:

(a) One way to measure the beam radius is to advance a knife edge into the beam with a micrometer and measure the power of the beam with a digital power meter downstream from the knife edge. Between what two percentages of power will the knife edge advance a distance equal to $2\omega$? (Make it symmetrical: i.e., $\Delta\%$ and $(100 - \Delta)\%$).

(b) A commercial instrument uses the scheme below to automate the spot measuring process. A cylindrical knife-edge rotates at angular speed $\Omega$. If the open aperture subtends an angle $\theta$ from the center and the radius of the tube is $R$, write down an expression for the "rise-time" of the signal at the detector as a function of the spot radius, where the rise-time is defined as being between the same two percentages as calculated above.

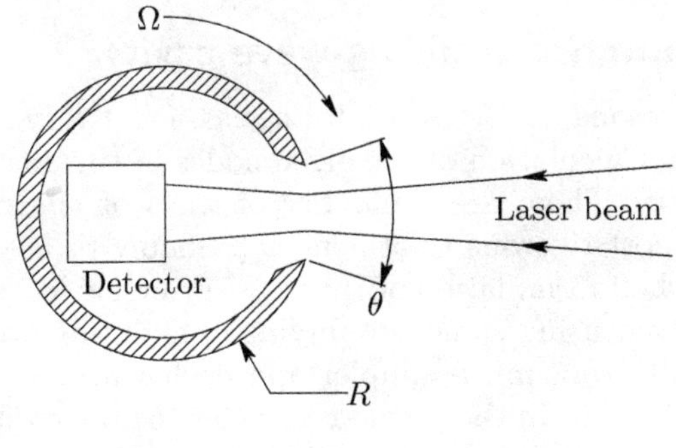

# 2

# Optical resonators – geometrical properties

## 2.1 Introduction

Optical cavities (resonators) have numerous applications, perhaps most notably as the feedback mechanism in laser oscillators. In the atomic physics laboratory, they are also used for optical spectrum analysis, a stable frequency reference for laser frequency stabilization and for enhancement of the power of fundamental beams in nonlinear frequency synthesis.

We will consider two variants of the optical resonator: the standing-wave cavity and the traveling-wave cavity. Both types are used in lasers but traveling-wave (or "ring") cavities are usually preferred for nonlinear frequency synthesis. Despite the apparent greater simplicity of the two-mirror standing-wave cavity compared to the four-mirror ring resonator, the analysis of the latter is actually no more difficult than that of the former. One problematic aspect of the ring resonator not shared by the standing-wave cavity is the inevitable presence of astigmatism, since a practical realization of a ring cavity requires that mirrors (and often intracavity crystals as well) be used off-axis. We will analyze this in some detail.

This chapter will discuss resonator eigenmodes, which depend only upon the cavity geometry and the wavelength of the radiation. The next chapter will cover cavity energy relationships, which depend upon the reflectivity of the cavity mirrors and various loss mechanisms. The latter discussion will introduce the concepts of cavity finesse, $Q$, photon lifetime, etc. This division between *geometrical* properties and *energy* relationships is perhaps analogous to the division in classical mechanics between kinematics and dynamics.

## 2.2 The two-mirror standing-wave cavity

The two-mirror resonator is essentially a classical Fabry–Pérot interferometer with spherical mirrors. Calculation of the eigenmodes of this instrument is complicated by the open geometry. There are a number of ways of analyzing an open, two-mirror resonator – the most rigorous treatment is probably to use formal diffraction theory. The Fresnel–Kirchoff formulation of Huygens' principle leads to two integral equations which can be solved using some simplifying assumptions to make the problem more tractable. We will use a much simpler approach which is based upon the Gaussian beam theory worked out in the previous chapter together with a simple and intuitively obvious self-consistency approach. Our treatment of course implicitly acknowledges diffraction, since the latter is built into the theory of Gaussian beams.

The two-mirror resonator consists of two spherical mirrors with radii of curvature $R_1$ and $R_2$ (Fig. 2.1). We are currently only interested in the field configurations which form stable eigenmodes of the cavity. Thus, we assume that there is some means (e.g., partially transmitting coatings) by which radiation can be injected into the space between the mirrors, but we are not interested at present in its details.

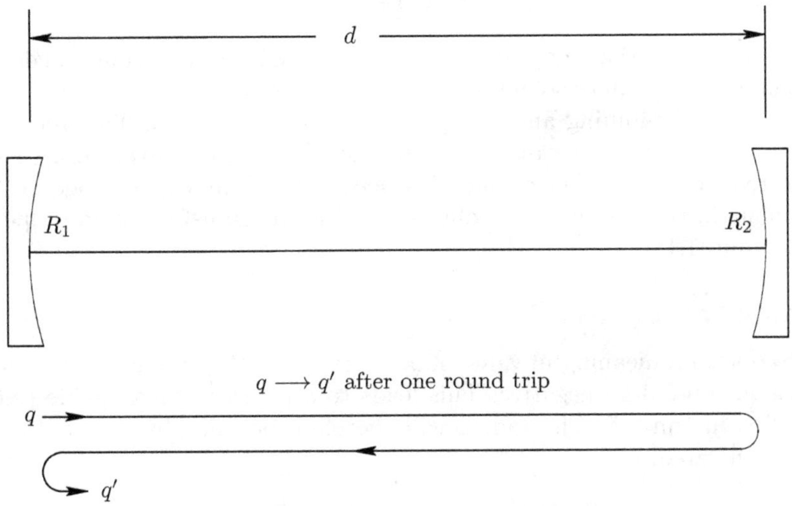


**Fig. 2.1** A two-mirror standing-wave cavity.

Our *self-consistent* approach requires that the complex beam parameter, $q$, replicates itself after one round trip in the cavity. If this happens, the field distribution also replicates itself after one round trip since $q$ contains everything there is to know about the Gaussian field (except for its *phase*, which will be discussed later when we determine the *resonant frequencies* of the cavity). In terms of the round trip ABCD matrix, self-consistency requires that

$$q = \frac{Aq+B}{Cq+D} \quad \text{or} \quad \frac{1}{q} = \frac{C+D\left(\frac{1}{q}\right)}{A+B\left(\frac{1}{q}\right)}. \tag{2.1}$$

Due to the way that $q$ is defined, it is more convenient to solve this equation for $q_1 = 1/q$. The resulting quadratic equation:

$$B\left(\frac{1}{q}\right)^2 + (A-D)\left(\frac{1}{q}\right) - C = 0, \tag{2.2}$$

has solutions:

$$q_1 = \frac{1}{q} = \frac{D-A}{2B} \pm \frac{1}{2B}\sqrt{(A-D)^2 + 4BC}. \tag{2.3}$$

Since the elements of the ABCD matrix are real, the first term must be associated with the real part of $1/q$, and the second term must be pure imaginary and be associated with the imaginary part of $1/q$. Thus, using eqn 1.10, one obtains:

$$R = \frac{2B}{D - A} \tag{2.4}$$

$$\omega = \sqrt{\frac{2\lambda|B|}{n\pi\sqrt{4-(A+D)^2}}}, \tag{2.5}$$

where we used the fact that the determinant of an ABCD matrix is unity in deriving the second equation. The values so obtained are of course valid only at the *reference point* in the cavity (the beginning and end points of the round trip). The reference point is completely arbitrary and would normally be chosen for convenience (a judicious choice can reduce the number of matrices needed to be multiplied together). Values at other points in the cavity can be obtained using the transformation properties of $q$ or eqns 1.17 and 1.18.

## 2.3 Stability

In order to obtain a meaningful value of $q$, we required that the expression under the square root in eqn 2.3 be negative. This leads to a physically reasonable (*real*) value for the radius in eqn 2.5. The radius will therefore be non-physical if the following inequality is violated:

$$\text{Stability criterion:} \quad |A + D| \le 2. \tag{2.6}$$

This inequality is called the *stability criterion* for the given cavity geometry. If it is violated, it will be impossible for the field to replicate itself after one round trip.

Some additional insight can be obtained by using a *ray matrix* approach and by replacing the two mirror cavity by an *equivalent* periodic sequence of lenses having the same *powers* (reciprocals of the focal length) as the mirrors. In this way, the ray trajectory will be *unfolded* and a stable cavity can be shown to have a periodic ray path which remains close to the axis while an unstable cavity will have a ray path which deviates increasingly from the cavity axis.

We can approach this analytically by writing down the overall ray matrix for $n$ *units of periodicity*, each of which consists of a pair of lenses as shown in Fig. 2.2. We require the $n$th power of an ABCD matrix representing one unit of periodicity. Using *Sylvester's theorem* to evaluate the matrix corresponding to $n$ periods yields

$$\begin{pmatrix} A & B \\ C & D \end{pmatrix}^n = \frac{1}{\sin\theta}\begin{pmatrix} A\sin n\theta - \sin(n-1)\theta & B\sin n\theta \\ C\sin n\theta & D\sin n\theta - \sin(n-1)\theta \end{pmatrix}, \tag{2.7}$$

where

$$\cos\theta = \frac{1}{2}(A+D) \Longrightarrow \theta = \cos^{-1}\left\{\frac{1}{2}(A+D)\right\}. \tag{2.8}$$

In order for $\theta$ to be *real*, the expression inside the curly brackets must be between $+1$ and $-1$, which leads to eqn 2.6. Thus, for a stable system with a real value of $\theta$, the

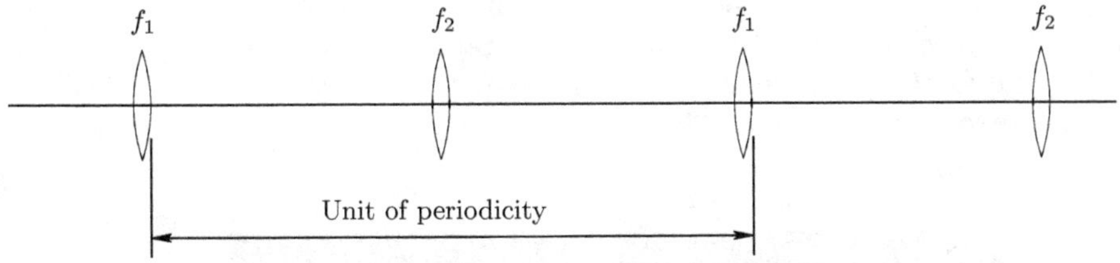


**Fig. 2.2** A lens sequence which is equivalent to a two-mirror cavity. Each *unit of periodicity* represents one round-trip path in the two-mirror cavity. The lenses are identified by their focal lenghts, $f_{1,2}$

matrix elements are simple sine and cosine functions of $n$, and a ray will have a periodic behavior. An unstable system will have an *imaginary* $\theta$ which yields *hyperbolic* sines and cosines and causes the ray to deviate exponentially from the axis.

We can determine the constraints on $R_1$ and $R_2$ required for stability by writing the round trip ABCD matrix as a product of four matrices. We take the (arbitrary) starting point just to the right of the left-hand mirror. Then we have a free-space propagation by a distance $d$, a reflection by the mirror whose radius of curvature is $R_2$, a second free-space propagation and a reflection by the mirror with radius $R_1$. We therefore obtain (writing the matrices in *reverse order*),

$$\begin{pmatrix} A & B \\ C & D \end{pmatrix} = \begin{pmatrix} 1 & 0 \\ -\frac{2}{R_1} & 1 \end{pmatrix} \begin{pmatrix} 1 & d \\ 0 & 1 \end{pmatrix} \begin{pmatrix} 1 & 0 \\ -\frac{2}{R_2} & 1 \end{pmatrix} \begin{pmatrix} 1 & d \\ 0 & 1 \end{pmatrix}. \tag{2.9}$$

In what follows, it is convenient to describe the cavity *geometry* in terms of two dimensionless parameters, $g_1$ and $g_2$:

$$g_1 \equiv 1 - \frac{d}{R_1} \tag{2.10}$$

$$g_2 \equiv 1 - \frac{d}{R_2}. \tag{2.11}$$

Then,

$$\begin{pmatrix} A & B \\ C & D \end{pmatrix} = \begin{pmatrix} 1 & 0 \\ \frac{2}{d}(g_1 - 1) & 1 \end{pmatrix} \begin{pmatrix} 1 & d \\ 0 & 1 \end{pmatrix} \begin{pmatrix} 1 & 0 \\ \frac{2}{d}(g_2 - 1) & 1 \end{pmatrix} \begin{pmatrix} 1 & d \\ 0 & 1 \end{pmatrix}$$

$$= \begin{pmatrix} 2g_2 - 1 & 2g_2 d \\ \frac{2}{d}(2g_1 g_2 - g_1 - g_2) & 4g_1 g_2 - 2g_2 - 1 \end{pmatrix}. \tag{2.12}$$

The *stability criterion* becomes:

$$|A + D| \leq 2 \Longrightarrow |4g_1 g_2 - 2| \leq 2. \tag{2.13}$$

We therefore obtain:

$$\text{Stability criterion:} \quad 0 \leq g_1 g_2 \leq 1. \tag{2.14}$$

This expression can be plotted (Fig. 2.3) on a two-dimensional coordinate system whose axes are $g_1$ and $g_2$. The plot indicates *regions of stability* bounded by the lines $g_1 = 0$ and $g_2 = 0$ together with the hyperbolas in eqn 2.14. The geometrical interpretation of $g_{1,2}$ should be clear: these parameters vary from 1 (plane mirrors) through 0 (*confocal*, where the mirror spacing equals the mirror radii) to $-1$ (spherical, where the mirrors are part of a sphere).

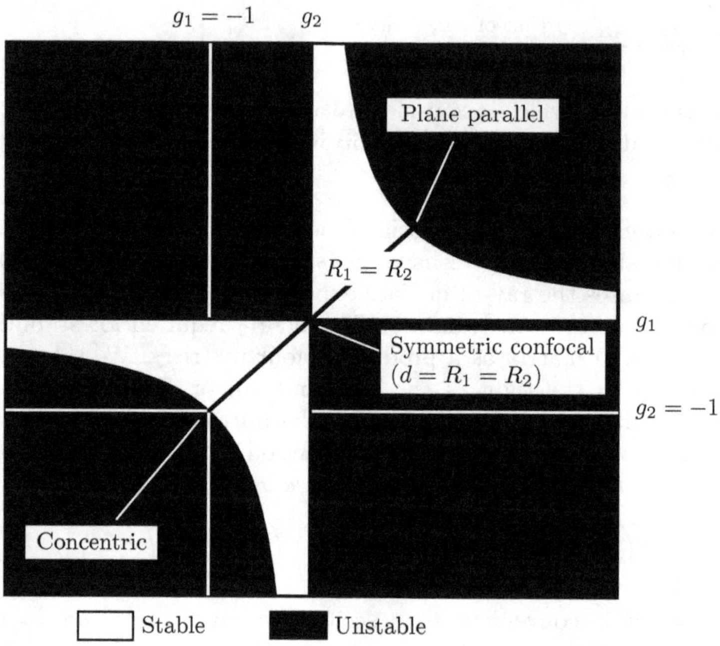


**Fig. 2.3** A *stability diagram* displaying regions of stability in terms of the parameters $g_{1,2}$ in a two-mirror cavity.

## 2.4 Solution for an arbitrary two-mirror stable cavity

The self-consistent solution for a two-mirror cavity will be obtained by first writing down the solution for $q_1 = 1/q$ using eqn 2.3 and the product matrix (eqn 2.12):

$$q_1 = \frac{1}{q} = \frac{g_1 - 1}{d} \pm \frac{1}{g_2 d}\sqrt{g_1 g_2 (g_1 g_2 - 1)}. \tag{2.15}$$

The spot size at the left-hand mirror (the reference point) can be obtained by using this result and eqn 1.10:

$$\omega^2 = \left(\frac{\lambda d}{n\pi}\right)\sqrt{\frac{g_2}{g_1(1 - g_1 g_2)}}. \tag{2.16}$$

The wavefront radius at the left-hand mirror can be obtained by substituting the elements of the ABCD matrix into eqn 2.4. It should come as no surprise that the result is $-R$, which is just the mirror radius (with a sign which is consistent with the *sign convention*). It would be a mistake, however, to conclude that this is always the case: in a ring cavity, the mirror radius and wavefront radius are different. The waist size can be obtained by substituting the imaginary part of $q_1$ from eqn 2.15 into eqn 1.25. After a bit of algebra, we get:

$$\omega_0^2 = \left(\frac{\lambda d}{n\pi}\right) \frac{\sqrt{g_1 g_2 (1 - g_1 g_2)}}{g_1 + g_2 - 2g_1 g_2}. \tag{2.17}$$

A few additional relations will be useful when we calculate the resonant frequencies of the cavity. These are the Rayleigh length, $z_R$, and the $z$-coordinates, $z_1$ and $z_2$, of mirrors 1 and 2 with respect to the waist (these all appear in the *Gouy phase* term for the resonant frequency). The Rayleigh length can be obtained from eqn 1.24 (or directly from eqn 2.17):

$$z_R = d\frac{\sqrt{g_1 g_2 (1 - g_1 g_2)}}{g_1 + g_2 - 2g_1 g_2}. \tag{2.18}$$

The distance from the waist to mirror 1 ($z_1$) can be obtained from eqn 1.23 (where two minus signs cancel, since we seek the *distance from the waist to the mirror*). The distance to mirror 2, $z_2$, is obtained from $z_1$ together with the relationship $z_2 - z_1 = d$. The results are:

$$z_1 = d\frac{g_2(g_1 - 1)}{g_1 + g_2 - 2g_1 g_2} \tag{2.19}$$

$$z_2 = d\frac{g_1(1 - g_2)}{g_1 + g_2 - 2g_1 g_2}. \tag{2.20}$$

It would be instructive at this point to analyze the two-mirror cavity for the very simple case of a *symmetric resonator*, where $R_1 = R_2 = R$. As we can see from Fig. 2.3, all symmetrical configurations between the plane mirror cavity and the spherical mirror cavity are stable. To find the spot sizes, we set $g_1 = g_2 = g = 1 - d/R$ and obtain:

$$\text{At mirror:} \quad \omega^2 = \left(\frac{\lambda d}{n\pi}\right)\sqrt{\frac{1}{1-g^2}} = \left(\frac{\lambda R}{n\pi}\right)\sqrt{\frac{d}{2R-d}} \tag{2.21}$$

$$\text{Waist:} \quad \omega_0^2 = \left(\frac{\lambda d}{2n\pi}\right)\sqrt{\frac{1+g}{1-g}} = \left(\frac{\lambda}{n\pi}\right)\sqrt{\frac{dR}{2} - \frac{d^2}{4}}. \tag{2.22}$$

A very common configuration is the one with $d = R$. This is the *confocal* resonator whose position on the stability diagram is halfway between the plane-mirror cavity and the spherical cavity (constrained to lie on the line tilted at 45° in Fig. 2.3). The values of the radius at the mirror and waist are

$$\omega_0 = \left(\frac{\lambda d}{2n\pi}\right)^{1/2} \tag{2.23}$$

$$\omega_{mirror} = \left(\frac{\lambda d}{n\pi}\right)^{1/2} = \sqrt{2}\omega_0. \tag{2.24}$$

Note that $d$ in this case is just the *confocal parameter*, which is perhaps the origin of its name.

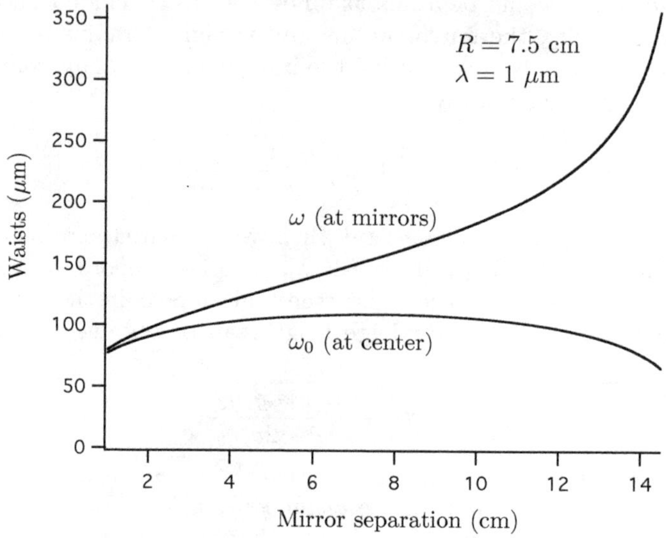


**Fig. 2.4** The spot sizes at the mirrors and waist sizes as a function of the mirror separation for a symmetrical two-mirror cavity.

The spot radii as a function of mirror separation are plotted in Fig. 2.4 for a cavity whose mirror radii are 7.5 cm. Notice that the cavity is unconditionally stable for small mirror separations (and $\omega$ approaches $\omega_0$ as the beam becomes collimated) but the spot size at the mirror increases very rapidly as one approaches the instability point (spherical cavity).

## 2.5 Higher-order modes

In deriving the solutions to the paraxial wave equation (eqn 1.5), we have considered cylindrically symmetric field configurations with a very simple Gaussian dependence on $r$. In order to describe an arbitrary field configuration, we need to find a complete set of functions, each one of which is a solution of the paraxial wave equation. We can then expand a general beam in a linear combination of these functions, which are called *modes*. The simple Gaussian beam studied in the previous chapter is only one (the lowest) of an infinite number of possible modes. We will now describe a complete set of solutions to the paraxial wave equation in *Cartesian coordinates*. One might expect a better coordinate system to be *cylindrical coordinates*, but the *eigenmodes* of the paraxial wave equation in the Cartesian system turn out to be more often observed in the laboratory than the eigenmodes in cylindrical coordinates.

After a great deal of algebra, one can show that the following functions are solutions of the paraxial wave equation:

$$\begin{aligned} u(x,y,z)_{nm} &= \frac{\omega_0}{\omega(z)} H_n(\sqrt{2}\frac{x}{\omega}) H_m(\sqrt{2}\frac{y}{\omega}) \\ &\quad \times \exp\left\{-i(kz - \Phi(m,n;z)) - i\frac{k}{2q}(x^2+y^2)\right\}, \end{aligned} \tag{2.25}$$

where

$$\frac{\omega_0}{\omega(z)} = \frac{1}{\sqrt{1+(z/z_R)^2}} \tag{2.26}$$

$$\Phi(n,m;z) = (n+m+1)\tan^{-1}(z/z_R) \qquad \text{(Gouy phase)}. \tag{2.27}$$

The indices $n$ and $m$ are integers, and the functions $H_n()$ and $H_m()$ are *Hermite polynomials* of order $n$ and $m$. The origin of the coordinate system ($z = 0$) is at the waist. When $n = m = 0$, this expression reduces to that of the *lowest-order* Gaussian beam, eqn 1.30.

We should note that, since all of the transverse coordinates (except for those which only effect the phase of $u$) in eqn 2.25 appear divided by $\omega(z)$, the *cross-sectional shape* of the beam is independent of the position along the beam and that the size of the profile is scaled by the *Gaussian* spot radius, $\omega(z)$. These field distributions are called *Hermite–Gaussian modes*. Since the free-space electric and magnetic fields are transverse to the $k$-vector, the modes are called $\text{TEM}_{mnq}$ (transverse electromagnetic). The indices $m$ and $n$ are defined above; $q$ is an axial mode number which will be defined shortly (and should be distinguished from the complex beam parameter – the notation is occasionally unfortunate). Plots of three different modes are shown in Fig. 2.5.

The Hermite–Gaussian modes have the following properties:

- They are all characterized by the *same* complex beam parameter, $q$, defined by

$$\frac{1}{q(z)} = \frac{1}{R(z)} - i\frac{\lambda}{n\pi\omega^2(z)}. \tag{2.28}$$

- They all satisfy the same ABCD rule as (lowest-order) Gaussian beams. The *mode numbers*, $m$ and $n$, are preserved under all of the transformations discussed in this book.
- The mode *shape* is independent of $z$ and scales with $\omega(z)$.
- The mode of index $m$ has a half-width, $x_m$ (in one transverse coordinate), where

$$x_m \approx \sqrt{m} \times \omega. \tag{2.29}$$

  Thus, $\omega$ is no longer the beam size in higher-order modes. A focused spot has its size *degraded* by $\sqrt{m}$.
- The maximum mode number ($m_{max}$) at a waist which will "fit" into an aperture of radius $a$ is

$$m_{max} \approx (a/\omega_0)^2. \tag{2.30}$$

  This *spatial filtering* behavior of small apertures allows one to filter out modes whose mode number (in either coordinate) is greater than $m_{max}$.
- The far-field half-angle spread, $\theta_m$, is also degraded compared to that ($\theta_0$) of the lowest-order mode:

$$\theta_m \approx \sqrt{m} \times \theta_0. \tag{2.31}$$

Since the definition of $q$ and the ABCD rule are not dependent upon the mode number and the mode numbers are not altered by reflection from spherical mirrors, the self-consistency procedure for determining the possible higher-order *cavity modes* is exactly the same as the procedure we have already outlined for the lowest-order Gaussian cavity mode. Thus, for every lowest-order Gaussian mode, there are an infinite number of higher-order modes with the same value of $q$. As we will shortly see, the resonant frequencies will in general be *different* for different modes.

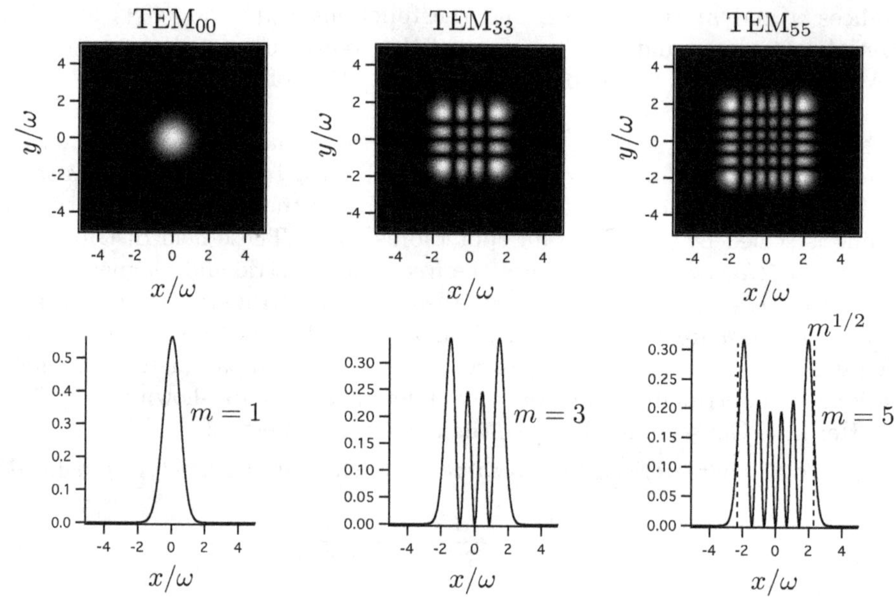


**Fig. 2.5** Three Hermite–Gaussian modes together with their intensity distributions in each dimension.

## 2.6 Resonant frequencies

We derived the possible cavity modes by insisting that the field *replicate* itself after one round trip. So far, we have only considered the replication of the field amplitude distribution, determined by the complex beam parameter, $q$. It is also necessary for the *phase* to be the same (modulo integral multiples of $2\pi$) after a round trip in order for the field to constructively interfere with itself after multiple passes (this is actually the only requirement when analyzing the classical Fabry–Pérot interferometer). From this condition, we can determine the resonant frequencies of the resonator. (In the following, we assume that the index of refraction, $n$, is unity.)

If the mirror separation is $d$, the round trip phase change, $\delta$, on axis can be determined from eqn 2.25:

$$\delta = 2kd - 2(n + m + 1)(\tan^{-1}(z_2/z_R) - \tan^{-1}(z_1/z_R)), \tag{2.32}$$

where $z_1$ and $z_2$ are the $z$-coordinates of the two mirrors measured from the waist. The two $\tan^{-1}$ functions can be combined using eqns 2.18, 2.19 and 2.20 together with the following trigonometric identity which holds for any $x_1$ and $x_2$:

$$\tan^{-1} x_1 - \tan^{-1} x_2 = \cos^{-1}\left\{\frac{1 + x_1 x_2}{\sqrt{(1 + x_1^2)(1 + x_2^2)}}\right\}. \tag{2.33}$$

After a fair amount of algebra, the result is:

$$\tan^{-1}(z_2/z_R) - \tan^{-1}(z_1/z_R) = \cos^{-1} \pm\sqrt{g_1 g_2}, \tag{2.34}$$

where $g_{1,2}$ are defined in eqns 2.10 and 2.11, and the plus is taken when $g_1, g_2 > 0$ while the minus is taken when $g_1, g_2 < 0$. A cavity resonance will occur when the round trip phase shift is an integral multiple of $2\pi$:

$$\delta = q(2\pi) \Longrightarrow \frac{\omega d}{c} - (n + m + 1)\cos^{-1} \pm\sqrt{g_1 g_2} = q\pi, \tag{2.35}$$

where $q$ is an integer. Solving for the resonant frequencies, $\nu_{nmq}$,

$$\nu_{nmq} = \left(q + (n + m + 1)\frac{\cos^{-1} \pm\sqrt{g_1 g_2}}{\pi}\right)\frac{c}{2d}. \tag{2.36}$$

For geometries which are close to planar, confocal or spherical, the Gouy phase terms are

$$\frac{\cos^{-1} \pm\sqrt{g_1 g_2}}{\pi} \approx \begin{cases} 0 & : \quad g_1, g_2 \to 1 \quad \text{(near-planar)} \\ 1/2 & : \quad g_1, g_2 \to 0 \quad \text{(near-confocal)} \\ 1 & : \quad g_1, g_2 \to -1 \quad \text{(near-spherical)} \end{cases} \tag{2.37}$$

The spacing between axial modes ($q$ values) is called the *free spectral range* and is given by:

$$\text{Free spectral range: FSR} = \frac{c}{2d}. \tag{2.38}$$

It is interesting to note that the mode spacing (both axial *and* transverse) in a confocal resonator is one half the free spectral range, since the Gouy phase term is 1/2. This geometry exhibits a great deal of degeneracy between transverse and axial modes and has the advantage of providing a spectrum which is largely independent of which modes are excited by the incoming radiation. As a result of this, the confocal geometry is preferred in commercial optical spectrum analyzers, since the alignment is relatively uncritical. Its free spectral range is still given by eqn 2.38, although manufacturers often (mistakenly) specify the mode spacing ($c/4d$) instead. The spherical and planar geometries share this decoupling between the spectrum and the alignment, but have the disadvantage of being on the edge of the stability region.

## 2.7 The traveling-wave (ring) cavity

There are a number of ways of constructing a traveling-wave cavity; we will describe here the *symmetrical bow-tie configuration* shown in Fig. 2.6. This cavity consists of four mirrors – two curved and two flat – and the mirrors are aligned so that the beam follows the zig-zag path shown in the figure. The curved mirrors have the same radii of curvature, and the configuration is symmetrical about a vertical line halfway between the mirrors. The equivalent sequence of lenses is shown in Fig. 2.7. We will use

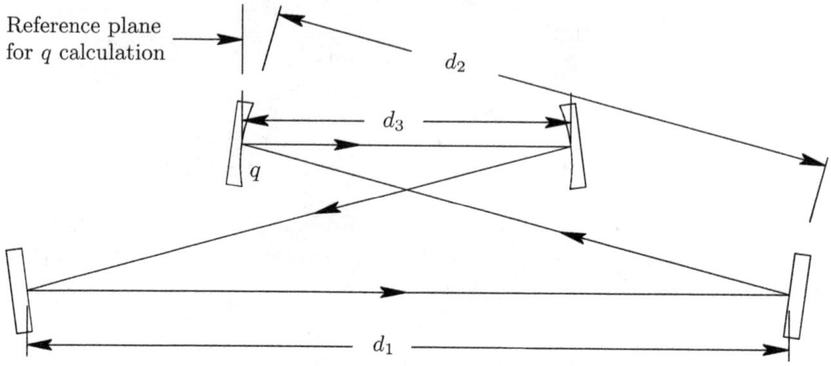


**Fig. 2.6** The symmetrical traveling wave bow-tie cavity.

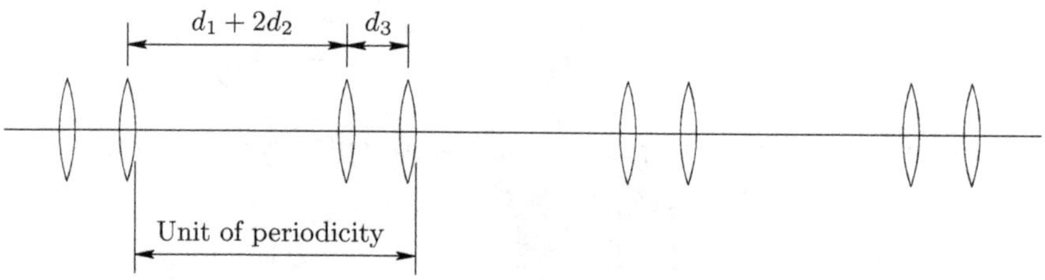


**Fig. 2.7** The lens sequence equivalent of a ring resonator

exactly the same procedure in analyzing the ring cavity as we used for the two-mirror standing-wave cavity. If we choose the reference point to be just to the right of the left-hand curved mirror, the round trip ABCD matrix is

$$\begin{pmatrix} A & B \\ C & D \end{pmatrix} = \begin{pmatrix} 1 & 0 \\ \frac{-2}{R} & 1 \end{pmatrix} \begin{pmatrix} 1 & d_1 + 2d_2 \\ 0 & 1 \end{pmatrix} \begin{pmatrix} 1 & 0 \\ \frac{-2}{R} & 1 \end{pmatrix} \begin{pmatrix} 1 & d_3 \\ 0 & 1 \end{pmatrix}. \tag{2.39}$$

From the figures, it should be apparent that there are only two important distances: the long path $d_1 + 2d_2$ and the shorter curved mirror separation, $d_3$. We will again introduce the two dimensionless parameters $g_1$ and $g_2$:

$$g_1 = 1 - \frac{d_1 + 2d_2}{R}, \tag{2.40}$$

$$g_2 = 1 - \frac{d_3}{R}, \tag{2.41}$$

and write the overall ABCD matrix in terms of them:

$$\begin{aligned} \begin{pmatrix} A & B \\ C & D \end{pmatrix} &= \begin{pmatrix} 1 & 0 \\ \frac{-2}{R} & 1 \end{pmatrix} \begin{pmatrix} 1 & R(1-g_1) \\ 0 & 1 \end{pmatrix} \begin{pmatrix} 1 & 0 \\ \frac{-2}{R} & 1 \end{pmatrix} \begin{pmatrix} 1 & R(1-g_2) \\ 0 & 1 \end{pmatrix} \\ &= \begin{pmatrix} 2g_1 - 1 & R(-2g_1g_2 + g_1 + g_2) \\ \frac{-4g_1}{R} & 4g_1g_2 - 2g_1 - 1 \end{pmatrix}. \end{aligned} \tag{2.42}$$

It should come as no surprise that the stability criterion is the same as in the two-mirror case:

$$|A + D| \leq 2 \Longrightarrow |4g_1g_1 - 2| \leq 2 \Longrightarrow 0 \leq g_1g_2 \leq 1. \tag{2.43}$$

Making the indicated substitutions, we get a constraint on both the minimum and maximum value of $d_3$ (unlike the two-mirror cavity which has no minimum stable mirror separation):

$$\text{Stability:} \qquad R \leq d_3 \leq \frac{R(d_1 + 2d_2)}{d_1 + 2d_2 - R}. \tag{2.44}$$

The stable range of $d_3$, $\Delta d_3$, can be written down as the following simple expression when $d_1 + 2d_2 \gg R$:

$$\text{Range:} \qquad \Delta d_3 \approx \frac{R^2}{d_1 + 2d_2} \qquad \text{when} \;\; d_1 + 2d_2 \gg R. \tag{2.45}$$

Before carrying out further analysis, we will establish a convention which distinguishes the two waists by using unprimed and primed variables: the unprimed variables refer to the upper waist (between the curved mirrors) and the primed quantities refer to the lower waist. The symmetry of the configuration being studied requires that the two waists be located exactly halfway between the respective pairs of mirrors.

We will use eqn 2.3 to determine the value of $q_1$ ($\equiv 1/q$) at the reference plane. The result is:

$$q_1 = \frac{2g_1(g_2 - 1)}{R(g_1 + g_2 - 2g_1g_2)} \pm i\frac{2\sqrt{g_1g_2(1 - g_1g_2)}}{R(g_1 + g_2 - 2g_1g_2)}. \tag{2.46}$$

The Rayleigh length, $z_R$, and waist radius, $\omega_0$, for the upper waist are then

$$z_R = -\frac{Im\{q_1\}}{|q_1|^2} = \frac{R\sqrt{g_1g_2(1 - g_1g_2)}}{2g_1} \tag{2.47}$$

$$\omega_0^2 = \frac{\lambda R\sqrt{g_1g_2(1 - g_1g_2)}}{2n\pi g_1}. \tag{2.48}$$

As a consistency check, one can evaluate the distance to the waist ($-Re\{q_1\}/|q_1|^2$) and confirm that it is indeed $d_3/2$.

The properties of the lower waist are determined by transforming the beam *backwards* past the left curved mirror. The transformation properties of $q_1$ are (eqn 2.1):

$$q_1' = \frac{C + Dq_1}{A + Bq_1} = \frac{2}{R} + q_1, \qquad \text{where} \quad \begin{pmatrix} A & B \\ C & D \end{pmatrix} = \begin{pmatrix} 1 & 0 \\ 2/R & 1 \end{pmatrix}. \tag{2.49}$$

Notice that we have used $+R$ instead of $-R$; this is required by the sign conventions for mirror radii when a beam travels *backwards*. The result for $q_1'$ is:

$$q_1' = \frac{2g_2(1 - g_1)}{R(g_1 + g_2 - 2g_1g_2)} \pm i\frac{2\sqrt{g_1g_2(1 - g_1g_2)}}{R(g_1 + g_2 - 2g_1g_2)}. \tag{2.50}$$

The lower Rayleigh length and waist radius are given by:

$$z'_R = -\frac{Im\{q_1\}}{|q_1|^2} = \frac{R\sqrt{g_1g_2(1-g_1g_2)}}{2g_2} \tag{2.51}$$

$$\omega_0'^2 = \frac{\lambda R\sqrt{g_1g_2(1-g_1g_2)}}{2n\pi g_2}. \tag{2.52}$$

Again, evaluating $Re\{q_1\}/|q_1|^2$ confirms that the lower waist is a distance $(d_1+2d_2)/2$ from the reference plane. An interesting consequence of the preceding calculations is a very simple relation between the two waist radii:

$$\frac{\omega_0}{\omega_0'} = \sqrt{\frac{g_2}{g_1}}. \tag{2.53}$$

Finally, the resonant frequencies are determined by writing down an equation (similar to eqn 2.32) for the round-trip phase change in a ring cavity. The result is:

$$\delta = kL - 2(n+m+1)\left(\tan^{-1}\left(\frac{d_1+2d_2}{2z'_R}\right) + \tan^{-1}\left(\frac{d_3}{2z_R}\right)\right). \tag{2.54}$$

Here, $L$ is the total round-trip distance ($= d_1+2d_2+d_3$). It is interesting (but probably not too surprising) that the two inverse tan functions can be combined to obtain exactly the same result as with the standing-wave cavity:

$$\tan^{-1}\left(\frac{d_1+2d_2}{2z'_R}\right) + \tan^{-1}\left(\frac{d_3}{2z_R}\right) = \cos^{-1}\pm\sqrt{g_1g_2}, \tag{2.55}$$

with the same treatment of the $\pm$ as in the standing-wave case. The resonant frequencies are then,

$$\nu_{nmq} = \left(q + (n+m+1)\frac{\cos^{-1}\sqrt{g_1g_2}}{\pi}\right)\frac{c}{L}, \tag{2.56}$$

and the *free spectral range* (FSR) is,

$$\text{FSR} = \frac{c}{L}. \tag{2.57}$$

If one uses the appropriate definitions for the $g_{1,2}$ parameters and round-trip distances, all of the important relations for the standing-wave and symmetric-ring cavities, except for the waist sizes, are identical. This should not be too surprising in view of the similarity of the ABCD matrices for the two cavity types. We should note that the ring-cavity symmetry requirement only mandates that the curved mirror radii be the same: all of the results hold for cavities with different diagonal lengths and identical radii provided that one changes the long path distance appropriately (from $d_1+2d_2$ to $L-d_3$). Finally, we will summarize the important properties for the two cavity types in Table 2.1.

A typical beam profile in a ring cavity is shown in Fig. 2.8. As can be seen, the waist between the curved mirrors is somewhat smaller than that between the flat mirrors. The former waist would normally be the location of a nonlinear crystal when the cavity is used for nonlinear enhancement, since the fields will be much larger at the smaller waist. A plot of the waist radii appears in Fig. 2.9. As one approaches the region of instability, the small waist radius decreases very rapidly.

**Table 2.1** Summary of cavity properties.

| Property | Standing wave | Symmetric ring |
|---|---|---|
| $g_1$ | $1 - \frac{d}{R_1}$ | $1 - \frac{d_1+2d_2}{R}$ |
| $g_2$ | $1 - \frac{d}{R_2}$ | $1 - \frac{d_3}{R}$ |
| Stability criterion | $0 \leq g_1 g_2 \leq 1$ | $0 \leq g_1 g_2 \leq 1$ |
| Path length | $2d$ | $L = d_1 + 2d_2 + d_3$ |
| Waist(s) | $\omega^2 = \left(\frac{\lambda d}{n\pi}\right) \frac{\sqrt{g_1 g_2 (1-g_1 g_2)}}{g_1+g_2-2g_1 g_2}$ | $\omega_0^2 = \left(\frac{\lambda R}{2n\pi}\right) \frac{\sqrt{g_1 g_2 (1-g_1 g_2)}}{g_1}$ <br> $\omega_0'^{\,2} = \left(\frac{\lambda R}{2n\pi}\right) \frac{\sqrt{g_1 g_2 (1-g_1 g_2)}}{g_2}$ |
| $\nu_{nmq}$ | $\left(q + (n+m+1)\frac{\cos^{-1} \pm\sqrt{g_1 g_2}}{\pi}\right) \frac{c}{2d}$ | $\left(q + (n+m+1)\frac{\cos^{-1} \pm\sqrt{g_1 g_2}}{\pi}\right) \frac{c}{L}$ |
| FSR | $\frac{c}{2d}$ | $\frac{c}{L}$ |

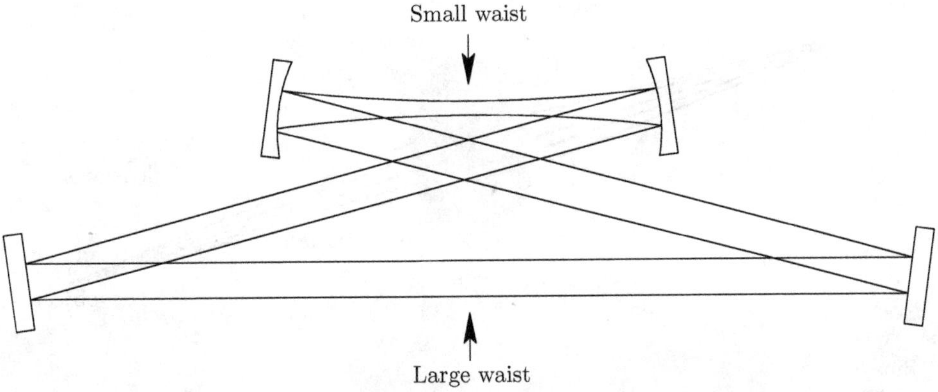


**Fig. 2.8** A typical beam profile in a ring cavity, showing the small waist between the curved mirrors and the larger waist between the flat mirrors.

## 2.8 Astigmatism in a ring cavity

The reader will have undoubtedly noticed that we have violated one of our assumptions in analyzing a ring cavity: the ring-cavity geometry is no longer cylindrically symmetric. From Fig. 2.6, it can be seen that the only place where this matters is where the beam is reflected from the two *curved* mirrors (reflection from the flat mirrors merely *folds* the geometry but introduces no aberrations). The off-axis reflection from the curved mirrors introduces the geometrical aberration called *astigmatism.*

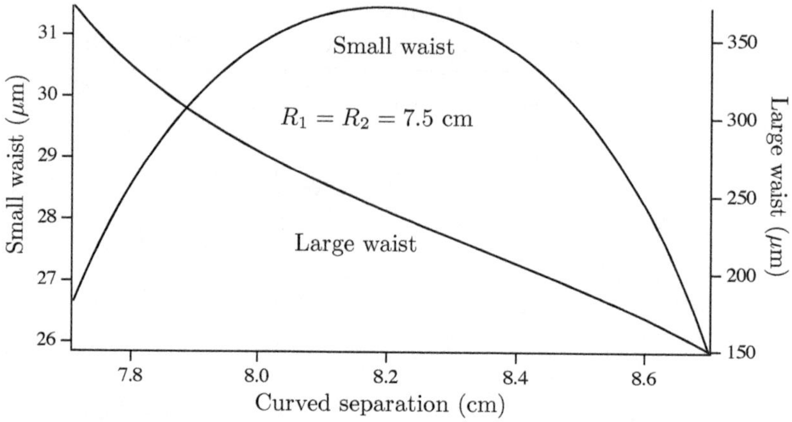


**Fig. 2.9** The small and large waists as a function of the curved mirror separation. The instability region is close to the boundaries of this plot, where the small waist is shrinking rapidly.

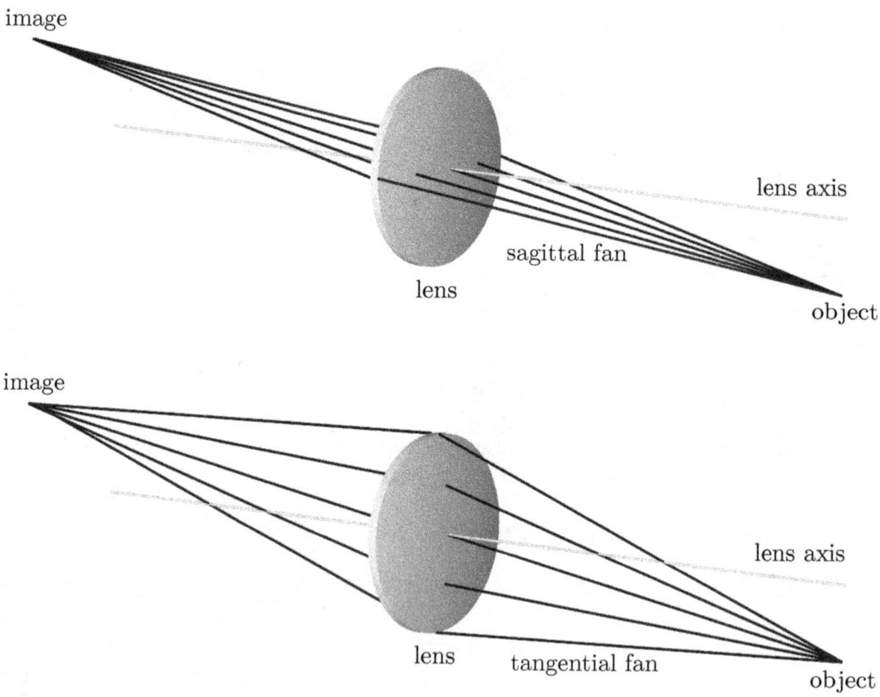


**Fig. 2.10** An illustration of the aberration of *astigmatism*, in which the tangential and sagittal rays need to be treated differently.

The aberration of astigmatism is illustrated in Fig. 2.10. Astigmatism occurs only for off-axis rays, and in describing such rays it is useful to define two planes: the *tangential plane*, which is parallel to the *plane of incidence* (the plane containing the axis of the system and the off-axis ray), and the *sagittal plane*, which is perpendicular to the tangential plane and also contains the off-axis ray. Astigmatism is due to the fact that, for an off-axis bundle of rays, the tangential rays will encounter a different geometry from the sagittal rays.

In the systems described in this book, the causes of astigmatism can always be considered to be optical elements which are tilted away from the symmetry axis of the system so that they break the cylindrical symmetry. We will only discuss *orthogonal*, or *simple, astigmatism* in which all of the tilt angles for a given optical system are in one of two orthogonal planes. More general astigmatism is beyond the scope of our treatment.

We treat astigmatism by considering each transverse coordinate separately but use exactly the same procedures as were used for the cylindrically symmetrical case. For orthogonal astigmatism, one can show that the paraxial wave equation (eqn 1.5) will be satisfied when individual paraxial equations for each transverse coordinate, identical in form to the original (with the same value of $k$), are satisfied. Each coordinate will therefore have its own ABCD matrix and its own value of $q$; the beam profile will now be *elliptical* since the two $\omega$ values are different (the sagittal and tangential wavefront "radii" will also be different). The Gouy phase terms for the transverse coordinates will be different from each other, and the phase shift for each coordinate will be $\pi/2$ (the total for both coordinates will still be $\pi$).

We will consider the astigmatism due to both tilted mirrors and tilted crystals (at Brewster's angle). The tilted mirror with radius of curvature $R$ is shown in Fig. 2.11. It

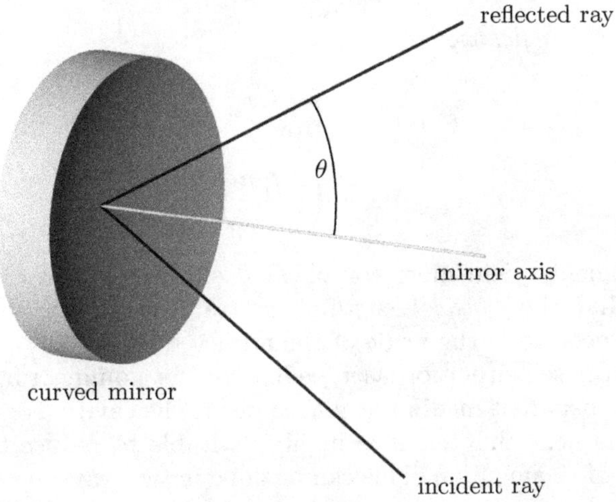


**Fig. 2.11** Off-axis reflection from a spherical mirror.

is well known (see, for example, Jenkins and White (1957)) that, for off-axis reflection of a mirror tilted at angle $\theta$, the *effective radius of curvature* of the mirror (and therefore the focal length) seen by a tangential bundle of rays will be different from that seen by a sagittal bundle. The two focal lengths are:

$$\text{Tangential focal length} = \frac{R\cos\theta}{2} \tag{2.58}$$

$$\text{Sagittal focal length} = \frac{R}{2\cos\theta}. \tag{2.59}$$

The tangential and sagittal ABCD matrices are:

$$M_T = \begin{pmatrix} 1 & 0 \\ \frac{-2}{R\cos\theta} & 1 \end{pmatrix}, \quad M_S = \begin{pmatrix} 1 & 0 \\ \frac{-2\cos\theta}{R} & 1 \end{pmatrix}. \tag{2.60}$$

A crystal of length $l$ and index of refraction, $n$, which is tilted by angle $\theta$ from normal incidence is shown in Fig. 2.12. One can show that the *effective length* of this crystal (the *length* that appears in the ABCD translation matrix, not the *optical length*) is different for tangential and sagittal rays. The resulting ABCD matrices,

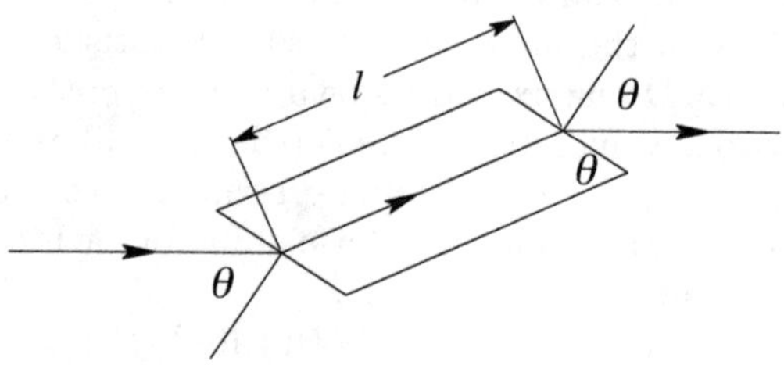


**Fig. 2.12** Off-axis transmission through a crystal.

when $\theta$ is *Brewster's angle*, are:

$$M_T = \begin{pmatrix} 1 & l/n^3 \\ 0 & 1 \end{pmatrix} \tag{2.61}$$

$$M_S = \begin{pmatrix} 1 & l/n \\ 0 & 1 \end{pmatrix}. \tag{2.62}$$

(Note: some treatments yield more complicated expressions since they use the *plate thickness* rather than the *physical length* of the transmitted ray, and these differ by a factor equal to the cosine of the angle of the refracted ray.)

Ring cavities are used either for laser feedback or for nonlinear frequency synthesis. In either case, a transparent medium is placed inside the cavity. For reasons which will be discussed in the next chapter, it is highly desirable to reduce the reflection (and absorption) losses of the medium. This can be done in two ways: by tilting the medium so that its surfaces are at Brewster's angle for a particular polarization, or by using a medium with antireflection coated surfaces. The Brewster approach allows one to *compensate* the astigmatism which will be an inescapable result of tilted cavity mirrors. The compensation can circularize the cavity mode between flat mirrors, which aids in coupling into the cavity, or it can circularize the mode *inside the crystal*, which might improve the efficiency of the nonlinear process for whose enhancement the cavity has been constructed.

In addition to introducing the possible cavity-coupling problems mentioned above, astigmatism will narrow the stability range of the cavity. This is shown in Fig. 2.13, where it is evident that the astigmatism results in a different range of stable curved mirror spacings in the $x$ and $y$ directions. The set of mirror spacings which results in both $x$ and $y$ stability is reduced.

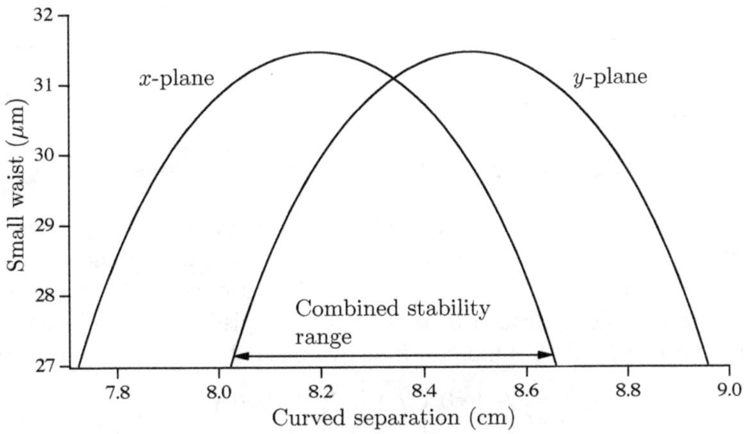


**Fig. 2.13** The reduction in the range of stable mirror spacings due to astigmatism.

## 2.9 Mode matching

It is generally desirable to excite only the lowest (Gaussian) mode in a cavity by judicious choice of the parameters of the input beam. In general, the coupled input beam can be expanded in an infinite series of cavity modes (which form a complete set), and each mode will appear with an amplitude given by the coefficient in the expansion. The problem becomes simplified when the input beam is a Gaussian and the desired cavity mode is the lowest: the internal beam is then completely described by the complex beam parameter and it suffices to match this parameter in the input beam. If one knows the size and location of the cavity mode waist (usually the lower waist in a ring cavity), the mode-matching problem becomes one of generating an input beam with the same waist radius and ensuring that the waists are at the same location. This can be done with one or two lenses.

The single-lens approach is illustrated in Fig. 2.14. We have two waists: $\omega_1$ is the waist of the input beam and $\omega_2$ is the cavity waist. We first define the quantity $f_0$:

$$f_0 = n\pi\omega_1\omega_2/\lambda = \sqrt{z_{R1}z_{R2}}, \tag{2.63}$$

where $z_{R1}$ and $z_{R2}$ are the Rayleigh lengths for waists $\omega_1$ and $\omega_2$. We then choose a lens whose focal length $f > f_0$ and adjust the distances $d_1$ and $d_2$ to satisfy

$$d_1 = f \pm \frac{\omega_1}{\omega_2}\sqrt{f^2 - f_0^2} \qquad (2.64)$$

$$d_2 = f \pm \frac{\omega_2}{\omega_1}\sqrt{f^2 - f_0^2}. \qquad (2.65)$$

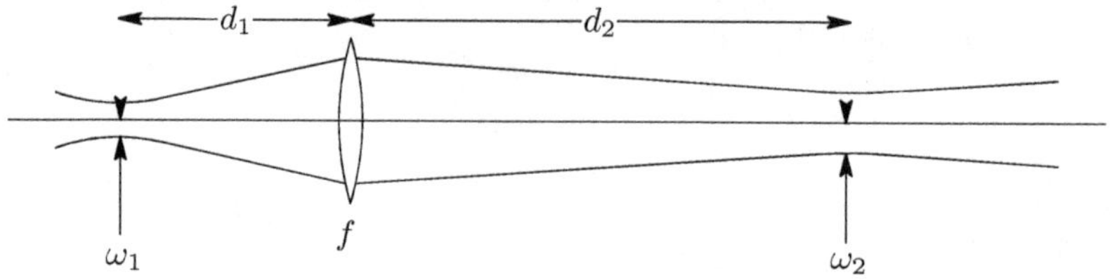


**Fig. 2.14** Mode matching with a single lens.

The disadvantage of this approach is that for a given lens the distances are fixed, and for fixed locations of the two waists it is not that easy in the usual laboratory environment to find a lens that will satisfy both equations.

A much more flexible approach to mode matching is to use two lenses. This situation is illustrated in Fig. 2.15. One can first evaluate the ABCD matrix up to lens 2:

$$\begin{pmatrix} A & B \\ C & D \end{pmatrix} = \begin{pmatrix} 1 & 0 \\ \frac{-1}{f_2} & 1 \end{pmatrix} \begin{pmatrix} 1 & d_3 \\ 0 & 1 \end{pmatrix} \begin{pmatrix} 1 & 0 \\ \frac{-1}{f_1} & 1 \end{pmatrix} \begin{pmatrix} 1 & d_1 \\ 0 & 1 \end{pmatrix}. \qquad (2.66)$$

Then the complex beam parameter at waist 2 is given by:

$$q_2 = \frac{Aq_1 + B}{Cq_1 + D} + d_2, \qquad (2.67)$$

and the desired beam properties at waist 2 are:

$$R_2 = 1/Re\{1/q_2\} \quad \text{and} \quad \omega_2 = \sqrt{\frac{-\lambda}{n\pi Im\{1/q_2\}}}. \qquad (2.68)$$

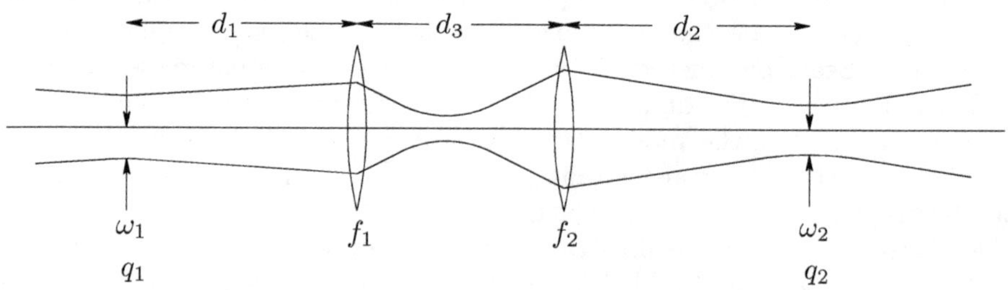


**Fig. 2.15** Mode matching with two lenses.

Although the calculations needed for two-lens mode matching are not too difficult, there is a very simple procedure which can quickly provide an approximate solution which is usually adequate. This approach is illustrated in Fig. 2.16.

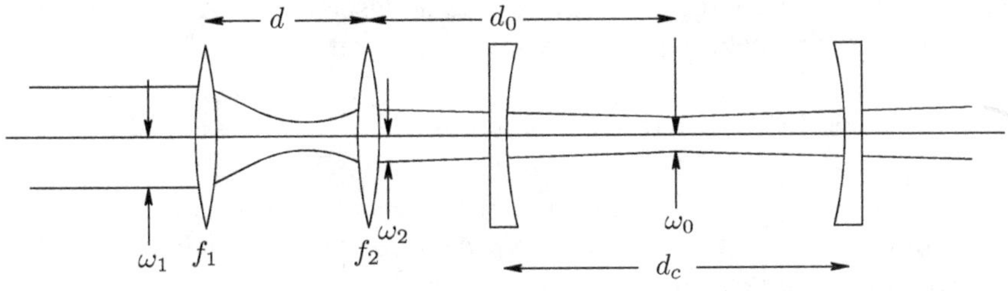


**Fig. 2.16** Illustration of a simplified approach to two-lens mode matching.

The simplified two-lens procedure is as follows:

| **Procedure** | **Example** |
|---|---|
| 1. Determine $\omega_0$ and $z_R$ from the cavity equations. | (Confocal) $d_c = 10$ cm, $\lambda = 1\ \mu$m, $\omega_0 = 126\ \mu$m, $z_R = 5$ cm |
| 2. Given $d_0$, the distance from waist to lens 2, determine $\omega_2$ using $\omega_2 = \omega_0\sqrt{1 + (d_0/z_R)^2}$. | $d_0 = 10$ cm, $\omega_2 = 282\ \mu$m |
| 3. Given input spot size ($\omega_1$), find two lenses whose focal lengths are in ratio $\omega_1/\omega_2$. | $\omega_1 = .05$ cm, ratio $\approx 1.8$, $f_1 = 1.8$ cm, $f_2 = 1$ cm |
| 4. Separate lenses by slightly more than $f_1 + f_2$ so waist is at center of cavity. | $f_1 + f_2 = 2.8$ cm |

## 2.10 Beam quality characterization: the $M^2$ parameter

The beams considered in this chapter and the previous one have been idealizations: either a pure Gaussian beam or a single-mode beam described by a Hermite Gaussian function. These idealized beams are rarely found in the real world. A Gaussian beam has the property of being a *minimum uncertainty* function: the product of the beam width and the diffraction angle is the minimum possible. For this reason, a Gaussian beam is in some sense an *ideal, diffraction-limited laser beam.* We seek a simple way of characterizing the departure of the beam characteristics (mainly the divergence) of an arbitrary beam from those of an ideal Gaussian beam. We will show that a very reasonable way of doing this is to use the so-called $M^2$ parameter.

Before describing the $M^2$ parameter, we should note that a measurement of the intensity profile at a *single place* is a completely inadequate way of characterizing a laser beam. There are illustrative examples in the literature of beam profiles which are almost indistinguishable from that of a Gaussian beam but actually contain *zero* $TEM_{00}$ (Gaussian) contribution. Such a beam might look like a Gaussian beam at one place, but it will *diverge much more rapidly* than an actual Gaussian beam. Two conclusions can be drawn from this. First, a beam is properly characterized by *both* its intensity profile and its divergence. Second, if a profile measurement is used for beam characterization, it must be done at *more than one place along the beam path* so that some measure of the divergence is obtained.

To obtain some motivation for the $M^2$ measurement, we will first consider the one-dimensional behavior of two actual beams: the diffracted beam from a plane wave which encounters a rectangular slit and a $TEM_{00}$ laser beam. A slit of width $2a$ will have a *far-field* electric field dependence which is given by

$$\text{Rectangular slit:} \quad E(x,z) \propto \frac{\sin(2\pi ax/z\lambda)}{2\pi ax/z\lambda}, \tag{2.69}$$

where $z$ is the propagation direction and $x$ is the transverse coordinate. It will be useful to examine the $z$-dependence of the *product of the near-field and far-field beam widths.* For this case, this product is

$$\text{Rectangular slit:} \quad \Delta x_0 \times \Delta x(z) = 0.5 \times \lambda z, \tag{2.70}$$

where $\Delta x_0 = a$ is the near-field half-width (the slit half-width) and $\Delta x(z)$ is the distance from $x = 0$ to the first null in the diffracted pattern. The Gaussian beam width as a function of $z$ is given by

$$\text{Gaussian beam:} \quad \omega^2(z) = \omega_0^2 \left(1 + \left(\frac{z}{z_R}\right)^2\right), \tag{2.71}$$

where $z = 0$ at the beam waist. The product of the near-field width $\omega_0$ and the far-field width $\omega$ is

$$\text{Gaussian beam:} \quad \omega_0 \times \omega(z) = \frac{1}{\pi} \times \lambda z, \tag{2.72}$$

where we have made the far-field measurement at many Rayleigh lengths from the waist, so that $\omega \gg \omega_0$. For these two cases, the product of the near-field and far-field widths is some constant times $\lambda z$. This numerical constant is a measure of the departure of the beam from the ideal Gaussian behavior: the larger it is, the greater is the departure.

There are a variety of definitions of beam width: the $1/e$ width, the width between first nulls, etc. Without getting too far into the justification of one width over another, we note that the $M^2$ parameter uses the *second moment* or *variance* widths, $\sigma_x$ and $\sigma_y$, defined by

$$\sigma_x^2 = \frac{\int_{-\infty}^{\infty} (x - x_0)^2 I(x,y)dxdy}{\int_{-\infty}^{\infty} I(x,y)dxdy} \tag{2.73}$$

$$\sigma_y^2 = \frac{\int_{-\infty}^{\infty} (y - y_0)^2 I(x,y)dxdy}{\int_{-\infty}^{\infty} I(x,y)dxdy}, \tag{2.74}$$

where $I(x,y)$ is the beam *intensity*, $x, y$ are the transverse coordinates and $x_0, y_0$ are at the *centroid* of the intensity function. It can be shown that the widths defined in this way obey a formula similar to that for a Gaussian beam

$$\sigma_x^2 = \sigma_{0x}^2 + \sigma_{\theta_x}^2 \times z^2 \tag{2.75}$$

$$\sigma_y^2 = \sigma_{0y}^2 + \sigma_{\theta_y}^2 \times z^2, \tag{2.76}$$

where $z = 0$ at the waist and $\sigma_{\theta_{x,y}}^2$ are the variances of the angular divergence of the beam in the $x$ and $y$ directions. It appears that these equations are rigorously satisfied only for the second moment definition of the beam width. For a Gaussian beam, $\sigma_x = \omega_x/2, \sigma_y = \omega_y/2$. In order to more conveniently compare an arbitrary beam to a *standard* Gaussian beam, the following width definitions are used:

$$W_x \equiv 2\sigma_x \tag{2.77}$$
$$W_y \equiv 2\sigma_y. \tag{2.78}$$

We are now in a position to define the $M^2$ parameter. An arbitrary beam has the divergence properties described by the following equations:

$$W_x^2(z) = W_{0x}^2 \left(1 + M_x^4 \times \left(\frac{z}{Z_{Rx}}\right)^2\right) \tag{2.79}$$
$$W_y^2(z) = W_{0y}^2 \left(1 + M_y^4 \times \left(\frac{z}{Z_{Ry}}\right)^2\right), \tag{2.80}$$

where $Z_{Rx}$ and $Z_{Ry}$ are *generalized Rayleigh lengths* defined by

$$\begin{aligned} Z_{Rx} &= \frac{n\pi W_{0x}^2}{\lambda} \\ Z_{Ry} &= \frac{n\pi W_{0y}^2}{\lambda} \end{aligned} \tag{2.81}$$

and $W_{0x}, W_{0y}$ are the widths at $z = 0$, while $n$ is the index of refraction. In analogy with a Gaussian beam, the near-field far-field products are

$$W_{0x} \times W_x(z) \approx M_x^2 \times \frac{z\lambda}{n\pi} \tag{2.82}$$
$$W_{0y} \times W_y(z) \approx M_y^2 \times \frac{z\lambda}{n\pi}, \tag{2.83}$$

where the far-field measurements are taken many "Rayleigh" lengths from the waist. Note that $M_{x,y}^2 = 1$ for a Gaussian beam and is greater than unity for non-Gaussian beams (since the Gaussian beam has the *minimum uncertainty* property). The $M^2$ parameters are a measure of "how many times diffraction limited" an arbitrary beam is, or, equivalently, how much faster a beam diverges than an equivalent Gaussian beam with the same "waist" size. Note that $M^2$ has little to do with the apparent shape of a laser beam: an astigmatic beam with an elliptical shape can still have a low $M^2$ since the latter measures the deviation from a diffraction-limited beam in only *a single plane* containing the propagation direction and one transverse coordinate. A practical means for measuring $M^2$ is left to the exercises.

## 2.11 Further reading

This chapter discusses the application of Gaussian beam theory to the geometrical aspects of standing-wave and symmetric-ring cavities. Much of the material recommended in the previous chapter will be applicable to the design of cavities, particularly the pioneering paper by Kogelnik and Lee (1966). Some of the ring-cavity theory is original to the author of this book, but excellent descriptions of ring cavities can be found in the early papers by members of the Boulder, Colorado NIST group, who used these cavities to enhance the generation of ultraviolet light from continuous-wave visible lasers. (See, for example, Hemmati (1983).)

## 2.12 Problems

(2.1) Due to manufacturing errors, a pair of mirrors (whose nominal radii should be 200 mm) have radii of 203 mm and 197 mm. There will be a problem if these mirrors are used in a "confocal" cavity with a mirror separation of 200 mm. What is the problem and what are the possible mirror spacings which solve it?

(2.2) You are given a flat plate of glass having index of refraction $n$ and parallel surfaces whose separation is $t$. This object is tilted with respect to the $z$-axis so that the angle between its *normal* and the $z$-axis is $\theta$. As a result of the tilt, the plate exhibits *astigmatism*: the behavior of a paraxial "fan" of rays in the tangential plane (the one containing the normal and $z$-axis) is different from the behavior of such a "fan" in the sagittal plane (the plane which contains the $z$-axis and is perpendicular to the tangential plane). Calculate the ABCD matrices between the entrance and exit points of the plate for both planes. Ignore the parallel displacement of the beam by the tilted plate: assume that the $z$-axis for the ray exiting the plate is displaced from the $z$-axis of the entering ray (and assume the rays make only small angles with respect to both axes).

(2.3) You are given a resonator consisting of a convex mirror ($R_1 < 0$) and a concave mirror ($R_2 > 0$) separated by a distance $d$. Find the values of $d$ for which the resonator is stable both for $|R_1| > R_2$ and $|R_1| < R_2$.

(2.4) Consider a near-planar symmetric resonator made with mirrors of radius $R$ and separation $d \ll R$. Obtain an approximate expression for the spot sizes on the mirrors and at the waist. Use your results to calculate the spot sizes when $R =$ 8 m and $d = 1$ m at a wavelength of 514 nm. (Assume $n = 1$.)

(2.5) Consider a near-concentric resonator composed of two mirrors of radius $R$ and separation $2R - \Delta d$. Give an approximate expression for the spot sizes on the mirrors and at the waist. Compare this to the result of problem 2.4. (Assume $n = 1$.)

(2.6) Using just a well-corrected lens and a device for measuring the spot size of a beam, describe a method to measure, *very approximately*, the $M^2$ value for an arbitrary beam.

(2.7) The mode-matching techniques for standing-wave cavities described in the text did not take account of the refractive power of the input mirror which acts like a weak negative lens. Assume that one has an approximately confocal cavity with mirror separation $d$. Approximately how far is the apparent waist position shifted by refraction from the input mirror? (Treat the input mirror as a thin plano-concave lens with refractive index $n$.) Assume $n = 1.5$ and determine the actual shift in terms of $d$.

# 3
# Energy relations in optical cavities

## 3.1 Introduction

In the previous chapter, we discussed the cavity field distributions which are the eigenmodes of the cavity. We required that the field distribution replicates itself after one round trip but said nothing about how many round trips the propagating field would make before it was attenuated, nor did we discuss how the field would be injected into the cavity (or extracted from the cavity). The discussion of cavity eigenmodes depended only on the resonator geometry and the wavelength of the radiation. In discussing the build-up of the field in the cavity, we need to consider everything which affects the power and energy of the internal field. Thus we need to be concerned with both the dissipative losses and the Fresnel reflection losses from the mirrors and intracavity elements. The study of cavity losses is intimately connected with the *frequency resolution* of cavities, since it determines how many times the beam can be made to interfere with itself. The discussion of energy relations in cavities is analogous to the study of *dynamics* in classical mechanics, while our previous study of cavity field configurations is analogous to the study of *kinematics*.

We will begin with a simple derivation of the relation between the field transmission and reflection coefficients based upon time reversal symmetry and, using this, proceed to derive the size and frequency dependence of the internal fields for standing-wave and traveling-wave cavities. We will then discuss resonator $Q$, finesse, photon lifetime and *impedance matching* into a cavity. Finally we will describe a novel technique for coupling radiation into a cavity using an intracavity transparent wedge, and will conclude with a discussion of some practical matters, such as the experimental determination of certain cavity parameters.

## 3.2 Reflection and transmission at an interface

An elegant derivation of the relation between the field transmission and reflection coefficients was made by the British physicist Sir George Gabriel Stokes (1819–1903). He assumed that there was *no dissipation* at the interface and therefore the process obeyed *time-reversal invariance*. His argument is illustrated in Fig. 3.1, which depicts a wave incident on a transparent surface.

Assume that the amplitude transmission coefficients are $t, t'$ and the reflection coefficients are $r, r'$, where the unprimed quantities are used when one approaches the interface from above and the primed quantities are used when the interface is approached from below. The three waves are shown on the left in the figure. The time-reversed case appears in the center plot. When we reverse time, we need to include three additional waves: the reflection from the bottom wave, and the transmission of both the bottom and top waves. All of the waves are shown in the plot on the right. In order for the right-hand plot to be equivalent to the time-reversed center plot, the two waves in the upper left must sum to $E$ and the two waves in the lower left must sum to zero. Thus,

$$E = Ett' + Err \quad \text{and} \quad 0 = Ert + Er't, \tag{3.1}$$

from which we conclude that

$$tt' + r^2 = 1 \quad \text{and} \quad r = -r.' \tag{3.2}$$

This shows that there is a sign change between the *amplitude* reflection coefficients on opposite sides of an interface and that the *intensity* transmission and reflection coefficients ($T$ and $R$) follow the usual rule (since $t = t'$):

$$R + T = 1, \quad \text{where} \quad R = r^2, T = t^2. \tag{3.3}$$

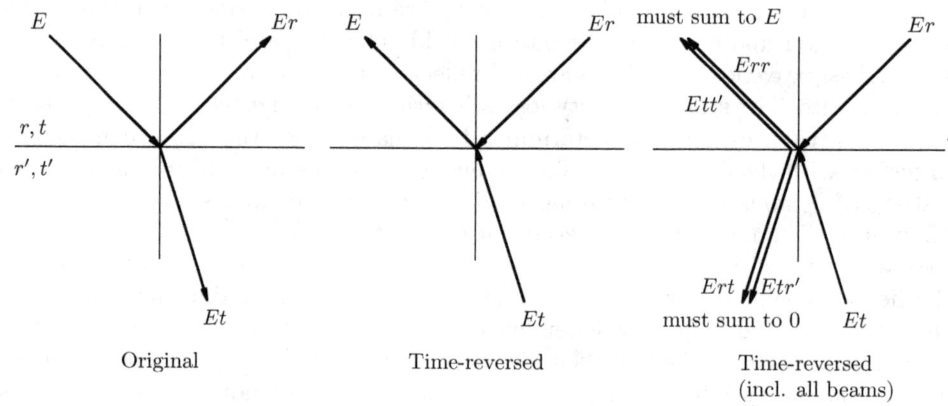


**Fig. 3.1** Original and time reversed behavior at interface.

## 3.3 Reflected fields from standing-wave cavity

We will use the above relations to determine the electric field of the wave which is reflected from a two-mirror standing-wave cavity. We will assume that the reflection and transmission from the mirrors are free from dissipative losses and will place a potentially lossy medium inside the cavity. The latter will model the Fresnel losses from an intracavity medium (such as a nonlinear crystal) as well as all of the *dissipative* losses of both the mirrors and this medium. The medium will be characterized by the *field* transmission coefficient, $t$. The mirrors have reflection and transmission coefficients, $r_1, r_2$ and $t_1, t_2$.

The total reflected field is calculated by summing the fields from all possible internal passes. The first three passes are shown in Fig. 3.2, where we assume a round-trip phase delay of $\delta$. We have used eqn 3.2, which requires that the sign from the first (external) reflection be opposite to the signs of the internal reflections. If we sum the first three passes, we obtain a reflected field, $E_r$, given by

$$\begin{aligned} E_r &= -E_0 r_1 + E_0 r_2 t_1^2 t^2 e^{-i\delta} + E_0 r_1 r_2^2 t_1^2 t^4 e^{-2i\delta} + \cdots \\ &= -E_0 r_1 + \frac{E_0 t_1^2}{r_1}\left(r_1 r_2 t^2 e^{-i\delta} + (r_1 r_2 t^2 e^{-i\delta})^2 + \cdots\right) \\ &= E_0 \frac{r_2 t^2 e^{-i\delta} - r_1}{1 - r_1 r_2 t^2 e^{-i\delta}}, \end{aligned} \tag{3.4}$$

where we have summed the geometrical series and used the relation $r_1^2 + t_1^2 = 1$.

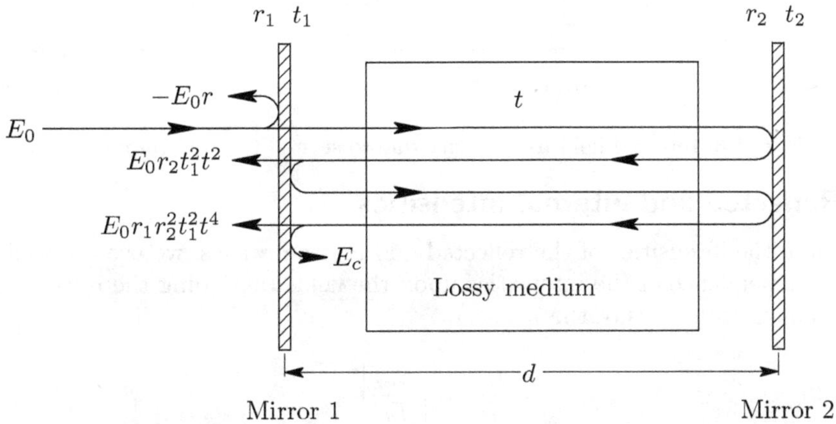


**Fig. 3.2** Reflected fields from cavity due to several round-trips inside the cavity.

## 3.4 Internal (circulating) field in a standing-wave cavity

Resonant cavities are often used to enhance the power incident on an intracavity crystal for nonlinear frequency synthesis. To aid in the analysis of such a system, we will calculate the internal, circulating field to the left of the internal medium in a standing-wave cavity; the remaining case of the transmitted field will be left as an exercise for the reader. We are only interested in the fields propagating in *the same direction as the incident wave*, since most applications (such as nonlinear optics enhancement) involving the internal field only work for a *traveling wave*. The actual field, of course, inside the cavity will be a *standing wave*: the superposition of two counter-propagating waves. As we did with the reflected fields, we sum the first three contributions to the internal field and use the formula for an infinite geometrical series to obtain the circulating field, $E_c$ (see Fig. 3.3):

$$\begin{aligned} E_c &= E_0 t_1 + E_0 r_1 r_2 t_1 t^2 e^{-i\delta} + E_0 r_1^2 r_2^2 t_1 t^4 e^{-2i\delta} + \cdots \\ &= E_0 t_1 \left(1 + r_1 r_2 t^2 e^{-i\delta} + (r_1 r_2 t^2 e^{-i\delta})^2 + \cdots\right) \\ &= \frac{E_0 t_1}{1 - r_1 r_2 t^2 e^{-i\delta}}. \end{aligned} \tag{3.5}$$

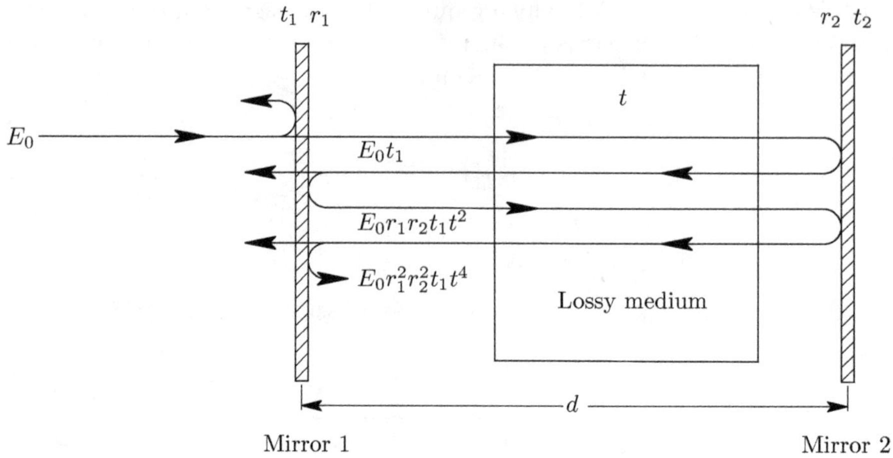


**Fig. 3.3** Internal field from cavity due to several internal round trips.

## 3.5 Reflected and internal intensities

To calculate the intensities of the reflected and internal waves, we use the well-known quadratic dependence of the intensities upon the fields (assuming the refractive index is the same on both sides of the mirror),

$$\frac{I_{r,c}}{I_0} = \left|\frac{E_{r,c}}{E_0}\right|^2, \tag{3.6}$$

to obtain

$$\text{Field:}\quad E_r = E_0 \frac{r_m e^{-i\delta} - r_1}{1 - r_1 r_m e^{-i\delta}} \tag{3.7}$$

$$\text{Intensity:}\quad I_r = I_0 \frac{(r_1 - r_m)^2 + 4r_1 r_m \sin^2 \delta/2}{(1 - r_1 r_m)^2 + 4r_1 r_m \sin^2 \delta/2} \tag{3.8}$$

for the reflected field and intensity and

$$\text{Field:}\quad E_c = E_0 \frac{t_1}{1 - r_1 r_m e^{-i\delta}} \tag{3.9}$$

$$\text{Intensity:}\quad I_c = I_0 \frac{t_1^2}{(1 - r_1 r_m)^2 + 4r_1 r_m \sin^2 \delta/2} \tag{3.10}$$

for the circulating field and intensity. In these expressions, we have defined the quantity $r_m \equiv r_2 t^2$ and used the relation $\cos\delta = 1 - 2\sin^2\delta/2$.

As mentioned above, the internal intensity is for the wave propagating in the same direction as the incident wave. The internal field will be a standing wave whose *power* will have a $\cos^2 kz$ dependence and whose *average power* will be twice that derived from our expression for the internal intensity, since we are ignoring the wave propagating in the other direction.

## 3.6 The resonant character of the reflected and circulating intensities

We have already found the relation between the round-trip phase ($\delta$) and the frequency of the beam (eqn 2.32). In the following, it will be simpler to consider the phase to be the independent variable and to investigate the behavior of the intensities near a *resonance*. We observe that the intensity formulas are periodic functions of $\delta$ with a period of $2\pi$. Thus, in terms of $\delta$, the *free spectral range* is $2\pi$.

The circulating intensity (eqn 3.10) has a maximum when $\delta = 0$ (modulo integral multiples of $2\pi$). The *half maximum intensity* occurs at a phase ($\delta_{1/2}$) which is a solution of:

$$(1 - r_1 r_m)^2 = 4 r_1 r_m \sin^2 \delta_{1/2}/2, \tag{3.11}$$

whose solution for $\delta_{1/2} \ll 1$ is:

$$\delta_{1/2} = 2 \sin^{-1}\left(\frac{1 - r_1 r_m}{2\sqrt{r_1 r_m}}\right) \approx \frac{1 - r_1 r_m}{\sqrt{r_1 r_m}}. \tag{3.12}$$

(We will shortly justify our assumption that $\delta_{1/2} \ll 1$.) The full width at half maximum (FWHM), $\Delta\nu_{1/2}$, of the circulating intensity at resonance is equal to:

$$\text{FWHM} \equiv \Delta\nu_{1/2} = 2\delta_{1/2} = \frac{2(1 - r_1 r_m)}{\sqrt{r_1 r_m}}. \tag{3.13}$$

One of the several definitions of the *finesse* (symbolized by $\mathcal{F}$) is that it is the ratio of the free spectral range of the cavity to the full width at half maximum of the *transmitted* (and internal) intensity resonances. Thus,

$$\text{Finesse: } \mathcal{F} = \frac{\text{free spectral range}}{\Delta\nu_{1/2}} = \frac{\pi\sqrt{r_1 r_m}}{1 - r_1 r_m} \quad (\text{FSR} = 2\pi). \tag{3.14}$$

The approximation ($\delta_{1/2} \ll 1$) made in deriving eqn 3.12 is justified by the fact that most useful cavities have a finesse $\gg 1$ and therefore $\delta_{1/2} \ll \pi$. For a lossless, empty cavity ($t = 1$) with identical mirrors ($r_1 = r_2$), eqn 3.14 agrees with the formula for finesse found in many optics textbooks:

$$\mathcal{F} = \frac{\pi\sqrt{r_1^2}}{1 - r_1^2} = \frac{\pi\sqrt{R}}{1 - R} \quad (t = 1, r_1 = r_2, R = r_1^2). \tag{3.15}$$

We conclude this section by obtaining the cavity electric fields and intensities as a function of the frequency of the laser beam. First, we note that the round-trip phase change, $\delta$, can be written as:

$$\delta = \frac{4\pi n d}{\lambda_0} + 2\epsilon = \frac{4\pi n d \nu}{c} + 2\epsilon, \tag{3.16}$$

where $\lambda_0$ is the vacuum wavelength of the radiation, $\nu$ is the frequency, $\epsilon$ is the phase change upon reflection from each mirror and $n$ is the index of refraction. Note that we have not included the Gouy phase term, since it has no frequency dependence and will merely shift all of the axial resonances by the same amount but not change the separation between them. For the same reason, we can set $\epsilon = 0$. From eqns 3.8 and 3.10, it should be obvious that the maximum circulating power and minimum reflected power both occur when $\delta/2 = q\pi$, where $q$ is an integer. From this we get our customary result that the axial resonance frequencies are given by $\nu = qc/2nd$. If we assume that the finesse is very high, we can expand the frequency around any of the resonances and obtain

$$\text{Circulating:} \quad E_c = E_0 \frac{t_1 \mathcal{F} \Delta\nu_{1/2}/\pi}{\Delta\nu_{1/2} + 2i(\nu - \nu_0)} \tag{3.17}$$

$$I_c = I_0 \frac{(t_1 \mathcal{F} \gamma/\pi)^2}{(\Delta\nu_{1/2})^2 + 4(\nu - \nu_0)^2}, \tag{3.18}$$

where $\nu_0$ is one of the resonance frequencies, and the FWHM ($\Delta\nu_{1/2}$) is just the free spectral range ($c/2nd$) divided by the finesse (eqn 3.14). We have assumed that $r_1, r_m \approx 1$ and have removed factors such as $r_1$ or $r_1 r_m$, but have retained differences such as $r_1 - r_m$ and $1 - r_1 r_m$.

The expressions for the reflected fields and intensities can also be written as a function of frequency under the same assumptions as for the internal field and intensity:

$$\text{Reflected:} \quad E_r = E_0 \frac{(r_m - r_1)\mathcal{F}\Delta\nu_{1/2}/\pi - 2i(\nu - \nu_0)}{\Delta\nu_{1/2} + 2i(\nu - \nu_0)} \tag{3.19}$$

$$I_r = I_0 \frac{(r_m - r_1)^2 (\mathcal{F}\Delta\nu_{1/2}/\pi)^2 + 4(\nu - \nu_0)^2}{(\Delta\nu_{1/2})^2 + 4(\nu - \nu_0)^2}. \tag{3.20}$$

## 3.7 Impedance matching

It is well known from electrical network theory that a generator delivers the maximum power to a load when the source impedance is equal to the complex conjugate of the load impedance. Although we have not defined impedance in the optical realm, inspection of eqn 3.10 indicates that there should be a set of parameters ($r_1, r_m$) for which the circulating power is a maximum at resonance ($\delta = 0$). By analogy with an electrical network, this optimum coupling is called the *impedance matched* case. We will determine the conditions for impedance matching.

First, it will be useful to state the significance of the parameter $r_m$. This is simply the fraction of the electric field which *remains* after one round trip, *excluding the input coupling*, $t_1$. We will define $L$ to be the *round-trip power loss*, again excluding the transmission through the input mirror. From the above,

$$L + r_m^2 = 1 \implies r_m^2 = 1 - L. \tag{3.21}$$

The expressions for the circulating and reflected power at resonance are (from eqns 3.8 and 3.10)

$$\text{Circulating:} \quad I_c = I_0 \frac{1 - r_1^2}{(1 - r_1 r_m)^2} \tag{3.22}$$

$$\text{Reflected:} \quad I_r = I_0 \frac{(r_1 - r_m)^2}{(1 - r_1 r_m)^2}, \tag{3.23}$$

where we have used $r_1^2 + t_1^2 = 1$. Differentiating the first with respect to $r_1$ and setting the result equal to zero yields an extremum (maximum) in the circulating power when $r_1 = r_m$. When this is the case, the second equation indicates that there will be *zero reflected power*. The *impedance-matched condition* is:

$$\text{Impedance matched:} \quad r_1 = r_m \Longrightarrow L = t_1^2. \tag{3.24}$$

Stated in words: *A cavity is impedance matched when the intensity transmission coefficient of the input-coupling mirror is equal to the sum of all of the other losses, excluding the input coupling.* When this happens, the internal circulating power will be a maximum and the reflection will be zero (at resonance).

It will be useful at this point to investigate the *impedance-mismatched* case. We start by observing (from eqn 3.10) that the ratio of $I_c$ under mismatched conditions to $I_c$ under matched conditions is (at resonance):

$$\frac{I_c}{I_{c,matched}} = \frac{(1 - r_1^2)(1 - r_m^2)}{(1 - r_1 r_m)^2}. \tag{3.25}$$

We will define the *mismatch parameter*, $\sigma$, to be:

$$\sigma \equiv \frac{1 - r_1}{1 - r_m} \approx \frac{1 - r_1^2}{1 - r_m^2} = \frac{t_1^2}{L} \quad \text{if } 1 - r_1 \ll 1 \text{ and } 1 - r_m \ll 1. \tag{3.26}$$

Thus, $\sigma = 1$ under matched conditions. From this, eqn 3.25 can be written as

$$\frac{I_c}{I_{c,match}} = \frac{2(1 + r_m)\sigma - (1 - r_m^2)\sigma^2}{(\sigma r_m + 1)^2} \approx \frac{4\sigma}{(\sigma + 1)^2} \qquad (r_m \approx 1). \tag{3.27}$$

A plot of the relative circulating power versus the mismatch parameter, $\sigma$, appears in Fig. 3.4. From the figure, it can be seen that an error in the input coupling by a factor of two (in either direction) reduces the circulating power by only 11%, which is fortunate in view of the difficulty in making mirrors with a precise transmission.

The reflected power on resonance will not be zero when the cavity is not impedance matched. Writing the formula for the reflected power (eqn 3.23) in terms of $\sigma$, one gets (when $r_m \approx 1$):

$$\frac{I_r}{I_0} = \frac{(r_1 - r_m)^2}{(1 - r_1 r_m)^2} = \left(\frac{\sigma - 1}{\sigma r_m + 1}\right)^2 \approx \left(\frac{\sigma - 1}{\sigma + 1}\right)^2. \tag{3.28}$$

Of course, $I_r/I_0 \approx 1$ far from resonance. Plots of the circulating and reflected intensity under non-matched conditions are shown in Fig. 3.5.

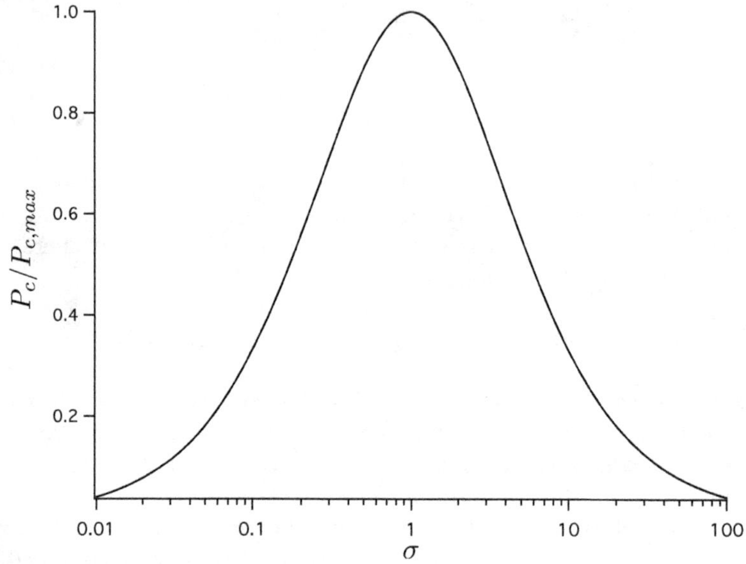


**Fig. 3.4** Relative circulating power versus mismatch parameter, $\sigma$.

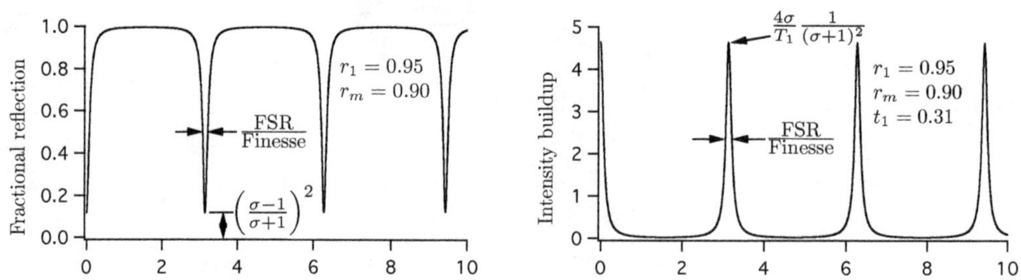


**Fig. 3.5** Spectrum of reflected and circulating power.

A few comments on impedance matching are perhaps in order before proceeding. First, under impedance-matched conditions, the *enhancement* ($I_c/I_0$) has a very simple representation and is simply connected to the *finesse*:

$$\text{Enhancement} = \frac{1}{T_1} = \frac{\langle\text{finesse}\rangle}{\pi} \quad \text{(impedance matched)} \tag{3.29}$$

Since $T_1 = L$ in a matched cavity, it is essential to reduce the losses as much as possible in order to obtain a significant enhancement. With a well-designed cavity, it is possible to obtain an enhancement of a bit greater than 100 when the cavity contains an internal nonlinear crystal. This has enormous implications for certain nonlinear processes: for example, second-harmonic generation, which depends quadratically on the fundamental power, can be enhanced by a factor of 10,000. Next, one might think that a practical method for experimentally identifying impedance matching is the absence of reflected intensity on resonance. While this is true, the converse is not: a non-zero reflected power at resonance does not necessarily mean that the cavity is *not* impedance matched. A finite reflection from an impedance-matched cavity can still be obtained when the cavity is not *mode matched*: not all of the incident (perhaps multimode) power is coupled into the single cavity mode and that which is not coupled will be reflected from the input mirror whether or not the cavity is impedance matched.

## 3.8 Fields and intensities in ring cavity

Our analysis of the symmetric-ring cavity is similar to that of the standing-wave cavity except that we require that all of the mirrors (except the *input coupler*) have nearly 100% reflectivity. The input coupler transmission and reflectivity are (to keep the notation consistent with the standing-wave cavity) $t_1$ and $r_1$. If we use the same definition of $r_m$ as with the standing-wave cavity, it should be evident that

$$r_m = t. \tag{3.30}$$

A schematic of the cavity appears in Fig. 3.6.

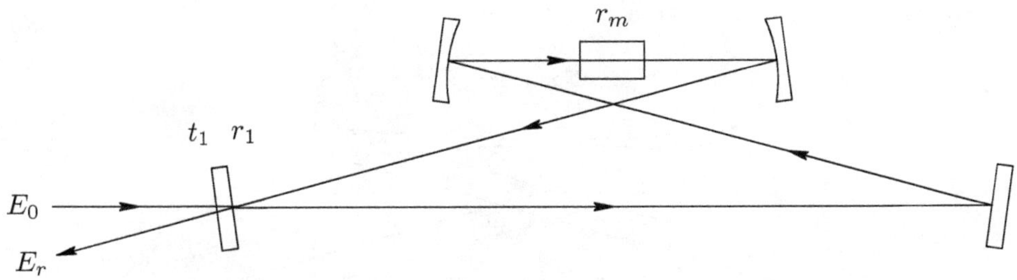


**Fig. 3.6** The ring cavity displaying items needed to calculate internal and reflected fields.

The analysis proceeds in exactly the same fashion as for the standing-wave cavity and begins by summing the fields for the first three round trips. The electric field of the reflected wave (with the usual definition of $\delta$ as the round-trip phase change) is

$$\begin{aligned} E_r &= -E_0 r + E_0 r_m t_1^2 e^{-i\delta} + E_0 r_1 r_m^2 t_1^2 e^{-2i\delta} + \cdots \\ &= -E_0 r + E_0 r_m t_1^2 e^{-i\delta}(1 + r_1 r_m e^{-i\delta} + \cdots) \\ &= E_0 \frac{r_m e^{-i\delta} - r_1}{1 - r_1 r_m e^{-i\delta}}. \end{aligned} \tag{3.31}$$

The circulating field is

$$\begin{aligned} E_c &= E_0 t_1 + E_0 r_1 r_m t_1 e^{-i\delta} + E_0 r_1^2 r_m^2 t_1 e^{-2i\delta} + \cdots \\ &= E_0 t_1 (1 + r_1 r_m e^{-i\delta} + r_1^2 r_m^2 e^{-2i\delta} + \cdots) \\ &= E_0 \frac{t_1}{1 - r_1 r_m e^{-i\delta}}. \end{aligned} \tag{3.32}$$

We observe that these are exactly the same results as for the standing-wave cavity. This should come as no surprise, since we were performing exactly the same operations and were careful to use the same definition of $r_m$ in both cases (although $r_m$ depends *differently* on the cavity parameters for the two cavity types).

From the preceding analysis, we should be safe in claiming that all of the results for the reflected and circulating fields and intensities of a standing-wave cavity, as well as all of the impedance-matching results, should apply unchanged to the ring cavity.

## 3.9 A novel "reflective" coupling scheme using a tilted wedge

The conventional way to couple light into a cavity is to use a partially transmitting mirror as one of the cavity mirrors and to couple light through it. The transmission of this *input-coupling mirror* will usually be determined by the impedance matching requirement. Cavities with extremely high finesses will use the slight transmission of ultra-high-reflecting cavity mirrors, but there are situations with medium-finesse cavities where one would like to couple an *adjustable* amount of light into the cavity. This might occur with a cavity containing a nonlinear crystal, where it is desirable to maximize the circulating power (by careful impedance matching). We will present a scheme (suggested by N. Yu and J. Torgerson) which uses *reflective* coupling to inject a variable amount of power into a ring cavity.

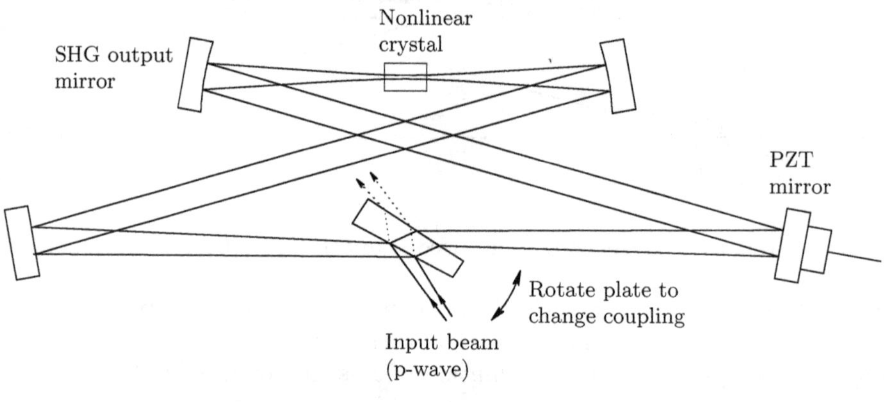


**Fig. 3.7** Diagram of ring cavity using "reflective" coupling.

The scheme is illustrated in Fig. 3.7. A slightly wedged element is inserted into the cavity and the light is coupled into the cavity by reflection from one of the surfaces (the near one in the figure). The loss from the other surface is nearly eliminated by using a polarization in the plane of the ring and ensuring that the angle of incidence on this surface is always near to Brewster's angle. A plot of the reflectivity from both surfaces of a 4° glass wedge appears in Fig. 3.8. For the plate depicted in the plot, the index of refraction is 1.45. As can be seen from the plot, adjusting the tilt angle by ±3° around the central angle of 56° will smoothly vary the coupling from 0.5% to 3.5% while keeping the loss at the Brewster surface below 0.1% (a value comparable to other losses in the cavity). One drawback to the scheme is that a change in the tilt angle will require readjustment of *all* of the other mirrors to preserve the cavity alignment. In practice, this has not been found to be a serious problem.

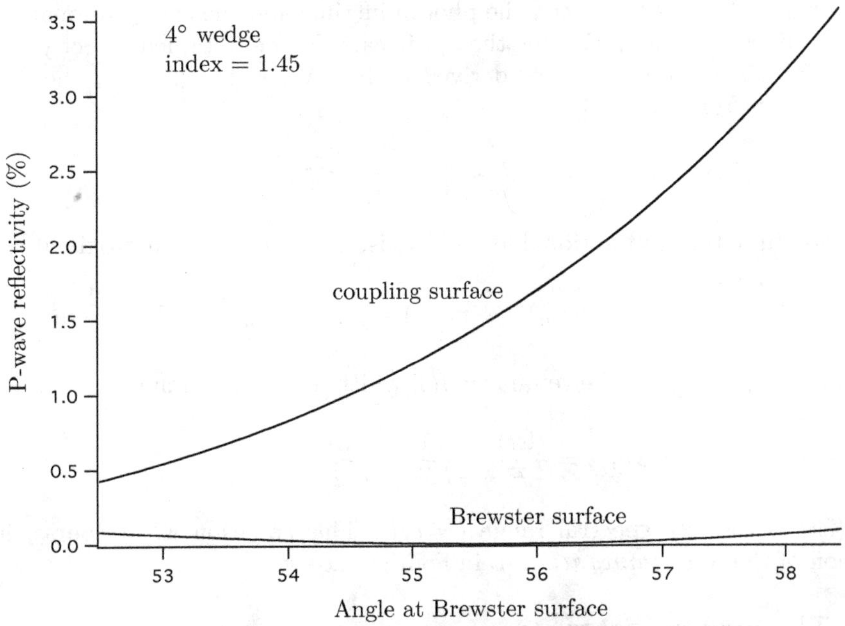


**Fig. 3.8** Reflectivity from "coupling" and "Brewster" surfaces as a function of tilt angle. The "p-wave" is polarized in the plane of the ring.

Using this approach, the analogue to the reflected field is the beam transmitted through the wedge. It is left as an exercise to show that all of the theory developed for transmissive coupling holds for reflective coupling, with some simple parameter substitutions.

## 3.10 Photon lifetime

As a result of the finite transmission of the cavity mirrors and the possibility of dissipative losses in the cavity, a wave injected into an optical cavity will decay with time. Although we have not introduced the concept of *photons* into our discussion, the *photon lifetime*, $t_c$, can be defined to be the average time a photon or injected beam will *live* inside the cavity. If $\mathcal{E}$ is the stored light energy in the cavity, the stored energy will satisfy

$$\frac{d\mathcal{E}}{dt} = -\frac{\mathcal{E}}{t_c}, \tag{3.33}$$

whose solution is

$$\mathcal{E} = \mathcal{E}_0 e^{-t/t_c}. \tag{3.34}$$

Introducing the quantity $L_{tot}$, which is the *fractional intensity loss per round trip*, the energy loss per round trip is $L_{tot}\mathcal{E}$ and the energy loss *per unit time* is $(c/l)L_{tot}\mathcal{E}$, where $l$ is the round-trip distance. The energy-loss equation can be rewritten as

$$\frac{d\mathcal{E}}{dt} = -\frac{cL_{tot}}{l}\mathcal{E} \Longrightarrow t_c = \frac{l}{cL_{tot}}. \tag{3.35}$$

Although the relation between the photon lifetime and the cavity resonance width can be obtained by simple Fourier theory, it can also be obtained directly from the relations for the finesse which we derived earlier. When $r_1, r_m \approx 1$, the finesse is approximately (from eqn 3.14):

$$\mathcal{F} \approx \frac{\pi}{1 - r_1 r_m}. \tag{3.36}$$

The round-trip intensity fractional loss ($L_{tot}$) is, using the same approximation,

$$L_{tot} = 1 - r_1^2 r_m^2 = (1 - r_1 r_m)(1 + r_1 r_m) \approx 2(1 - r_1 r_m) \Longrightarrow \mathcal{F} = \frac{2\pi}{L_{tot}}. \tag{3.37}$$

Using these two relations, the resonance full width at half maximum, $\Delta\nu_{1/2}$, is:

$$\Delta\nu_{1/2} = \frac{\langle \text{fsr} \rangle}{\mathcal{F}} = \frac{c}{l}\frac{1}{\mathcal{F}} = \frac{c}{l}\frac{L_{tot}}{2\pi} = \frac{1}{2\pi t_c}, \tag{3.38}$$

where $\langle \text{fsr} \rangle$ is the free spectral range ($= c/l$). This equation is, of course, just an expression of the *uncertainty relation* in Fourier theory.

## 3.11 The quality factor, $Q$

The quality factor, $Q$, is a familiar quantity from electrical circuit theory and is usually defined as the *frequency resolution* of a resonant circuit. There is an equivalent and much more useful definition of $Q$ which allows it to be applied to *all resonant phenomena.* This definition is:

$$Q \equiv \frac{2\pi \times \text{stored energy}}{\text{energy loss per cycle}}. \tag{3.39}$$

If the loss term is expressed as a *loss rate* instead of a loss per cycle, the definition becomes:

$$Q \equiv \frac{\omega \times \text{stored energy}}{\text{energy loss rate}} = -\frac{\omega\mathcal{E}}{d\mathcal{E}/dt} = \omega t_c, \tag{3.40}$$

where the last identity follows from eqn 3.33 and $\omega$ is the resonant frequency.

Using eqn 3.38, we can now show that the above definitions of $Q$ are equivalent to the familiar notion of $Q$ as a measure of *frequency resolution.* If $\nu = \omega/2\pi$ is the resonant frequency in Hz,

$$Q = 2\pi\nu t_c = 2\pi\nu\left(\frac{1}{2\pi\Delta\nu_{1/2}}\right) = \frac{\nu}{\Delta\nu_{1/2}}. \tag{3.41}$$

## 3.12 Relation between $Q$ and finesse

We first recall that the finesse ($\mathcal{F}$) is defined as

$$\mathcal{F} = \frac{\langle \text{fsr} \rangle}{\Delta\nu_{1/2}} = \frac{c}{l\Delta\nu_{1/2}} = \frac{2\pi c t_c}{l}. \tag{3.42}$$

Using the frequency-resolution definition of $Q$,

$$Q = \frac{\nu}{\Delta\nu_{1/2}} = \frac{\nu l \mathcal{F}}{c} = \frac{l\mathcal{F}}{\lambda}. \tag{3.43}$$

In words, the $Q$ is *equal to the finesse times the number of wavelengths in a round-trip* (or the number of *half-waves* between the mirrors in a standing-wave cavity).

We saw earlier that the energy in a cavity decays exponentially with a characteristic time equal to $t_c$. Both the *circulating power* and the power coupled out through any cavity mirror will, of course, have the same exponential behavior with time. Thus, after $n$ round-trips, the power will be

$$P = P_0 e^{-t/t_c} = P_0 e^{-\frac{nl}{ct_c}}, \tag{3.44}$$

where $P_0$ is the power at $t = 0$. The power is therefore reduced by a factor of $e^{-1}$ when $n = ct_c/l = \mathcal{F}/2\pi$ (eqn 3.42). From this, we see that *the power in the cavity will be reduced by* $1/e$ *after* $\mathcal{F}/2\pi$ *round-trips.* From the point of view of interferometry, one can think of a cavity as a device which *causes* $\approx \mathcal{F}/2\pi$ beams to interfere with one other, where the interference is due to the *folding* of the path rather than to the presence of multiple slits or lines, as in a grating. This is similar to multiple beam interference experiments, where the appropriate width is reduced by a factor equal to the *number of beams that interfere.* In a cavity, this number is very roughly equal to the finesse divided by $2\pi$.

## 3.13 Alternative representation of cavity loss

In anticipation of our treatment of *laser gain*, it will be useful to represent the internal cavity loss as being due to a *uniformly distributed absorption coefficient*, $\alpha$, which excludes mirror Fresnel reflection losses. Thus for a standing-wave cavity, we require that $\alpha$ satisfy

$$t^4 = e^{-\alpha l}. \tag{3.45}$$

If the round-trip fractional energy loss is $L_{tot}$, the *fraction of the power remaining after one round trip* is $e^{-L_{tot}}$. From this, we obtain

$$e^{-L_{tot}} = r_1^2 r_2^2 t^4 = r_1^2 r_2^2 e^{-\alpha l} \Longrightarrow L_{tot} = \alpha l - 2\ln(r_1 r_2). \tag{3.46}$$

For the case of a ring cavity, $r_2 = 1$ and the exponent of the $t$ term would be 2, since the beam traverses the internal medium once per round trip. Equation 3.46 can be considered either as a formula for $L_{tot}$ or a definition of $\alpha$. In the former case, we allow for the intriguing possibility that a *negative* $\alpha$ can make $L_{tot} = 0$, in which case there will be *no damping of the internal field.* This self-sustained cavity field is of course the result of a *coherent laser gain* and will be discussed in detail in Chapter 5.

## 3.14 Experimental determination of cavity parameters

An experimentalist is interested, for obvious reasons, in measuring the various cavity parameters which have been discussed in this and the preceding chapter. Certain cavity properties, such as the free spectral range, are trivial to obtain, since they depend very simply on the cavity geometry. The *dynamic* properties – those which pertain to the energy relations in a cavity – are a little more difficult to determine. We will briefly discuss techniques used to determine the *enhancement* and *finesse*.

The most straightforward way to measure the finesse is to use its definition as the ratio of the free spectral range to the full width at half maximum of a cavity resonance. This is a reasonable procedure for cavities with low to modest finesse ($\leq 200$), if one has access to a fairly stable tunable laser at the appropriate wavelength. This technique is not easy to apply to cavities with extremely high finesse, since the stability requirement placed on the laser might be too great. Furthermore, it does not measure the enhancement directly, and the relation between enhancement and finesse is simple only when the cavity is impedance matched. We will present two techniques: one which can be used to directly measure the enhancement in a *build-up cavity* and the other which can be used to measure the finesse of a cavity whose finesse is extremely high.

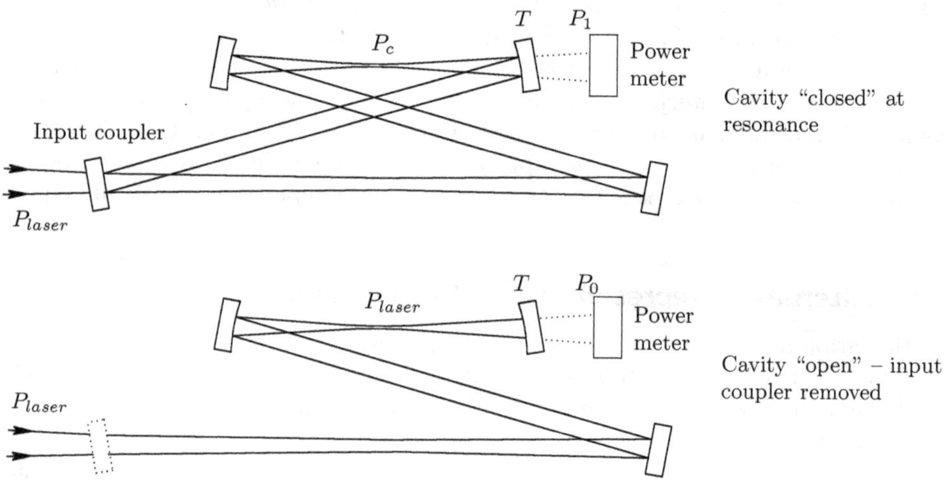


**Fig. 3.9** Technique for measuring cavity enhancement by making two different power measurements of light leaking through a cavity mirror, whose (small) transmission is $T$. The enhancement is $P_1/P_0$.

We will discuss the enhancement measurement first, since it is simpler. It is conveniently employed to evaluate a cavity which is used to enhance nonlinear processes such as second-harmonic generation or sum-frequency mixing. The technique is illustrated in Fig. 3.9. The procedure is to place a sensitive power meter behind one of the cavity mirrors and measure the *leakage* of radiation through this mirror. Two power measurements are made. One, which we will call $P_1$, is done with the cavity at resonance and aligned (it is most convenient if the cavity is *locked* to the laser frequency so that the power can be measured over some modest time period, though it can also be done by *slowly* scanning the laser with the cavity *free-running*). If we assume that $T$ is the *transmission* of the cavity mirror through which the leakage is measured,

$$P_1 = TP_c, \tag{3.47}$$

where $P_c$ is the circulating power. The other power measurement is made by removing the input coupler (keeping the laser power the same). If we call this measured power, $P_0$, and the laser power, $P_{laser}$, then

$$P_0 = TP_{laser}, \tag{3.48}$$

and we assume that the reflectivity of all of the cavity mirrors is $\approx 1$. The enhancement is simply the ratio of these two power measurements:

$$\text{Enhancement} \equiv \frac{P_c}{P_{laser}} = \frac{P_1}{P_0}. \tag{3.49}$$

The finesse of a cavity can also be determined by applying a very short pulse of light at the resonant frequency of the cavity and observing the exponential decay of the intracavity power as it leaks out through one of the mirrors. A common experimental setup is shown in Fig. 3.10. The laser is tuned near to the cavity resonance and the cavity is allowed to drift into resonance. After the intracavity power reaches a certain threshold (as measured by a detector which samples the leakage through one of the cavity mirrors), a very fast optical switch (usually an *acousto-optic modulator*; see Chapter 15) quickly switches off the beam. A digital sampling oscilloscope stores a curve of the leakage power as a function of time, and the finesse is determined by fitting an exponential to the decay curve using

$$P(t) = P_0 e^{-t/t_c} = P_0 e^{-(2\pi c/l\mathcal{F})t}. \tag{3.50}$$

This approach is called the *ring-down* technique for obvious reasons. It is important

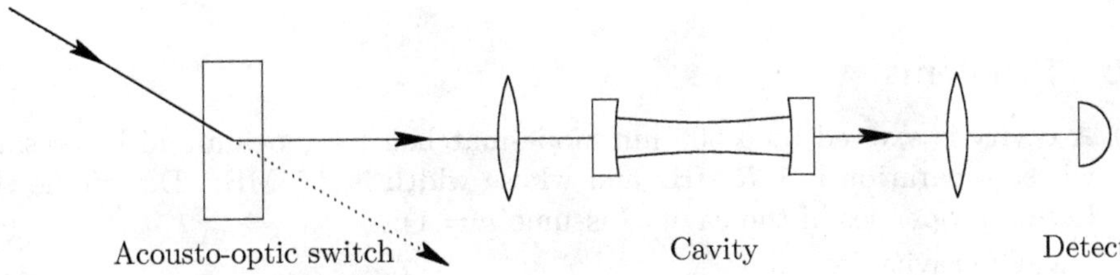


**Fig. 3.10** A commonly used setup for measuring cavity finesse using the *ring-down method.*

that the fall time of the optical switch is short compared to the photon storage time to avoid distorting the exponential decay curve. As one can see, the method works very well for cavities with extremely high finesse (approaching $10^6$ using mirrors with a power loss in the part-per-million range) since the storage times are then relatively large and the time delays in the electronics have a smaller effect on the results. A typical curve appears in Fig. 3.11, where the rapid filling is shown for $t < 0$ and the exponential decay is shown for $t > 0$.

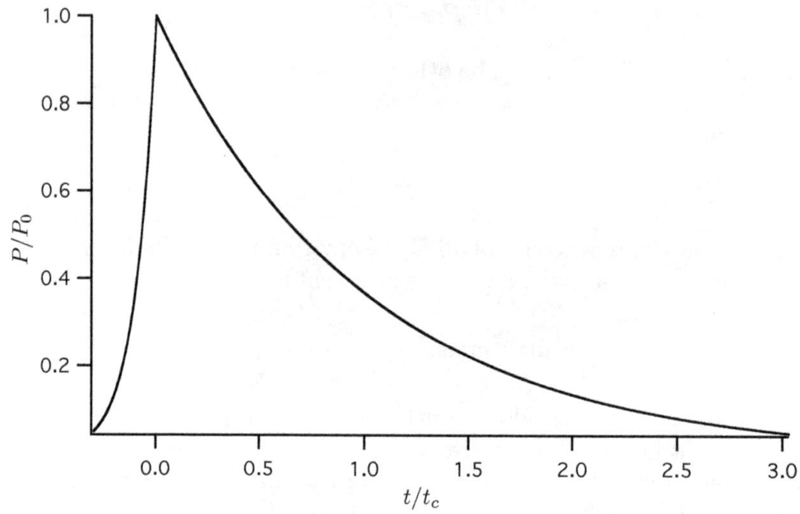


**Fig. 3.11** The signal obtained using the *ring-down* method. The cavity rapidly fills when $t < 0$ and exponentially decays when $t > 0$.

## 3.15 Further reading

This chapter discusses the dynamical aspects of optical cavities. The references mentioned in the previous two chapters are still useful, as well as a number of excellent treatises on optics, such as Born and Wolf (1980) and Hecht (2002). In addition, a discussion of impedance matching into a cavity occurs in only one place, to the author's knowledge: the paper by Ashkin, Boyd and Dziedzic (1966). The publication on the use of a wedge for variable reflective coupling appears in the paper by Torgerson and Nagourney (1999).

## 3.16 Problems

(3.1) A cavity is excited by a 500 nm mode-matched laser beam and has resonances whose separation is 125 MHz and whose width is 2.5 MHz. Determine the following properties of the cavity (assume $n = 1$):
  a The cavity length
  b The cavity finesse
  c The cavity $Q$
  c The photon lifetime

(3.2) Calculate the *transmission* of the electric field and intensity for a cavity having mirrors with reflectivity $r_1, r_2$ (and transmission $t_1, t_2$), separation $d$ and containing a medium with transmission $t$. This is just the case not discussed in the text.

(3.3) For the three-mirror ring cavity depicted in the diagram, calculate the free spectral range, finesse, $Q$, and photon lifetime. As shown, the three arms are 20 cm long and have the indicated *field* reflectivities. Assume $\lambda = 600$ nm.

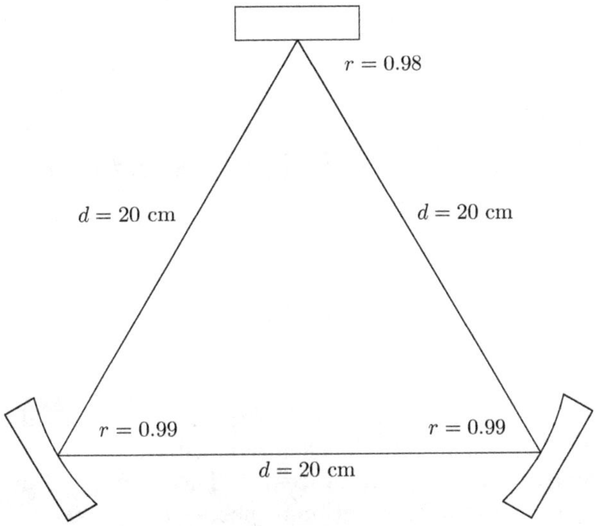


(3.4) A *thin etalon* is sometimes used in a laser cavity to help achieve single-mode operation. Such a device consists of a thin plate of quartz which is operated near normal incidence and tilted about an axis normal to the laser beam for tuning purposes. If the index of refraction is 1.5 and the thickness is 0.2 cm, find the linewidth of the etalon and an expression for the frequency as a function of the tilt angle (this dependence is not necessarily linear). The tilt angle is the angle between the normal to the plate and the input $k$-vector.

(3.5) Calculate the internal and *transmitted* fields and intensities for a ring cavity which uses the *reflective* coupling approach discussed in this chapter. Assume that the mirrors have 100% reflectivity and that there is an internal medium with transmission $r_m$. Neglect the effects of the surface at Brewster's angle and assume that the *coupling* surface has *field* reflectivity and transmission $r$ and $t$. The internal fields and intensities should be calculated just to the left of the coupling surface.

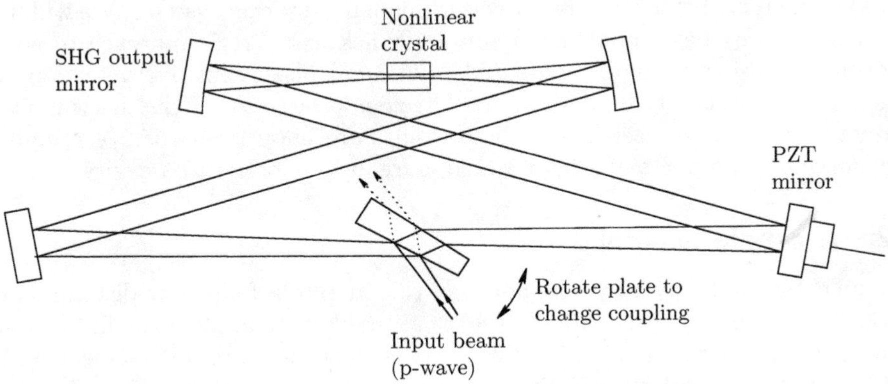

# 4 Optical cavity as frequency discriminator

## 4.1 Introduction

Earlier in this book, we mentioned some of the uses for optical cavities in an atomic physics laboratory, such as the feedback mechanism in lasers, spectrum analyzers and enhancement devices for nonlinear processes. These applications are fairly simple to understand and do not require additional pieces of apparatus to perform their functions. The use of a cavity as a frequency reference element, however, requires that one have a means of generating an *error signal* which will correct the laser frequency when it differs from that of an exact cavity resonance. The correction signal is ideally a linear function of the frequency deviation from resonance and is called a *discriminant.* This application requires some additional equipment and warrants some additional discussion. The use of a cavity as a *passive frequency discriminator* will be discussed in this chapter. In Chapter 13 we will discuss *active* discriminators derived from atomic or molecular resonances.

Two techniques will be discussed in detail: the Pound–Drever–Hall (PDH) scheme, which uses a frequency-modulated signal reflected from a cavity, and the passive polarization scheme, which analyzes the polarization of the reflected signal from a cavity containing a polarizing element. The former scheme is usually used for locking a laser to a cavity, and the latter for the reverse process: locking a cavity to a laser (usually for enhancement of a nonlinear process). We will briefly touch on the use of a cavity in transmission, which historically preceded the PDH method and is still found in some commercial dye or Ti–sapphire laser frequency-locking systems. We will illustrate the principles by beginning with a simple "quick and dirty" approach to generate a discriminant using a length-modulated cavity and phase-sensitive detection. We will conclude the discussion by deriving the *frequency response* of the discriminator to a frequency-modulated laser beam; these results are important when one analyzes the servomechanism aspects of a laser which is frequency locked to a cavity.

## 4.2 A simple example

A simple and probably very common way to generate a frequency discriminant when a commercial optical spectrum analyzer is available is to apply a small, low-frequency sinusoidal voltage to the cavity's piezo-electric length-controlling element and detect the transmitted signal using a phase-sensitive detector whose reference is a sample of the modulation voltage. A block diagram of the apparatus appears in Fig. 4.1. The

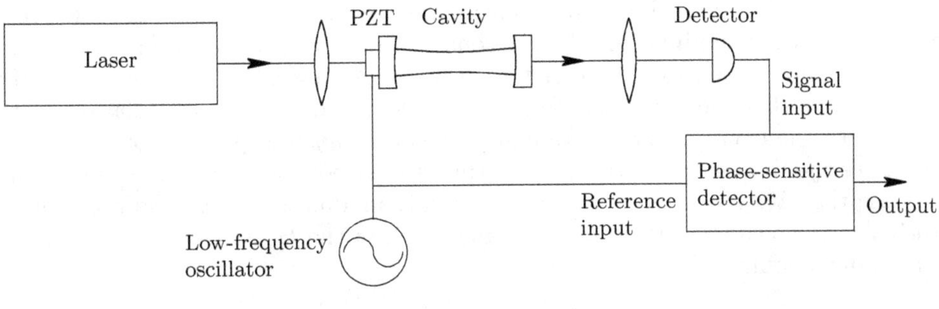


**Fig. 4.1** Block diagram of "simple" apparatus used to generate a discriminant.

cavity will have its resonance frequency modulated by the applied voltage. As can be seen from Fig. 4.2, when the laser is below resonance, the transmission characteristic of the cavity will convert the modulation of the cavity's *resonant frequency* into amplitude modulation with a certain phase, and when the laser is above resonance the converted amplitude modulation will have the opposite phase. At exact resonance, there will be no signal at the modulation frequency (there will be one at *twice* this frequency). Thus, one will get a curve with the appropriate shape for a discriminant when one plots the output of a phase-sensitive detector against the laser frequency – if the modulation *excursion* is small compared to the cavity width, the signal will essentially be proportional to the *first derivative* of the cavity transmission lineshape function. The phase-sensitive detector can be *defined* as a device which produces an output which is jointly proportional to the amplitude of the input and the cosine of the phase difference between the signal and some reference. The operation of *multiplication* by a constant sinusoidal reference followed by low-pass filtering is a reasonable model of a phase-sensitive detector.

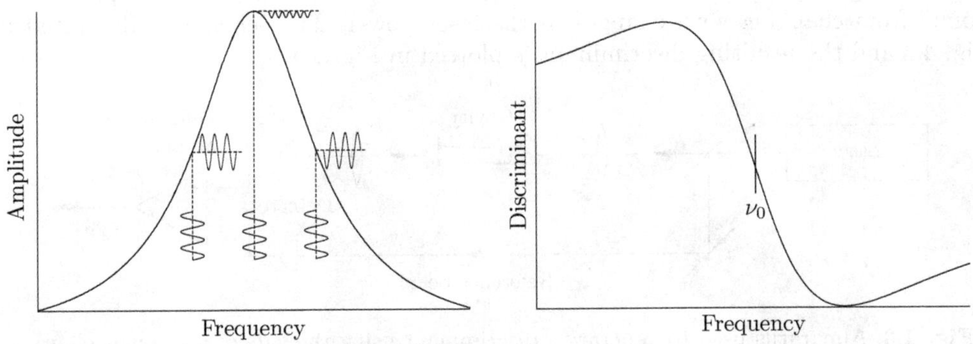


**Fig. 4.2** Illustration of simple discriminator function. Below resonance (left), the output is in phase with modulation, above resonance the output is 180° out of phase and on resonance the output is at twice the frequency. The output of the phase-sensitive detector is the *first derivative* of the resonance curve for small deviation, as shown on the right.

The functional form of the resonance can be easily obtained if we assume that the modulation frequency is much lower than the inverse of the photon storage time –

this allows us to consider the cavity transmission to be a quasi-static process even in the presence of the modulation of its resonance frequency. It can then be described by an expression proportional to that for the circulating intensity (eqn 3.18). If we further assume that the size of the *frequency excursion* due to the modulation is small compared to the cavity FWHM, we can perform a *Taylor expansion* of the function which gives the transmitted intensity in terms of the frequency. The length modulation will cause the cavity resonance to be sinusoidally modulated. From the point of view of the cavity, one can consider the frequency, $\nu(t)$, of the *laser beam* to be modulated by the same amount:

$$\nu(t) = \nu_0 + \delta\nu \sin(\nu_m t), \tag{4.1}$$

where $\delta\nu$ is the maximum deviation of the cavity frequency due to the modulation, $\nu_m$ is the (low) modulation frequency and $\nu_0$ is the laser frequency. Performing the expansion about $\nu_0$, the intensity at the detector ($I_d$) will have following time dependence:

$$I_d(t) = I_d(\nu_0) + \left.\frac{\partial I_d}{\partial \nu}\right|_{\nu_0} (\delta\nu \sin(\nu_m t)). \tag{4.2}$$

The time-dependent intensity will generate a time-dependent photocurrent in the detector, and the signal at the output of the phase-sensitive detector will simply be proportional to the first derivative with respect to frequency of the Lorentzian lineshape function given in eqn 3.18 and plotted in Fig. 4.2.

## 4.3 Side of resonance discriminant

Historically, one of the first successful methods of generating an *optical* discriminant was also the simplest. It uses the side of a Lorentzian resonance, which is fairly linear for small excursions from the "lock point". A sample of the laser intensity is subtracted from the signal due to transmission through a cavity to prevent the zero point (lock point) from changing with changes in the laser power. The scheme is illustrated in Fig. 4.3 and the resulting discriminant is plotted in Fig. 4.4.

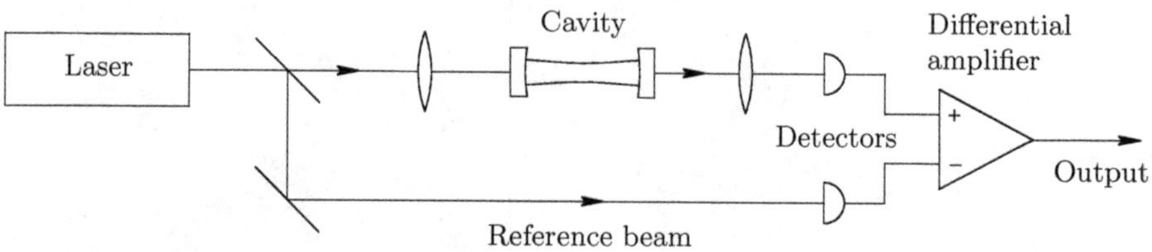


**Fig. 4.3** Apparatus used to generate a discriminant using the side of a resonance curve.

When used as a reference for laser stabilization, this scheme has a number of problems, perhaps the principal one being the time delays resulting from the transmission through the cavity. These delays are on the order of the photon lifetime in the cavity and can greatly limit the bandwidth of the stable servo system used for frequency locking. The successor to this approach – the Pound–Drever–Hall scheme – uses the reflected beam from the cavity and avoids this problem.

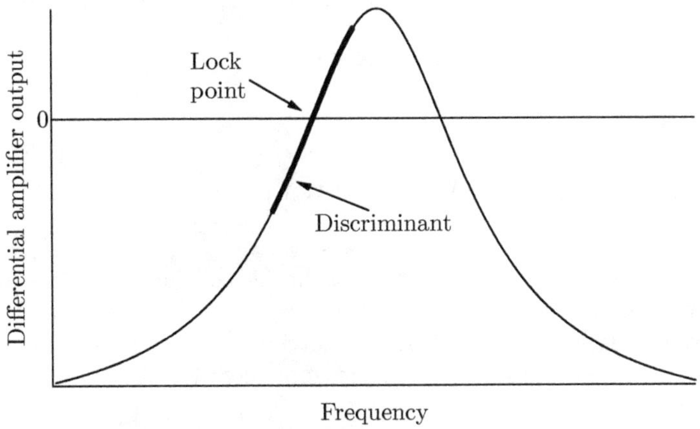


**Fig. 4.4** Plot of discriminant generated using the side of a resonance. Of course, the portion above the Lorentzian resonance could have also been used.

## 4.4 The manipulation of polarized beams: the Jones calculus

We will take a short diversion to summarize a convenient formalism for describing and manipulating *fully polarized* beams, since the next two methods used to generate a discriminant make extensive use of polarization-manipulating devices. This formalism is called the *Jones calculus*. We should emphasize that this approach is only applicable to fully polarized waves – partially polarized light is described by other formalisms, such as the *Stokes parameters*.

A polarized beam is represented by a two-element column vector (called a *Jones vector*) giving the Cartesian components of the electric field as a function of time:

$$\boldsymbol{E}(t) = \begin{pmatrix} E_x(t) \\ E_y(t) \end{pmatrix}. \tag{4.3}$$

The components of the field are *complex numbers*. They are often *normalized* to unity complex norm ($\boldsymbol{E} \cdot \boldsymbol{E}^* = 1$), though this is not always done. Usually, common complex factors (including the time dependence of a complex sinusoid) are factored out and removed. Thus, the four common polarizations are (normalized):

$$\text{Horizontal:} \quad \boldsymbol{E}_h = \begin{pmatrix} 1 \\ 0 \end{pmatrix} \tag{4.4}$$

$$\text{Vertical:} \quad \boldsymbol{E}_v = \begin{pmatrix} 0 \\ 1 \end{pmatrix} \tag{4.5}$$

$$\text{Right circular:} \quad \boldsymbol{E}_h = \frac{1}{\sqrt{2}} \begin{pmatrix} 1 \\ -i \end{pmatrix} \tag{4.6}$$

$$\text{Left circular:} \quad \boldsymbol{E}_h = \frac{1}{\sqrt{2}} \begin{pmatrix} 1 \\ i \end{pmatrix}. \tag{4.7}$$

The *Jones matrices* represent the *linear transformation* which a Jones vector undergoes when it passes through an optical device. If the Jones matrix is $\mathcal{A}$ and the incident and transmitted Jones vectors are $\boldsymbol{E}_i$ and $\boldsymbol{E}_t$ respectively, then

$$\boldsymbol{E}_t = \mathcal{A}\boldsymbol{E}_i, \tag{4.8}$$

or writing the elements out,

$$\begin{pmatrix} E_{tx} \\ E_{ty} \end{pmatrix} = \begin{pmatrix} a_{11} & a_{12} \\ a_{21} & a_{22} \end{pmatrix} \begin{pmatrix} E_{ix} \\ E_{iy} \end{pmatrix}. \tag{4.9}$$

The key to the utility of the Jones calculus is the fact that one can represent a complex device as a *sequence of simple devices*, and the overall Jones matrix can be obtained by multiplication of the Jones matrices of the simple components. If the beam encounters devices represented by Jones matrices $\mathcal{A}_1, \mathcal{A}_2 \cdots \mathcal{A}_n$, the overall Jones matrix of the sequence is

$$\mathcal{A} = \mathcal{A}_n \mathcal{A}_{n-1} \cdots \mathcal{A}_1, \tag{4.10}$$

where one must write the matrices in the *reverse order* from that in which the devices are encountered by the beam. (One should note the similarity to the treatment of a composite ABCD matrix for cascaded systems.)

The Jones matrices of some common devices are:

$$\text{Horizontal polarizer:} \quad \begin{pmatrix} 1 & 0 \\ 0 & 0 \end{pmatrix} \tag{4.11}$$

$$\text{Vertical polarizer:} \quad \begin{pmatrix} 0 & 0 \\ 0 & 1 \end{pmatrix} \tag{4.12}$$

$$\lambda/4\text{, vertical fast axis:} \quad \begin{pmatrix} 1 & 0 \\ 0 & i \end{pmatrix} \tag{4.13}$$

$$\lambda/4\text{, horizontal fast axis:} \quad \begin{pmatrix} 1 & 0 \\ 0 & -i \end{pmatrix}, \tag{4.14}$$

where the last two devices are *quarter-wave plates* – devices which advance or retard the phase of a beam by $\pi/2$ radians.

Since the Jones vector is a conventional two-dimensional vector, it can be evaluated in a *rotated coordinate system* by applying the familiar rotation matrix in the $xy$ plane:

$$\boldsymbol{E}' = S(\theta)\boldsymbol{E}, \tag{4.15}$$

where $\boldsymbol{E}'$ is the Jones vector in the rotated system and $S(\theta)$ is the rotation matrix for an angle of $\theta$:

$$S(\theta) = \begin{pmatrix} \cos\theta & \sin\theta \\ -\sin\theta & \cos\theta \end{pmatrix}. \tag{4.16}$$

In the rotated frame,

$$\boldsymbol{E}_t' = \mathcal{A}'\boldsymbol{E}_i', \tag{4.17}$$

where $\mathcal{A}'$ is the Jones matrix in the rotated frame.

Since

$$\boldsymbol{E}'_i = S(\theta)\boldsymbol{E}_i \tag{4.18}$$
$$\boldsymbol{E}'_t = S(\theta)\boldsymbol{E}_t, \tag{4.19}$$

multiplying eqn 4.8 on the left by $S$ and using the identity $S^{-1}S = 1$, one has

$$\boldsymbol{E}'_t = S\mathcal{A}S^{-1}\boldsymbol{E}'_i, \tag{4.20}$$

which implies that the *Jones matrix in the rotated system* is

$$\mathcal{A}' = S\mathcal{A}S^{-1}, \tag{4.21}$$

where the inverse of the *orthogonal matrix* $S$ is $S^\dagger$.

Using this, one can immediately write down the Jones matrices for rotated objects, such as

$$\text{Linear polarizer at } -45^\circ\text{:} \quad \frac{1}{2}\begin{pmatrix} 1 & 1 \\ 1 & 1 \end{pmatrix} \tag{4.22}$$

$$\text{Linear polarizer at } 45^\circ\text{:} \quad \frac{1}{2}\begin{pmatrix} 1 & -1 \\ -1 & 1 \end{pmatrix}, \tag{4.23}$$

and, in general, the Jones matrix for a linear polarizer at angle $\theta$ to the $x$-axis is:

$$\mathcal{A} = \begin{pmatrix} \cos^2\theta & -\sin\theta\cos\theta \\ -\sin\theta\cos\theta & \sin^2\theta \end{pmatrix}. \tag{4.24}$$

It should be kept in mind that *a rotation of the coordinate system by angle $\theta$ is equivalent to a rotation of the apparatus by angle $-\theta$.*

## 4.5 The polarization technique

The *polarization* technique generates a discriminant by placing a polarization-sensitive element in the optical cavity and observing the *interference* between two reflected beams with orthogonal polarizations; the orientation of the intracavity element is chosen so that the two polarizations have very different intracavity losses. The signal is actually the *difference* between the intensities of the *two opposite circular polarization components* of the reflected beam. We will show that this does indeed generate a useful discriminant by using our cavity theory and some shortcuts from the Jones calculus. Since a potentially lossy intracavity element is used, the technique is not employed in the highest-finesse cavities (where the losses are in the parts-per-million range) but finds frequent application in build-up cavities used in nonlinear optics where there is often a polarization-sensitive element already in place in the cavity.

The apparatus for generating a polarization signal is shown in Fig. 4.5, where it is used in a ring cavity containing a nonlinear crystal whose faces are at Brewster's angle for polarization in the plane of the figure. Since the *lossless* polarization of the Brewster surfaces is in the ring plane, the electric field of the incoming beam ($E_0$) is tilted slightly by an angle of $\theta$ with respect to the $x$-axis. The reflected beam

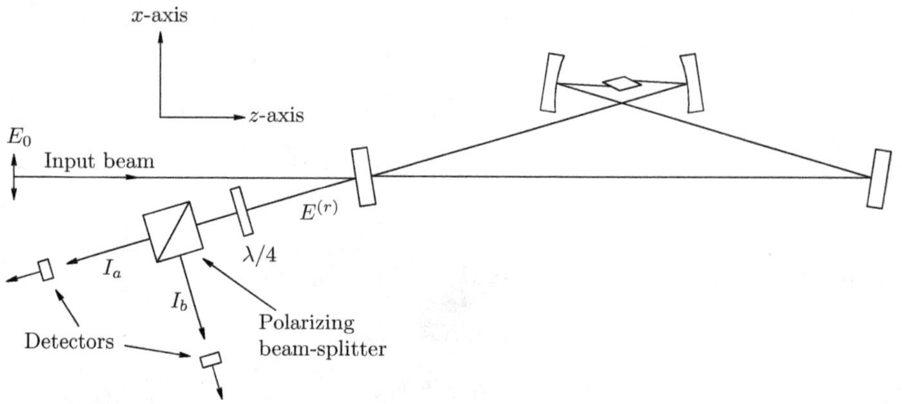


**Fig. 4.5** The polarization technique used with a ring cavity. The $x$-axis is in the plane of the ring; the beam $I_b$ is in the ring plane and the incoming field, $E_0$, makes an angle $\theta$ with the $x$-axis. The $y$-axis is perpendicular to the plane of the ring. The *fast axis* of the quarter-wave plate makes an angle of 45° to the $x$-axis.

electric field, $\boldsymbol{E}^{(r)}$, has two components: $E_x^{(r)}$ and $E_y^{(r)}$. The *polarizing beam-splitter* is just a polarizer with two outputs: polarization along the $x$-axis (beam with intensity $I_a$) passes straight through it without loss, and polarization along the $y$-axis ($I_b$) is reflected by 90°. The Jones matrix for the quarter-wave plate can be obtained from eqn 4.14 after rotating by 45° (eqn 4.21). This can be shown to be:

$$\lambda/4 \text{ at } 45^\circ: \quad \mathcal{A} = \frac{1-i}{2}\begin{pmatrix} 1 & -i \\ -i & 1 \end{pmatrix}. \tag{4.25}$$

The Jones vectors for beams $I_a$ and $I_b$ are

$$\text{Beam a:} \ \boldsymbol{E}_a = \frac{1-i}{2}\begin{pmatrix} 1 & 0 \\ 0 & 0 \end{pmatrix}\begin{pmatrix} 1 & -i \\ -i & 1 \end{pmatrix}\begin{pmatrix} E_x^{(r)} \\ E_y^{(r)} \end{pmatrix} \tag{4.26}$$

$$\text{Beam b:} \ \boldsymbol{E}_b = \frac{1-i}{2}\begin{pmatrix} 0 & 0 \\ 0 & 1 \end{pmatrix}\begin{pmatrix} 1 & -i \\ -i & 1 \end{pmatrix}\begin{pmatrix} E_x^{(r)} \\ E_y^{(r)} \end{pmatrix}. \tag{4.27}$$

The signal at each detector is obtained by taking the square modulus of the fields:

$$I_a \propto |E_x^{(r)} - iE_y^{(r)}|^2 \tag{4.28}$$
$$I_b \propto |E_x^{(r)} + iE_y^{(r)}|^2, \tag{4.29}$$

where the proportionality constants are the same for $I_a$ and $I_b$. The overall signal $(I_a - I_b)$ is due to an interference between the real part of one linear polarization component and the imaginary part of the other.

We will assume that the finesse of the cavity is very low for the $y$-polarized wave. This is justified by the fact that this polarization has a $\approx 20\%$ *Fresnel* loss from the crystal (for a refractive index of 1.5), so that the finesse of the $y$-polarization in an

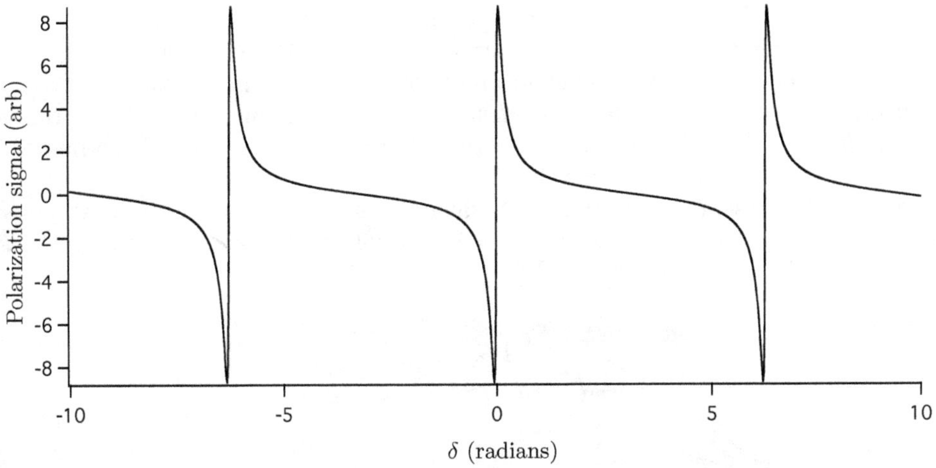


**Fig. 4.6** Plot of polarization signal versus $\delta$.

impedance-matched cavity must be less than about 15, while that of the $x$-polarization can be 300 or more. This assumption allows us to ignore the frequency dependence and write the field reflection of the $y$-component simply as

$$E_y^{(r)} = -rE_{0y}, \tag{4.30}$$

where $r$ is the field reflectivity of the input mirror. Under impedance-matched conditions, the $x$-polarization field reflection is (from eqn 3.7):

$$E_x^{(r)} = E_{0x}\frac{r(e^{-i\delta}-1)}{1-r^2e^{-i\delta}}. \tag{4.31}$$

If $\theta$ is the angle that the input field makes with the $x$-axis, then the input field components are

$$E_{0x} = E_0\cos\theta \tag{4.32}$$
$$E_{0y} = E_0\sin\theta. \tag{4.33}$$

The intensities are then

$$I_{ab} \propto E_0^2\left|\frac{r(e^{-i\delta}-1)}{1-r^2e^{-i\delta}}\cos\theta \pm ir\sin\theta\right|^2, \tag{4.34}$$

where the minus sign is for $I_a$ and the plus is for $I_b$. The desired signal is proportional to $I_a - I_b$:

$$\text{Signal} \propto 4E_0^2r^2t^2\cos\theta\sin\theta\frac{\sin\delta}{(1-r^2)^2+4r^2\sin^2\delta/2}, \tag{4.35}$$

where $t$ is the field transmission of the input mirror. The independent variable, $\delta$, is the round-trip phase change of the intracavity beam. A plot of the discriminant appears in Fig. 4.6. Normally, one would use the smallest practical $\theta$ in the interests of coupling most of the power into the low-loss polarization where it would be effectively enhanced by the cavity.

## 4.6 Frequency modulation

As an aid in understanding the Pound–Drever–Hall (PDH) technique, we will discuss the frequency modulation of a monochromatic optical field. We will consider the modulated field in both the *frequency* and *time* domains, and will start with a graphical aid: the *phasor* model of an oscillating signal. This model will also help in describing the PDH approach to generating a discriminant.

Figure 4.7 represents a *narrowband* frequency-modulated wave using both phasors and the frequency domain (the term *narrowband* will be defined shortly). The phasor

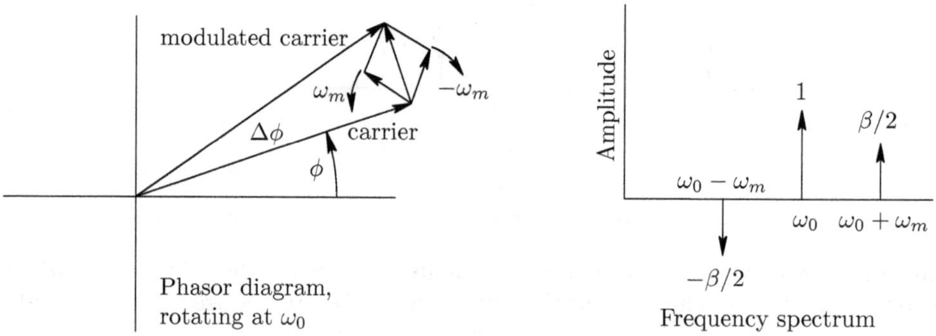


**Fig. 4.7** Representation of frequency-modulated wave using phasors (left) and the frequency domain (right).

model represents a sinusoidal signal by a uniformly rotating vector with constant length. To aid in visualization, the coordinate system is assumed to rotate at the same angular speed as the wave so that the phasor is fixed in space. The amplitude of the wave is simply the length of the vector, and the *instantaneous phase* is the angle it makes with the $x$-axis. At any time, one can generate a physically realistic function by projecting the phasor onto either the $x$- or the $y$-axis. Phasors have the nice property of following the rules of vector addition: one can superpose any number of fields simply by adding the phasors representing them as vectors.

One begins with an unmodulated sinusoidal signal at angular frequency $\omega_0$; this is called the *carrier*. Frequency modulation is obtained by adding two small phasors (called *sidebands*) of equal length and rotating in opposite directions at angular frequency (in the rotating frame) $\omega_m$; they are phased so that they are perpendicular to the carrier phasor when they overlap. As can be seen in the diagram, the resultant of the three phasors is one whose *phase* varies sinusoidally at the rotation speed of the sideband phasors but whose amplitude is essentially constant. This is *phase modulation*, which is identical to frequency modulation for a single sinusoidal modulating signal. If the amplitude of the modulating phasors is small compared to that of the carrier, only two are needed to preserve the constancy of the resultant amplitude. In this case, the size of the phase modulation (called the *phase deviation*) is much less than 1 radian and the modulation is called *narrowband*. The two sidebands are at $\omega_0 + \omega_m$ and $\omega_0 - \omega_m$. Thus, there are three components in the spectral representation of the phase-modulated carrier as shown on the right side of Fig. 4.7. The 180° difference between the two sidebands can be understood by considering the relative sideband phase when both sideband phasors are lined up along the carrier: they are 180° out of phase at that instant of time.

Some further insight into the modulation process can be obtained by considering *amplitude modulation*. Figure 4.8 displays the phasors and spectrum of an amplitude-modulated carrier. As can be seen from the plots, amplitude modulation is identical to

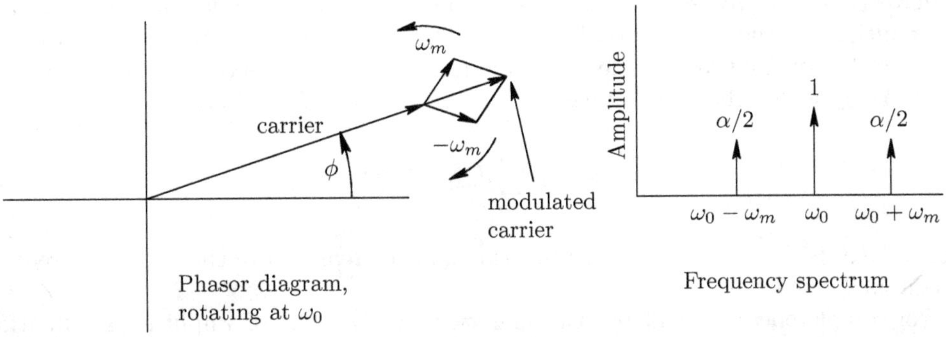


**Fig. 4.8** Representation of amplitude modulated wave using phasors (left) and the frequency domain (right).

frequency modulation except that both sidebands are shifted in phase by 90° relative to the carrier so that they modulate the amplitude of the carrier, keeping its phase constant. In fact, one can convert between amplitude and frequency modulation by using a scheme that rotates the sideband phase by 90° without changing the phase of the carrier (this is the idea behind *phase contrast microscopy*). Both sidebands have the same phase in the spectral representation.

The analytic representation of a phase-modulated carrier is completely straightforward. If we represent an unmodulated monochromatic wave by its time-dependent electric field:

$$E(t) = E_0 \cos(\phi(t)) = E_0 \cos(\omega_0 t), \tag{4.36}$$

the time-dependent phase $\phi(t)$ is just $\omega_0 t$ and increases linearly with time. In order to modulate this phase, we add a time-dependent modulation term $\Delta\phi(t)$. If this modulation is sinusoidal with a maximum of $\beta$ radians, then the phase-modulated field is

$$E(t) = E_0 \cos(\phi(t) + \Delta\phi(t)) = E_0 \cos(\omega_0 t + \beta \sin \omega_m t), \tag{4.37}$$

where the *phase modulation deviation is* $\beta$ radians. The equivalence of this to *frequency modulation* should now be clear if we define the *instantaneous frequency* ($\omega(t)$) to be the *time derivative of the total phase*

$$\omega(t) = \frac{d}{dt}(\phi(t) + \Delta\phi(t)) = \omega_0 + \beta\omega_m \cos \omega_m t. \tag{4.38}$$

The quantity $\beta$ is called the *modulation index* regardless of whether one is describing frequency or phase modulation. The carrier frequency ($\omega_0$) is now being modulated sinusoidally and the maximum frequency shift is $\beta\omega_m$. This latter quantity is called the frequency modulation *deviation*, $\Delta\omega$. The criterion for narrowband modulation is that $\beta \ll 1$ or $\Delta\omega \ll \omega_m$. The narrowband modulated field can be easily expanded using standard trigonometric identities and taking account of the fact that $\beta \ll 1$,

$$E(t) = E_0 \left[\cos\omega_0 t - (\beta/2)\cos(\omega_0 - \omega_m)t + (\beta/2)\cos(\omega_0 + \omega_m)t\right], \quad (4.39)$$

where the 180° phase difference between the sidebands is now clearly shown. If $\beta > 1$, we have *wideband frequency modulation*, which requires additional sidebands at integral multiples of the modulation frequency. These sidebands are present to essentially preserve the constant amplitude of the modulated carrier when the phase deviation is large. Analytically, the field is given by an expansion in *Bessel functions*:

$$E(t) = E_0 \sum_{-\infty}^{\infty} J_n(\beta) e^{i(\omega_0 + n\omega_m)t}, \quad (4.40)$$

where $J_n(\beta)$ is the $n$th order Bessel function and the real part of the expansion reduces to eqn 4.39 when $\beta \ll 1$.

For completeness, we will give the analytic representation of amplitude modulation:

$$\begin{aligned} E(t) &= E_0(1 + \alpha\cos\omega_m t)\cos\omega_0 t \\ &= E_0\left[\cos\omega_0 t + (\alpha/2)\cos(\omega_0 - \omega_m)t + (\alpha/2)\cos(\omega_0 + \omega_m)t\right], \end{aligned} \quad (4.41)$$

where $\alpha$ is the *index of amplitude modulation*. In communication practice, $\alpha$ is sometimes represented as a *per cent*, where 100% amplitude modulation is considered to be the largest index which can be detected without distortion.

## 4.7 The Pound–Drever–Hall approach

The Pound–Drever–Hall method has become the standard method for frequency locking a laser to a very stable cavity and has achieved spectacular performance – lasers with short term linewidths of well under 1 Hz are now routine. It is the optical version of the very old microwave *Pound stabilizer*, which used almost an identical approach in its later incarnations (earlier versions did not use frequency modulation). The technique is characterized by the use of both the reflected wave from a cavity and frequency modulation to generate the discriminant. A schematic of the apparatus appears in Fig. 4.9.

The laser beam is first frequency (or phase) modulated at a frequency ($\omega_m$) which is somewhat larger than the cavity linewidth, usually with an *electro-optic modulator* (which will be described in Chapter 15). A convenient way of efficiently accessing the reflected beam is to use a polarizing beam-splitter and quarter-wave plate whose fast axis is at 45° to the plane of the figure. The wave plate converts the light into circular polarization and the reflected wave is converted back to linear polarization which is orthogonal to the input beam, causing the reflected beam to be deflected by 90° by the polarizing beam-splitter. The cavity will convert the frequency modulation into amplitude modulation resulting in a signal at $\omega_m$ at the output of the detector. The discriminant is obtained by using the mixer to coherently extract the component of

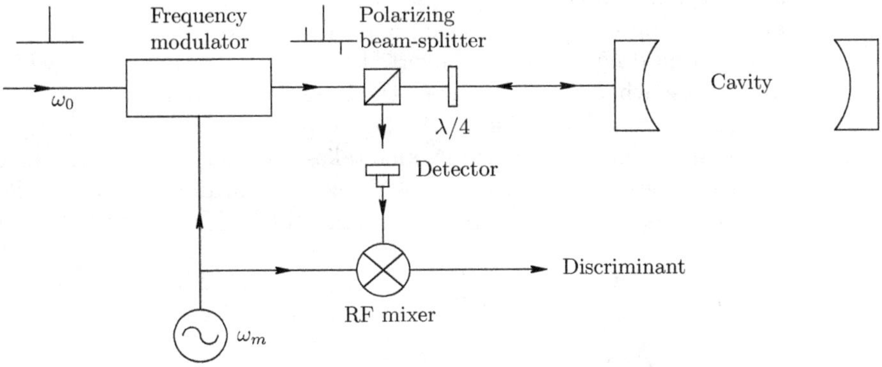


**Fig. 4.9** Schematic of Pound–Drever–Hall apparatus.

the signal at $\omega_m$. The mixer performs the algebraic operation of *multiplication*, where it *coherently* shifts the frequency of the signal spectrum both up and down by the reference frequency. The up-shifted output (at $2\omega_m$) is removed by a low-pass filter and the downshifted DC part is the discriminant.

Some insight into the scheme can be obtained from Fig. 4.10, where we consider the effect of a cavity on the phasors which represent a frequency-modulated beam. When

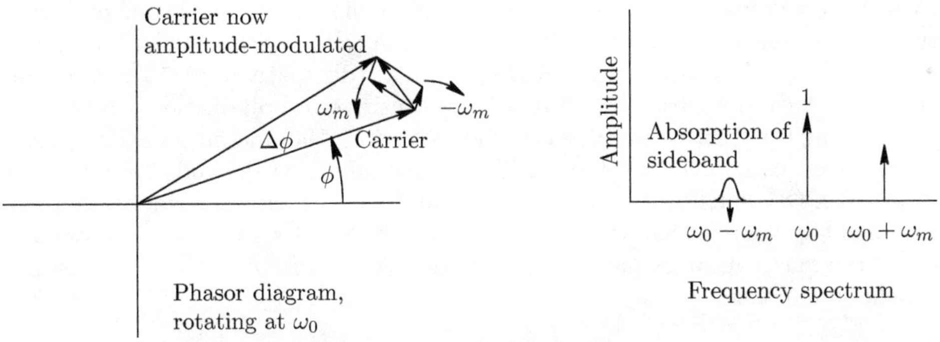


**Fig. 4.10** Influence of cavity on phasor diagram and spectrum of frequency-modulated beam.

the lower sideband is over the cavity resonance, both the amplitude and the *phase* of the reflected component at $\omega_0 - \omega_m$ will be changed. The amplitude reduction is shown on the left side of the figure (the phase change is zero at the exact resonance of the lower sideband). From the phasor diagram, one can easily see that the delicate balance between the sidebands and the carrier has been disrupted, and the resultant phasor will now be amplitude modulated at $\omega_m$ as the sidebands rotate in opposite directions. If one studies the diagram, it should be plausible that the case of the upper sideband being over the cavity resonance will generate an identical amplitude modulation which is 180° out of phase with the first. Thus, one will generate a signal which is of one sign when the laser is below resonance and of the opposite sign when the laser is above resonance: i.e., a discriminant. In summary, the cavity reflection disrupts the amplitude and phase relations among the sidebands and carrier of a frequency modulated beam, causing the frequency modulation to be converted to amplitude modulation which can be extracted by a mixer whose reference is at the modulation frequency.

Before proceeding to the derivation of the signal, it would be useful to say a few words about the polarization manipulation which takes place as the beam traverses the apparatus. Using the Jones calculus, the incident Jones vector undergoes the following transformations up to just before it enters the beam-splitter for the second time:

$$\frac{1-i}{2}\begin{pmatrix}1 & i\\ i & 1\end{pmatrix}\begin{pmatrix}-1 & 0\\ 0 & -1\end{pmatrix}\frac{1-i}{2}\begin{pmatrix}1 & i\\ i & 1\end{pmatrix}\begin{pmatrix}1 & 0\\ 0 & 0\end{pmatrix}\begin{pmatrix}E_{0x}\\ E_{0y}\end{pmatrix}. \tag{4.42}$$

The rightmost Jones matrix represents the polarization due to the polarizing beam-splitter and the matrix to the left of this one is due to the quarter-wave plate at 45°. The reflection from the cavity is represented by a minus identity matrix, since the (off-resonance) reflection causes a 180° phase shift for both field components. Finally, the leftmost matrix is due to a second pass through the quarter-wave plate. Performing the indicated multiplications,

$$\text{Polarization before second beam-splitter pass:} \quad \begin{pmatrix}0\\ -E_{0x}\end{pmatrix}. \tag{4.43}$$

This is of course the desired polarization for a 90° deflection by the beam-splitter. One is thus able to detect the beam reflected from the cavity without loss. The reflected beam can also be extracted using a conventional (dissipative) 50/50 beam-splitter, which might have the advantage of being less sensitive to polarization errors.

In deriving the discriminant for the Pound–Drever–Hall method, we start with the frequency-modulated field, where we will consider only two sidebands but with a modulation index ($\beta$) which can be unity or greater. The higher-order sidebands produce discriminant features which are separated from the main resonance by integral multiples of the modulation frequency, but we are only interested in the main resonance. The field is (eqn 4.40)

$$E(t) = E_0\left\{J_0(\beta)e^{i\omega_0 t} + J_1(\beta)(e^{i(\omega_0+\omega_m)t} - e^{i(\omega_0-\omega_m)t})\right\}, \tag{4.44}$$

where we used the fact that $J_{-n}(\beta) = (-1)^n J_n(\beta)$, and the modulation frequency is $\omega_m$.

The complex reflectivity of the cavity is taken from eqn 3.19 under impedance-matched conditions and expressed as a function of angular frequency, $\omega$,

$$E_r \equiv E_0 F(\omega) = -E_0 r\frac{2i(\omega-\omega')/\delta\omega}{1+2i(\omega-\omega')/\delta\omega}, \tag{4.45}$$

where $\omega'$ is the cavity resonant frequency and $\delta\omega$ is the full width at half maximum of the cavity resonance. We will assume that $|r| \approx 1$ in the following. Since there are three frequency components incident on the cavity and reflection from the cavity is a *linear operation*, by the *principle of superposition* the reflected signal is simply the sum of the reflections from each component taken individually:

$$\begin{aligned} E_r/E_0 =& J_0(\beta)F(\omega_0)e^{i\omega_0 t} \\ &+ J_1(\beta)\left(F(\omega_0+\omega_m)e^{i(\omega_0+\omega_m)t} - F(\omega_0-\omega_m)e^{i(\omega_0-\omega_m)t}\right). \end{aligned} \tag{4.46}$$

If we write the detected signal at the output of the photodetector amplifier as $V_{out} = GP_r \propto |E_r|^2$, where $P_r$ is the reflected power and the detector responsivity is $G$,

$$\begin{aligned} V_{out} =& 2GP_0J_0(\beta)J_1(\beta) \\ &\times \{\, \mathrm{Re}\,[F(\omega_0)F^*(\omega_0+\omega_m) - F^*(\omega_0)F(\omega_0-\omega_m)] \cos\omega_m t \\ &+ \mathrm{Im}\,[F(\omega_0)F^*(\omega_0+\omega_m) - F^*(\omega_0)F(\omega_0-\omega_m)] \sin\omega_m t \,\}, \end{aligned} \tag{4.47}$$

where $P_0$ is the incident power and we have left out the squared terms, which are constant in time, and have also left out the cross terms between the sidebands. It turns out that only the coefficient of the $\sin\omega_m t$ term gives a signal having the desired *dispersion shape* (the $\cos\omega_m t$ term generates a series of *Lorentzians*). The final signal is therefore

$$\begin{aligned} V_{out} =& 2GP_0J_0(\beta)J_1(\beta) \\ &\times \{\mathrm{Im}\,[F(\omega_0)F^*(\omega_0+\omega_m) - F^*(\omega_0)F(\omega_0-\omega_m)]\} \sin\omega_m t. \end{aligned} \tag{4.48}$$

The mixer will remove the $\sin\omega_m t$ term by shifting it down to zero frequency (DC). Substituting for $F(\omega)$ and defining $\Delta\omega \equiv \omega_0 - \omega'$

$$\begin{aligned} V'_{out} =& \frac{32GP_0J_0(\beta)J_1(\beta)\omega_m^2}{\delta\omega^3} \\ &\times \frac{\Delta\omega - 4\frac{\Delta\omega(\Delta\omega+\omega_m)(\Delta\omega-\omega_m)}{\delta\omega^2}}{\left(1+4\left(\frac{\Delta\omega}{\delta\omega}\right)^2\right)\left(1+4\left(\frac{\Delta\omega+\omega_m}{\delta\omega}\right)^2\right)\left(1+4\left(\frac{\Delta\omega-\omega_m}{\delta\omega}\right)^2\right)}, \end{aligned} \tag{4.49}$$

where $V'_{out}$ is the signal after being demodulated by a mixer whose local oscillator frequency is $\omega_m$. This expression is plotted in Fig. 4.11. If one is only interested in the central resonance ($\Delta\omega \leq \delta\omega$) and $\omega_m \gg \delta\omega$, this expression can be greatly simplified as

$$V'_{out} = \frac{8GP_0J_0(\beta)J_1(\beta)}{\delta\omega} \frac{-\Delta\omega}{1+4\left(\frac{\Delta\omega}{\delta\omega}\right)^2} \qquad \Delta\omega \leq \delta\omega, \quad \omega_m \gg \delta\omega, \tag{4.50}$$

which is a conventional *dispersion signal.* The slope at resonance can be easily determined by differentiation:

$$\frac{\partial V'_{out}}{\partial\omega_0} = -8GP_0J_0(\beta)J_1(\beta)\frac{1}{\delta\omega}. \tag{4.51}$$

The positive and negative peaks occur when $\Delta\omega = \pm\delta\omega/2$. The actual slope can be shown to be twice the value obtained by dividing the difference between the maximum and the minimum by $\delta\omega$. Finally, the expression $J_0(\beta)J_1(\beta)$ can be shown to be at its maximum when $\beta = 1.08$.

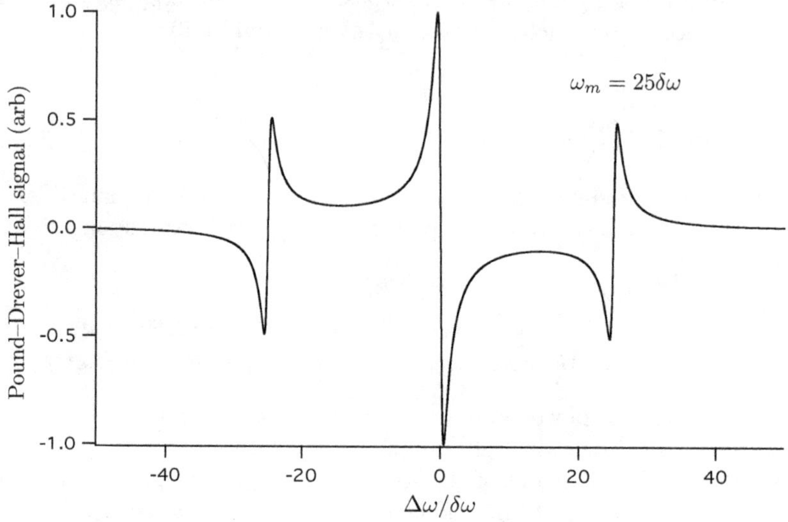


**Fig. 4.11** Signal from Pound–Drever–Hall apparatus plotted for $\beta = 1.08$.

## 4.8 Frequency response of a cavity-based discriminator

In this chapter, we have discussed the use of an optical cavity as a frequency discriminator, with the understanding that it will be used in some kind of frequency servo system. The analysis of the servo loop requires knowledge of the frequency response of each element in the loop, including the discriminator. The discussion of the latter can be a bit confusing, since the independent variable (the item being controlled) is itself a frequency and we are seeking the *frequency response of a frequency.* Operationally, we can apply some frequency modulation at $\omega_N$ to the laser and derive the amplitude and phase of the response of the discriminator as a function of this modulation frequency. Viewing the desired result as a response to a *frequency-modulated laser* should avoid the confusion.

In order to analyze the frequency response of the PDH discriminant, we will assume that the laser frequency is locked to the cavity resonant frequency using the PDH scheme. We will then generate a small sinusoidal *frequency dither* at $\omega_N$ and determine the amplitude and phase of the discriminator output at $\omega_N$:

$$\text{Frequency dither input signal:} \quad \Delta\omega(t) = \Delta\omega_0 \cos\omega_N t. \tag{4.52}$$

The phase modulation and modulation index of the input signal are

$$\text{Phase modulation:} \quad \Delta\phi(t) = \frac{\Delta\omega_0}{\omega_N}\sin\omega_N t = \alpha\sin\omega_N t \tag{4.53}$$

$$\text{Modulation index:} \quad \alpha = \frac{\Delta\omega_0}{\omega_N}. \tag{4.54}$$

The calculation proceeds in much the same way as the discriminant calculation except that we now have *two phase-modulation inputs*: the one at $\omega_m$ needed to implement the PDH approach and the "noise" input at $\omega_N$. Since the noise signal is very small, we can treat it as a *narrowband* modulation. The doubly modulated field is

$$\begin{aligned}
E(t) &= E_0 \exp(i\omega_0 t + i\beta \sin \omega_N t + i\alpha \sin \omega_N t) \\
&= E_0 \left[ J_0(\beta) e^{i\omega_0 t} + J_1(\beta)(e^{i(\omega_0+\omega_m)t} - e^{i(\omega_0-\omega_m)t}) \right] \\
&\times \left[ 1 + \frac{\alpha}{2}(e^{i(\omega_0+\omega_N)t} - e^{i(\omega_0-\omega_N)t}) \right] \\
&= E_0 \left[ J_0(\beta) e^{i\omega_0 t} + \frac{J_0(\beta)\alpha}{2}(e^{i(\omega_0+\omega_N)t} - e^{i(\omega_0-\omega_N)t}) \right. \\
&\left. + J_1(\beta)(e^{i(\omega_0+\omega_m)t)} - e^{i(\omega_0-\omega_m)t}) \right],
\end{aligned} \tag{4.55}$$

where we have left out terms at $\omega_m \pm \omega_N$ since they will not contribute to the signal at $\omega_m$. Using the principle of superposition to obtain the reflected field,

$$\begin{aligned}
\frac{E_r}{E_0} = &\left[ J_0(\beta) e^{i\omega_0 t} F(\omega_0) + \frac{J_0(\beta)\alpha}{2} \{ e^{i(\omega_0+\omega_N)t} F(\omega_0 + \omega_N) \right. \\
&\left. - e^{i(\omega_0-\omega_N)t} F(\omega_0 - \omega_N) \} + 2ie^{i\omega_0 t} J_1(\beta) \sin \omega_m t \right],
\end{aligned} \tag{4.56}$$

where $F(\omega_0 \pm \omega_m) = 1$, since the sidebands lie well outside the cavity width. Since $\omega_0 = \omega'$ (laser on resonance), we can considerably simplify this expression using $F(\omega_0) = 0$ and $F(\omega_0 + \omega_N) = F^*(\omega_0 - \omega_N)$ to obtain

$$\begin{aligned}
\frac{E_r}{E_0} = &\left[ iJ_0(\beta)\alpha e^{i\omega_0 t} \mathrm{Im}\{ e^{i\omega_N t} F(\omega_0 + \omega_N) \} \right. \\
&\left. + 2ie^{i\omega_0 t} J_1(\beta) \sin \omega_m t \right].
\end{aligned} \tag{4.57}$$

Proceeding as before to determine $V_{out}$, we calculate the square modulus and consider only the terms with a $\sin \omega_m t$ time dependence, since the mixer and low-pass filter will isolate these terms:

$$\begin{aligned}
V_{out}(\omega_m) &\propto 4\alpha J_0(\beta) J_1(\beta) \mathrm{Im}\{ F(\omega_0 + \omega_N) e^{i\omega_N t} \} \sin \omega_m t \\
&= 4\alpha J_0(\beta) J_1(\beta) |F(\omega_0 + \omega_N)| \cos(\omega_N t + \phi) \sin \omega_m t,
\end{aligned} \tag{4.58}$$

where

$$\phi = -\tan^{-1} \frac{\mathrm{Re}\{F(\omega_0 + \omega_N)\}}{\mathrm{Im}\{F(\omega_0 + \omega_N)\}} \tag{4.59}$$

and the slightly unusual expression for the phase is chosen so that the response is a cosine rather than a sine for easier comparison to the cosine "noise" input signal ($\Delta\omega(t)$). The output can be seen as a "beating" between the reflected sidebands at $\omega_m$ and those at $\omega_N$ (the reflected carrier is zero on resonance). Finally, dropping the constant multiplicative terms (except for $\alpha$, which is proportional to the *amplitude of the input signal*), the output of the mixer is

$$V_{mixer} \propto |F(\omega_0 + \omega_N)|\alpha \cos(\omega_N t + \phi)$$
$$= \frac{2\omega_N/\delta\omega}{\sqrt{1 + 4\left(\frac{\omega_N}{\delta\omega}\right)^2}} \alpha \cos(\omega_N t + \phi). \tag{4.60}$$

This can be rewritten as follows to emphasize the *transfer function* aspect of the equation:

$$V_{mixer} \propto \left(\frac{1}{\sqrt{1 + 4\left(\frac{\omega_N}{\delta\omega}\right)^2}}\right) \Delta\omega_0 \cos(\omega_N t + \phi) \tag{4.61}$$

$$\phi = -\tan^{-1}\frac{2\omega_N}{\delta\omega}. \tag{4.62}$$

Thus, an input frequency signal $\Delta\omega_0 \cos \omega_N t$ is multiplied by the bracketed factor to obtain the output amplitude, and its phase is shifted by $\phi$ to obtain the output phase. The *transfer function* of the discriminator is identical to that for a Butterworth low-pass filter consisting of a series $R$ and shunt $C$ having time constant $\tau = RC$. It is easy to show that the transfer function of this circuit is:

$$G(\omega) = \frac{1}{1 + i\omega\tau}. \tag{4.63}$$

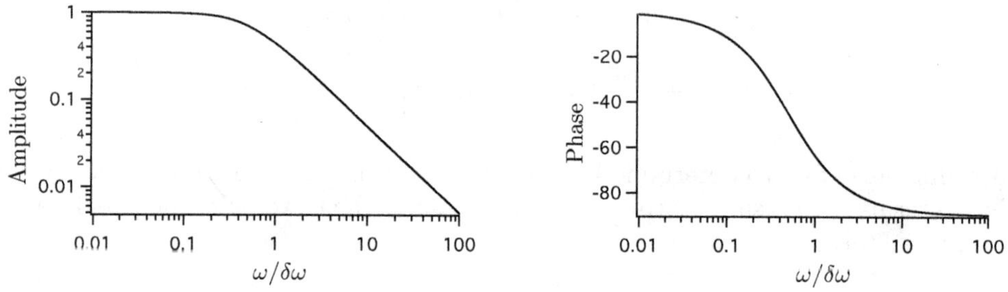


**Fig. 4.12** Magnitude and phase of the PDH transfer function. The $-3$ db point occurs at $\delta\omega/2$.

The magnitude and phase of $G(\omega)$ are:

$$|G(\omega)| = \frac{1}{\sqrt{1 + \omega^2\tau^2}} \tag{4.64}$$

$$\phi = -\tan^{-1}\omega\tau. \tag{4.65}$$

If we let $\tau = 2/\delta\omega$ the equivalence is established.

A plot of the magnitude and phase of the transfer function appears in Fig. 4.12. The plot displays the characteristic amplitude and phase characteristic of a single-pole low-pass filter. For the purposes of servo theory, the transfer function has one *pole* (zero of the denominator) which occurs at the −3 db point of the amplitude plot and the point where the phase lags by −45°. This point occurs at

$$\text{Pole location } (-3 \text{ db point}):\quad \omega_N = \frac{\delta\omega}{2}. \tag{4.66}$$

Below this frequency the output of the discriminator is more or less proportional to the input signal; above it, the discriminator behaves like an integrator with its characteristic 90° phase lag and 6 db/octave amplitude roll-off. This lag can be compensated in the electronics by placing a *zero* (a *differentiator*) at the same frequency.

## 4.9 Further reading

The best references on the three methods for generating a discriminant appear in the original papers describing the techniques. The side-of-line approach was described in the paper by Helmcke, Lee and Hall (1973); the Pound–Drever–Hall (PDH) scheme was originally described in the paper by Drever, Hall et al. (1983) and the polarization approach was described in the paper by Hansch and Couillaud (1980). There is a very large literature on the PDH method due to its use in the LIGO project, whose purpose is the detection of gravitational waves. Most papers on the PDH technique do not discuss its frequency response; one of the few exceptions is the paper by Day, Gustafson and Byer (1992).

## 4.10 Problems

(4.1) Determine the Jones matrix for a polarizing beam-splitter whose transmission axis is rotated (about the propagation direction) by angle $\theta$ from the $x$-direction. A polarizing beam-splitter is a device (in the form of a cube) which transmits one component of light and reflects the orthogonal component (see figure).

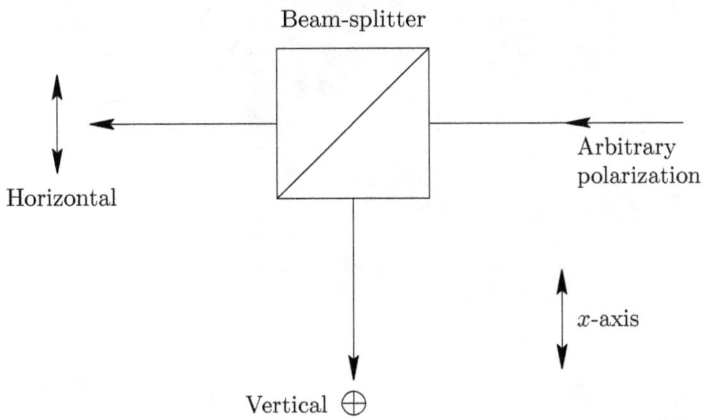

(4.2) Show, using Jones matrices, that a circular analyzer's behavior is independent of rotations about the propagation axis. A circular analyzer consists of a quarter-wave plate and polarizer with the fast axis of the wave plate at 45° to the transmission polarization direction of the polarizer. It transmits one type of circular polarization (right or left) and rejects the other.

(4.3) All of the discriminants discussed in this chapter can have a problem with offsets: any *additive* constant signal at the discriminator output will shift the *lock point* (the zero crossing) away from the exact cavity resonance. Calculate the size of the lock point shift (in frequency units) as a function of the cavity width and peak fringe height for the Pound–Drever–Hall discriminant.

(4.4) The offsets mentioned in the previous problem can be generated by inadvertent *amplitude modulation* from the electro-optic modulator which is used to phase modulate the beam. Calculate the approximate size of the offset due to a small amplitude modulation whose index is $\alpha$.

# 5

# Laser gain and some of its consequences

## 5.1 Introduction

The crucial element common to all lasers is the ability of a specially prepared medium to amplify optical radiation. We will describe this phenomenon from two viewpoints. First, we will derive the *optical Bloch equations*, based upon a semiclassical treatment of a two-level atomic system and a radiation field. As a visualization aid, we will discuss the formally identical *classical Bloch equations*, which describe the behavior of a magnetic moment in a combined static and rotating magnetic field. It will be shown that laser gain is proportional to the imaginary part of the linear susceptibility, which will be obtained from the semiclassical treatment. The second viewpoint will use a slightly modified form of Einstein's A and B coefficients adapted to describing interactions between atoms and *monochromatic* fields. Various damping and broadening mechanisms will be introduced in a reasonable but non-rigorous fashion and the crucial distinction between *homogeneous* and *inhomogeneous* broadening will be introduced.

Our approach will be reasonably complete but not as exhaustive as other books on the subject. The goal is not to prepare the reader for a career in the design of lasers but rather to develop an appreciation in the reader of the several laser systems described in later chapters.

## 5.2 The wave equation

We will use the *semiclassical* approach to the interaction of an atom with radiation: the atom will be treated quantum-mechanically but the radiation will be treated classically as a time-dependent term in the Hamiltonian. We start with the Schrödinger equation

$$i\hbar\frac{\partial\psi}{\partial t} = (H_0 + H_I(t))\psi, \tag{5.1}$$

where $H_0$ is the Hamiltonian of the atom *in the absence* of radiation and $H_I(t)$ is the time-dependent term which completely describes the interaction of the atom with radiation.

The atom has two *non-degenerate* levels labeled 1 (*ground state*) and 2 (*excited state*) with level 2 having the larger energy. The wavefunction, $\psi(\boldsymbol{r}, t)$, is expanded in eigenstates of the unperturbed Hamiltonian $H_0$ with time-dependent expansion coefficients. The rapidly varying time dependence of the eigenstates is factored out so that the coefficients $c_1(t)$ and $c_2(t)$ vary slowly with time. Thus,

$$\psi(\boldsymbol{r},t) = c_1(t)|1\rangle e^{-i\omega_1 t} + c_2(t)|2\rangle e^{-i\omega_2 t}, \qquad |c_1|^2 + |c_2|^2 = 1, \tag{5.2}$$

where

$$H_0|1\rangle = E_1|1\rangle \tag{5.3}$$
$$H_0|2\rangle = E_2|2\rangle \tag{5.4}$$
$$\omega_{1,2} = E_{1,2}/\hbar. \tag{5.5}$$

Substituting $\psi(\boldsymbol{r},t)$ (eqn 5.2) into Schrödinger's equation (eqn 5.1) and making use of the *orthonormality* of $|1\rangle$ and $|2\rangle$ yields two coupled equations:

$$\begin{aligned} i\hbar\dot{c}_1 &= \langle 1|H_I(t)|2\rangle e^{-i\omega_0 t} c_2 \\ i\hbar\dot{c}_2 &= \langle 2|H_I(t)|1\rangle e^{i\omega_0 t} c_1, \end{aligned} \tag{5.6}$$

where $\omega_0 = \omega_2 - \omega_1$ and we have used $\langle 1|H_I|1\rangle = \langle 2|H_I|2\rangle = 0$ for states of definite parity, since the presumed (electric dipole) interaction term, $H_I$, has *odd parity*.

## 5.3 The interaction term

In the presence of an electromagnetic field, the Hamiltonian is:

$$H = \frac{1}{2m}(\boldsymbol{p} - e\boldsymbol{A})^2 + e\phi + V = H_0 + H_I, \qquad H_0 = \frac{\boldsymbol{p}^2}{2m} + V, \tag{5.7}$$

where $\boldsymbol{p}$ is the canonical momentum, $V$ is the potential energy exclusive of the radiation field, and $\phi$ and $\boldsymbol{A}$ are the scalar and vector potentials defined by:

$$\boldsymbol{E} = -\nabla\phi - \frac{\partial \boldsymbol{A}}{\partial t} \tag{5.8}$$
$$\boldsymbol{B} = \nabla \times \boldsymbol{A}. \tag{5.9}$$

Due to the way they are defined, the potentials $\phi$ and $\boldsymbol{A}$ are somewhat arbitrary. The ability to change the potentials without changing the fields is called *gauge invariance*, and the constraints on a particular choice of potentials is called a *gauge*. In the absence of sources (charges and currents), one can use the *Coulomb gauge*, which is defined by the equation

$$\text{Coulomb gauge:} \quad \nabla \cdot \boldsymbol{A} = 0. \tag{5.10}$$

Since there are no sources, we can make a gauge transformation which eliminates the scalar potential:

$$\phi = 0. \tag{5.11}$$

Then, the interaction term is

$$H_I = \frac{e}{m}\boldsymbol{A} \cdot \boldsymbol{p} + \frac{e^2}{2m}\boldsymbol{A}^2 \approx \frac{e}{m}\boldsymbol{A} \cdot \boldsymbol{p}, \tag{5.12}$$

where we have used $\nabla \cdot \boldsymbol{A} = 0$ and have ignored the $\boldsymbol{A}^2$ term, which is generally much smaller than the $\boldsymbol{A} \cdot \boldsymbol{p}$ term.

We will assume that the field is given by a plane traveling wave whose vector potential is:

$$\boldsymbol{A}(\boldsymbol{r},t) = \boldsymbol{A}_0 \cos(\boldsymbol{k}\cdot\boldsymbol{r} - \omega t). \tag{5.13}$$

The electric field is

$$\boldsymbol{E} = \boldsymbol{E}_0 \sin(\boldsymbol{k}\cdot\boldsymbol{r} - \omega t), \qquad \boldsymbol{E}_0 = \omega \boldsymbol{A}_0. \tag{5.14}$$

The interaction term is currently expressed as a function of $\boldsymbol{p}$; we can use the following identity to convert matrix elements of $\boldsymbol{p}$ into matrix elements of $\boldsymbol{r}$:

$$\langle i|\boldsymbol{p}|j\rangle = im\omega_{ij}\langle i|\boldsymbol{r}|j\rangle, \tag{5.15}$$

where $\omega_{ij} = (E_i - E_j)/\hbar$. This identity is obtained by taking matrix elements of the commutator $[H_0, \boldsymbol{r}]$ between energy eigenstates $i$ and $j$ and using the identity $[r_i, p_j] = i\hbar\delta_{ij}$. From eqns 5.12 and 5.15, the matrix elements of $H_I$ are

$$\langle i|H_I|j\rangle = ie\boldsymbol{E}_0 \cdot \langle i|\boldsymbol{r}\cos(\boldsymbol{k}\cdot\boldsymbol{r} - \omega t)|j\rangle, \tag{5.16}$$

where we are assuming that we are in the vicinity of a *resonance* and thus substitute $\omega$ for $\omega_{ij}$. In evaluating the matrix element of the cosine term, one integrates over the atom, whose spatial extent is $\approx 0.1$ nm. The spatial period of the cosine function is the wavelength of optical radiation, $\approx$ 500 nm. Thus, one can make the *dipole approximation* and assume that $\boldsymbol{k}\cdot\boldsymbol{r} \ll 1$ over the atom. Then, ignoring a constant phase term ($i$), the desired *dipole matrix element* is

$$\langle i|H_I|j\rangle = \langle i|e\boldsymbol{E}_0 \cdot \boldsymbol{r}|j\rangle \cos(\omega t). \tag{5.17}$$

## 5.4 The rotating-wave approximation

The interaction matrix element can be written in terms of a *Rabi frequency*, $\Omega$, which is defined as:

$$\Omega \equiv \frac{1}{\hbar}\langle 1|e\boldsymbol{r}\cdot\boldsymbol{E}_0|2\rangle = \frac{e}{\hbar}\int \psi_1^*(\boldsymbol{r})\boldsymbol{r}\cdot\boldsymbol{E}_0\psi_2(\boldsymbol{r})d^3\boldsymbol{r}. \tag{5.18}$$

The coupled equations (eqns 5.6) are now

$$\begin{aligned} i\dot{c}_1 &= \Omega\cos(\omega t)e^{-i\omega_0 t}c_2 \\ i\dot{c}_2 &= \Omega^*\cos(\omega t)e^{i\omega_0 t}c_1. \end{aligned} \tag{5.19}$$

These equations depend upon two (very high) optical frequencies, $\omega$ and $\omega_0$, and upon two slowly varying quantities, $c_1$ and $c_2$. These disparate time dependences can be resolved if we can find a way of eliminating the rapidly oscillating terms. This can be done by expanding the cos() in exponentials and throwing away the very-high-frequency $e^{\pm i(\omega+\omega_0)t}$ terms. The coupled equations then become

$$\begin{aligned} i\dot{c}_1 &= \frac{\Omega}{2}e^{i\delta t}c_2 \\ i\dot{c}_2 &= \frac{\Omega^*}{2}e^{-i\delta t}c_1, \end{aligned} \tag{5.20}$$

where $\delta = \omega - \omega_0$. Ignoring the high-frequency terms is called the *rotating-wave approximation*. It is equivalent to replacing the *oscillating* term $\cos\omega t$ with a (rotating) complex exponential, $\cos\omega t \pm i\sin\omega t$ $(= e^{\pm i\omega t})$, where the sign is chosen to make all of the factors in the equations slowly varying. The validity of this will be discussed later on, and we will see what this means *graphically* when we consider the *classical Bloch equations*. Finally, we eliminate all of the explicit time dependence on the right-hand side by transforming the coefficients $c_{1,2}$ into coefficients $\tilde{c}_{1,2}$ using a matrix:

$$\begin{pmatrix} \tilde{c}_1 \\ \tilde{c}_2 \end{pmatrix} = \begin{pmatrix} e^{-i\delta t/2} & 0 \\ 0 & e^{i\delta t/2} \end{pmatrix} \begin{pmatrix} c_1 \\ c_2 \end{pmatrix}. \tag{5.21}$$

The coupled equations are now:

$$\begin{aligned} i\dot{\tilde{c}}_1 &= \frac{1}{2}(\delta\tilde{c}_1 + \Omega\tilde{c}_2) \\ i\dot{\tilde{c}}_2 &= \frac{1}{2}(\Omega\tilde{c}_1 - \delta\tilde{c}_2). \end{aligned} \tag{5.22}$$

Since $\delta \ll \omega_0$, the $\tilde{c}_{1,2}$ coefficients are still slowly varying with time.

## 5.5 Density matrix of two-level system

There are three approaches to solving the original coupled equations (eqns 5.6). One can use a perturbation approach, where the field is so weak that $c_2$ never has any appreciable amplitude and integrate the second equation assuming that $c_1 \approx 1$ (this is done *after* damping terms are added to the equations). The time integral of a function of the form $\cos(\omega t)e^{\pm i\omega_0 t}$ has two terms: one with a resonant denominator and one without. Invoking the *rotating-wave approximation* consists in ignoring the term with the non-resonant denominator.

The perturbation approach is not appropriate when studying lasers because the excited-state amplitude can be quite large. A solution for arbitrary excitation can be obtained by combining the two equations into a single second-order differential equation. An alternative approach which is also valid for strong excitation is to take the transformed equations (eqns 5.22) and form a *density matrix*. We will discuss the density matrix in this section.

Density matrices are generally used in quantum statistics where one has a statistical *ensemble* and can assign a probability to each possible wavefunction of the ensemble. In our case, we have a *pure state* whose wavefunction is known. However, there are some advantages to using a density matrix even then. The density operator, $\rho$, for a pure state $\psi$ is defined formally as

$$\rho = |\psi\rangle\langle\psi| \tag{5.23}$$

and is obviously Hermitian. Substituting from eqns 5.2 and 5.21, we obtain

$$\begin{aligned} \rho = |\tilde{c}_1|^2|1\rangle\langle 1| &+ \tilde{c}_1\tilde{c}_2^*|1\rangle\langle 2|e^{i\omega t} \\ &+ \tilde{c}_2\tilde{c}_1^*|2\rangle\langle 1|e^{-i\omega t} + |\tilde{c}_2|^2|2\rangle\langle 2| \end{aligned} \tag{5.24}$$

whose matrix representation (also denoted by $\rho$) is

$$\rho = \begin{pmatrix} |\tilde{c}_1|^2 & \tilde{c}_1\tilde{c}_2^* e^{i\omega t} \\ \tilde{c}_2\tilde{c}_1^* e^{-i\omega t} & |\tilde{c}_2|^2 \end{pmatrix}. \tag{5.25}$$

It will be demonstrated later on that the density matrix of a two-level system transforms under rotations in the same way as any operator operating on a spin-1/2 particle. Thus, in a frame rotating at angular rate $\omega$, the density matrix, $\tilde{\rho}$, is given by

$$\tilde{\rho} = R\rho R^{-1} = \begin{pmatrix} |\tilde{c}_1|^2 & \tilde{c}_1\tilde{c}_2^* \\ \tilde{c}_2\tilde{c}_1^* & |\tilde{c}_2|^2 \end{pmatrix}, \tag{5.26}$$

where the rotation matrix for a spin-1/2 particle rotating at angular rate $\omega$ is

$$R = \begin{pmatrix} e^{i\omega t/2} & 0 \\ 0 & e^{-i\omega t/2} \end{pmatrix}. \tag{5.27}$$

Thus, by transforming to a rotating coordinate system, we have eliminated the time dependence at optical frequencies of the density matrix elements, all of which are now slowly varying (in the rotating frame). When we discuss the classical Bloch equations, we will also perform a rotation at the frequency of the rotating magnetic field.

The diagonal and off-diagonal elements of $\rho$ have interesting interpretations. The diagonal elements $\rho_{ii}$ give the *probability* of finding the atom in state $i$. These are sometimes called the *populations* of the state $i$, though this term is perhaps more appropriate when one multiplies by the number density of atoms. The off-diagonal elements are the *coherences*, which are used when one takes the expectation value of some non-diagonal operator such as the dipole moment of the atom. The populations vary slowly with time, and the coherences have a time dependence at optical frequencies (in the *laboratory* frame).

From the definition of $\tilde{\rho}_{ij}$ together with the coupled differential equations (eqns 5.22), one can obtain the following differential equations for $\tilde{\rho}$:

$$\begin{aligned} \frac{d\tilde{\rho}_{12}}{dt} &= \frac{d\tilde{\rho}_{21}^*}{dt} = -i\delta\tilde{\rho}_{12} + \frac{i\Omega}{2}(\tilde{\rho}_{11} - \tilde{\rho}_{22}) \\ \frac{d\tilde{\rho}_{22}}{dt} &= -\frac{d\tilde{\rho}_{11}}{dt} = \frac{i\Omega}{2}(\tilde{\rho}_{21} - \tilde{\rho}_{12}). \end{aligned} \tag{5.28}$$

These equations can be rewritten in a compact form if we define the following quantities:

$$u \equiv \tilde{\rho}_{12} + \tilde{\rho}_{21} = 2\mathrm{Re}\{\tilde{\rho}_{12}\} \tag{5.29}$$

$$v \equiv -i(\tilde{\rho}_{12} - \tilde{\rho}_{21}) = 2\mathrm{Im}\{\tilde{\rho}_{12}\} \tag{5.30}$$

$$w \equiv \tilde{\rho}_{11} - \tilde{\rho}_{22} = \text{Population difference} \tag{5.31}$$

$$\boldsymbol{R} \equiv u\hat{x} + v\hat{y} + w\hat{z} = \text{Bloch vector} \tag{5.32}$$

$$\boldsymbol{W} \equiv \Omega\hat{x} + \delta\hat{z} = \text{Effective field.} \tag{5.33}$$

Substituting these into eqns 5.28 yields the *optical Bloch equations (without damping)*:

$$\dot{u} = \delta v \tag{5.34}$$

$$\dot{v} = -\delta u + \Omega w \tag{5.35}$$

$$\dot{w} = -\Omega v. \tag{5.36}$$

These equations can be combined into a single vector equation:

$$\frac{d\boldsymbol{R}}{dt} = \boldsymbol{R} \times \boldsymbol{W}. \tag{5.37}$$

It should be kept in mind that the optical Bloch equation is just the Schrödinger equation (in the so-called *interaction representation*) for an irradiated two-level system expressed in terms of quantities (*populations* and *coherences*) which are more familiar to the experimentalist than is the wavefunction. We will discuss the significance of this equation by deriving a formally identical *classical* equation which describes the behavior of a magnetic dipole in a combined static and rotating magnetic field.

## 5.6 The classical Bloch equation

Consider a *magnetic moment* $\boldsymbol{\mu}$ in a vertical magnetic field, $\boldsymbol{H}_0 = H_0\hat{\boldsymbol{z}}$ (Fig. 5.1). Equating the time rate of change of the angular momentum, $\boldsymbol{L}$, to the torque,

$$\frac{d\boldsymbol{L}}{dt} = \gamma \boldsymbol{L} \times \boldsymbol{H}_0, \tag{5.38}$$

where $\boldsymbol{\mu} = \gamma \boldsymbol{L}$, and $\gamma$ is the *gyromagnetic ratio* (ratio of the magnetic moment to the angular momentum). It is straightforward to show that the magnetic moment precesses around the static field at an angular rate equal to $\omega_L = \gamma H_0$, called the *Larmor frequency*.

We now add a small magnetic field, $\boldsymbol{H}_1$, *rotating* in the $xy$-plane at angular rate $\omega$. It is convenient to transform to a *frame which is rotating at the same rate*, so that $\boldsymbol{H}_1$ is fixed in the new frame (along the $x$-axis, $\boldsymbol{H}_1 = H_1\hat{\boldsymbol{x}}$). We use the relationship between the time derivative of a vector in a *fixed frame* and one in a *frame rotating* at $\omega$ (this can be derived by performing an infinitesimal rotation and using the definition of the derivative):

$$\frac{d}{dt} = \frac{\partial}{\partial t} + \boldsymbol{\omega} \times \tag{5.39}$$

where $d/dt$ is in the fixed frame and $\partial/\partial t$ is in the rotating frame. Thus,

$$\frac{d\boldsymbol{L}}{dt} = \frac{\partial \boldsymbol{L}}{\partial t} + \boldsymbol{\omega} \times \boldsymbol{L} \tag{5.40}$$

and, substituting into eqn 5.38,

$$\frac{\partial \boldsymbol{L}}{\partial t} = \gamma \boldsymbol{L} \times \left( \left( H_0 - \frac{\omega}{\gamma} \right) \hat{\boldsymbol{z}} + H_1 \hat{\boldsymbol{x}} \right) \qquad (\boldsymbol{\omega} = -\omega \hat{\boldsymbol{z}}) \tag{5.41}$$

is the equation for the rate of change of $\boldsymbol{L}$ in the *rotating frame*. The quantity in the outer brackets is called the *effective field*, since it is the field in the rotating frame about which the magnetic moment precesses. The *classical Bloch equation* is obtained by identifying this quantity as $\boldsymbol{W}/\gamma$:

$$\frac{\partial \boldsymbol{L}}{\partial t} = \boldsymbol{L} \times \boldsymbol{W}, \tag{5.42}$$

which is the same equation as the one derived earlier for a two-level quantum-mechanical system.

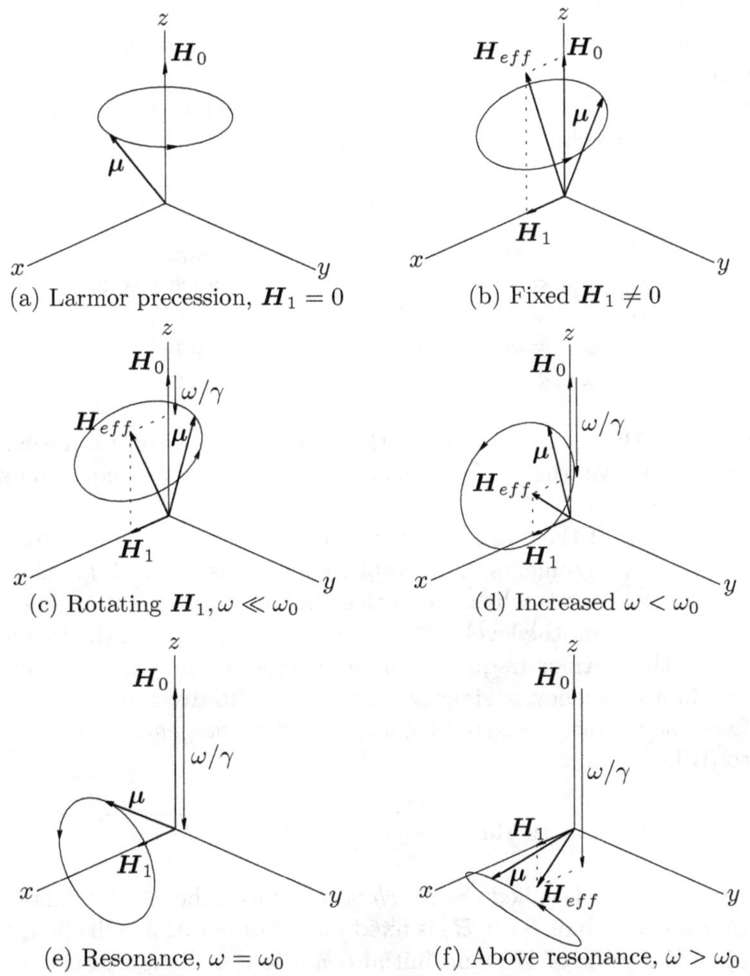


**Fig. 5.1** Magnetic moment, $\boldsymbol{\mu}$, precessing in various magnetic fields. The addition of a rotating $\boldsymbol{H}_1$ field ((c) through (f)) in a frame rotating with $\boldsymbol{H}_1$ shortens the static field (in the rotating frame) and the moment now precesses around $\boldsymbol{H}_{eff}$. At *resonance* ((e)), the *effective field* is along the $x$-axis and the oscillating $z$-component of $\boldsymbol{\mu}$ has its maximum amplitude.

The interpretation of the classical equation is straightforward (see Fig. 5.1). With no $\boldsymbol{H}_1$ field, the magnetic moment precesses about $\boldsymbol{H}_0$ at the Larmor frequency, $\omega_L$. When one adds a *fixed* $\boldsymbol{H}_1$ field, the precession is now around the combined fields $H_0\hat{\boldsymbol{z}} + H_1\hat{\boldsymbol{x}}$. If one allows $\boldsymbol{H}_1$ to rotate at angular rate $\omega$, the time derivative transformation causes $\boldsymbol{H}_0$ to shorten by $(\omega/\gamma)\hat{\boldsymbol{z}}$. When the rotation is at the Larmor frequency, the $z$-component of the magnetic field will be reduced to zero (in the rotating frame) and the magnetic moment will precess around the horizontal $\boldsymbol{H}_1$ field at the *Rabi frequency*, $\gamma H_1$. If one is measuring the $z$-component of the magnetic moment, it will have a sinusoidal time dependence at the Rabi frequency. This is the *resonant* behavior that is sought after in magnetic resonance experiments, where the Rabi frequency was first introduced.

We now make the following *equivalences* between the classical quantities and those used in the quantum-mechanical optical Bloch equation:

| Classical | | Quantum | | |
|---|---|---|---|---|
| $\gamma H_1$ | $\Longleftrightarrow$ | $\Omega$ | Rabi frequency | |
| $\gamma H_0$ | $\Longleftrightarrow$ | $\omega_0$ | Resonance frequency | (5.43) |
| $\boldsymbol{L}$ | $\Longleftrightarrow$ | $\boldsymbol{R}$ | Bloch vector | |
| $\boldsymbol{W}/\gamma$ | $\Longleftrightarrow$ | $\boldsymbol{W}$ | Effective field | |
| $\delta = \omega_0 - \omega$ | $\Longleftrightarrow$ | $\delta = \omega - \omega_0$ | Frequency detuning, | |

with the result that $\boldsymbol{W} = \delta\hat{\boldsymbol{z}} + \Omega\hat{\boldsymbol{x}}$ in both the classical and quantum-mechanical cases. The reason for calling $\boldsymbol{W}$ the *effective field* in the optical Bloch equation should now be clear.

The interpretation of the two-level atom irradiated by a monochromatic field follows from the above assignments. The Rabi frequency is the rate at which the atom cycles between the two levels when the optical frequency is exactly resonant with the energy separation between the levels. The resonance frequency of the two-level system is identified with the Larmor frequency of the magnetic moment (with no $\boldsymbol{H}_1$ field). Although the Rabi frequency is strictly only defined at resonance, the off-resonance precession frequency is often referred to, and is called the *generalized Rabi frequency*. It is equal to $|W|$:

$$\text{Generalized Rabi frequency} = |W| = \sqrt{\delta^2 + \Omega^2}. \tag{5.44}$$

The precessing vector is called the *Bloch vector*. From the Bloch equation, it should be obvious that the length of $\boldsymbol{L}$ (or $\boldsymbol{R}$) is fixed but its orientation will change with time, depending upon the value of $\boldsymbol{W}$ (and initial conditions). The projection of the vector on the $z$-axis is the $z$-component of the angular momentum in magnetic resonance experiments and the *population inversion*, $\rho_{11} - \rho_{22}$, in a two-level atom. The population inversion will play a central role in laser theory: it is just the excess probability of finding the atom in the ground state over the probability of finding it in the excited state. Thus, a Bloch vector pointing up will correspond to an atom in the ground state and a Bloch vector pointing down will correspond to an atom in the excited state. (From the way we have defined the population inversion, a *negative* inversion corresponds to the excited-state population being *greater* than that of the ground state.) The $x$- and $y$-components of the Bloch vector are proportional to the real and imaginary parts of the coherence element, $\rho_{12}$. We will show that these components are proportional to the quadrature components of an *oscillating dipole moment* induced in the atom by the optical field. In the classical case, the horizontal projection of the Bloch vector is the horizontal component of the oscillating magnetic dipole moment.

The classical model also provides a rationale for the *rotating-wave approximation*. In most magnetic resonance experiments, one uses an *oscillating field* rather than a rotating one. An oscillating field can be written as the sum of two counter-rotating fields:

$$\begin{aligned}\boldsymbol{H}(t) &= H_0\hat{\boldsymbol{x}}\cos\omega t = \boldsymbol{H}_+(t) + \boldsymbol{H}_-(t)\\ \boldsymbol{H}_+(t) &= \frac{1}{2}H_0\left(\hat{\boldsymbol{x}}\cos\omega t + \hat{\boldsymbol{y}}\sin\omega t\right)\\ \boldsymbol{H}_-(t) &= \frac{1}{2}H_0\left(\hat{\boldsymbol{x}}\cos\omega t - \hat{\boldsymbol{y}}\sin\omega t\right),\end{aligned} \tag{5.45}$$

where it should be clear that $\boldsymbol{H}_+$ and $\boldsymbol{H}_-$ rotate in opposite senses. At resonance, one of these fields will rotate exactly in step with the Larmor precession of the magnetic moment and the rotating frame will rotate so that this component is fixed. In the rotating frame, the other component will rotate at $2\omega_L$ and will be very far-off resonance and will therefore only slightly perturb the magnetic moment (the idea of *far-off resonance* will be clarified when we discuss radiative damping and a *resonance width*). In the two-level atom, the rotation is not in coordinate space but is in the complex plane: it corresponds to the use of complex exponentials ($e^{\pm i\omega t}$) rather than sines and cosines to represent the oscillatory time dependence of the fields. Just as in the magnetic resonance case, the oscillatory field can be resolved into two counter-rotating complex exponentials, and the *non-resonant* (rapidly oscillating) term will have very little effect on the Bloch vector (the off-resonance perturbation is sometimes called an *AC Stark shift* or a *Bloch–Siegert shift*).

## 5.7 Connection between two-level atom and spin-1/2 system

In the preceding sections, we showed an equivalence between a *quantum-mechanical* two-level system and a *classical* magnetic moment whose evolution in time in a magnetic field is described by the Bloch equations. In this section, we will provide some additional motivation for using the optical Bloch equations by discussing the close connection between a two-level atomic system and a *quantum-mechanical* spin-1/2 system.

In the process of deriving the optical Bloch equations, we had found it useful to rotate the $2\times 2$ density matrix of a two-level system using a rotation matrix (eqn 5.27) for a spin-1/2 particle; we now provide some justification for this transformation. We will begin by demonstrating that the rotation of a spin-1/2 wavefunction using a $2\times 2$ matrix $R_z(\theta)$ is equivalent to a rotation in three-dimensional space about the $z$-axis by the angle $\theta$.

We will assume familiarity with the fundamentals of quantum-mechanical angular momentum in all that follows. The angular momentum states of a spin-1/2 system will be represented by column vectors (spinors) having two elements:

$$\begin{aligned}|+1/2\rangle &\longrightarrow \begin{pmatrix}1\\0\end{pmatrix}\\ |-1/2\rangle &\longrightarrow \begin{pmatrix}0\\1\end{pmatrix}\end{aligned} \tag{5.46}$$

where $|\pm 1/2\rangle$ are the eigenstates of $S_z$, the $z$-component of the spin angular momentum, with eigenvalues $\pm\hbar/2$. It is convenient to write the matrix representation of the spin angular momentum operator as a vector $\boldsymbol{S}$ whose components are proportional to the three *Pauli spin matrices*, $\sigma_{x,y,z}$:

$$S_i = \frac{\hbar}{2}\sigma_i, \qquad i = x, y, z, \tag{5.47}$$

where the Pauli matrices are

$$\begin{aligned}\sigma_x &= \begin{pmatrix} 0 & 1 \\ 1 & 0 \end{pmatrix} \\ \sigma_y &= \begin{pmatrix} 0 & -i \\ i & 0 \end{pmatrix} \\ \sigma_z &= \begin{pmatrix} 1 & 0 \\ 0 & -1 \end{pmatrix}.\end{aligned} \tag{5.48}$$

These matrices satisfy both a commutation and anticommutation relation:

$$\begin{aligned}[\sigma_i, \sigma_j] &\equiv \sigma_i\sigma_j - \sigma_j\sigma_i = 2i\epsilon_{ijk}\sigma_k \qquad & i, j, k = x, y, z \\ \{\sigma_i, \sigma_j\} &\equiv \sigma_i\sigma_j + \sigma_j\sigma_i = 2\delta_{ij}I \qquad & I = \begin{pmatrix} 1 & 0 \\ 0 & 1 \end{pmatrix},\end{aligned} \tag{5.49}$$

where $\epsilon_{ijk}$ is zero if any indices are repeated, is $+1$ for any cyclic permutation of the indices $x, y, z$ and $-1$ otherwise. For a system whose angular momentum is $\boldsymbol{J}$, an arbitrary state $|\alpha\rangle$ will be transformed by a rotation about the $z$-axis into a state $|\alpha'\rangle$ according to

$$|\alpha'\rangle = e^{-iJ_z\theta/\hbar}|\alpha\rangle, \tag{5.50}$$

where $\theta$ is the rotation angle. If $j = 1/2$, the rotation matrix, $R_z(\theta)$, is

$$\text{Rotation of spin-1/2 system:} \qquad R_z(\theta) = \begin{pmatrix} e^{-i\theta/2} & 0 \\ 0 & e^{i\theta/2} \end{pmatrix}. \tag{5.51}$$

This result can be derived using the *generalized Euler formula*,

$$e^{-i\boldsymbol{\sigma}\cdot\boldsymbol{n}\theta} = I\cos\theta - i\boldsymbol{\sigma}\cdot\boldsymbol{n}\sin\theta \qquad \boldsymbol{n} = \text{a unit vector}, \tag{5.52}$$

which can be obtained from the series expansion of the exponential together with eqns 5.49. The rotation matrix is *unitary*, which implies that:

$$R_z^{-1}(\theta) = R_z^{\dagger}(\theta) = R_z(-\theta) = \begin{pmatrix} e^{i\theta/2} & 0 \\ 0 & e^{-i\theta/2} \end{pmatrix}. \tag{5.53}$$

So far, we have determined how a *spin-1/2 system* behaves under a rotation. In all that follows, the system will be stationary and we will be *rotating the coordinate system*. This is accomplished by changing $\theta$ to $-\theta$ in the rotation matrix:

$$\text{Rotation of coordinate system:} \qquad R_z(\theta) = \begin{pmatrix} e^{i\theta/2} & 0 \\ 0 & e^{-i\theta/2} \end{pmatrix}. \tag{5.54}$$

If $M$ is some matrix which operates on spinors, we would like to determine how it is transformed under the rotation, $R_z(\theta)$, of the coordinate system. To do so, we begin with a matrix equation in which a spinor is operated on by $M$:

$$\begin{pmatrix} c \\ d \end{pmatrix} = \begin{pmatrix} M_{11} & M_{12} \\ M_{21} & M_{22} \end{pmatrix} \begin{pmatrix} a \\ b \end{pmatrix}. \tag{5.55}$$

The spinors are transformed under rotations using $R_z(\theta)$:

$$\begin{aligned} \begin{pmatrix} a' \\ b' \end{pmatrix} &= R_z(\theta) \begin{pmatrix} a \\ b \end{pmatrix} \\ \begin{pmatrix} c' \\ d' \end{pmatrix} &= R_z(\theta) \begin{pmatrix} c \\ d \end{pmatrix}, \end{aligned} \tag{5.56}$$

where the primed quantities are those after the rotation. Since the coordinate system is being rotated, we require that eqn 5.55 be preserved under the rotation; that is, we would like the equation to be the same when we use primed quantities. Using eqns 5.56, we have:

$$R_z^{-1}(\theta) \begin{pmatrix} c' \\ d' \end{pmatrix} = \begin{pmatrix} M_{11} & M_{12} \\ M_{21} & M_{22} \end{pmatrix} R_z^{-1}(\theta) \begin{pmatrix} a' \\ b' \end{pmatrix}. \tag{5.57}$$

Multiplying on the left by $R_z(\theta)$, we obtain the matrix $M'$ in the rotated coordinate system:

$$\begin{pmatrix} M'_{11} & M'_{12} \\ M'_{21} & M'_{22} \end{pmatrix} = R_z(\theta) \begin{pmatrix} M_{11} & M_{12} \\ M_{21} & M_{22} \end{pmatrix} R_z^{-1}(\theta). \tag{5.58}$$

In order to examine the transformation properties of $M$ under a rotation about the $z$-axis, we first express $M$ as a linear combination of the identity matrix ($I$) and the three Pauli matrices ($\sigma_i$). Writing $M$ in this fashion,

$$M = \tfrac{1}{2}(I + \boldsymbol{A} \cdot \boldsymbol{\sigma}), \tag{5.59}$$

where the 1/2 is there to force the trace (sum of diagonal elements) to be unity, in anticipation of identifying $M$ as a density matrix. The quantity $\boldsymbol{A}$ is a three-dimensional vector whose transformation properties will now be determined. We first write eqn 5.58 in terms of $\boldsymbol{A}$ and $\boldsymbol{\sigma}$:

$$\begin{aligned} \tfrac{1}{2}(I + \boldsymbol{A}' \cdot \boldsymbol{\sigma}) &= \tfrac{1}{2} R_z(\theta)(I + \boldsymbol{A} \cdot \boldsymbol{\sigma}) R_z^{-1}(\theta)) \\ \Longrightarrow \boldsymbol{A}' \cdot \boldsymbol{\sigma} &= R_z(\theta) \boldsymbol{A} \cdot \boldsymbol{\sigma} R_z^{-1}(\theta), \end{aligned} \tag{5.60}$$

where we have cancelled the term involving the identity on both sides (the identity matrix is unchanged under a rotation). Note that the Pauli matrices on the left are unprimed: they are constant matrices (eqn 5.48). The right-hand side consists of three terms, which correspond to the three components of $\boldsymbol{A}$. Using eqn 5.54, the three terms are:

$$\begin{aligned} A_x R_z(\theta) \sigma_x R_z^{-1}(\theta) &= A_x \begin{pmatrix} 0 & e^{i\theta} \\ e^{-i\theta} & 0 \end{pmatrix} = A_x(\sigma_x \cos\theta - \sigma_y \sin\theta) \\ A_y R_z(\theta) \sigma_y R_z^{-1}(\theta) &= A_y \begin{pmatrix} 0 & -ie^{i\theta} \\ ie^{-i\theta} & 0 \end{pmatrix} = A_y(\sigma_y \cos\theta + \sigma_x \sin\theta) \\ A_z R_z(\theta) \sigma_z R_z^{-1}(\theta) &= A_z \sigma_z. \end{aligned} \tag{5.61}$$

Since the Pauli matrices (together with $I$) are *linearly independent*, we can use eqn 5.61 and equate the coefficients of each $\sigma_i$ on both sides of eqn 5.60 to obtain a relationship between $\boldsymbol{A}$ and $\boldsymbol{A}'$:

$$\begin{aligned} A'_x &= A_x \cos\theta + A_y \sin\theta \\ A'_y &= -A_x \sin\theta + A_y \cos\theta \\ A'_z &= A_z. \end{aligned} \tag{5.62}$$

This is the well-known transformation rule for the components of a vector when the coordinate system is rotated about the $z$-axis by angle $\theta$. We have thus demonstrated that the rotation of a spinor in its abstract two-dimensional space using $R_z(\theta)$ is equivalent to a three-dimensional rotation of a coordinate system about the $z$-axis. One can obtain formulas for the elements of $M$ in terms of the components of $\boldsymbol{A}$ using eqn 5.59:

$$\begin{pmatrix} M_{11} & M_{12} \\ M_{21} & M_{22} \end{pmatrix} = \frac{1}{2}\begin{pmatrix} 1 + A_z & A_x - iA_y \\ A_x + iA_y & 1 - A_z \end{pmatrix}. \tag{5.63}$$

Solving for the components of $\boldsymbol{A}$:

$$\begin{aligned} A_x &= M_{12} + M_{21} \\ A_y &= i(M_{12} - M_{21}) \\ A_z &= M_{11} - M_{22}. \end{aligned} \tag{5.64}$$

Observe that this is exactly the same relationship (other than an unimportant sign of $A_y$) as we obtained for the Bloch vector in terms of the Hermitian density matrix in eqn 5.32. For a continuous rotation at rate $\omega$, we can replace $\theta$ by $\omega t$ and obtain the rotation matrix (eqn 5.27) that we used in our earlier treatment of the density matrix:

Continuous rotation of coordinate system: $$R_z(\omega t) = \begin{pmatrix} e^{i\omega t/2} & 0 \\ 0 & e^{-i\omega t/2} \end{pmatrix}. \tag{5.65}$$

In summary, the similarity transformation $(R_z(\omega t)\rho R_z^{-1}(\omega t))$ of the density matrix $(\rho)$ is completely equivalent to the continuous rotation in three dimensions of the coordinate system of the Bloch vector about the $z$-axis at rate $\omega$.

## 5.8 Radiative and collision-induced damping

The semiclassical treatment so far can actually describe most of the aspects of a laser – in particular, the *coherent amplification* which results from *stimulated emission.* Although spontaneous decay can be patched into the semiclassical theory using the *Fermi golden rule*, a satisfactory treatment requires the quantization of the radiation field. We will use a *phenomenological approach* to introduce spontaneous decay in a possibly ad hoc (but reasonable) manner: we will add *damping terms* to the differential equations satisfied by the density matrix. It turns out that it is much easier to do this in a natural way to the density matrix equations than to the differential equations for the coefficients, $c_i(t)$, since the behavior of the populations and coherences under various damping mechanisms is easier to describe. We will also modify the Bloch equations to account for *collision-induced damping* in a low-pressure gas.

We will begin with the observation that, in the absence of radiation, an atom with an excited-state *population* ($\tilde{\rho}_{22}$) will spontaneously decay at a rate ($\gamma$) which is proportional to the population:

$$\frac{d\tilde{\rho}_{22}}{dt} = -\gamma\tilde{\rho}_{22}, \tag{5.66}$$

whose solution is

$$\tilde{\rho}_{22}(t) = \tilde{\rho}_{22}(0)e^{-\gamma t}. \tag{5.67}$$

Since we call level 1 the "ground state", we will assume that it does not decay.

What is the corresponding effect on the coherence elements, $\tilde{\rho}_{12}$ and $\tilde{\rho}_{21}$? We can answer this by using the following relation, which follows from the definition of the density matrix:

$$|\tilde{\rho}_{12}(t)| = \sqrt{\tilde{\rho}_{11}(t)\tilde{\rho}_{22}(t)}. \tag{5.68}$$

We have already established that $\tilde{\rho}_{22}(t)$ decays exponentially at rate $\gamma$, and $\tilde{\rho}_{11}$ does not decay. Thus, in the absence of a radiation field,

$$\tilde{\rho}_{12}(t) = \tilde{\rho}_{12}(0)e^{-\gamma t/2}, \tag{5.69}$$

and therefore $\tilde{\rho}_{12}$ decays at a rate $\gamma/2$. We will display the most recent density matrix equations of motion here for reference:

$$\begin{aligned}\dot{u} &= \delta v \\ \dot{v} &= -\delta u + \Omega w \\ \dot{w} &= -\Omega v.\end{aligned}$$

Recalling that

$$\begin{aligned}u &= 2Re\{\tilde{\rho}_{12}\} \\ v &= 2Im\{\tilde{\rho}_{12}\} \\ w &= \tilde{\rho}_{11} - \tilde{\rho}_{22},\end{aligned}$$

the following modified density matrix equations will have the correct damping properties when the field is turned off:

$$\begin{aligned}\dot{u} &= \delta v - \frac{\gamma}{2}u \\ \dot{v} &= -\delta u + \Omega w - \frac{\gamma}{2}v \\ \dot{w} &= -\Omega v - \gamma(w-1).\end{aligned} \tag{5.70}$$

The term in brackets in the last equation is due to the fact that $w$ relaxes to $w = 1$ (ground state), while the coherences $(u, v)$ relax to zero.

The steady-state solution of eqns 5.70 can be obtained by setting all of the time derivatives equal to zero and solving the three resulting linear equations. The results are:

$$u = \frac{\Omega\delta}{\delta^2 + \Omega^2/2 + \gamma^2/4} \tag{5.71}$$

$$v = \frac{\Omega\gamma/2}{\delta^2 + \Omega^2/2 + \gamma^2/4} \tag{5.72}$$

$$w = \frac{\delta^2 + \gamma^2/4}{\delta^2 + \Omega^2/2 + \gamma^2/4}. \tag{5.73}$$

The value of $\rho_{22}$ ($= \tilde{\rho}_{22}$) comes up in the study of *fluorescence*, which is proportional to $\rho_{22}$:

$$\rho_{22} = \frac{1-w}{2} = \frac{\Omega^2/4}{\delta^2 + \Omega^2/2 + \gamma^2/4}. \tag{5.74}$$

This is a simple *Lorentzian*. We see here the presence of the *resonance width*: all of these expressions have denominator $\delta^2 + \Omega^2/2 + \gamma^2/4$ and, in particular, the full width at half maximum of the expression for $\rho_{22}$ versus the detuning, $\delta$, is just $\gamma$ (when $\Omega \ll \gamma$). We also notice, in these equations, the phenomenon of *saturation*: when $\Omega \geq \gamma$ the resonance width of $\rho_{22}$ will increase:

$$\text{Full width at half maximum} = 2\sqrt{\Omega^2/2 + \gamma^2/4} = \sqrt{2\Omega^2 + \gamma^2}. \tag{5.75}$$

Finally, the unsaturated excited-state population on resonance is $(\Omega/\gamma)^2$, while the saturated population on resonance approaches 1/2. We will have more to say about saturation later.

The treatment of collisions in a low-pressure gas can be quite involved; we will touch on the main points. We will start with the result from *kinetic theory* that the probability, $p(t)dt$, that an atom has a period $t$ which is free from collisions is:

$$p(t)dt = (1/\tau_{col})\exp(-t/\tau_{col})dt, \tag{5.76}$$

where $dt$ is a very small observation time and $\tau_{col}$ is the *average time between collisions* (which depends upon the *type of collision*).

There are two kinds of collisions: *elastic*, which conserve energy in the atom, and *inelastic*, which do not. Elastic collisions therefore *do not result in transitions* but instead *randomize* to some extent the *phase* of the atomic wavefunction. Inelastic collisions result in transitions, almost always to a lower energy level. It turns out that inelastic collisions have the same effect on the Bloch vector as spontaneous decay, and the *rate for inelastic collisions can be added to that for spontaneous decay.* Thus, the total damping rate, $\gamma_{total}$, is

$$\gamma_{total} = \gamma_{spont} + \gamma_{inelast}. \tag{5.77}$$

Elastic collisions are observed far more frequently and their treatment is slightly more complicated. We need to find a relationship between the collision rate, $1/\tau_{col}$, and the damping parameter, $\gamma_{elast}$, for elastic collisions. Our model will be the radiation *emitted by an ensemble of n atoms*, where we assume that the effect of collisions on absorption will be the same as on emission. A model of the emission process which is consistent with the semiclassical theory is that the atom emits a series of wavetrains whose average length is $c\tau_{coll}$ and whose phase is completely random. We will evaluate the *autocorrelation function* of the emitted electric field and use the *Wiener–Khintchine* theorem to obtain the *power spectrum*, whose frequency dependence will give us our desired result.

The electric field is

$$\begin{aligned} E(t) &= E_1(t) + E_2(t) + \cdots + E_n(t) \\ &= E_0 e^{-i\omega_0 t}\left\{e^{i\phi_1(t)} + e^{i\phi_2(t)} + \cdots + e^{i\phi_n(t)}\right\}, \end{aligned} \tag{5.78}$$

where the $\phi_i(t)$ are the *random* phases of the wavetrains. The autocorrelation function is then

$$\langle E(t)^* E(t+\tau)\rangle = E_0^2 e^{-i\omega_0\tau}\langle\sum_i^n\sum_j^n e^{i(\phi_i(t+\tau)-\phi_j(t))}\rangle \tag{5.79}$$

$$= E_0^2 e^{-i\omega_0\tau}\sum_i^n\langle e^{i(\phi_i(t+\tau)-\phi_i(t))}\rangle \tag{5.80}$$

$$= n\langle E_i(t)^* E_i(t+\tau)\rangle, \tag{5.81}$$

where the second equation is a result of the *cross terms* averaging to zero and the third is due to the fact that all of the wavetrains are *statistically* equivalent. The autocorrelation function on the last line is that due to a *single atom.*

The salient feature of our model of elastic collisions is that the *phase of the emitted wavetrain is completely randomized by the collision.* Thus, after each collision, the contribution of an atom drops to zero. The single-atom autocorrelation function, $\langle E_i(t)^* E_i(t+\tau)\rangle$, is therefore proportional to the probability that the time between collisions is *longer than* $\tau$ (collision times shorter than $\tau$ yield a zero autocorrelation since the two electric field factors are then *totally uncorrelated*). Using the probability distribution (eqn 5.76), we therefore have

$$\langle E_i(t)^* E_i(t+\tau)\rangle = E_0^2 e^{-i\omega_0\tau}\langle e^{i[\phi_i(t+\tau)-\phi_i(t)]}\rangle \tag{5.82}$$

$$= E_0^2 e^{-i\omega_0\tau}\int_\tau^{-\infty} p(t)dt \tag{5.83}$$

$$= E_0^2 e^{-i\omega_0\tau-\tau/\tau_{coll}}, \tag{5.84}$$

and the autocorrelation function for all $n$ atoms is

$$\langle E(t)^* E(t+\tau)\rangle = nE_0^2 e^{-i\omega_0\tau-\tau/\tau_{coll}}. \tag{5.85}$$

The *power spectrum* is obtained by invoking the Wiener–Khintchine theorem, which states: *the power spectrum of a signal is equal to the Fourier transform of its autocorrelation function.* Thus, the power spectrum, $F(\omega)$, is proportional to

$$F(\omega) \propto \frac{1}{(\omega - \omega_0)^2 + (1/\tau_{coll})^2)}. \tag{5.86}$$

The *linewidth* (full width at half maximum) of this function is $2/\tau_{coll}$. We observe that the power spectrum has the same Lorentzian frequency dependence as the population provided that we use the following relation between the *elastic collision width* and the collision time:

$$\gamma_{elast} = \frac{2}{\tau_{coll}}. \tag{5.87}$$

The next question is how to incorporate $\gamma_{elast}$ into the density matrix equations. Since only the *coherences* and not the *populations* are involved in elastic collisions, it should be plausible that the damping be included *only* in the $\rho_{12}$ (i.e., $u$ and $v$) equation (once more demonstrating the utility of the density matrix since it is not obvious how one would do this using the coefficients, $c_i(t)$). We will define the total damping (due to radiation and collisions of both types) as $\gamma'$, where

$$\gamma' = \gamma_{total} + \gamma_{elast} = \gamma_{spont} + \gamma_{inelast} + \gamma_{elast}. \tag{5.88}$$

Thus, the density matrix equations are

$$\dot{u} = \delta v - \frac{\gamma'}{2} u \tag{5.89}$$

$$\dot{v} = -\delta u + \Omega w - \frac{\gamma'}{2} v \tag{5.90}$$

$$\dot{w} = -\Omega v - \gamma_{total}(w - 1), \tag{5.91}$$

where it will be recalled that $\gamma_{total}$ is the combined inelastic collision and radiative damping rate. The steady-state solutions to these equations are

$$u = \frac{\Omega\delta}{\delta^2 + \frac{1}{4}\gamma'^2 + \frac{1}{2}(\gamma'/\gamma_{total})\Omega^2} \tag{5.92}$$

$$v = \frac{\frac{1}{2}\Omega\gamma'}{\delta^2 + \frac{1}{4}\gamma'^2 + \frac{1}{2}(\gamma'/\gamma_{total})\Omega^2} \tag{5.93}$$

$$w = \frac{\delta^2 + \frac{1}{4}\gamma'^2}{\delta^2 + \frac{1}{4}\gamma'^2 + \frac{1}{2}(\gamma'/\gamma_{total})\Omega^2} \tag{5.94}$$

and

$$\rho_{22} = \frac{\frac{1}{4}(\gamma'/\gamma_{total})\Omega^2}{\delta^2 + \frac{1}{4}\gamma'^2 + \frac{1}{2}(\gamma'/\gamma_{total})\Omega^2}. \tag{5.95}$$

It is perhaps worth noting that the two kinds of collision-induced relaxations (elastic and inelastic) were first observed in magnetic resonance experiments where they were associated with the relaxation times $T_2$ and $T_1$. The relaxation due to elastic collisions is sometimes called *transverse relaxation*, and it occurs with characteristic time $T_2$ $(= \tau_{coll})$. It is also occasionally referred to as *spin–spin relaxation*, since in magnetic resonance it is due to an interaction between spins which causes the component perpendicular (transverse) to the magnetic field to randomize while the longitudinal component is not affected. Conversely, *longitudinal relaxation* in magnetic resonance occurs with characteristic time $T_1$ ($= 1/\gamma_{inelast}$) and is also called *spin–lattice* relaxation since the component of the spin along the magnetic field (and therefore the energy) relaxes, due (in a solid) to the interaction between the spin and the surrounding lattice. Using the optical Bloch model, longitudinal relaxation is seen to be the relaxation of the population (the $z$-component of the Bloch vector) while transverse relaxation is due to the randomization of the atomic wavefunction's *phase* (the $x$- and $y$-components of the Bloch vector).

## 5.9 The atomic susceptibility and optical gain

We will now calculate the *atomic susceptibility* for a classical wave traveling through a linear, isotropic medium whose polarizability will be described using the semiclassical theory from the first part of this chapter. This will lead in a fairly natural way to *laser gain.*

We will start with the *material relations* among the quantities $\boldsymbol{D}$ (electric displacement), $\boldsymbol{E}$ (electric field) and $\boldsymbol{P}$ (polarization):

$$\boldsymbol{D} = \epsilon_0 \boldsymbol{E} + \boldsymbol{P} \tag{5.96}$$

$$\boldsymbol{P} = \epsilon_0 \chi \boldsymbol{E}, \tag{5.97}$$

where $\chi$ is the *susceptibility* of the medium. We will consider the polarization to consist of two parts: that due to the atomic transition and that from all other sources:

$$\begin{aligned} \boldsymbol{D} &= \epsilon_0 \boldsymbol{E} + \boldsymbol{P} + \boldsymbol{P}_{transition} \\ &= \epsilon \boldsymbol{E} + \epsilon_0 \chi(\omega) \boldsymbol{E} \\ &= \epsilon' \boldsymbol{E}, \end{aligned} \tag{5.98}$$

where

$$\epsilon'(\nu) = \epsilon \left[1 + \frac{\epsilon_0}{\epsilon} \chi(\omega)\right] \tag{5.99}$$

and $\chi(\omega)$ is due *only* to the resonant processes in the atoms.

A monochromatic wave propagating in the $z$-direction through the medium is described by

$$\boldsymbol{E}(z,t) = \boldsymbol{E}_0 e^{i(\omega t - k'z)}, \tag{5.100}$$

where $k'$ is the wave vector in the medium and is given by:

$$k' = \frac{\omega}{c'} = \omega\sqrt{\mu\epsilon'} \approx k\left(1 + \frac{\epsilon_0}{2\epsilon}\chi\right), \qquad |\chi| \ll 1, \tag{5.101}$$

where $c'$ is the speed of light in the medium and $k$ is the wave vector in free space. The frequency-dependent *atomic susceptibility* is a complex quantity:

$$\chi = \chi' - i\chi'', \tag{5.102}$$

where the minus sign is used to be consistent with other treatments. The propagation constant is now

$$k' \approx k\left[1 + \frac{\chi'(\omega)}{2n^2}\right] - i\frac{k\chi''(\omega)}{2n^2}, \tag{5.103}$$

where $n = \sqrt{\epsilon/\epsilon_0}$ is the *index of refraction* of the medium (exclusive of the resonant processes). Substituting this into the equation for the traveling wave,

$$\boldsymbol{E}(z,t) = \boldsymbol{E}_0 e^{i(\omega t - kz)}\left\{e^{(i\Delta k + \gamma(\omega)/2)z}\right\}, \tag{5.104}$$

where

$$\Delta k = -\frac{k\chi'(\omega)}{2n^2} \tag{5.105}$$

$$\gamma(\nu) = -\frac{k\chi''(\omega)}{n^2}. \tag{5.106}$$

The electric field of the wave will undergo a *phase change* given by $\Delta kz$ and will be fractionally *attenuated* (or amplified) by $\gamma/2$ per unit length (the factor of 2 is present since $\gamma$ is actually an *intensity* absorption coefficient). We see that the attenuation or gain of the wave is given by an expression which is proportional to the imaginary part of the atomic susceptibility, which we will now derive in terms of the atomic quantities.

We will first modify the density matrix equations to allow for an *equilibrium population* which is different from that of our original treatment, where we assumed that the atom will relax into the ground state ($w = 1$). We now allow for the possibility that some external agency (called a *pumping mechanism*) can maintain an arbitrary (and probably *non-thermal*) population difference, $w_0$, in the absence of radiation. The density matrix equations are now

$$\dot{u} = \delta v - \frac{\gamma'}{2}u \tag{5.107}$$

$$\dot{v} = -\delta u + \Omega w - \frac{\gamma'}{2}v \tag{5.108}$$

$$\dot{w} = -\Omega v - \gamma_{total}(w - w_0), \tag{5.109}$$

where in the third equation it should be evident that, in the absence of radiation, the equilibribrium value of $w$ is $w_0$. The steady-state solutions are

$$u = \frac{\Omega\delta w_0}{\delta^2 + \frac{1}{4}\gamma'^2 + \frac{1}{2}(\gamma'/\gamma_{total})\Omega^2} \tag{5.110}$$

$$v = \frac{\frac{1}{2}\Omega\gamma' w_0}{\delta^2 + \frac{1}{4}\gamma'^2 + \frac{1}{2}(\gamma'/\gamma_{total})\Omega^2} \tag{5.111}$$

$$w = \frac{(\delta^2 + \frac{1}{4}\gamma'^2)w_0}{\delta^2 + \frac{1}{4}\gamma'^2 + \frac{1}{2}(\gamma'/\gamma_{total})\Omega^2}. \tag{5.112}$$

Since we seek the atomic susceptibility, we must calculate the atomic polarization induced by the radiation. The polarization of each atom is obtained by taking the expectation value of the polarization operator, $\boldsymbol{p} = -e\boldsymbol{r}$. If we use the original wavefunction (eqn 5.2),

$$\begin{aligned}\langle \boldsymbol{p} \rangle &= \left(c_1^* e^{i\omega_1 t}\langle 1| + c_2^* e^{i\omega_2 t}\langle 2|\right) \boldsymbol{p} \left(c_1 e^{-i\omega_1 t}|1\rangle + c_2 e^{-i\omega_2 t}|2\rangle\right) \\ &= \boldsymbol{p}_{12}(\rho_{12} e^{i\omega_0 t} + \rho_{21} e^{-i\omega_0 t}) \\ &= \boldsymbol{p}_{12}(\tilde{\rho}_{12} e^{i\omega t} + \tilde{\rho}_{21} e^{-i\omega t}) \\ &= 2\boldsymbol{p}_{12}(\mathrm{Re}\{\tilde{\rho}_{12}\}\cos\omega t - \mathrm{Im}\{\tilde{\rho}_{12}\}\sin\omega t) \\ &= \boldsymbol{p}_{12}(u\cos\omega t - v\sin\omega t),\end{aligned} \tag{5.113}$$

where we assume that $\boldsymbol{p}_{12} \equiv p$ is real (we will henceforth ignore the vectorial nature of the dipole moment). The last equation confirms our earlier claim that the coherences are proportional to the components of an oscillating (actually rotating) induced atomic dipole moment. Using this result, we can now write down the *macroscopic dipole moment*, $P = N\langle p \rangle$, where $N$ is the number of atoms per unit volume:

$$P = -pN\Omega w_0 \frac{-\frac{1}{2}\gamma' \sin\omega t + \delta\cos\omega t}{\delta^2 + \frac{1}{4}\gamma'^2 + \frac{1}{2}(\gamma'/\gamma_{total})\Omega^2} E_0, \tag{5.114}$$

and we use an overall minus sign since $p < 0$ for an electron. We extract the imaginary component of $\chi$ by first considering the time-dependent polarization,

$$\begin{aligned}P(t) &= \mathrm{Re}\{\epsilon_0 \chi E_0 e^{i\omega t}\} \\ &= E_0\epsilon_0(\chi' \cos\omega t + \chi'' \sin\omega t),\end{aligned} \tag{5.115}$$

from which we obtain

$$\chi''(\omega) = \frac{p\Omega\gamma'\Delta N_0}{2\epsilon_0 E_0} \frac{1}{\delta^2 + \frac{1}{4}\gamma'^2 + \frac{1}{2}(\gamma'/\gamma_{total})\Omega^2}, \tag{5.116}$$

where we define the zero-field population difference $Nw_0 \equiv \Delta N_0$. Using the expression for the Rabi frequency, we can simplify this equation:

$$\chi''(\omega) = \frac{p^2\gamma'\Delta N_0}{2\hbar\epsilon_0} \frac{1}{\delta^2 + \frac{1}{4}\gamma'^2 + \frac{1}{2}(\gamma'/\gamma_{total})\Omega^2}. \tag{5.117}$$

The population difference of the ensemble of atoms is called the *inversion* and is represented by $\Delta N$, which is obtained by multiplying the equation for $w$ by $N$:

$$\Delta N = Nw = \frac{(\delta^2 + \frac{1}{4}\gamma'^2)\Delta N_0}{\delta^2 + \frac{1}{4}\gamma'^2 + \frac{1}{2}(\gamma'/\gamma_{total})\Omega^2}. \tag{5.118}$$

From this, we can further simplify the equation for $\chi''$:

$$\chi''(\omega) = \frac{p^2\gamma'\Delta N}{2\hbar\epsilon_0} \frac{1}{\delta^2 + \frac{1}{4}\gamma'^2}. \tag{5.119}$$

We note that $\Delta N$ (unlike $\Delta N_0$) is the inversion *in the presence of the optical field.* We further note that this expression for $\chi''$ seems to have lost the term in the denominator which is responsible for saturation. However, the saturation behavior is now due to the factor $\Delta N$, which clearly changes from the field-free value $\Delta N_0$ when $\Omega$ becomes comparable to $\gamma'$. Finally, the $\chi''$ equation can be written in terms of a *normalized lineshape function*, $g(\omega)$, whose integral over all frequencies is unity:

$$\chi''(\omega) = \frac{p^2 \Delta N \pi}{\hbar \epsilon_0} g(\omega), \tag{5.120}$$

where

$$g(\omega) = \frac{\gamma'/2\pi}{(\omega - \omega_0)^2 + \frac{1}{4}\gamma'^2}. \tag{5.121}$$

The absorption coefficient, $\gamma(\omega)$, is obtained by substituting the expression for $\chi''$ into eqn 5.106:

$$\gamma(\omega) = -\frac{kp^2 \Delta N \pi}{n^2 \hbar \epsilon_0} g(\omega), \tag{5.122}$$

where it is hoped that the distinction between the damping constant $\gamma'$ and the absorption coefficient $\gamma(\omega)$ is clear, despite the confusing notation. The effect of $\gamma(\omega)$ on a traveling wave is

$$\boldsymbol{E}(z) = \boldsymbol{E}(0) e^{\gamma(\omega) z/2}. \tag{5.123}$$

From the way that $\Delta N$ is defined $(N(\tilde{\rho}_{11} - \tilde{\rho}_{22}))$, it will be *positive* when the atom is in thermal equilibrium at a modest temperature and most of the population is in the ground state. Thus, there will be *absorption* of the light as it passes through the medium. However, if the excited-state population is greater than that of the ground state, we have a *population inversion* and $\Delta N$ will be *negative* with the result that there will be *coherent optical gain*: the amplitude of the wave will *increase* as the wave traverses the medium.

We can further simplify eqn 5.122 by replacing $p^2$ with an expression derived from the *rate of spontaneous emission* (see, for example, Loudon (1983)):

$$\gamma_{spont} = \frac{1}{\tau_{spont}} = \frac{\omega^3 n^3 D^2}{3\pi \epsilon_0 \hbar c^3}, \tag{5.124}$$

where $D^2 = |\langle 1|ex|2\rangle|^2 + |\langle 1|ey|2\rangle|^2 + |\langle 1|ez|2\rangle|^2$. Since the three matrix elements in $D^2$ are equal (there is no preferred direction), $D^2 \Rightarrow 3|\langle 1|ex|2\rangle|^2 = 3p^2$, and the expression for $\gamma_{spont}$ is

$$\gamma_{spont} = \frac{\omega^3 n^3 p^2}{\pi \epsilon_0 \hbar c^3}. \tag{5.125}$$

Substituting the value of $p^2$ into the expression for $\gamma(\omega)$ yields the formula for *laser gain*:

$$\text{Laser gain:} \quad \gamma(\omega) = -\frac{\lambda^2 \Delta N}{4n^2 \tau_{spont}} g(\omega) = -\frac{\lambda^2 \Delta N}{8\pi n^2 \tau_{spont}} g(\nu), \tag{5.126}$$

where the last equality comes from $g(\omega) = g(\nu)/2\pi$. A *laser amplifier* consists of a medium in which $\Delta N < 0$ with the result that a traveling wave will increase in amplitude as it traverses the medium. Since $\Delta N$ *depends upon the intensity of the wave*, the dependence of the field amplitude (and intensity) on $z$ will not necessarily be a simple exponential. One must go back to the differential equation for the intensity, $I(z)$,

$$\frac{dI(z)}{dz} = \gamma(\omega)I(z), \tag{5.127}$$

from which we get the *laser gain equation*:

$$\text{Laser gain equation:} \quad \frac{dI(z)}{dz} = -\frac{\lambda^2 \Delta N}{8\pi n^2 \tau_{spont}} g(\omega)I(z). \tag{5.128}$$

## 5.10 The Einstein A and B coefficients

The most important result of the recent sections, the laser gain equation, can also be derived from the phenomenological treatment due to Einstein. According to this theory, there are three possible processes which an atom can undergo in the presence of radiation. These are: *spontaneous emission, absorption and stimulated emission.* If the *energy spectral density* (energy per unit volume per unit frequency interval) is $\rho(\nu)$, the three processes are described by three parameters, $A_{21}, B_{21}$ and $B_{12}$, in three rate equations:

$$\begin{aligned} \text{Spontaneous emission:} \quad & \frac{dN_2}{dt} = -A_{21}N_2 \\ \text{Absorption:} \quad & \frac{dN_2}{dt} = -\frac{dN_1}{dt} = +B_{12}N_1\rho(\nu) \\ \text{Stimulated emission:} \quad & \frac{dN_2}{dt} = -\frac{dN_1}{dt} = -B_{21}N_2\rho(\nu), \end{aligned} \tag{5.129}$$

where it is assumed the atom has two levels, labeled 1 and 2, with number densities in each level $N_1$ and $N_2$. The photon emitted by stimulated emission is assumed to have the same values of $\boldsymbol{k}$, polarization and phase as the incident photon. The three processes are shown schematically in Fig. 5.2.

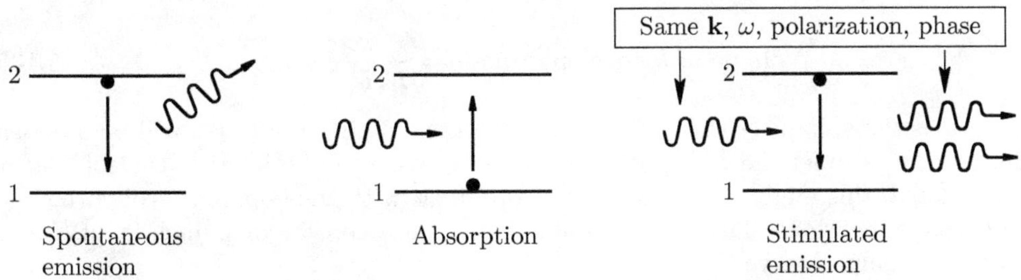


**Fig. 5.2** The three possible radiative interactions in a two-level atom according to the Einstein theory.

Before solving these rate equations, it will be useful to derive the *Planck formula* for the spectral density, $\rho(\nu)$, of radiation which is in thermal equilibrium with a *black body* at temperature, $T$. Planck's hypothesis was that the electromagnetic field is *quantized* – the energy in the radiation field is carried by *photons*, whose energy is $h\nu$, where $\nu$ is the frequency. Since the radiation is in thermal equilibrium with a thermal reservoir (the black body) at temperature $T$, the probability of finding $n$ photons at frequency $\nu$ is given by a *Boltzmann factor*:

$$\text{Probability of } n \text{ photons} \propto e^{-nh\nu/k_BT}. \tag{5.130}$$

From this, the average number of photons at frequency $\nu$ is

$$\bar{n} = \frac{\sum_n ne^{-nh\nu/k_BT}}{\sum_n e^{-nh\nu/k_BT}}. \tag{5.131}$$

This is evaluated by defining $X$

$$X \equiv \sum_n e^{-h\nu/k_BT}. \tag{5.132}$$

Then, the value of $\bar{n}$ is

$$\bar{n} = \left(\frac{-k_BT}{h}\right)\frac{1}{X}\frac{\partial X}{\partial \nu} = \frac{1}{e^{h\nu/k_BT}-1}. \tag{5.133}$$

This is the number of photons per *mode* at $\nu$. Since we are seeking a *spectral density*, we need to multiply $\bar{n}$ by the *number of modes per frequency interval*. In order to do this, we need to model the modes of the radiation field. There are two useful models. One can consider the field to be a series of *standing waves* in a rectangular cavity of dimensions $L_x \times L_y \times L_z$ or to be a series of *traveling waves* satisfying *periodic boundary conditions*. Both models yield the same results; we will choose the cavity since it is based upon slightly more familiar concepts. In order to satisfy the boundary conditions at the walls, the possible values of $\boldsymbol{k}$ are

$$\boldsymbol{k} = \left(\frac{\pi l}{L_x}, \frac{\pi m}{L_y}, \frac{\pi n}{L_z}\right) \qquad l, m, n \text{ integers}. \tag{5.134}$$

Each mode occupies a volume of $k$-space given by:

$$\text{Volume of } k\text{-space per mode} = \frac{\pi^3}{L_xL_yL_z} = \frac{\pi^3}{V}, \tag{5.135}$$

where $V$ is the volume of the cavity. Since it turns out that the result will be independent of $V$, we can let the dimensions of the cavity become infinite after the calculation is completed; this might avoid conceptual problems with the apparent arbitrariness of the cavity dimensions. The modes would then form a *continuum*, which is, of course, what is actually observed.

In order to calculate the number of modes between $k$ and $k + dk$, we consider a spherical shell in *the first octant of k-space*:

$$\text{Volume of spherical shell} = \tfrac{1}{8} \times 4\pi k^2 dk. \tag{5.136}$$

The factor of $\frac{1}{8}$ is there in order to avoid counting the modes twice: changing the sign of any component of $\boldsymbol{k}$ yields a value *from the same mode* (since the modes are *standing waves*; this would not be the case with *traveling waves*). The number of modes between $k$ and $k + dk$ is obtained by dividing the volume of the shell by the volume per mode and multiplying the result by 2, since there are *two possible polarizations per mode*:

$$\text{Number of modes between } k \text{ and } k + dk = \frac{Vk^2 dk}{\pi^2}. \tag{5.137}$$

We can convert this to the number of modes between $\nu$ and $\nu + d\nu$ (using $k = 2\pi\nu/c$):

$$\text{Number of modes between } \nu \text{ and } \nu + d\nu = V \frac{8\pi\nu^2 d\nu}{c^3}. \tag{5.138}$$

The spectral density is given by the product of three factors: the number of photons per mode, the energy per photon ($h\nu$) and the number of modes per frequency interval. Evaluating this product and dividing by $V$ (since we want the energy density) yields the Planck formula:

$$\rho = \frac{8\pi h\nu^3}{c^3} \frac{1}{e^{h\nu/k_B T} - 1}. \tag{5.139}$$

The rate equations (eqn 5.129) are solved by adding together the three sources for $dN_2/dt$ and setting the sum equal to zero (since the populations are at *equilibrium*). The result is:

$$\text{At equilibrium:} \quad 0 = \frac{dN_2}{dt} = -\frac{dN_1}{dt} = -A_{21}N_2 + B_{12}N_1\rho(\nu) - B_{21}N_2\rho(\nu). \tag{5.140}$$

Since the levels are in *thermal equilibrium* with the radiation, the ratio $N_2/N_1$ is equal to a Boltzmann factor:

$$\frac{N_2}{N_1} = \frac{g_2}{g_1} e^{-h\nu/k_B T}, \tag{5.141}$$

where $g_i$ is the *multiplicity* of the $i$th state: the number of levels having the same energy. Solving eqn 5.140 for $N_2/N_1$ and setting it equal to the Boltzmann factor, one can obtain an equation for $\rho(\nu)$:

$$\rho(\nu) = \frac{A_{21}}{B_{21}} \frac{1}{\frac{B_{12}g_1}{B_{21}g_2} e^{h\nu/k_B T} - 1}. \tag{5.142}$$

This equation is formally similar to the Planck formula (eqn 5.139) and is identical to it if

$$g_2 B_{21} = g_1 B_{12} \tag{5.143}$$

$$\frac{A_{21}}{B_{21}} = \frac{8\pi h\nu^3}{c^3}. \tag{5.144}$$

Using these important relationships among $A_{21}$, $B_{21}$, and $B_{12}$, one can obtain the rate that the level 2 population changes due to absorption and stimulated emission (ignoring spontaneous emission):

$$\frac{dN_2}{dt} = (B_{12}N_1 - B_{21}N_2)\rho(\nu) = A_{21} \frac{c^3}{8\pi h\nu^3} \left( \frac{g_2}{g_1} N_1 - N_2 \right) \rho(\nu). \tag{5.145}$$

Before proceeding further, we note that we are attempting to use Einstein's theory with *monochromatic* radiation instead of the very broadband radiation for which the theory was originally intended. To adapt the theory to monochromatic radiation, we first observe that the rates involving the $B$ coefficient are written as:

$$\langle\text{Transition rate}\rangle = B \times \langle\text{spectral energy density}\rangle. \tag{5.146}$$

When we derived the transition rates using quantum mechanics, the parameter which we now call the $B$ coefficient had a spectral width due to damping processes. The quantity $\rho(\nu)$ also has a spectral width: in the classical use of the theory, the width of the $B$ coefficient is much *smaller* than that of the radiation and the latter can be ignored. For monochromatic radiation, the quantity $\rho(\nu)$ can be represented as a *delta function*:

$$\rho(\nu) = \rho_{\nu'}\delta(\nu - \nu'). \tag{5.147}$$

In order to allow both spectrally broad and spectrally narrow radiation in a single theory, we therefore modify the theory so that the transition rates are given by the *frequency integral* of the product of three factors:

$$\text{Transition rate} = \int Bg(\nu)\rho(\nu)d\nu, \tag{5.148}$$

where we have factored the $B$ coefficient into a constant factor and a lineshape function, $g(\nu)$, whose integral over all frequencies is unity. Thus, in the classical case of a broad $\rho$ and narrow $B$, the integral is $B\rho$, as required by the original Einstein theory. For the case of monochromatic radiation, we have

$$\text{Transition rate} = Bg(\nu)\rho_\nu. \tag{5.149}$$

Note that $\rho_\nu$ is no longer a spectral density, but instead is simply an *energy density* of the radiation (energy per unit volume). For a plane wave, the intensity, $I_\nu$, is

$$I_\nu = c\rho_\nu. \tag{5.150}$$

We can now write the rate equation for $N_2$ in terms of $I_\nu$ and the wavelength $\lambda$:

$$\frac{dN_2}{dt} = A_{21}\frac{\lambda^2}{8\pi}g(\nu)\left(\frac{g_2}{g_1}N_1 - N_2\right)\frac{I_\nu}{h\nu}, \tag{5.151}$$

where we note that the last factor $(I_\nu/h\nu)$ is the *photon flux*.

This equation is an expression for a rate which is equal to some quantity times a photon flux. The multiplier of the flux and population is therefore a *cross section*, $\sigma(\nu)$:

$$\sigma(\nu) = A_{21}\frac{\lambda^2}{8\pi}g(\nu). \tag{5.152}$$

For *radiative damping only*, the normalized lineshape function $g(\nu)$ is

$$g(\nu) = \frac{\Delta\nu/2\pi}{(\nu-\nu_0)^2 + \frac{1}{4}(\Delta\nu)^2}. \tag{5.153}$$

The Einstein $A_{21}$ coefficient is $1/\tau$, where $\tau$ is the radiative lifetime. Therefore, $\Delta\nu = A_{21}/2\pi$ and at *resonance* $g(\nu_0) = 4/A_{21}$. The result is that the resonant cross section is

$$\text{At resonance:} \quad \sigma(\nu_0) = \frac{\lambda^2}{2\pi}. \tag{5.154}$$

Interpreting $\sigma$ as a cross-sectional area across which a photon will cause an excitation, this result means that an atom's influence extends out about one wavelength, despite the atomic diameter being more than 1000 times smaller (the same is true of radio antennas).

If we multiply eqn 5.151 by $h\nu$, we obtain the *rate of change of the internal energy density of the atoms* as a result of their interaction with the radiation. Thus,

$$\frac{dW}{dt} = A_{21}\frac{\lambda^2 \Delta N g(\nu)}{8\pi} I_\nu \qquad \Delta N \equiv \frac{g_2}{g_1}N_1 - N_2, \tag{5.155}$$

where the atoms' internal energy density is $W$. In the absence of *sources of radiation*, one can write a *continuity equation* for the intensity (this actually is a direct result of the *Poynting theorem*):

$$\nabla \cdot \boldsymbol{S} + \frac{\partial \rho}{\partial t} = 0, \tag{5.156}$$

where $\boldsymbol{S}$ is the Poynting vector (the vector of the intensity) and $\rho$ is the electromagnetic energy density. If there is a source,

$$\nabla \cdot \boldsymbol{S} + \frac{\partial \rho}{\partial t} = -\frac{dW}{dt}, \tag{5.157}$$

where the minus sign comes from the fact that an *increase* in the energy of the atoms comes about as a result of a *decrease* in the energy of the radiation. Substituting for $dW/dt$ and setting $d\rho/dt = 0$ (since we are interested in the equilibrium situation),

$$\frac{dI_\nu}{dz} = -A_{21}\frac{\lambda^2 \Delta N g(\nu)}{8\pi} I_\nu, \tag{5.158}$$

where we are assuming wave propagation in the $z$-direction. This is the same as the quantum-mechanical result (since $A_{21} = 1/\tau_{spont}$ and we assumed $n = 1$ in the derivation).

## 5.11 Doppler broadening: an example of inhomogeneous broadening

In earlier sections, we discussed two broadening mechanisms: radiative (natural) broadening and collision (elastic and inelastic) broadening. These two mechanisms have this in common: *at any time, the distribution of possible absorption or emission frequencies is the same for all atoms.* This type of broadening is called *homogeneous*. Conversely, if *at any time the distribution of absorption/emission frequencies is different for different atoms*, the broadening is called *inhomogeneous*. The presence of two categories of broadening has profound implications for the spectrum of the light emitted by a laser, as we shall discover later.

There are a number of possible sources of inhomogeneous broadening. Perhaps the most obvious is in a crystal, where, due to a random distribution of impurities or other inhomogeneities, different atoms will have slightly different local environments and therefore will have different transition frequencies. The most common inhomogeneous broadening in gas lasers is due to the distribution of atomic velocities and consequent distribution of transition frequencies because of the *Doppler shift.* We will discuss *Doppler broadening* in this section.

A collection of atoms in thermal equilibrium at temperature $T$ in an atomic vapor will have a *Maxwellian* distribution of velocities, where the probability $(dN/N)$ of the atom's velocity along the $z$-axis being between $v_z$ and $v_z + dv_z$ is

$$\frac{dN}{N} = \left(\frac{M}{2\pi k_B T}\right)^{1/2} e^{-\frac{Mv_z^2}{2k_B T}} dv_z, \tag{5.159}$$

where $M$ is the atomic mass and $k_B$ is the Boltzmann constant. As a result of the motion of the atoms, their transition frequencies will experience a first-order Doppler shift:

$$\nu = \nu_0 \left(1 - \frac{v_z}{c}\right), \tag{5.160}$$

where $\nu_0$ is the rest frequency of the atom and we assume that the laser beam is directed in the $z$-direction. Combining these two equations, we get a normalized Doppler line shape, $g(\nu)$:

$$g(\nu) = \left(\frac{4\ln 2}{\pi}\right)^{1/2} \frac{1}{\Delta\nu_D} e^{-4\ln 2\left(\frac{\nu-\nu_0}{\Delta\nu_D}\right)^2}, \tag{5.161}$$

where the full width at half maximum, $\Delta\nu_D$, is given by:

$$\begin{aligned} \Delta\nu_D &= \left(\frac{8k_B T\ln 2}{Mc^2}\right)^{1/2} \nu_0 \\ &\approx 7.2 \times 10^{-7}\sqrt{\frac{T}{A}}\nu_0 \end{aligned} \tag{5.162}$$

and $A$ is the atomic mass in amu.

A plot of a Gaussian and Lorentzian line with the same width and height appears in Fig. 5.3. The lineshapes are quite similar within one half width of resonance, but the much longer *tail* of the Lorentzian line should be evident. Since homogeneous broadening usually results in a Lorentzian, and inhomogeneous broadening usually results in a Gaussian, one might say that *all broadening is homogeneous sufficiently far from resonance.*

When there is more than one broadening mechanism, the *combined* lineshape is the *convolution* of the two component lineshapes:

$$g_{combined}(\nu) = \int_{-\infty}^{\infty} g_1(\nu - \nu')g_2(\nu')d\nu', \tag{5.163}$$

where $g_1(\nu)$ and $g_2(\nu)$ are the two original lineshape functions. For similar broadening mechanisms, the combined linewidth is a simple function of the original linewidths.

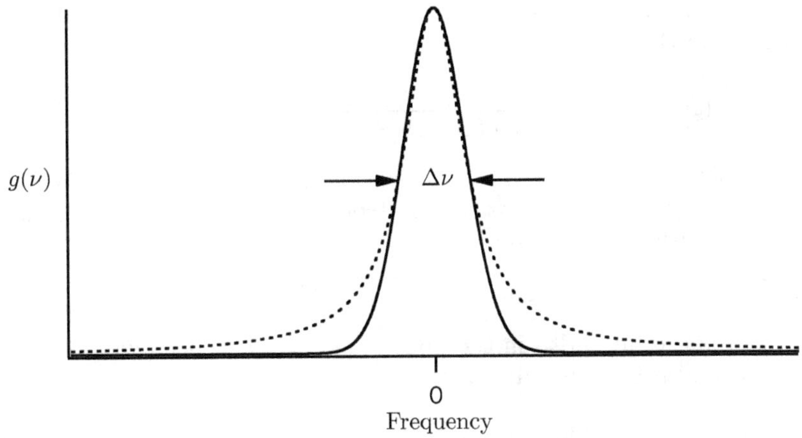


**Fig. 5.3** A Lorentzian (dashed) and Gaussian (solid) with the same linewidth and amplitude.

We saw earlier that for radiative and collision broadening (or two types of collision broadening), the rates and therefore the widths simply add:

$$\text{Homogeneous (Lorentzian):} \quad \Delta\nu_{total} = \Delta\nu_1 + \Delta\nu_2 \tag{5.164}$$

and the combined lineshape function is also a Lorentzian. It is well known that the convolution of a Gaussian with another Gaussian is also a Gaussian whose width is the quadratic sum of the original widths:

$$\text{Inhomogeneous (Gaussian):} \quad \Delta\nu_{total} = \sqrt{\Delta\nu_1^2 + \Delta\nu_2^2}. \tag{5.165}$$

The case of *different types* of broadening mechanisms is a little more complicated. When the linewidth of one is much greater than the other, one can simply use the dominant function, although far from resonance the Lorentzian always dominates. When the widths are similar, one obtains a *Voigt profile*, which is not surprisingly similar to both a Lorentzian and a Gaussian. (See, for example, Liu (2001) for a discussion of the Voigt profile together with a simple analytical expression for evaluating it.)

## 5.12 Comments on saturation

Now that we have touched on the notion of saturation and have discussed the difference between homogeneous and inhomogeneous broadening, we would like to illustrate this difference using a simple "thought" experiment. The experiment introduces the phenomenon of "hole burning", which we will describe in the limiting case of an inhomogeneous width which is much larger than the homogeneous width. First, we will derive some simple consequences of saturation.

Earlier in this chapter, we saw that the linewidth of a transition would increase when the Rabi frequency, $\Omega$, was comparable to or greater than the homogeneous linewidth, $\gamma'$. We will examine the effects of a large Rabi frequency by first writing the gain constant (eqn 5.126) in terms of $\Delta N_0$ instead of $\Delta N$ in order to display clearly the dependence on the Rabi frequency:

$$\gamma(\omega) = -\frac{\lambda^2\gamma'\Delta N_0}{16\pi^2\tau_{spont}}\frac{1}{(\omega-\omega_0)^2+\frac{1}{4}\gamma'^2+\frac{1}{2}(\gamma'/\gamma_{total})\Omega^2}. \tag{5.166}$$

In order to simplify the discussion, we will assume that there is only radiative damping ($\gamma_{total} = \gamma' = \gamma$). When $\Omega \geq \gamma$, there will be two consequences. First, as already mentioned, the width will increase:

$$\Delta\omega = \Delta\omega_0\sqrt{1+2(\Omega/\gamma)^2}, \tag{5.167}$$

where $\Delta\omega_0 = \gamma$ is the linewidth in the limit of zero $\Omega$. The peak gain (or absorption) at *resonance* will also be reduced:

$$\gamma(\omega_0) = \gamma(\omega_0)|_{\Omega=0}\frac{1}{1+2(\Omega/\gamma)^2}. \tag{5.168}$$

It would be instructive at this point to observe the behavior of the inversion ($\Delta N$) as a function of the Rabi frequency by rewriting the equation for $\Delta N$ (eqn 5.118):

$$\Delta N = \Delta N_0\frac{(\omega-\omega_0)^2+\frac{1}{4}\gamma^2}{(\omega-\omega_0)^2+\frac{1}{4}\gamma^2+\frac{1}{2}\Omega^2}. \tag{5.169}$$

From this, we see clearly that $\Delta N$ varies from $\Delta N_0$ when $\Omega = 0$ to something approaching zero when $\Omega \gg \gamma$. The fact that, in the absence of pumping, $\Delta N$ is never negative leads to the general principle that *in the absence of an external pumping mechanism, it is impossible to invert a two-level population by continuously irradiating the atoms at or near their resonant frequency.* As the intensity increases, the populations will tend to be *equalized* and the population difference (the inversion) will approach zero.

We can rewrite eqns 5.168 and 5.167 as:

$$\gamma(\omega_0) = \gamma(\omega_0)|_{\Omega=0}\frac{1}{1+\frac{I}{I_s}} \tag{5.170}$$

$$\Delta\omega = \Delta\omega_0\sqrt{1+\frac{I}{I_s}}, \tag{5.171}$$

where $I$ is the intensity and $I_s$ is the *saturation intensity* at resonance and is equal to

$$I_s = \frac{\pi\hbar\omega_0}{\lambda^2\tau_{spont}}, \tag{5.172}$$

where we used $I = \epsilon_0 c|E|^2/2$ and eqn 5.125 to obtain this result and assumed that $n = 1$. The saturation intensity can also be obtained by recognizing that the onset of saturation occurs when *the excitation rate is equal to the spontaneous decay rate:*

$$\text{Saturation:} \quad \left(\frac{I_s}{\hbar\omega_0}\right)\sigma = \left(\frac{I_s}{\hbar\omega_0}\right)\left(\frac{\lambda^2}{2\pi}\right) = \frac{1}{\tau_{spont}} \Longrightarrow I_s = \frac{2\pi\hbar\omega_0}{\lambda^2\tau_{spont}}, \tag{5.173}$$

which a similar to eqn 5.172.

Equation 5.170 gives the gain reduction due to saturation at *resonance*; we can obtain the off-resonance result by factoring eqn 5.166 into the product of a Lorentzian and a second intensity-dependent factor

$$\gamma(\omega) = \gamma(\omega)|_{\Omega=0} \frac{1}{1 + \frac{I}{I_s}\bar{g}(\omega - \omega_0)}, \tag{5.174}$$

where $\bar{g}(\omega - \omega_0)$ is a Lorentzian normalized to unity at resonance:

$$\bar{g}(\omega - \omega_0) \equiv \frac{\gamma^2}{\gamma^2 + 4(\omega - \omega_0)^2}. \tag{5.175}$$

The population (eqn 5.169) can also be factored in a similar fashion:

$$\Delta N = \Delta N_0 \frac{1}{1 + \frac{I}{I_s}\bar{g}(\omega - \omega_0)}. \tag{5.176}$$

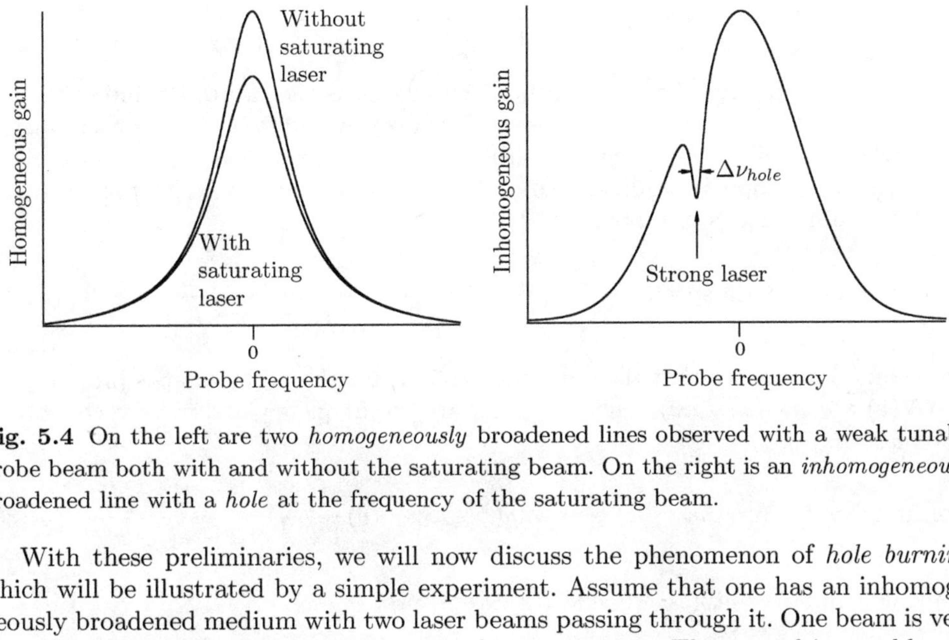


**Fig. 5.4** On the left are two *homogeneously* broadened lines observed with a weak tunable probe beam both with and without the saturating beam. On the right is an *inhomogeneously* broadened line with a *hole* at the frequency of the saturating beam.

With these preliminaries, we will now discuss the phenomenon of *hole burning*, which will be illustrated by a simple experiment. Assume that one has an inhomogeneously broadened medium with two laser beams passing through it. One beam is very intense and is tuned slightly to one side of the resonance. The second is tunable and weak enough to not alter the populations of the atoms. The former is called the *saturating* beam and the latter is called the *probe* beam. If one has a means of measuring the *gain* of the probe as one tunes it over the inhomogeneous line, one will obtain a curve similar to the one on the right of Fig. 5.4. The saturating beam will *eat a hole out of the inhomogeneous profile*; the hole is located at the frequency of the strong beam.

The reason for this behavior is that the inhomogeneously broadened medium can be divided up into a large number of *independent, spectrally narrow packets*, each of which is homogeneously broadened and whose center frequencies cover the entire inhomogeneous width. When a strong saturating field appears at some frequency within the inhomogeneous width, only the packet resonant with the field will be saturated; the other packets will not be affected since they are not resonant with the field. Thus, the saturated packet will have a smaller $\Delta N$ and therefore a smaller gain and a *hole* in the inhomogeneous gain curve will appear.

An approximate analytical expression can be obtained for the gain experienced by the probe beam in the presence of the saturating beam by assuming that the inhomogeneous broadening is due to the Doppler effect. For atoms moving with speed $v$ along the $z$-axis, the Doppler shift for the saturating laser (which propagates in the $z$-direction) is

$$\Delta\omega'_{Doppler} = \frac{v}{c}\omega' = k'v, \tag{5.177}$$

where $k'$ is the wave vector of the fixed frequency saturating laser and $\omega'$ is its frequency. The population difference induced by the saturating laser is therefore (from eqn 5.176)

$$\Delta N(v) = \Delta N_0 \frac{1}{1 + \frac{I}{I_s}\bar{g}(\omega' - \omega_0 - k'v)}, \tag{5.178}$$

where $I$ is the intensity of the saturating laser and $I_s$ is the saturation intensity. The velocity distribution of the atoms is given by $N(v)$, where $N(v)dv$ is the number of atoms whose velocity component along the laser beam is between $v$ and $v + dv$. Since the unsaturated population difference, $\Delta N_0$, is proportional to $N(v)$, the saturated population difference is given by:

$$\text{For atoms with speed } v\text{:} \quad \Delta N(v) \propto N(v) \frac{1}{1 + \frac{I}{I_s}\bar{g}(\omega' - \omega_0 - k'v)}. \tag{5.179}$$

From eqn 5.126, we see that the gain of the atoms for the probe laser is proportional to $\Delta N(v) \times g(\omega - \omega_0 - kv)$, where $\omega$ and $k$ are the frequency and wave vector of the probe laser.

$$\text{For atoms with speed } v\text{:} \quad \gamma(\omega) \propto N(v) g(\omega - \omega_0 - kv) \frac{1}{1 + \frac{I}{I_s}\bar{g}(\omega' - \omega_0 - k'v)} \tag{5.180}$$

To include all of the atoms, we *sum (integrate)* over $v$:

$$\begin{aligned} \gamma(\omega) &\propto \int_{-\infty}^{\infty} N(v) g(\omega - \omega_0 - kv) \frac{1}{1 + \frac{I}{I_s}\bar{g}(\omega' - \omega_0 - k'v)} dv \\ &\propto N((\omega - \omega_0)/k) \frac{1}{1 + \frac{I}{I_s}\bar{g}(\omega - \omega')}, \end{aligned} \tag{5.181}$$

where we have treated the $g(\omega - \omega_0 - kv)$ factor as a delta function, selecting velocities which satisfy $\omega - \omega_0 = kv$. (In the argument of the lineshape function, $\bar{g}(\omega' - \omega_0 - k'v)$, very little error is introduced by setting $k'$ equal to $k$.) A simple expression can be obtained in the limit that $I$ is somewhat smaller than $I_s$ by expanding the second factor,

$$\gamma(\omega) \propto N((\omega - \omega_0)/k) \left\{ 1 - \frac{I}{I_s} \bar{g}(\omega - \omega') \right\}. \tag{5.182}$$

When $\omega - \omega' \gg \gamma$, we have a simple unmodified Doppler lineshape proportional to $N((\omega - \omega_0)/k)$. At frequencies near $\omega'$, the second term in the curly brackets will produce a *hole* in the population difference and therefore a hole in the gain expression, as shown in Fig. 5.4.

If one performs this experiment with a *homogeneously broadened* medium, the results will be very different. This case is illustrated by the curves on the left side of of Fig. 5.4. Here, the effect of the strong saturating beam is to lower the gain over the *entire line* rather than over a small portion of it as in the inhomogeneous case. These two very different behaviors will alter the spectrum of light emitted by lasers whose amplifying media are subject to the two types of broadening.

## 5.13 Further reading

The author's favorite book on the semiclassical laser theory is Loudon's *The Quantum Theory of Light* (1983). Virtually all of the previously mentioned books on laser theory and Gaussian beams use an approach similar to ours; perhaps one of the most readable is Yariv's book on quantum electronics (1989). Finally, an excellent atomic physics text which describes the optical Bloch equations is Foot (2005); we used this text for much of our treatment of this topic.

## 5.14 Problems

(5.1) For a system in thermal equilibrium, calculate the temperature at which the spontaneous and stimulated emission rates are equal for $\lambda = 500$ nm and the wavelength at which these rates are equal for a temperature of 4000 K.

(5.2) Derive the *Wien law* by determining the product of the temperature and wavelength at the peak of Planck's formula for the spectral density of blackbody radiation. This product should depend upon $h$ and $k$ (Boltzmann constant). Make reasonable approximations if necessary.

(5.3) A He–Ne laser produces light at 632.8 nm due to transitions in neon atoms. Assuming pure $^{20}$Ne at a temperature of $\approx 300°$C, answer the following questions.
   (a) What is the Doppler linewidth (FWHM) of the transition?
   (b) How many *blackbody* modes per unit volume couple to this transition?
   (c) How many $\text{TEM}_{00q}$ modes are within the Doppler full width of this transition if the mirror spacing is 1 m?

(5.4) All resistors generate thermal noise, called "Johnson noise", whose RMS voltage is given by $V_{Johnson} = \sqrt{4k_B TR\Delta\nu}$ where $R$ is the resistance, $T$ is the absolute temperature, $k$ is Boltzmann's constant and $\Delta\nu$ is the electrical bandwidth of the system. This noise can be considered to be blackbody radiation in a one-dimensional "cavity" where $h\nu \ll k_B T$. Derive this formula using the following steps.

- Use the procedure for determining the spectral density of blackbody radiation that we derived in the text. (Three factors need to be multiplied together.)
- The number of modes can be determined using the following model. Assume that the resistor $R$ is connected to a transmission line (a *one-dimensional* enclosure) whose characteristic impedance is also $R$. The only things you need to know about characteristic impedance are that a transmission line terminated in it will *not experience any reflections* and will have a resistance at the input equal to the characteristic impedance, $R$. Thus, the transmission line's length is irrelevant (which aids the model, since the length does not appear in the result).
- Once you get the correct spectral density of the energy, calculate the *power* by determining the *time* for the energy emitted by one resistor to travel down the line and be absorbed by the other. Then use the remaining known parameters $(R, \Delta\nu)$ to obtain the voltage.

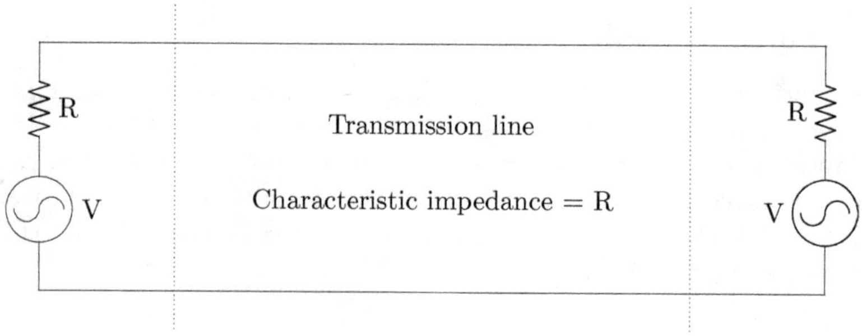


- The above diagram displays the appropriate model. One needs a resistor $R$ at *both* ends to avoid reflections (both resistors have Johnson noise).

# 6

# Laser oscillation and pumping mechanisms

## 6.1 Introduction

This chapter will combine the results of Chapters 3 and 5 by placing an amplifying medium in a resonant cavity and evaluating the behavior of the resulting *laser oscillator*. First, we will determine the conditions for laser oscillation and the expected power output of the laser. Next, we will discuss optical *pumping* techniques for three-level and four-level lasers. Finally, the spectral characteristics of laser radiation will be considered. This will include the *frequency pulling* of the empty cavity modes, the multimode behavior of an inhomogeneously broadened laser (including *spatial hole burning*) and the ultimate linewidth (Schawlow–Townes limit) of a laser. Although it is slightly off topic, the remarkable behavior of the photon occupation number above threshold will also be discussed. The fairly subtle distinction between the photon statistics of a coherent laser beam and that of thermal radiation will be treated in some detail. For most of the chapter, the laser will be modeled by a gain medium inside a *traveling-wave cavity*, since the spatially nonuniform field in a standing-wave cavity is somewhat more difficult to treat.

## 6.2 The condition for laser oscillation

In Chapter 3, the round-trip fractional intensity loss in a cavity was represented by the quantity $L_{tot}$ so that the *intensity remaining after one round trip* is $e^{-L_{tot}}$. The internal intensity transmission is $t^2$ for a traveling-wave cavity and $t^4$ for a standing-wave cavity. The internal loss was considered to be distributed throughout the cavity and was therefore described by an absorption coefficient, $\alpha$. The fractional intensity remaining after one round trip due to absorption is $e^{-\alpha l}$, where $l$ is the round-trip distance. Equating $e^{-L_{tot}}$ to the combined absorption effects and Fresnel reflectivity, one has:

$$e^{-L_{tot}} = r_1^2 r_2^2 e^{-\alpha l}, \tag{6.1}$$

where $r_1$ and $r_2$ are the field reflectivities of the cavity mirrors. Although most lasers using either type of cavity have only one partially transmitting mirror, we will continue to allow for the possibility of two partially transmitting mirrors. The fractional round-trip loss is

$$L_{tot} = \alpha l - 2\ln(r_1 r_2). \tag{6.2}$$

In Chapter 5, the fractional intensity gain (or loss) per unit length due to resonant atomic processes was represented by $\gamma(\omega)$, which satisfies the equation

$$\frac{dI_\nu}{dz} = \gamma(\omega)I_\nu. \tag{6.3}$$

In a *laser amplifier*, the intensity increase after one pass can be substantial and the $z$-dependence of $\gamma(\omega)$ is therefore not exponential, since $\gamma(\omega)$ depends upon the intensity via the quantity $\Delta N$. The single-pass gain of a *laser oscillator* is usually much smaller and we will assume that the intensity has an exponential dependence on $z$:

$$I_\nu = I_{\nu 0}e^{\gamma(\omega)z}, \tag{6.4}$$

where we are assuming the gain medium is in a *traveling-wave cavity.* This expression is formally the same as that which describes absorption in a cavity containing an absorbing medium; we will thus include the possibility of gain by modifying $\alpha$:

$$\alpha \Longrightarrow \alpha - \gamma(\omega) \tag{6.5}$$

so that the expression for $L_{tot}$ is now

$$L_{tot} = (\alpha - \gamma(\omega))l - 2\ln(r_1 r_2). \tag{6.6}$$

When $L_{tot} = 0$, any field inside the cavity will be indefinitely sustained, since there are no net losses. We thus have the condition for *laser oscillation:*

$$\text{Laser oscillation condition:} \quad L_{tot} = 0 \Longrightarrow \gamma(\omega) = \alpha - \frac{2}{l}\ln(r_1 r_2), \tag{6.7}$$

where, of course, this happens *only at the resonant frequencies of the cavity.* Essentially, this equation tells us that the gain must compensate for both the distributed loss, $\alpha$, and the losses in cavity intensity due to the radiation leaking out through the partially transmitting cavity mirrors. It is important to realize that the compensation must be *exact:* if it is too small the oscillations will die out and if it is too large the power will increase indefinitely, an obviously unphysical condition. As the gain constant is increased (via an external pumping mechanism), the laser will reach *threshold* when the gain is exactly equal to the loss. This is called the *threshold gain*, $\gamma_t(\omega)$. The gain can not be larger than this: it is *clamped* to $\gamma_t$. Of course, by increasing the pumping, the power output will increase, but the gain will remain the same. This seeming paradox will be resolved in the next section.

## 6.3 The power output of a laser

In the last section, we showed that self-sustained laser oscillation will occur when the gain is equal to the loss:

$$\gamma_t = \alpha - \frac{2}{l}\ln(r_1 r_2). \tag{6.8}$$

As the zero-field inversion, $\Delta N_0$, is made more negative (via increased pumping), the zero-field gain ($\gamma_0$) will increase, but the saturated gain will remain clamped to $\gamma_t$ (eqn 5.170):

$$\gamma_t = \gamma_0 \frac{1}{1 + \frac{I_\nu}{I_s}}. \tag{6.9}$$

Again, we assume a *traveling-wave* cavity, since then $I_\nu$ (the internal intensity) will be spatially uniform. We see that as the unsaturated gain, $\gamma_0$, increases, $I_\nu$ will be increased just enough to keep $\gamma_t$ constant. Therefore, the output power will increase with increased pumping. This equation also explains how laser oscillation is *initiated.* One increases the pumping until $\gamma > \gamma_t$. Then, any noise (from spontaneous emission) or other disturbance that has Fourier components at the laser frequency will be amplified and the resulting intensity will grow until *saturation reduces the gain to* $\gamma_t$, at which point the intensity will be constant. Thus, the *clamping mechanism of the gain is saturation.*

Combining these equations and solving for $I_\nu$,

$$I_\nu = I_s\left(\frac{\gamma_0}{\gamma_t} - 1\right) = I_s\left(\frac{\gamma_0 l}{\alpha l - 2\ln(r_1 r_2)} - 1\right). \tag{6.10}$$

Assume that only one mirror (the *output coupler*, mirror 1) has any appreciable transmission. Then,

$$-2\ln(r_1 r_2) = -\ln r_1^2 = -\ln(1-T) \approx T, \tag{6.11}$$

where $T$ is the (intensity) transmission of the output coupler and we assume that $T \ll 1$. If we call the round-trip loss (excluding transmission, $T$) $L_i$, then $L_i = \alpha l$ and

$$I_\nu = I_s\left(\frac{g_0}{L_i + T} - 1\right), \tag{6.12}$$

where $g_0 \equiv \gamma_0 l$ is the *unsaturated gain per round trip.*

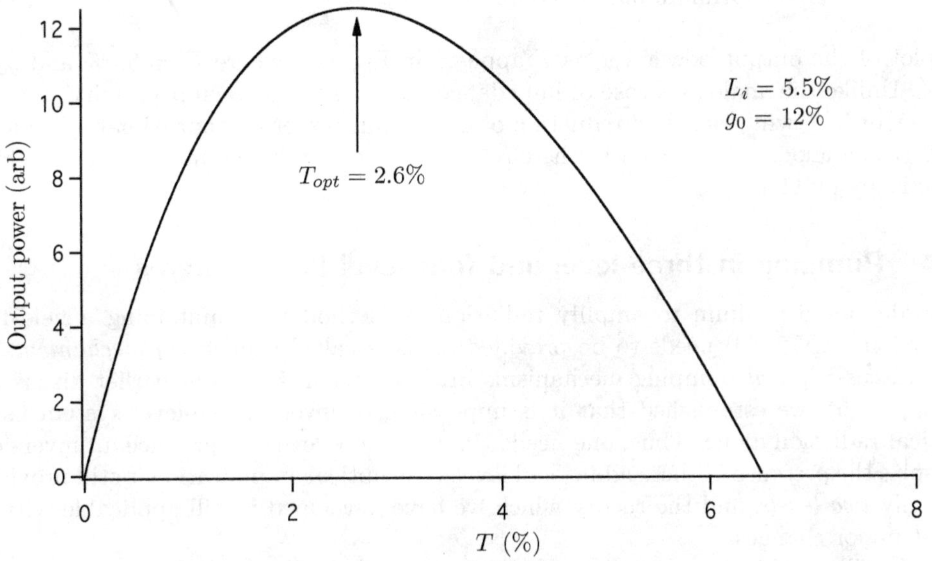


**Fig. 6.1** Output power versus output coupling.

The quantity $I_\nu$ is the internal circulating intensity, some of which will be absorbed by the atoms and some of which will stimulate emission. In order to determine the power output, we need to multiply $I_\nu$ by the *saturated gain*, $\gamma_t$. This will yield the *net* *power per unit volume, $P_e$, provided by the amplifying medium.* The *internal power* is obtained by multiplying this by the *mode volume*, $V_m$, which is the volume overlap between the amplifying medium and the radiation. Finally, the *external power* is obtained by multiplying the internal power by $T/(L_i + T)$. Putting all of these factors together, the useful output power, $P_0$, is

$$P_0 = P_e V_m \frac{T}{L_i + T} = \gamma_t I_\nu V_m \frac{T}{L_i + T} = \left(\frac{L_i + T}{l}\right) I_\nu V_m \left(\frac{T}{L_i + T}\right), \qquad (6.13)$$

where the last step comes from equating the gain $(\gamma_t l)$ to the total loss $(L_i + T)$ and solving for $\gamma_t$. Finally, substituting for $I_\nu$, yields

$$\text{Available power from a laser:} \quad P_0 = \frac{V_m I_s}{l}\left(\frac{g_0}{L_i + T} - 1\right) T. \qquad (6.14)$$

We should note that we could have also obtained this result by multiplying the internal intensity, $I_\nu$ (eqn 6.12), by the effective *cross-sectional area* $(V_m/l)$ and the mirror transmission, $T$.

The only conveniently available parameter which can be adjusted to maximize the power output is the output coupler transmission, $T$. One can determine the optimum value of $T$ by differentiating and setting the derivative equal to zero. The result is:

$$\text{Optimum output coupling:} \quad T_{opt} = -L_i + \sqrt{g_0 L_i}. \qquad (6.15)$$

A plot of the output power versus $T$ appears in Fig. 6.1, where $L_i = 5.5\%$ and $g_0 = 12\%$. Unlike the analogous case of impedance matching into an empty cavity, a factor of 2 error in $T$ will result in a reduction of the output power by approximately a factor of 2 (the analogous reduction in the circulating intensity in the impedance-matching case is about 11%).

## 6.4 Pumping in three-level and four-level laser systems

In order for a medium to amplify radiation, a method for maintaining a field-free inversion $(\Delta N_0 < 0)$ needs to be provided. This is called a *pumping mechanism*; we will discuss *optical* pumping mechanisms in this section. From our earlier discussion (Chapter 5), we established that it is impossible to invert a two-level system using optical radiation alone. Thus, one needs three or more levels to produce an inversion. Despite the presence of these additional levels, the optical amplification is still provided by only *two levels*, and the theory which we have developed is still applicable with at most minor changes.

We will consider laser media with three and four levels and determine the population inversion due to pumping using the *rate equations*, which describe the relationships among the *populations* and their rates of change. As with the Einstein $A$ and $B$ coefficients, a theory which uses the rate equations needs additional ad hoc assumptions, which only appear naturally using quantum mechanics. The added terms are intuitively obvious and fairly easy to evaluate using the theory already developed.

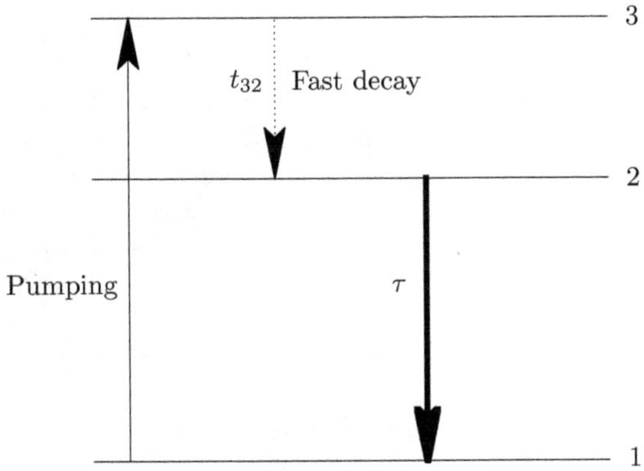


**Fig. 6.2** Level structure of a three-level laser.

Figure 6.2 depicts a three-level atomic system. For simplicity, we assume that all of the levels have a *multiplicity* of 1. The excitation rate of level 2 due to the laser radiation is given by eqn 5.151:

$$\frac{dN_2}{dt} = A_{21}\frac{\lambda^2}{8\pi}g(\nu)(N_1 - N_2)\frac{I_\nu}{h\nu} = W(\nu)(N_1 - N_2), \tag{6.16}$$

which defines the quantity $W(\nu)$ as the *stimulated transition rate.* The rate equations can be created for each level, $N_i$, by equating $dN_i/dt$ to an expression in which one *adds* the rates of the *sources* to $N_i$ and subtracts the rates of the *sinks* from $N_i$. Thus, an exponential decay (at rate $1/\tau$) from level $i$ to level $j$ requires a term $-N_i/\tau$ in the equation for $N_i$ and a term $N_i/\tau$ in the equation for $N_j$. We will represent the *optical pumping rate* by $R$, so that level 1 is depleted at rate $-RN_1$ and level 2 is increased at rate $RN_1$. Level 3 is just an intermediate in the pumping process and does not need a rate equation as long as $t_{32} \ll \tau$ so that the pumping can be considered to be directly to level 2. The rate equations for levels 1 and 2 are

$$\frac{dN_1}{dt} = -RN_1 + \frac{N_2}{\tau} + (N_2 - N_1)W(\nu) \tag{6.17}$$

$$\frac{dN_2}{dt} = RN_1 - \frac{N_2}{\tau} - (N_2 - N_1)W(\nu) \tag{6.18}$$

$$N_T = N_1 + N_2, \tag{6.19}$$

where $N_T$ is the *total* number density of atoms. These equations can be solved in the *steady-state* by setting the derivatives equal to zero and solving the resulting algebraic equations. Noting that the equations for $N_1$ and $N_2$ are the same, we have only *two* independent equations which can be solved for the inversion. The result is

$$\Delta N = \frac{1/\tau - R}{1/\tau + R + 2W(\nu)}N_T. \tag{6.20}$$

We are mainly interested in the inversion ($\Delta N_t$) at *threshold*, where $W(\nu) = 0$:

$$\text{Three-level laser:} \quad \Delta N_t = \frac{1/\tau - R_t}{1/\tau + R_t} N_T, \tag{6.21}$$

where $R_t$ is the *threshold pumping rate*. We see that with the addition of level 3, we can now invert levels 1 and 2 simply by making $R > 1/\tau$. The ground state still retains a considerable population (the Boltzmann factor $\approx 1$) and one needs to pump *fairly hard* to obtain a significant inversion. If we add a level somewhat *below* level 1, we can avoid this and greatly increase the ease of obtaining an inversion. This is the *four-level* laser scheme.

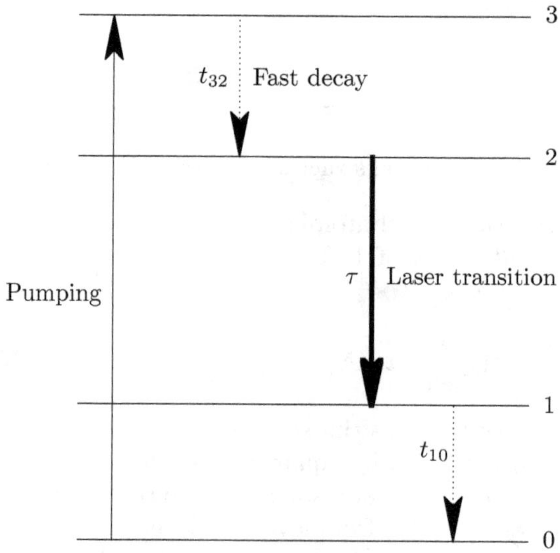


**Fig. 6.3** Level structure of a four-level laser.

A four-level laser system is shown in Fig. 6.3. We have taken the three-level system and added level 0 below the lower amplification level (level 1) with an energy gap sufficient to make the Boltzmann factor for level 1 much less than one. As in the three-level system, the decay time from level 1 to level 0 is much less than $\tau$. The three rate equations are (excluding one for level 3 for the same reasons as for the three-level system)

$$\frac{dN_0}{dt} = -RN_0 + \frac{N_1}{t_{10}} \tag{6.22}$$

$$\frac{dN_1}{dt} = -\frac{N_1}{t_{10}} + \frac{N_2}{\tau} + (N_2 - N_1)W(\nu) \tag{6.23}$$

$$\frac{dN_2}{dt} = RN_0 - \frac{N_2}{\tau} - (N_2 - N_1)W(\nu) \tag{6.24}$$

$$N_T = N_0 + N_1 + N_2. \tag{6.25}$$

Solving for $\Delta N$ at threshold in the steady state (again realizing that the equations are not all independent),

$$\text{Four-level laser:} \quad \Delta N_t = -\frac{R_t}{R_t + 1/\tau} N_T, \tag{6.26}$$

where we used $t_{10} \ll \tau$. Notice that at the pumping rate needed to barely reach threshold in a three-level system ($R = 1/\tau$), the four-level system is already significantly inverted ($\Delta N = -\frac{1}{2} N_T$). If one solves for the threshold pumping rates, the ratio of the pumping rates for a threshold inversion for the two laser systems is

$$\text{At threshold:} \quad \frac{(R_t)_{four-level}}{(R_t)_{three-level}} = \frac{\Delta N_t}{N_T + \Delta N_t} \ll 1. \tag{6.27}$$

Thus, we see that by adding a level somewhat below the lower amplification level we have tremendously eased the pumping requirements.

## 6.5 Laser oscillation frequencies and pulling

Laser oscillation occurs only at the resonant frequencies of the enclosing cavity, since these are the frequencies at which the considerations of loss and gain apply (the *feedback* necessary for oscillation is not available at other frequencies). When we derived the quantum-mechanical gain constant, we found that it is proportional to the *imaginary part* of the complex susceptibility, $\chi$. We also found that the *real part* of the susceptibility produced a *phase shift*. This phase shift will alter the cavity resonances from their values in the absence of the amplifying medium. To summarize the results from Chapter 5, a wave propagating in the amplifying medium will be described by

$$\boldsymbol{E}(z,t) = \boldsymbol{E}_0 e^{i(\omega t - kz)} \left\{ e^{(i\Delta k + \gamma(\omega)/2)z} \right\}, \tag{6.28}$$

where

$$\Delta k = -\frac{k\chi'(\omega)}{2n^2} \tag{6.29}$$

$$\gamma(\nu) = -\frac{k\chi''(\omega)}{n^2}. \tag{6.30}$$

In Chapter 2, we found that the resonant frequencies of an empty cavity (one not containing an amplifying medium) are determined by

$$nkl - 2(m + n + 1)\cos^{-1} \pm\sqrt{g_1 g_2} = 2q\pi, \tag{6.31}$$

where $l$ is the round-trip distance, $g_{1,2}$ are geometrical parameters, $q$ is an integral *axial mode index* and the $n$ before the $k$ is the index of refraction. The integers $m$ and $n$ multiplying the $\cos^{-1}$ are transverse mode indices. This expression is valid for both standing-wave and traveling-wave cavities. In the presence of an amplifying medium, this equation is changed to

$$nkl\left[1 + \frac{\chi'}{2n^2}\right] - 2(m + n + 1)\cos^{-1} \pm\sqrt{g_1 g_2} = 2q\pi, \tag{6.32}$$

due to the $\Delta k$ term in the susceptibility. The resonant frequencies are

$$\nu = \left(q + (n+m+1)\frac{\cos^{-1}\sqrt{g_1 g_2}}{\pi}\right)\frac{c}{nl}\left(1 + \frac{\chi'}{2n^2}\right)^{-1}. \tag{6.33}$$

The above equation can be used to determine the resonant frequencies, but it is more useful to express them in terms of the empty cavity resonance frequencies. This change in the resonant frequencies from those of the empty cavity is called *frequency pulling.* We begin by rewriting the above equation in terms of the *empty cavity resonance frequency,* $\nu_{qnm}$,

$$\nu\left[1 + \frac{\chi'}{2n^2}\right] = \nu_{qnm}, \tag{6.34}$$

where the empty cavity is still described by a refractive index, $n$. A *dispersion relationship* exists between $\chi'$ and $\chi''$

$$\chi' = \frac{2(\nu_0 - \nu)}{\Delta\nu}\chi'', \tag{6.35}$$

where $\Delta\nu$ is the width of the atomic resonance whose center frequency is $\nu_0$. This relation is a general result but can be derived from eqns 5.114 and 5.115. We now have

$$\nu\left[1 - \frac{\nu_0 - \nu}{\Delta\nu}\frac{\gamma(\nu)}{k}\right] = \nu_{qnm}, \tag{6.36}$$

where the expression for $\gamma(\omega)$ (eqn 6.30) was used. If the pulling effect is small and $\nu$ is very close to $\nu_{qnm}$,

$$\nu \approx \nu_{qnm} - (\nu_{qnm} - \nu_0)\frac{c\gamma(\nu_{qnm})}{2\pi n\Delta\nu}. \tag{6.37}$$

At threshold, we have $\gamma_t(\nu_{qnm})l = L_{tot}$ (round-trip gain equals loss) and from eqn 3.37 the finesse is $2\pi/L_{tot}$. Thus,

$$\frac{c\gamma(\nu_{qnm})}{2\pi n} = \frac{c}{\mathcal{F}ln} = \Delta\nu_{1/2}, \tag{6.38}$$

where $\Delta\nu_{1/2}$ is the *empty cavity linewidth.* Putting these equations together, we have the *frequency pulling relationship*

$$\text{Frequency pulling:} \quad \nu = \nu_{qnm} - (\nu_{qnm} - \nu_0)\left(\frac{\Delta\nu_{1/2}}{\Delta\nu}\right). \tag{6.39}$$

Thus, the resonance frequencies will be shifted from their empty cavity values by the product of the ratio of the cavity linewidth to the gain linewidth (usually small) and the detuning from the gain resonance frequency. The pulling will always be toward $\nu_0$.

## 6.6 Inhomogeneous broadening and multimode behavior

When we placed an amplifying medium in a resonant cavity, we discovered that under certain conditions a self-sustained field is possible: the laser oscillates. The frequencies at which oscillation is possible are the resonant frequencies of the *cavity,* modified (*pulled*) by the presence of the amplifying medium. The actual cavity mode(s) at which oscillation takes place is determined by the requirement that the gain is equal to the loss. The question arises: is it possible for a laser to oscillate *simultaneously* at more than one cavity mode frequency? The answer depends upon the nature of the broadening in the amplifying medium.

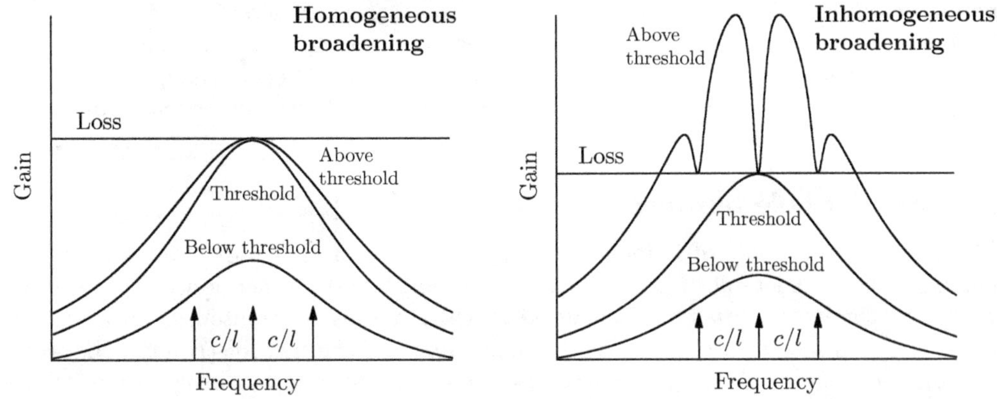


**Fig. 6.4** Gain curves for homogeneously and inhomogeneously broadened lasers for three pumping rates: below threshold, at threshold and above threshold.

A set of gain versus frequency curves appears in Fig. 6.4 for both homogeneously and inhomogeneously broadened lasers. The behavior of the curves for different pumping conditions illustrates how an inhomogeneously broadened laser can oscillate in more than one cavity mode simultaneously. In the figure, the cavity is adjusted so that one mode is always at the peak of the gain curve.

The curve on the left in the figure shows a homogeneously broadened gain curve at three pumping rates: below threshold, exact threshold and above threshold. As the pumping increases, the gain curve will increase in both amplitude and width until *threshold* is reached where the peak gain just equals the loss. The peak gain is *clamped* at this level and the laser will oscillate at the frequency corresponding to the peak. Further increases in the pumping rate will cause the gain curve to widen, but the peak will still be clamped to $\gamma_t$ and the unique oscillation frequency will be the same. All of this is a consequence of the behavior of the homogeneous gain curve under various levels of saturation, as discussed in Chapter 5. One can therefore state the general principle that *at any particular time, a homogeneously broadened laser is inherently single-mode.* There might be *technical noise* (mechanical fluctuations in the medium or cavity causing frequency fluctuations of a non-fundamental nature) and this might cause a *mode hop*, but the laser will still oscillate in only one mode at a time.

Below threshold, the inhomogeneously broadened gain curve behaves in a similar fashion to the homogeneously broadened curve, as shown on the right side of the figure. Again, at exact threshold, the laser will oscillate at the peak of the gain curve. Above threshold the behavior is very different from the homogeneous case. Since the model of the inhomogeneously broadened medium is a series of homogeneously broadened packets distributed in frequency over the inhomogeneous linewidth, each *cavity* mode can oscillate independently of the others provided that the gain at the mode frequency is greater than or equal to the loss. The plot shows a number of *holes* at several cavity modes with the bottom of the hole clamped to the loss as required by the condition for laser oscillation. For all frequencies *between cavity modes*, the gain can be greater than the loss since no clamping can occur at frequencies not corresponding to a cavity mode (possible laser oscillation can only occur at a cavity mode). It is as though a *large number of independent lasers exist* whose peak gain curve frequencies are distributed over the inhomogeneous linewidth. Thus, the general principle is that *an inhomogeneously broadened medium can sustain laser oscillation at a number of modes within the inhomogeneous linewidth.*

## 6.7 Spatial hole burning

Before leaving the topic of inhomogeneous broadening, a very important form of inhomogeneous broadening in a *standing-wave laser* should be mentioned. It is due to the *spatial inhomegeneity* which is caused by the standing wave intensity distribution inside the gain medium. Because of the dependence of the field distribution between the mirrors on the mode frequency, the laser can oscillate simultaneously in more than one mode. The phenomenon is called *spatial hole burning* and is illustrated in Fig. 6.5.

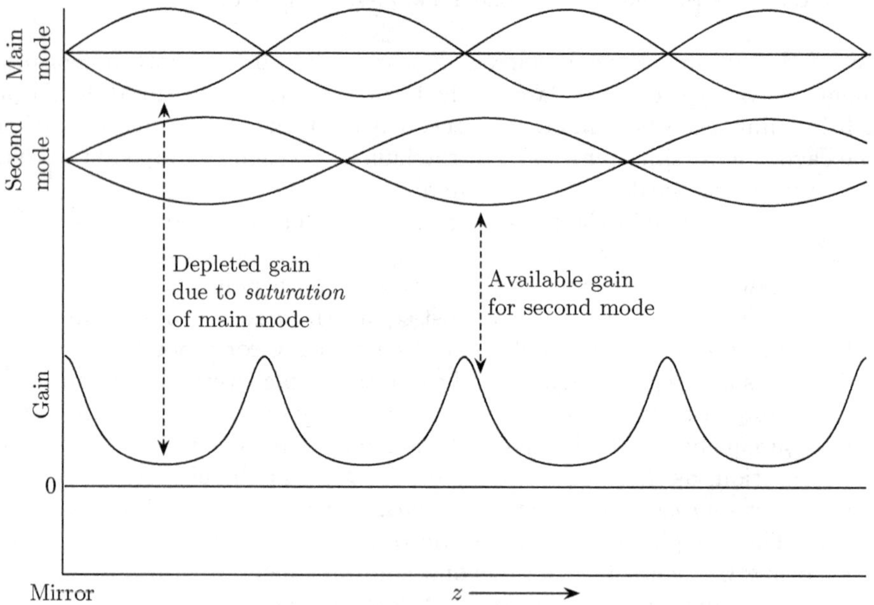


**Fig. 6.5** Illustration of *spatial hole burning.* The bottom curve is the saturated gain due to the standing-wave field of the main mode. A second mode standing-wave field will have a different spatial distribution and can also oscillate due to the availability of gain.

In a standing-wave cavity, if the intensity of the internal field in *each direction* is $I_\nu$, the gain saturation relationship is

$$\gamma(z) = \gamma_0 \frac{1}{1 + \frac{2I_\nu \sin^2 kz}{I_s}}, \tag{6.40}$$

which explicitly shows the $z$-dependence of the saturated gain. The presence of a spatially periodic gain is clearly depicted in Fig. 6.5 and is the reason that this phenomenon is called *spatial hole burning.* (The hole burning formerly discussed is called *spectral hole burning.*) In a standing wave, the clamping of the gain to the loss *at each point* is no longer a reasonable condition for laser operation. Instead, one requires that the rate of energy gain by the field due to stimulated emission be equal to the rate of energy loss *integrated over the entire length of the amplifying medium.* The intensity increase due to the gain is

$$\text{Intensity increase:} \quad \int_0^l I\gamma(z)dz = 2\gamma_0 I_\nu \int_0^l \frac{\sin^2(kz)dz}{1 + (2I_\nu/I_s)\sin^2(kz)}, \tag{6.41}$$

where $I = 2I_\nu \sin^2 kz$ is the total intensity in a standing wave. The power output in a standing-wave laser can be calculated by equating the intensity increase to the integrated loss and performing a fair bit of algebra. One will find that the output power is as much as 30% smaller than in a traveling-wave cavity due to the standing-wave intensity pattern.

The possibility of multimode oscillation in a standing-wave laser containing a homogeneously broadened medium can be explained by Fig. 6.5. The fundamental mode (at the top of the figure) gives rise to the periodic gain curve at the bottom of the figure. The maximum gain points are located at the *nodes* in the standing-wave pattern and the minimum gain points are at the *antinodes* of the field. The figure also shows a different cavity mode whose nodes and antinodes are displaced considerably from those of the fundamental mode. As a result of this *different field distribution*, the second cavity mode has *gain available at places where its field is strong* and can therefore oscillate at the same time as the fundamental mode. Thus, we have possible multimode behavior due to *spatial inhomogeneities* as well as *spectral inhomogeneities.* This form of multimode behavior is most easily cured by using a traveling-wave cavity instead of a standing-wave cavity: the resulting laser is usually called a *ring laser.* Another *treatment* is to place the gain medium *very close* to one of the standing-wave cavity mirrors. This will force the different field distribution to be *displaced considerably in frequency* from the fundamental mode where it can be more easily suppressed by intracavity etalons or other frequency-discriminating elements.

## 6.8 Some consequences of the photon model for laser radiation

The semiclassical theory treats the radiation field classically and is capable of describing the most important aspects of laser oscillation, such as the gain which is due to stimulated emission. There are, however, some aspects of laser operation which are best described using a *photon model*, which requires a fully quantum mechanical treatment of the field. We will postulate the existence of photons in a somewhat ad hoc manner and derive a number of important laser beam properties from this model. The first of the three topics using the photon model is an examination of the occupation number, $\bar{n}$, above threshold.

The determination of the occupation number requires the steady-state solution of two rate equations: one for the atoms and one for the photons. We first encountered photons when we derived the *Planck formula* for the spectral density of blackbody radiation. We postulated that the energy in the radiation field is an integral multiple of $h\nu$ or, alternatively, the field at frequency $\nu$ is composed of an integral number of *photons*, each of which has energy $h\nu$. From this, the intensity, $I_\nu$, of a unidirectional wave propagating inside the amplifying medium is

$$I_\nu = c\frac{nh\nu}{V_m}, \tag{6.42}$$

where $V_m$ is the *mode volume* of the field and $n$ is the number of photons. The rate equation for the upper amplifying level of a four-level laser is

$$\begin{aligned} \frac{dN_2}{dt} &= RN_0 - \frac{N_2}{\tau} - (N_2 - N_1)\left(\frac{\lambda^2}{8\pi}\frac{I_\nu}{h\nu}A_{21}g(\nu)\right) \\ &= RN_0 - \frac{N_2}{\tau} - (N_2 - N_1)\frac{K}{V_m}n \end{aligned} \tag{6.43}$$

$$\text{where} \quad K = \frac{\lambda^3}{4\pi^2}A_{21}\frac{\nu}{\Delta\nu}, \tag{6.44}$$

and we have defined $K$ so that $KnN_2$ is the *stimulated emission rate at resonance.* We will assume that $t_{10}$ is so small that level 1 has negligible population ($N_1 \approx 0$). Then, multiplying by $V_m$:

$$V_m\frac{dN_2}{dt} = RV_mN_0 - \frac{N_2V_m}{\tau} - KnN_2. \tag{6.45}$$

The rate equation for the photon number, $n$, requires one additional assumption: the decay rate for *spontaneous emission* is the same as that for stimulated emission with one photon per mode. Thus, the stimulated rate $KnN_2$ becomes $K(n+1)N_2$ when we add spontaneous decay. If the *photon decay rate* (due to cavity losses) is $1/t_c$, the photon number rate equation is

$$\frac{dn}{dt} = K(n+1)N_2 - \frac{n}{t_c}. \tag{6.46}$$

These two rate equations are readily solved for $n$ in the steady state (they are quadratic equations since $nN_2$ appears in both):

$$n = \frac{V_m}{2K\tau}\left\{\left(\frac{R}{R_t} - 1\right) + \left[\left(\frac{R}{R_t} - 1\right)^2 + \frac{4R}{R_t}\frac{K\tau}{V_m}\right]^{1/2}\right\}, \tag{6.47}$$

where $R_t$ is the threshold pumping rate:

$$\text{At threshold:} \quad R_t = \frac{1}{N_0K\tau t_c}. \tag{6.48}$$

A plot of the photon occupation number from below to above threshold appears in

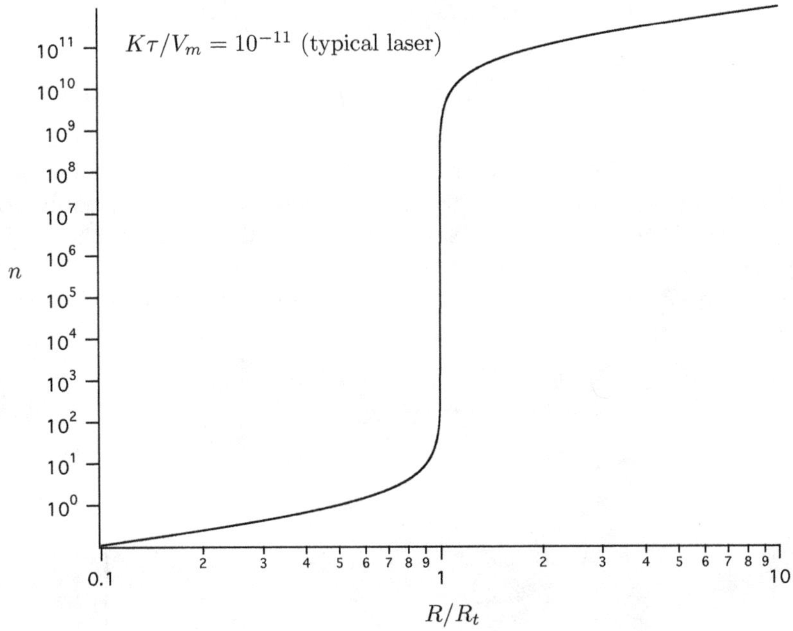


**Fig. 6.6** Photon occupation number, $n$, both above and below threshold.

Fig. 6.6. The remarkable jump in $n$ above threshold is akin to a *phase transition*. Since photons are bosons, it is a *Bose–Einstein condensation* of photons: the sudden appearance of an enormous number of particles in the *same quantum state*.

The energy spectral density of a *thermal* (blackbody) field is described by the Planck formula. The number of photons per mode (occupation number) is:

$$\bar{n} = \frac{1}{e^{h\nu/k_B T} - 1}. \tag{6.49}$$

We see that in order to approach one photon per mode of radiation at 500 nm, the temperature of the radiation source would need to be about 5000 K (the temperature of the surface of the sun). If the photon field from a laser were described by the same statistics as a thermal field (this is *not* an appropriate description, as we shall see later), the temperature of the source would need to be about $10^{10}$ K – clearly a stupendous temperature. This is one of the several remarkable features of the radiation from a laser; we will discuss others in the following sections.

## 6.9 The photon statistics of laser radiation

Before considering the statistics of laser radiation, we will briefly discuss *temporal coherence*. The temporal coherence of a wave is determined with the help of the *normalized first-order correlation function*, $g_1(\tau)$, which is defined as

$$g_1(\tau) = \frac{\langle E^*(t)E(t+\tau)\rangle}{\langle E^*(t)E(t)\rangle} \tag{6.50}$$

$$\text{where } \langle E^*(t+\tau)E(t)\rangle = \frac{1}{T}\int_0^T E^*(t+\tau)E(t)dt, \tag{6.51}$$

where $T$ is a large but finite averaging time. Essentially, $g_1(\tau)$ is a measure of the correlation between a wave and the same wave delayed by $\tau$, or, equivalently, the correlation between the wave and the same wave a distance $c\tau$ from the original position along the propagation direction. Its magnitude can be obtained from a *Michelson interferometer* (Fig. 6.7). The intensity at the detector, $I_d$, is related to that at the

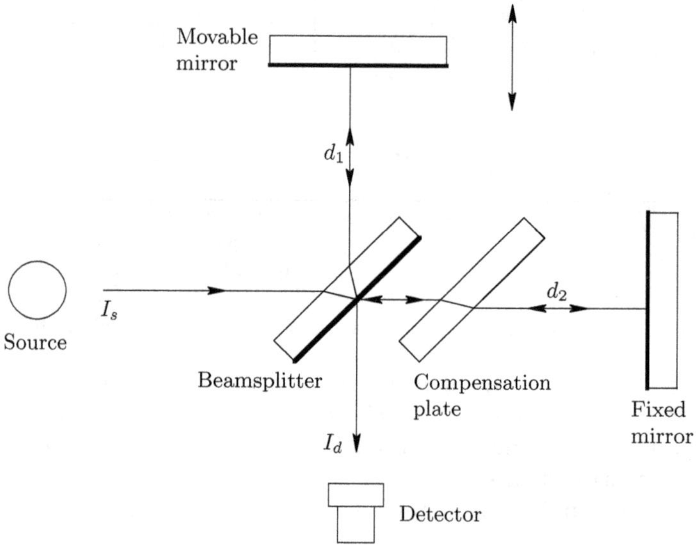


**Fig. 6.7** A Michelson interferometer used to measure temporal coherence.

source, $I_s$, by

$$I_d = \tfrac{1}{2}I_s[1 + \text{Re}\{g_1(\tau)\}], \tag{6.52}$$

where $\tau = 2(d_1 - d_2)/c$. The *fringe visibility*, $V(\tau)$, is

$$V(\tau) \equiv \frac{I_{max} - I_{min}}{I_{max} + I_{min}} = |g_1(\tau)|. \tag{6.53}$$

We define the *coherence time*, $\tau_c$, to be *the value of $\tau$ at which $|g_1(\tau)|$ and therefore the fringe visibility in a Michelson interferometer is reduced to one half of its maximum* (the maximum occurs when $\tau = 0$). We can relate $\tau_c$ to the *bandwidth*, $\delta\nu$, of the radiation using the following approximate argument:

Define $\Delta \equiv |d_1 - d_2| = c\tau/2$. The fringe peaks will appear when $\Delta = n\lambda$, where $\lambda$ is the average wavelength of the radiation. If there is a finite bandwidth, $\delta\nu$, of the radiation, the fringe peaks will be smeared by $\delta\Delta = -\Delta(\delta\nu/\nu)$. The fringes will disappear when $\delta\Delta \approx \lambda/2$, which implies that $-\Delta(\delta\nu/\nu) = -(c\tau_c/2)(\delta\nu/\nu) = \lambda/2$ and therefore $\tau_c \approx 1/\delta\nu$. A more careful calculation yields the following result:

$$\text{Coherence time: } \tau_c = \frac{1}{2\pi\delta\nu} \tag{6.54}$$

$$\text{Coherence length: } l_c \equiv c\tau_c. \tag{6.55}$$

We will show later on that the bandwidth of laser radiation can be well below 1 Hz; thus the coherence time can be 1 s or more and the coherence length can be *greater than* 300,000 km!

Light sources can be divided into two distinct classes, depending upon the photon statistics of the radiation they emit. Conventional light sources based upon a thermal emitter or a discharge lamp are said to emit *chaotic light.* Included in this category is a laser below threshold where the spontaneous emission will be amplified by the gain medium. The other type of source is a *laser* above threshold whose radiation is said to be *coherent.* The photon statistics of chaotic radiation is very different from that of coherent radiation.

We will begin our photon statistics study by assuming that one has an apparatus which counts photons during fixed time intervals of duration $T$. The quantity $P_n(t, t')$ is the probability of counting $n$ photons in an interval from $t$ to $t + t'$ and $p(t)dt$ is the probability of counting *one* photon in a very small interval, $dt$ (it is assumed there is negligible probability of counting *two or more* photons in this interval). Let the range of variable $t'$ be between 0 and $T$ so that $t + t'$ lies in the interval between $t$ and $t + T$. We can obtain a differential equation for $P_n(t, t')$ by considering the two *uncorrelated* ways of counting $n$ photons during the interval from $t$ to $t + t' + dt'$, where $dt'$ is a very small interval:

1. $n - 1$ photons between $t$ and $t + t'$ and 1 photon in $dt'$
2. $n$ photons between $t$ and $t + t'$ and 0 photons in $dt'$

Adding the two probabilities, we obtain

$$P_n(t, t' + dt') = P_{n-1}(t, t')p(t')dt' + P_n(t, t')(1 - p(t')dt'). \tag{6.56}$$

We can rearrange the terms to make the left side a derivative:

$$\frac{P_n(t, t' + dt') - P_n(t, t')}{dt'} = \frac{dP_n(t, t')}{dt'} = (P_{n-1}(t, t') - P_n(t, t'))p(t'). \tag{6.57}$$

If $\bar{I}$ is the photon emission rate, averaged over one optical cycle, and $\alpha$ is the *quantum efficiency* of the photodetector, then

$$p(t)dt = \alpha\bar{I}dt \tag{6.58}$$

and

$$\frac{dP_n(t, t')}{dt'} = (P_{n-1}(t, t') - P_n(t, t'))\alpha\bar{I}(t). \tag{6.59}$$

It can be shown (by *substitution*) that the solution to this equation satisfying the boundary condition $P_{n\neq 0}(t, 0) = 0$ is

$$P_n(t,T) = \frac{[X(t,T)T]^n e^{-X(t,T)T}}{n!}, \tag{6.60}$$

where

$$X(t,T) = \frac{\alpha}{T}\int_t^{t+T} \bar{I}(t')dt' \tag{6.61}$$

is the photon *counting rate* averaged over $T$. The usual practice is to average a large number of measurements over time. Then, the *average* value of $P_n(t,T)$ is called $P_n(T)$ and is given by

$$P_n(T) = \left\langle \frac{[X(t,T)T]^n e^{-X(t,T)T}}{n!} \right\rangle_t, \tag{6.62}$$

where the subscript $t$ denotes averaging over $t$.

The key to the difference between the two types of statistics lies in the behavior with time of the quantity $\bar{I}(t)$. If $\bar{I}(t)$ is independent of $t$, then $X(t,T)$ is independent of both $t$ and $T$ and

$$X(t,T)T = \alpha \bar{I} T = \bar{n}. \tag{6.63}$$

The calculation of the time average is trivial (since there is *no* time dependence), yielding the *Poisson distribution*

$$\text{Poisson distribution (constant } \bar{I}\text{):} \quad P_n(T) = \frac{\bar{n}^n e^{-\bar{n}}}{n!}. \tag{6.64}$$

From the way that $\bar{I}(t)$ is defined, $\bar{n}$ is clearly the *average number of photons* registered by the counting apparatus during interval $T$. This result is valid for a classical stable wave and also for the so-called quantum-mechanical *coherent state* of the radiation field (indeed, the latter is just a sum over $n$-photon states with coefficients equal to the square root of $P_n$, so Poisson statistics are an immediate consequence when one squares the coefficient). It is also valid for a *chaotic* source if *one counts for a period greater than the coherence time*, $t_c$, since the fluctuations will average out and $\bar{I}(t)$ can be considered constant. Thus, the distinction between the different statistics depends upon *the behavior of $\bar{I}(t)$ over counting periods which are short compared to the coherence time.*

A chaotic wave will be described by an $\bar{I}(t)$ which depends upon time for counting periods less than $t_c$. In order to calculate the average over $t$ as required by eqn 6.62, we need to know the *distribution of intensities*, $p[\bar{I}(t)]$, which is the probability that a measurement will yield a value between $\bar{I}(t)$ and $\bar{I}(t) + d\bar{I}(t)$. This can be obtained from the following considerations. We will model the chaotic source as a group of similar atoms emitting waves whose electric fields have the same frequency and amplitude but random phases:

$$\begin{aligned} E(t) &= E_1(t) + E_2(t) + \cdots + E_n(t) \\ &= E_0 e^{-i\omega_0 t}(e^{i\phi_1(t)} + e^{i\phi_2(t)} + \cdots + e^{i\phi_n(t)}) \\ &= E_0 e^{-i\omega_0 t}\beta(t) e^{i\phi(t)}. \end{aligned} \tag{6.65}$$

The quantities $\beta(t)$ and $\phi(t)$ are the time-varying amplitude and phase which result from the addition of $n$ complex numbers with unity magnitude and random phases. The intensity is

$$\bar{I}(t) = \tfrac{1}{2}\epsilon_0 c|E(t)|^2 = \tfrac{1}{2}\epsilon_0 c\beta^2(t)E_0^2. \tag{6.66}$$

The distribution of intensities reduces to finding the distribution of the quantity $\beta(t)$. The appropriate (normalized) probability can be obtained from *random walk* theory and is given by

$$p[\beta(t)] = (1/\pi n)e^{-\beta^2(t)/n}, \tag{6.67}$$

since we are adding a series of vectors of fixed length and random direction, which is equivalent to a random walk. From eqn 6.66, one can convert this into a distribution of intensities,

$$p[\bar{I}(t)] = (1/\bar{I})e^{-\bar{I}(t)/\bar{I}}, \qquad \bar{I} = \tfrac{1}{2}\epsilon_0 c E_0^2 n. \tag{6.68}$$

The appropriate value of $X(t, T)$ for chaotic light is

$$\text{Chaotic light:} \quad X(t, T) = \alpha\bar{I}(t), \tag{6.69}$$

since, for $T \ll t_c$, the intensity is essentially *constant* during the counting interval, but varies from one counting interval to the next. The value of $P_n(T)$ for chaotic light is obtained by integrating the product of the distribution of intensities and $P_n(t, T)$,

$$P_n(T) = \frac{1}{\bar{I}}\int_0^\infty e^{-\bar{I}(t)/\bar{I}}\frac{[\alpha\bar{I}(t)T]^n e^{-\alpha\bar{I}(t)T}}{n!}d\bar{I}(t). \tag{6.70}$$

This integral is actually easier to evaluate than it appears, and the result is

$$\text{Chaotic source (Bose–Einstein distribution):} \quad P_n(T) = \frac{\bar{n}^n}{(1+\bar{n})^{1+n}}, \tag{6.71}$$

where $\bar{n}$ is again $\alpha\bar{I}T$.

Our procedure for calculating $P_n(T)$ for chaotic sources was very general and was based upon some simple photon-counting considerations together with the assumption that the source intensity was not constant for times much shorter than the coherence time. For the *particular case* of a thermal (blackbody) source, it is very straightforward to obtain the same result. The probability of observing $n$ photons in thermal equilibrium with the source is just the normalized Boltzmann factor

$$P_n = \frac{e^{-nh\nu/k_BT}}{\sum_m e^{-mh\nu/k_BT}}, \tag{6.72}$$

where we have dropped the $T$ dependence in $P_n$. The numerator can be evaluated from the expression for $\bar{n}$ obtained during the derivation of Planck's formula, and the denominator is just a geometrical series which can also be expressed as a function of $\bar{n}$. The result is

$$\text{Blackbody radiation:} \quad P_n = \frac{\bar{n}^n}{(1+\bar{n})^{n+1}}. \tag{6.73}$$

Finally, we will evaluate the *variance* for the two distributions. The variance, $\Delta n^2$, is defined as the average of $(n - \bar{n})^2$ and is given by

$$\text{Variance:} \quad \langle \Delta n^2 \rangle = \langle n^2 \rangle - \bar{n}^2. \tag{6.74}$$

It will be left as an exercise for the reader to verify that the variances of the two distributions are:

$$\text{Chaotic source:} \quad P_n = \frac{\bar{n}^n}{(1+\bar{n})^{n+1}}, \qquad \langle \Delta n^2 \rangle = \bar{n}^2 + \bar{n} \tag{6.75}$$

$$\text{Laser source:} \quad P_n = \frac{\bar{n}^n e^{-\bar{n}}}{n!}, \qquad \langle \Delta n^2 \rangle = \bar{n}. \tag{6.76}$$

We thus have the familiar result that the RMS fluctuation in count rates is equal to $\sqrt{\bar{n}}$ for almost all sources encountered in the laboratory. Most chaotic sources have linewidths of many MHz and would require counting times of less than a microsecond to observe their departure from Poisson statistics. Figure 6.8 displays an exception: it

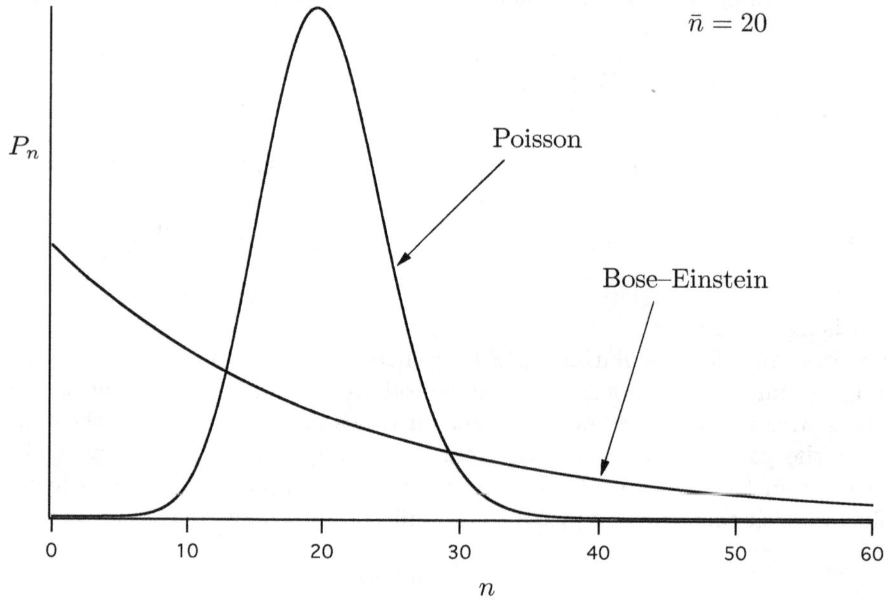


**Fig. 6.8** Comparison of coherent (Poisson) and chaotic (Bose–Einstein) distributions for $\bar{n} = 20$. These are theoretical curves, but the two very different distributions have been observed in the laboratory (Freed 1966).

shows the difference in the distributions for $\bar{n} = 20$. These two very different distributions were observed in the laboratory for a laser both above and below threshold (Freed 1966).

## 6.10 The ultimate linewidth of a laser

From our calculation of the occupation number ($\bar{n}$) of the radiation from a laser above threshold, we discovered that $\bar{n}$ is extremely high and therefore the radiation will be an excellent approximation to a classical, stable, monochromatic wave. Our approach ignored the effects of *spontaneous emission*, since this process does not fit naturally into the framework of a semiclassical treatment. When one includes spontaneous emission, the randomness of the spontaneous photons will add some uncertainty to the phase and frequency of the electric field, and the laser radiation will acquire a finite linewidth. We will give here a simplified derivation of the *Schawlow–Townes* expression for the ultimate linewidth of a laser.

Our approach will be to consider the atoms and radiation to be *one composite quantum system*, described by a quality factor, $Q$, which is defined in terms of the *stored energy* and *energy loss rate* (eqn 3.39). The stored energy will be in the *field*.

We begin with the rate equation for the photon number, which is reproduced here:

$$\frac{dn}{dt} = K(n+1)N_2 - \frac{n}{t_c}, \qquad K = \frac{\lambda^3}{4\pi^2} A_{21} \frac{\nu}{\Delta\nu}, \tag{6.77}$$

where $N_2$ is the upper laser level population. Let us assume that we have a four-level laser and that the lower laser level has negligible population. In the steady state ($dn/dt = 0$) and well above threshold ($n \gg 1$), $N_2$ is

$$\text{Steady state, well above threshold:}\ N_2 = \left(\frac{n}{n+1}\right)\frac{1}{Kt_c} \rightarrow \frac{1}{Kt_c}. \tag{6.78}$$

From the rate equation, the *spontaneous rate* (the "1" in the brackets) into a mode is

$$\text{Spontaneous rate} = KN_2 = \frac{1}{t_c} \tag{6.79}$$

from the equation immediately above, and the *spontaneous power into the mode* is obtained by multiplying by $h\nu$:

$$\text{Spontaneous power} = \frac{h\nu}{t_c}. \tag{6.80}$$

If we assume that the *output coupling* is the only loss, the output power, $P_o$, is

$$P_o = \frac{nh\nu}{t_c}, \tag{6.81}$$

since $t_c$ is the rate that the energy leaks out through the output coupler. The *stored mode energy* is $nh\nu$:

$$\text{Stored energy} = nh\nu = t_c P_o. \tag{6.82}$$

The $Q$ factor is given by

$$Q = \frac{\nu}{\delta\nu} = \omega \times \frac{\text{stored energy}}{\text{spontaneous energy emission rate}} = \frac{(2\pi\nu)t_c P_o}{(h\nu/t_c)} = \frac{2\pi t_c^2 P_o}{h}. \quad (6.83)$$

Solving for $\delta\nu$,

$$\delta\nu \approx \frac{2\pi(\Delta\nu_{1/2})^2 h\nu}{P_o}, \quad (6.84)$$

where we used the cavity expression, $t_c = 1/2\pi\Delta\nu_{1/2}$. Other treatments include a factor $N_2/\Delta N$ which is unity in our case. The important parameter is the *cavity linewidth* ($\Delta\nu_{1/2}$), which can be quite large in the extremely short cavities found in semiconductor lasers whose ultimate linewidths can therefore be in the tens of MHz (there is actually another *linewidth-enhancing* factor in semiconductor lasers). At the other extreme, a conventional laboratory continuous-wave dye or Ti–sapphire laser can have an ultimate linewidth of less than one Hz. It should be stressed again that the Schawlow–Townes linewidth is a *lower bound*: other (usually remediable) factors such as mechanical vibrations and thermal effects will invariably make the observed linewidth much larger than that derived from this expression. The increased linewidth from the latter factors is called *technical noise*. When we discuss *laser frequency stabilization*, we will show how one can remove the technical frequency noise and even reduce the laser linewidth to below the Shawlow–Townes limit.

## 6.11 Further reading

Much of the discussion of the photon statistics is based upon Loudon's book on the quantum theory of light (1983). The discussions of the laser oscillation condition and pulling comes largely from Yariv (1989), and the discussions of the Schawlow–Townes limit and photon statistics are similar to those in Milonni and Eberly (1988).

## 6.12 Problems

(6.1) The *threshold* pumping power is measured with two different output couplers whose power transmissions are $T_1$ and $T_2$. If the corresponding threshold pumping powers are $P_1$ and $P_2$ and everything else is the same, write down a formula for the *internal losses* of the laser (assume that all the other mirrors are totally reflecting). For simplicity, assume that the laser is a ring laser with very low losses.

(6.2) Calculate the shift due to frequency pulling in a He–Xe laser operating at the 3.51 $\mu$m transition of Xe. Assume that the transition is Doppler broadened with a FWHM of 200 MHz and that the laser is a standing-wave cavity with single-pass logarithmic loss $\rho$ (defined as $\rho = -\ln(1-T)$) of 0.5 and length of 30 cm. Estimate the pulling when the cavity resonance is detuned 50 MHz from the gain curve center. (Use the expression for *homogeneous broadening* derived in the text; this is less than a factor of two different from the *inhomogeneous* broadening result.)

(6.3) Determine the quantum limit to the linewidth for the following lasers:

(a) A He–Ne laser ($\lambda = 632.8$ nm) with a cavity length of 30 cm, total loss per pass of 1% (including output coupling) and power output of 1 mW.
(b) A GaAs/GaAlAs diode laser ($\lambda = 850$ nm) with uncoated facets (refractive index 3.5), a cavity length of 120 $\mu$m, loss coefficient ($\alpha$) of 10 cm$^{-1}$ and power output of 5 mW.
(c) A ring dye laser ($\lambda = 590$ nm) with a total path length of 75 cm, loss factor ($\alpha l$) of 0.05, output coupling of 3% and power of 100 mW.

(6.4) (a) Estimate the bandwidth and coherence time of white light, assuming that it comprises wavelengths from 400 nm to 700 nm.
(b) Show that the coherence length of white light is on the order of the wavelength.
(c) Using a Michelson interferometer, is it possible to observe "white-light" fringes (that is, an observable fringe pattern with the white light defined above)?

(6.5) Calculate the coherence time and the coherence length (both of these are related to the "longitudinal coherence") for the following sources (this comparison should illustrate one of the main differences between *thermal* light and laser light):
(a) A mercury source emitting at 546 nm with a bandwidth of $\Delta\lambda \approx 0.01$ nm.
(b) A YAG laser at 1064 nm with a bandwidth $\Delta\nu \approx 10$ kHz.

(6.6) Show that the average intensity ($I_d$) observed at the detector in a Michelson interferometer depends upon the intensity at the source ($I_s$) and the normalized first-order correlation function, $g_1(\tau)$, and is given by the following relationship: $I_d = \frac{1}{2} I_s [1 + \mathrm{Re}\{g_1(\tau)\}]$

(6.7) Show that the variance ($\langle \Delta n^2 \rangle$) of the photon number for a Poisson distribution is given by the following equation:
$\langle \Delta n^2 \rangle \equiv \langle (n - \bar{n})^2 \rangle = \bar{n}$

(6.8) Show that the variance ($\langle \Delta n^2 \rangle$) of the photon number for a thermal (Bose–Einstein) distribution is given by the following equation:
$\langle \Delta n^2 \rangle \equiv \langle (n - \bar{n})^2 \rangle = \bar{n}^2 + \bar{n}$

# 7 Descriptions of specific CW laser systems

## 7.1 Introduction

In this chapter, we will describe several continuous-wave laser systems, aided by the theory formulated in the previous two chapters. Due to its dominance, the ubiquitous diode laser warrants at least one chapter of its own. Several of the laser systems described in this chapter are currently being replaced by simpler and less expensive diode and diode-pumped solid-state systems.

In the category of lasers whose days are probably numbered in the atomic physics laboratory are argon ion lasers and dye lasers. Some commercial realizations of these lasers will be described. The diode-pumped frequency-doubled $Nd:YVO_4$ laser and the Ti–sapphire laser (pumped by the former) will be discussed. Finally, the formerly ubiquitous He–Ne laser (currently supplanted by red diode lasers) will be described, in part for historical reasons and also because it employs an interesting and unique excitation transfer mechanism for pumping. We will start with the He–Ne laser.

## 7.2 The He–Ne laser

A schematic of an early helium–neon (He–Ne) laser appears in Fig. 7.1. The laser

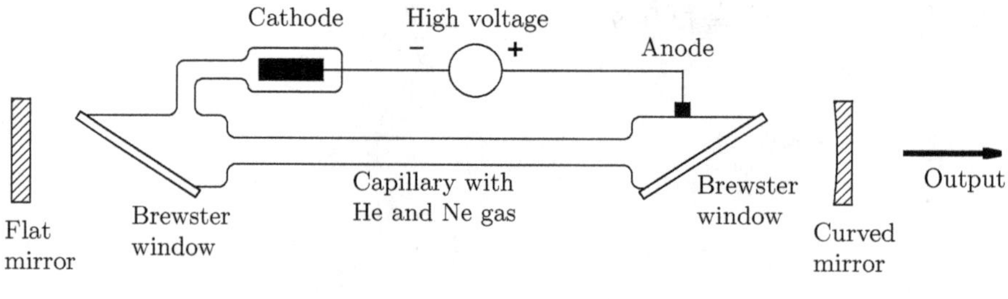


**Fig. 7.1** Schematic of helium–neon laser.

consists of a glass capillary filled with a mixture of helium and neon gases and with windows inclined at Brewster's angle on both ends. A DC glow discharge is maintained by a high-voltage power supply connected to an anode and cathode placed inside the tube. We shall shortly demonstrate that there is an optimum current which produces the maximum output power, and the discharge current is set to this value (the voltage–current characteristic depends on the pressure of the gases and the capillary dimensions). It has been found that the optimum performance is obtained when the helium partial pressure is about five times that of neon and the product of the total pressure and capillary diameter is about 4 Torr-mm. Thus, for a typical laser with a 2 mm diameter capillary, about 2 Torr of helium and 0.4 Torr of neon is used. The laser emits radiation at a number of wavelengths and the optimum parameters have a slight wavelength dependence.

The pumping mechanism can be understood by reference to Fig. 7.2, which shows the energy diagrams of both neutral helium and neutral neon. The principal pumping

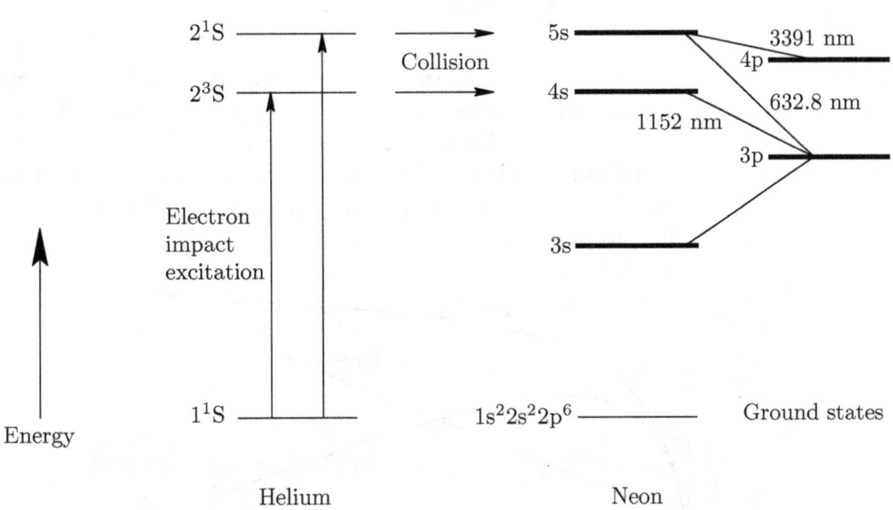


**Fig. 7.2** Level diagram of helium and neon as an aid in explaining the pumping mechanism in He–Ne laser. The excited levels in neon are actually *groups* of levels (composed of different *terms*) whose outer electron state is specified in the diagram. The *s*-states have four levels and the *p*-states have ten.

mechanism is a two-step process which relies on the near resonance between the $2^3S$ and $2^1S$ levels in helium and the $4s$ and $5s$ levels in neon. The energetic electrons in the discharge excite the helium atoms into the *metastable* $2^3S$ and $2^1S$ levels, where they remain for $10^{-4}$ seconds and $5 \times 10^{-6}$ seconds respectively. These lifetimes are much shorter than expected for the $2^3S$ and $2^1S$ states, since they are fairly rapidly quenched by the *discharge*.

Once the helium atoms are excited to the metastable states, they transfer their excitation, *via collisions*, to the $4s$ and $5s$ states of neon. This process is greatly enhanced by the accidental near resonance between these levels and the metastable levels in helium. This step *pumps* the neon atoms into their *upper lasing level.* From this point, the neon atoms behave like a four-level laser system, since the $S$ to $P$ transition rate ($\approx 10^{-7}$ /s) responsible for laser amplification is somewhat smaller than the rate ($\approx 10^{-8}$ /s) of the $3p, 4p$ to $3s$ transitions, which deplete the lower laser level. The figure displays three possible laser frequencies; the most common one is

632.8 nm (there is another one at 543 nm which is not shown; it is due to a transition from the 5*s* state to a different term in the 3*p* configuration).

The reason that there is an optimum discharge current is the following. Let the ground-state population in helium be $N_1$, the metastable population be $N_2$ and the discharge current be $I$. If $\alpha_1$ and $\alpha_2$ are constants, the helium metastable-state *excitation rate* is $N_1\alpha_1 I$ and the *de-excitation rate* is $N_2(\alpha_2 + \alpha_3 I)$, where the constant term is due to other de-excitations, such as wall collisions. At equilibrium, the two rates are equal and the metastable-state helium population is

$$N_2 = N_1 \frac{\alpha_1 I}{\alpha_2 + \alpha_3 I}. \tag{7.1}$$

We assume that, via collisions, the excited-state neon populations will be proportional to the helium metastable populations. Thus, the upper laser level (in neon) will *saturate* with increasing current. However, the lower lasing level is increasingly populated as the current increases, due to electron bombardment. The result will be a *peak* in the population difference and, therefore, gain. This phenomenon is depicted in Fig. 7.3, which displays the populations and gain in neon as a function of the current. The peak

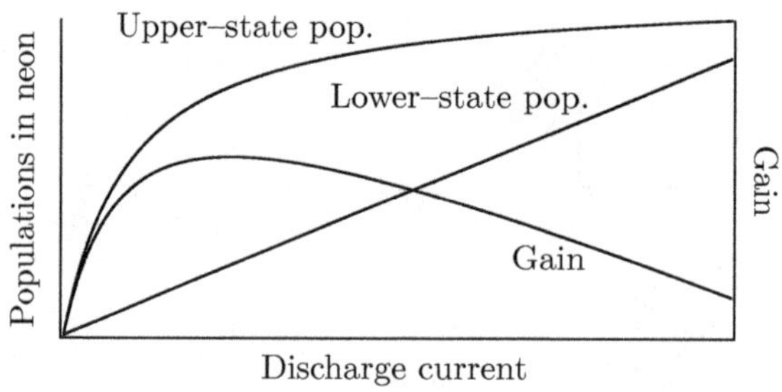


**Fig. 7.3** Saturation of upper-state population and gain maximum in neon in a He–Ne laser.

gain occurs at a discharge current of some tens of milliamps in a He–Ne laser, thus restricting the maximum power of such lasers to several milliwatts at most. Despite the much lower efficiency of an argon laser, the latter can produce 1000 times more power, since the gain does not saturate with increased current.

## 7.3 The argon-ion laser

The argon-ion laser is another system which is pumped by an electrical discharge which both ionizes the argon gas and populates the upper laser level in the ion. It is an extraordinarily inefficient system but it can generate much more power than a He–Ne laser on a number of wavelengths in the blue and ultraviolet regions of the spectrum. A simplified cross-sectional drawing of a commercial argon ion laser tube appears in Fig. 7.4.

We will shortly show that the gain in an argon-ion laser is proportional to the *square* of the current density in the discharge, and therefore a considerable effort is made to confine the discharge as much as possible. An ingenious technique for accomplishing

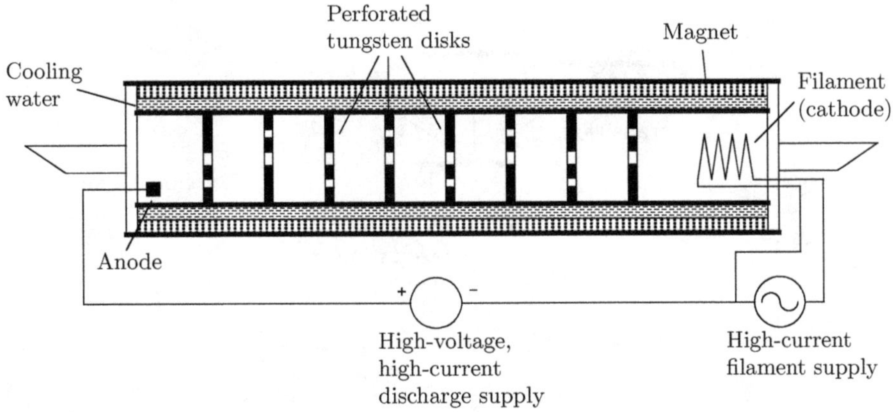


**Fig. 7.4** Simplified cross-sectional view of argon-ion laser tube with associated power supplies.

this uses a series of perforated tungsten disks whose outer edge is in contact with a cooling water jacket. The central hole in the disks confines the discharge to a column whose diameter is equal to that of the hole and the excellent thermal conductivity of the tungsten removes the considerable heat generated by the discharge. The additional holes in the disks serve as a gas return path, required to prevent large axial pressure gradients due to the *pumping* from the discharge. The holes are deliberately misaligned between adjacent disks (shown as *staggered* in the figure) to provide a larger path and to ensure that the discharge uses only the central holes. A filament is used to increase the area of the cathode and reduce its work function, thereby increasing the ionization rate and protecting the cathode. Most commercial argon lasers have an external argon reservoir which is provided with an automatically operated solenoid valve to keep the argon pressure relatively constant (the discharge will remove argon from the tube over time). A relatively large magnetic field (about 1000 gauss), generated by a solenoid surrounding the tube, will cause the electrons to spiral around the field lines and more effectively ionize the argon atoms.

The level diagram of $Ar^+$ appears in Fig. 7.5. The *zero* of the energy scale is set equal to the energy of the neutral argon ground state. This shows more clearly the total energy needed to excite the upper laser level of the ion. Unlike in the He–Ne laser, the laser levels in $Ar^+$ are directly excited by electron bombardment in the discharge. Due to the exponential dependence of the electron energy distribution, which is proportional to $e^{-E/k_BT}$, a two-step pumping process is much faster than a single-step process, even for the relatively *hot* electrons in a discharge. A single-step process would require an electron with more than 35 eV while the two-step process needs two $\approx$ 18 eV electrons. The first electron ionizes the neutral argon and the second electron excites the ion to the upper laser level. There are actually three ways that the second excitation can proceed. The ion can be directly excited to the upper laser level (from the ion ground state), the ion can be excited to higher energy levels followed by radiative decay to the upper laser level, and the ion can be excited to a

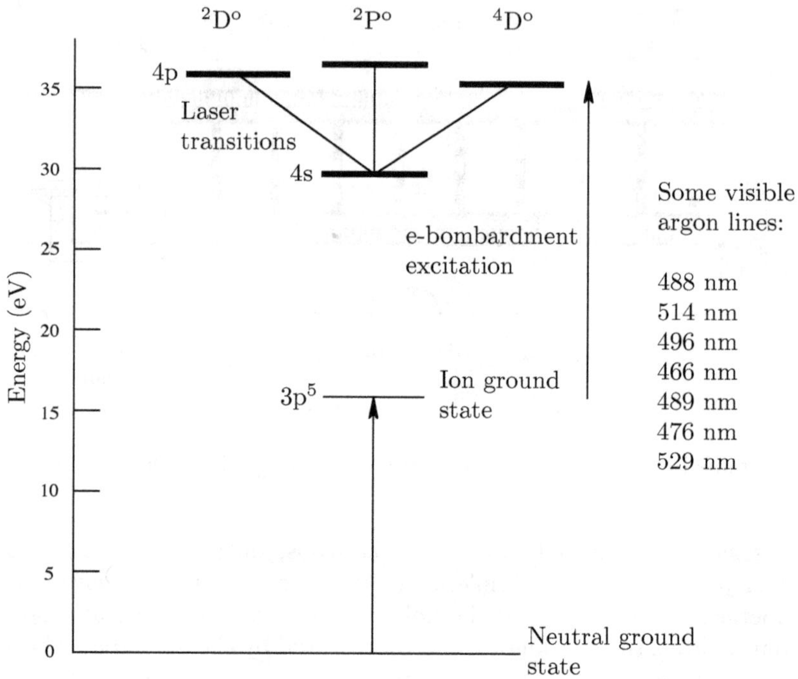


**Fig. 7.5** Levels of $Ar^+$ referred to ground state of neutral argon.

metastable level, where a collision with another electron will transfer the ion to the upper laser level. The first two mechanisms proceed at a rate which is proportional to the electron density, $N_e$, in the discharge. Since the ionization rate is also proportional to the electron density, the total excitation rate is proportional to the electron density squared:

$$\frac{dN_2}{dt} \propto N_e^2 \propto J^2, \tag{7.2}$$

where $J$ is the current density (which is proportional to $N_e$). From this, we see why argon-ion laser designers go to great lengths to increase the current density, for example, by using perforated disks and strong magnetic fields as discussed above. Since the excitation rate to the upper laser level is proportional to the current squared and the quenching rate from either laser level is linear in the current, the pumping rate does not *saturate*, as it does in the He–Ne laser.

As is the case with neon, the lifetime of the upper laser level ($\approx 10^{-8}$ s) is about ten times longer than that of the lower laser level, so one can apply four-level laser theory. One can define the laser *quantum efficiency* as the ratio of the energy of the stimulated photon to the energy needed to excite the upper laser level. The quantum efficiency of an argon laser is very low: about 7%. However, since the argon pumping scheme does not saturate, one can obtain a significant amount of power by using a very large discharge current. Typical argon lasers can operate with currents of 60 A or more, while He–Ne lasers are limited to currents in the tens of milliamps range. The actual overall efficiency of an argon laser is much less than its quantum efficiency. At a discharge current of 60 A, the discharge voltage is about 500 V while the power output is only about 20 W; this yields an overall efficiency of less than 0.001 and explains why water cooling is absolutely necessary in *large-frame* argon-ion lasers.

## 7.4 The continuous-wave organic dye laser

It is often desirable in the laboratory to have a laser which is tunable over a wide range of wavelengths. In addition to the obvious utility of such systems, they also make very good candidates for *mode-locked* operation, as we shall see later. There are two practical widely tunable systems in the visible and near-infrared: those using an organic dye as the active medium and those using a titanium–sapphire crystal. Although they have some superficial similarities, they will be considered separately.

The active medium in a dye laser is an organic dye dissolved in a solvent (which actually has a surprisingly important role in determining the laser characteristics). The dye molecules have a complicated level structure which can be modeled as a simple diatomic molecule, as shown in Fig. 7.6. The figure displays the *vibrational* levels in

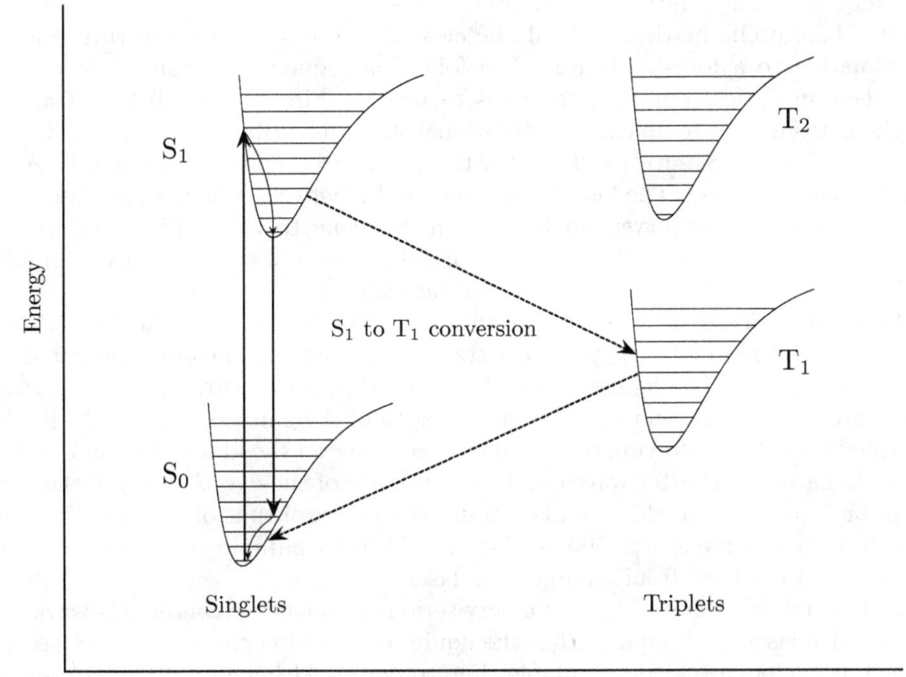


**Fig. 7.6** Simplified model of organic dye level structure.

a diagram whose horizontal coordinate is the intranuclear distance. Each potential-energy curve corresponds to a different *electronic* level, and the vibrational levels have a fine structure of rotational levels, which are not shown. The minima of the potential energy curves for the different *electronic* levels are at different intranuclear distances; this is necessary for laser behavior, as we will see below. The wide tunability of the dye laser is due to the rich vibration-rotational level structure which is further broadened by the presence of the liquid solvent.

Four sets of *electronic levels* are shown in the diagram: two *singlets* ($S_1$ and $S_2$) and two *triplets* ($T_1$ and $T_2$). The laser transitions are between the two singlet levels and they are very fast (nanosecond lifetimes), since the transitions are *allowed.* The transitions between singlets and triplets are *forbidden* and therefore have very long *radiative* lifetimes. When the dye is excited to a higher electronic level, the various vibrational levels *very quickly* thermalize (in picoseconds), with the maximum population appearing at the potential minimum. Any molecule which ends up in a triplet state will be removed from the laser amplification process until it decays back to the singlet levels (which can take a good fraction of a second). An important rule for determining the behavior of the vibrational levels during an electronic transition is the *Franck–Condon principle*, which states that electronic transitions will proceed along a *vertical line* in diagrams such as Fig. 7.6. The reason for this is that the electronic transition time is much shorter than the vibrational period, and therefore the intranuclear distance will not change very much during an electronic transition.

With these preliminaries, it should be clear that the dye level structure is a good approximation to a four-level laser. The following sequence of transitions will take place when an optical pumping source is turned on. First there will be a transition from the bottom of $S_0$ to an excited vibrational state in $S_1$ (due to the Franck–Condon principle). Then the $S_1$ levels will quickly thermalize to populate the upper laser level (near the bottom of $S_1$). The laser transition will be between this level and an excited vibrational level in the lower singlet system, $S_0$. Due to the *different intranuclear distances* in $S_0$ and $S_1$, the laser transition will be very distinct from the pumping transition and a four-level laser system will be realized.

The absorption and emission curves for a popular dye are shown in Fig. 7.7. The absorption curve is due to the pumping transitions, and the emission curve is due to the laser transitions. The figure clearly shows the shift between the two curves, which corroborates our model, and the $\approx 70$ nm breadth of the emission curve, which allows good tunability. The unfortunate partial overlap between the absorption and emission curves will harm the shorter-wavelength performance of the dye. A curve of the triplet absorption is also shown; this can also reduce the power output of the laser. The peak of the absorption curve is near 500 nm, which is ideal for pumping with the very strong 488 nm or 514 nm lines from an argon-ion laser.

The long triplet decay lifetime is a very serious problem for continuous-wave operation of a dye laser, and requires that the liquid dye solution circulate in a closed loop so that the triplets are given an ample time to decay. Although the triplet-to-singlet transitions are radiatively forbidden, non-radiative transitions promoted by the solvent can quickly populate the triplet levels and the dye needs to flow fairly rapidly to prevent this from significantly reducing the laser power. The commonly adopted scheme is to produce a thin *dye stream* (about 0.1 mm thickness) by forcing the dissolved dye through a jet under pressure from a circulating pump (Fig. 7.8). The stream is positioned at the waist between two concave laser cavity mirrors and oriented at Brew-

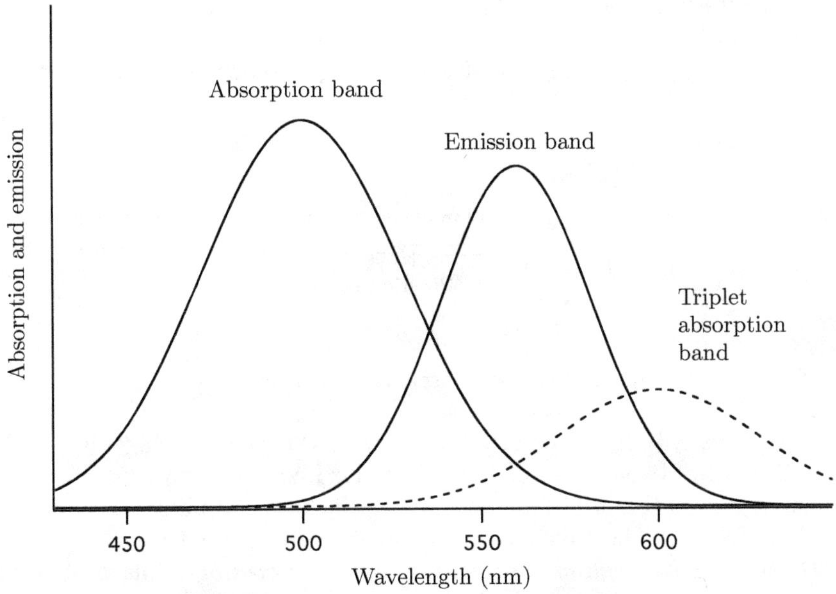


**Fig. 7.7** Absorption and emission curves for a common dye.

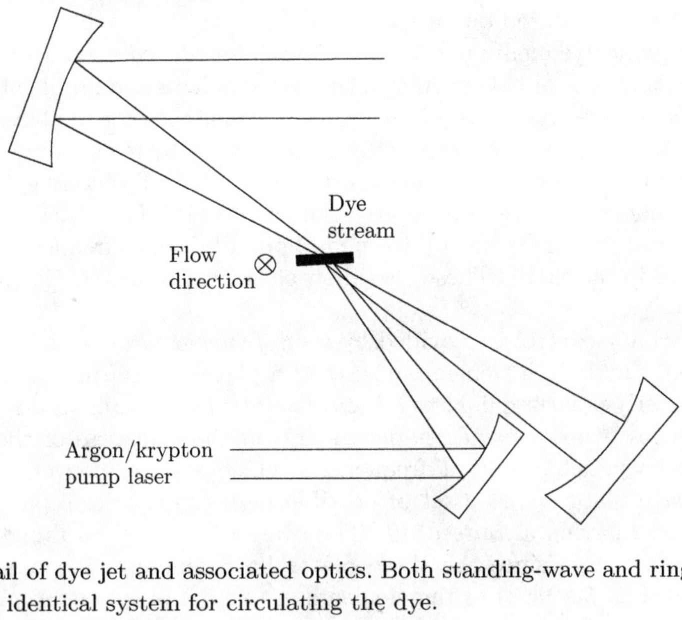


**Fig. 7.8** Detail of dye jet and associated optics. Both standing-wave and ring lasers employ essentially the identical system for circulating the dye.

ster's angle to minimize the Fresnel loss. With a typical dye stream speed of 10 m/s and spot size of 10 $\mu$m, a dye molecule spends only about 1 $\mu$s in the laser cavity; this is usually shorter than the time for significant triplet build-up. A disadvantage of such large dye stream velocities is that a bubble will cause disturbances in $\approx 1$ $\mu$s and therefore generate frequency noise in the MHz region, which requires very fast servo systems to remove.

A complete commercial ring dye laser system is shown in Fig. 7.9. The mirror

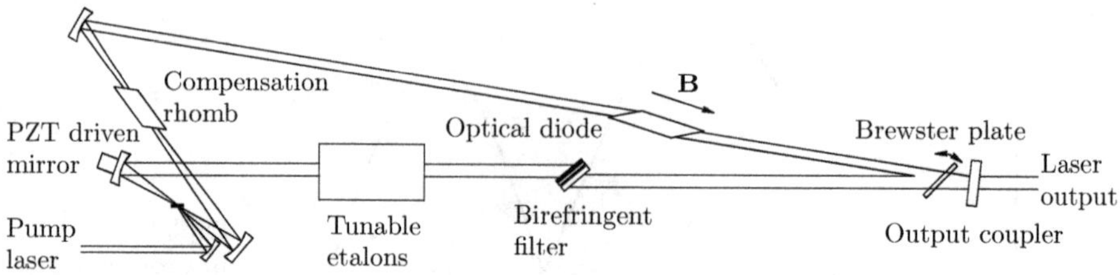


**Fig. 7.9** Schematic of ring dye laser with electronic frequency controls.

configuration is essentially a distorted version of the bow-tie ring discussed earlier with three instead of two curved mirrors. The third curved mirror generates a second small waist which can be used for intracavity frequency doubling (the smallest waist is at the dye stream). The cavity astigmatism is compensated partly by the dye jet and mostly by a *compensation rhomb* (or a *doubling crystal* in place of the rhomb) whose surfaces are at Brewster's angle. The tendency of the laser to operate in both directions is eliminated with an *optical diode*, which uses the *Faraday effect* to rotate the polarization in a manner which makes one direction much more lossy than the other (this will be elaborated on later).

Since the organic dye medium is *homogeneously broadened* and a ring configuration is used (eliminating spatial hole burning), the ring dye laser should be intrinsically single mode. While this is true *at any time*, there is a tendency for the laser to *mode hop* due to mechanical disturbances, especially those due to hydrodynamical fluctuations in the dye stream. This tendency is increased by the spectrally broad gain curve which allows several modes to have very nearly the same gain. To increase the frequency selectivity and reduce this tendency to mode hop, additional frequency-selective elements are placed in the cavity. These also allow electronic control of the laser frequency to some extent.

Coarse frequency control is provided by a *birefringent filter*, which is essentially a three-plate *Lyot filter*. Each element consists of a plate of birefringent material which converts the linearly polarized light to *elliptical polarization* at all but a series of equally spaced frequencies, whose spacing is dependent on the thickness of the plate. Since laser operation is possible only at frequencies which are *not* converted to elliptical polarization, each plate *allows* a set of possible laser frequencies separated by a *free spectral range* and having a *finite width*. The overall free spectral range of the three-plate filter is very large since the three plates have different, incommensurate free spectral ranges. The *fast axes* of the elements are in the plane of the plates, and the frequency of the filter is changed by rotating the plates about an axis normal to their surface.

Fine frequency control is provided by changing the cavity length using two intracavity elements: the piezo-electric (PZT) controlled mirror can provide relatively rapid frequency changes (at a rate of up to about 15 kHz) and a galvanometrically driven plate at Brewster's angle (the *Brewster plate*) can change the path length at a somewhat lower rate by rotating, as shown in the figure. Mode hops are reduced using a pair of *etalons* (simple Fabry–Pérot filters having low finesse) with considerably different widths. The etalon with the smaller width is electronically locked to the laser frequency and acts like a *tracking filter*. The laser can be frequency locked to an external cavity using these electronic frequency controls, and its stabilized linewidth is about 1 MHz. Although a given dye might be tunable only over a range of 50 to 100 nm, a large number of dyes are available covering the entire visible and near-infrared region of the spectrum and generating from several milliwatts to more than a watt of narrowband radiation.

Dye lasers have numerous difficulties. Perhaps their greatest problem is their sheer inefficiency when one includes the pump laser. One needs from 30 kW to 50 kW from the mains (with all of its attendant problems such as the need for water cooling and the increasing cost of electrical power) to generate as little as a few milliwatts of narrowband radiation (this inefficiency becomes even more stark when the visible light is *frequency doubled* into the ultraviolet). The finite lifetime and high replacement cost of the pumping laser tubes are well known to most atomic physicists. The finite lifetime of dyes is another nuisance, together with the need to periodically handle potentially toxic dyes during a dye change. Finally, a complete dye laser and argon pump laser can cost more than $200,000. Frequency-doubled and diode-pumped solid-state lasers for dye pumping are more efficient and have less costly maintenance than argon lasers, though their initial cost is similar. The next laser to be discussed – the titanium–sapphire laser – avoids some of the problems of a liquid dye solution, but real advances in the utility of laboratory laser systems probably depend upon the rapidly increasing sophistication of diode and fiber lasers.

## 7.5 The titanium–sapphire laser

The titanium-sapphire laser is continuously tunable from about 700 nm to 1020 nm using a single crystal, making it the most widely tunable laser available. It is usually pumped with the 532 nm line of a diode-pumped, frequency-doubled CW Nd:$YVO_4$ laser and can deliver more than one watt of narrowband radiation over most of its range. The active medium is a crystal of sapphire (aluminum oxide, $Al_2O_3$) *doped* with about 1% by weight of $TiO_3$. The titanium atoms replace the aluminum atoms at some sites and behave as though they were triply ionized ($Ti^{3+}$).

The level structure of $Ti^{3+}$ appears in Fig. 7.10. The $Ti^{3+}$ ion has a single active 3d electron available for laser gain; the other electrons participate in the bonding with the nearby oxygen atoms. Each $Ti^{3+}$ ion is surrounded by six oxygen ions and the 3d level is split by the crystal field into a number of levels in two groups, labeled $^2E$ and $^2T_2$. The laser transitions take place between these two groups of electronic states. The electronic states are split into a large number of vibrational (phonon) levels, and one can draw potential-energy diagrams as a function of the titanium–oxygen separation; these resemble the curves for an organic dye. The resemblance to the dye is further enhanced by the strong coupling between the electronic states ($^2E$ and $^2T$) and the Ti–O distance (crystal lattice), resulting in a horizontal shift between the lower and upper laser levels. This shift is much larger than in a dye, and the

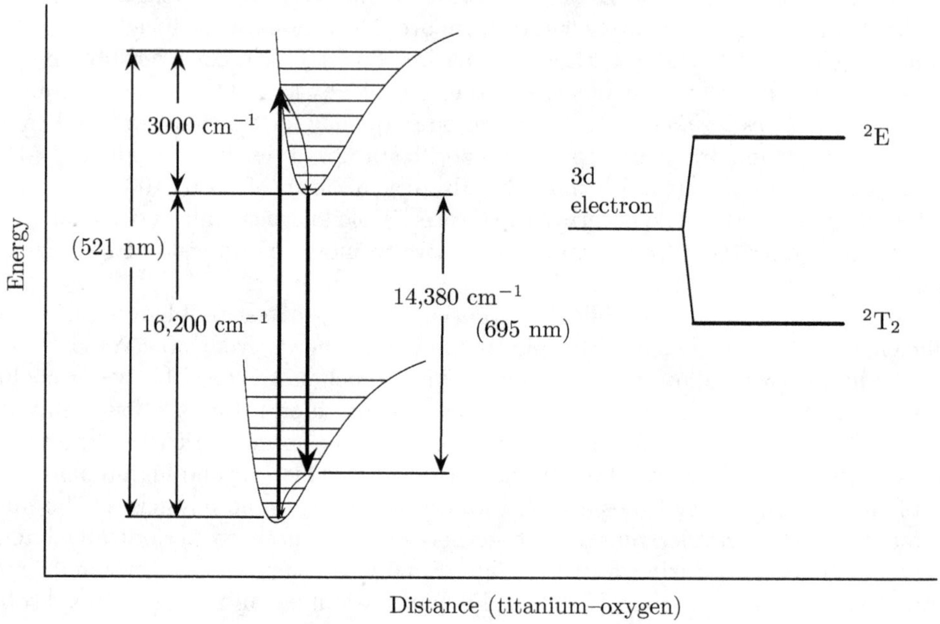


**Fig. 7.10** Level structure of active $Ti^{3+}$ ions in sapphire crystal. The heavy lines on the right represent several levels.

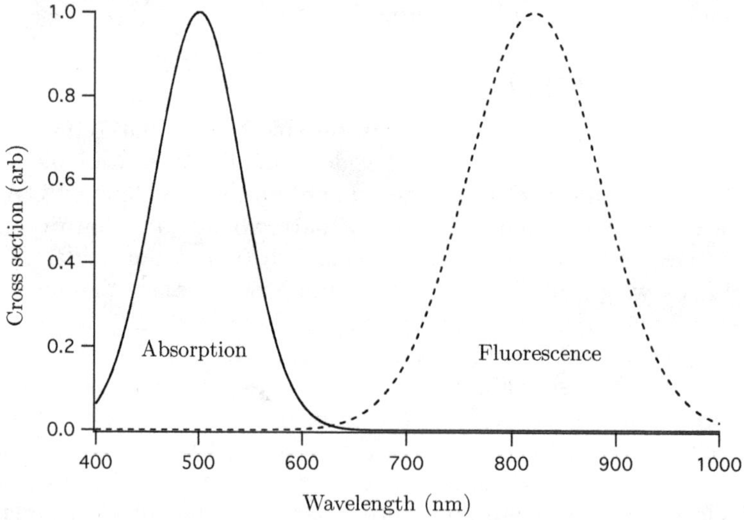


**Fig. 7.11** Absorption and emission in a titanium–sapphire crystal.

absorption and emission curves have very little overlap (Fig. 7.11). The analysis of the laser transitions in a titanium–sapphire system is very much like that in a dye molecule, and the Franck–Condon principle is still valid. The displacement between the ground and excited-state manifolds and the extremely short (picosecond) thermalization times within each manifold allow four-level laser theory to be used. The lifetime of the upper laser level is about 3.9 μs and, unlike in the dye, there are no non-radiative transitions to contend with (there is no analogue to the dye laser *triplet* states in a titanium–sapphire laser). As can be seen in Fig. 7.10, the pumping wavelength should be about 521 nm. Although current practice is to use the 532 nm line from a high-power frequency-doubled Nd:$YVO_4$ laser as a pump, one can also use the 514 nm or 488 nm lines from an argon-ion laser.

A rough schematic of a commercial titanium–sapphire laser appears in Fig. 7.12. The laser is very similar to the ring dye laser discussed earlier. One significant difference

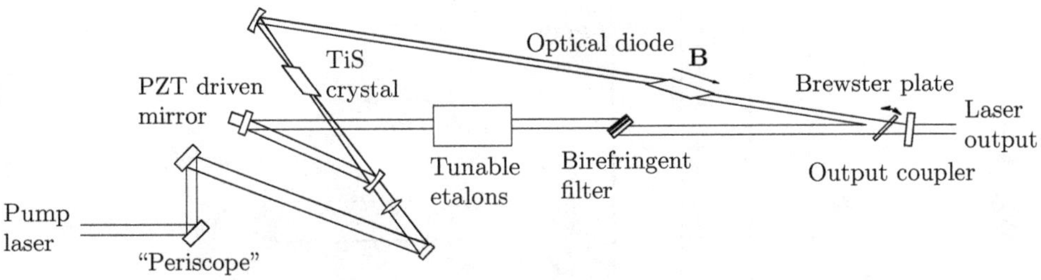


**Fig. 7.12** Commercial realization of a titanium–sapphire laser.

is the presence of the titanium–sapphire crystal at approximately the place of the compensation rhomb in the dye laser. Since the gain of titanium–sapphire is much less than that of most dyes, the titanium–sapphire crystal is much longer than the dye film thickness (≈ 200 times longer) and the crystal is *collinearly* pumped. The pumping optics consist of a *periscope* to steer the pump beam, a lens and convex mirror to focus it and a dichroic cavity mirror (transmissive at the pump wavelength, reflective at the laser wavelength) to allow the beam to impinge on the crystal. The frequency control elements are exactly the same as in the ring-dye laser.

Although titanium–sapphire lasers are very useful in spectroscopy (and their high power allows a practical extension of their range via frequency doubling), their principal current use is in extremely narrow pulse (femtosecond) *mode-locked* systems. This application is facilitated by the very large width of the gain curve. This topic will be discussed in detail in Chapter 11.

## 7.6 The CW neodymium–yttrium–aluminum–garnet (Nd:YAG) laser

A very common continuous-wave laser uses a crystal of *yttrium–aluminum–garnet* ($Y_3Al_5O_{12}$, abbreviated *YAG*) doped with about 1% $Nd^{3+}$, which replaces the Y and is responsible for the laser gain. Since all of the $Nd^{3+}$ ions experience similar fields from the host lattice, the laser transition is only slightly inhomogeneously broadened (unlike the Nd:glass laser, in which the host is amorphous glass and the inhomogeneous broadening is very great). At room temperature, the laser has two principal lines:

1064 nm (the stronger) and 946 nm. A level diagram of $Nd^{3+}$ in a YAG host appears in Fig. 7.13. (The atomic physics of these transitions is discussed in some detail in Chapter 10.)

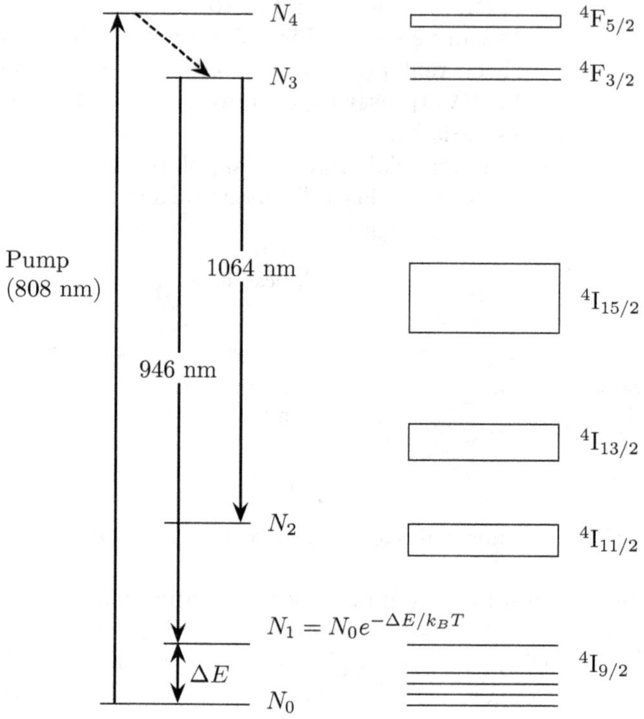


**Fig. 7.13** Level structure of the Nd:YAG system.

The YAG laser is pumped with 808 nm radiation which excites the $Nd^{3+}$ from the ground state to the $^4F_{5/2}$ level, where it decays very quickly to the $^4F_{3/2}$ upper laser level via a non-radiative transition. The laser transitions to the lower-lying $^4I$ levels are forbidden in isolated $Nd^{3+}$ ions by the selection rules for dipole transitions ($\Delta J = 0, \pm 1$), but the perturbations from the crystal fields allow these transitions to take place. The lifetime of the $^4F_{3/2}$ state is about 230 $\mu$s. The lower laser level for the 1064 nm line ($^4I_{11/2}$) has a lifetime of about 30 ns and the thermal population is very small, since the energy is about 10 times $k_B T$. Thus, the laser follows classical four-level theory for this line.

The laser can also emit radiation at 946 nm using the $^4I_{9/2}$ state as the lower laser level. Unfortunately, this level has appreciable population at room temperature and the level structure for this transition is a *quasi-three-level system.* The cross section for this transition is also about ten times smaller than that for the 1064 nm line. For these reasons, the 946 nm line is somewhat weaker than the 1064 nm line.

The CW YAG laser can be pumped with a linear discharge lamp. This lamp is optically coupled to the YAG crystal by placing the lamp and crystal at the two foci of an *elliptical* cavity. To remove the considerable heat generated by the lamp, cooling water is allowed to flow between the lamp and YAG rod (both of which are sealed to avoid contact with the water). The water is usually deionized to reduce its tendency to coat the reflecting surface of the elliptical cavity. These lamp-pumped lasers can generate more than 10 W of CW power at 1064 nm. The YAG laser can also be pumped with the 808 nm radiation from a high-power laser diode, and two such lasers will be discussed in the next two sections.

## 7.7 The YAG non-planar ring oscillator: a novel ring laser geometry

There are two interesting variants of the continuous-wave Nd:YAG laser. One, which will be discussed in the next section, is the diode-pumped high-power frequency-doubled $Nd:YVO_4$ laser, which is increasingly replacing argon-ion lasers as pumping sources, particularly for titanium–sapphire lasers. The other is the so-called *monolithic non-planar ring oscillator*, which is an exceptionally simple and frequency stable source of up to 1 W at 1064 nm (and about 500 mW at 946 nm).

The YAG non-planar ring oscillator (NPRO) consists of a single crystal of Nd:YAG material, cut in such a manner that it supports an *internal* circulating beam which does not lie in a single plane. A drawing of the laser showing the internal path of the beam appears in Fig. 7.14. The beam is confined by total internal reflection on

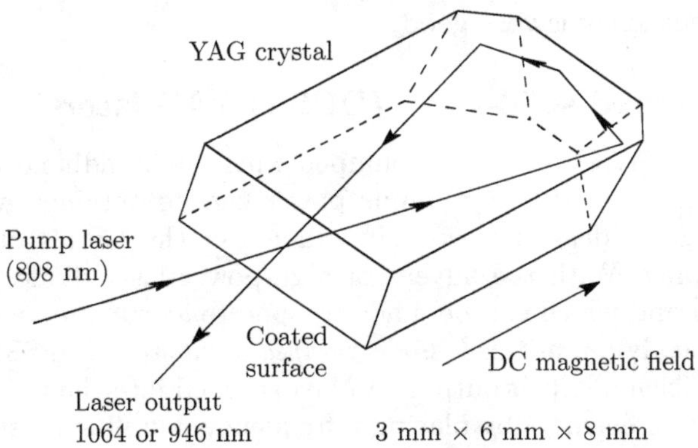


**Fig. 7.14** The Nd:YAG non-planar ring oscillator.

all of the facets except the output face, where the output coupling is determined by the reflection from a surface with a multilayer coating. The pump beam at 808 nm is directed at the point where the laser beam emerges; the pump is quickly absorbed. The direction and beam parameters of the pump beam are adjusted to ensure that only the lowest mode ($TEM_{00}$) is excited.

Unidirectional operation is obtained using the Faraday effect, produced by a magnetic field. The magnetic field is chosen to be largely along the internal propagation direction of the circulating beam. In general, optical diodes based upon Faraday rotation have three elements: a *reciprocal* polarization rotator, a *non-reciprocal* rotator (using the Faraday effect) and a polarizer. The rotations produced by the reciprocal and non-reciprocal elements are approximately equal and add for one beam direction and cancel for the other (desired) direction. With the addition of a polarizer, the two directions have different losses, and oscillation will take place in the direction with lower loss. In the NPRO, the reciprocal rotation is provided by the *non-planar* path inside the YAG crystal and the non-reciprocal rotation is due to the Faraday effect. The multilayer coating on the output coupling facet serves as the (partial) polarizer. The inventors of the YAG NPRO used the Jones calculus to analyze the integral optical diode, and the calculations are quite complicated.

The key advantage of the monolithic NPRO is its extraordinary freedom from technical frequency noise. This is entirely due to the laser resonator being constructed from a single piece of Nd:YAG material, which is intrinsically very rigid. The single-direction ring oscillation eliminates the possibility of spatial hole burning, facilitating single-frequency operation. The *free-running* (i.e., without active frequency stabilization) short-term linewidth of the NPRO was measured by beating two independent lasers together and is about 3 kHz. This is still much greater than the Shawlow–Townes linewidth (less than 1 Hz) but is much less than the free-running linewidth of most single-frequency lasers constructed from discrete components.

The YAG NPRO is also quite efficient, due to its large quantum efficiency and the high efficiency of the high-power laser diodes used for pumping. A commercial realization of the laser produces about 1 W (at 1064 nm) from a pair of pumping sources which together provide 3 W. The pump sources are about 50% efficient, so the overall electrical efficiency is very good.

## 7.8 Diode-pumped solid-state (DPSS) YAG lasers

In earlier times, CW YAG lasers were pumped with the broadband light from a high-pressure discharge lamp. Although ample power can be obtained with this arrangement, there is a great deal of waste, since much of the lamp's output contributes little to the pumping. With the advent of high-power laser diodes, whose electrical efficiency is $\approx$50% and which can be made to operate at 808 nm, laser manufacturers began providing neodymium-based *diode-pumped solid-state* (DPSS) lasers with high power and high efficiency. Their output can be extended into the visible with little loss in output power (or efficiency) by placing a frequency doubling crystal inside the cavity. Neodymium-based frequency-doubled sources are currently available whose output power in the green (532 nm) is greater than 15 W, and these are rapidly supplanting argon-ion lasers for certain laser pumping applications (particularly for pumping titanium–sapphire lasers). These systems draw less than 2 kW from the mains, which is about twenty times less than a comparable argon-ion laser.

A schematic of a commercial frequency-doubled DPSS laser appears in Fig. 7.15. This laser uses $Nd^{3+}$ in an yttrium orthovanadate ($YVO_4$) crystal. The level structure is similar to that of Nd:YAG, with a 1064 nm line and a few other lines at somewhat different wavelengths from the analogous lines in YAG. The Nd:$YVO_4$ medium has a greater gain and greater pump bandwidth but provides similar performance to YAG

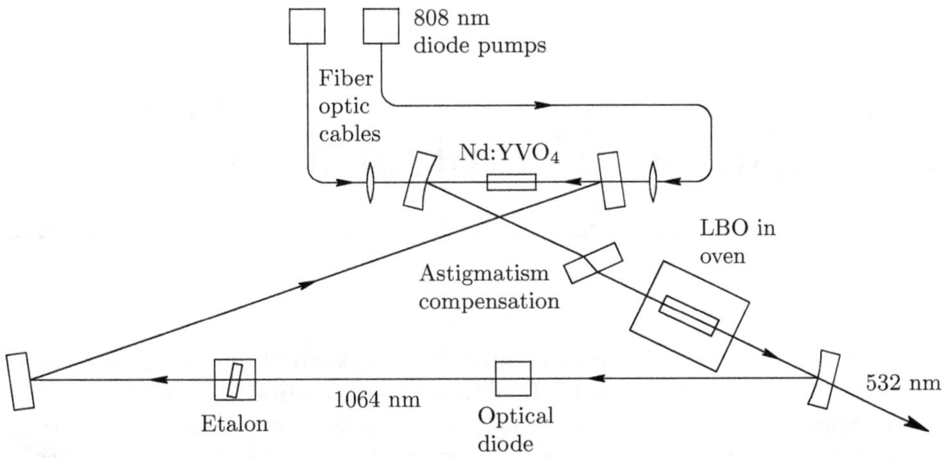


**Fig. 7.15** The Nd:$YVO_4$ diode-pumped solid-state laser with intracavity frequency doubling.

at high powers (it has a lower threshold than Nd:YAG, providing better low-power performance). Its principal advantage over YAG is a larger pump bandwidth, making the laser performance less sensitive to drifts in the pump diode wavelength.

A ring configuration is used to prevent spatial hole burning, and a frequency doubling crystal (lithium triborate, LBO) is placed at a waist between the two curved mirrors. Two high-power diode sources collinearly pump the laser crystal from both ends. The pump lasers are arrays arranged in *diode bars* and the pump radiation is delivered to the laser crystal using fiber-optic cables and focusing lenses. The pump diodes together with their cooling apparatus are located in a remote power supply to reduce the cooling requirements for the laser cavity. All but one of the cavity mirrors are *dichroic*, transmitting at the pump wavelength (such as 808 nm or 532 nm) and reflecting at 1064 nm. A conventional optical diode, consisting of a reciprocal rotator and a Faraday rotator, is used to ensure single-direction operation.

## 7.9 Further reading

Excellent descriptions of actual lasers can be found in Svelto (2004), Verdeyen (1995), Yariv (1989) and Milonni and Erberly (1988). A brief description of the YAG NPRO appears in an article in *Science* by Byer (1988), and a detailed theory (using the Jones calculus) is in the paper by Nilsson, Gustafson and Byer (1989). It goes without saying that a great deal of current information about commercial laser systems is available on the worldwide web, where one can obtain the original patents and occasionally useful sales brochures.

# 8 Laser gain in a semiconductor

## 8.1 Introduction

There is little doubt that the most common laser system is the semiconductor laser diode, which appears in CD and DVD players, laser pointers, supermarket scanners, fiberoptic communications, etc. The analogy between semiconductor lasers and transistors is quite illuminating, and the ultimate replacement of most traditional discharge-pumped lasers (analogous to vacuum tubes) by diode lasers is assured. Almost all of the important laser sources currently used in atomic physics laboratories are either semiconductor diodes or are pumped by diodes.

The theory of semiconductor laser gain will be given here. The most important results of the solid-state physics of semiconductors will be presented first, as a background. Obviously, the topic of laser-diode physics could easily fill a book several times as thick as this one. Following our earlier guideline, just enough theory will be provided to give the reader an appreciation of the several kinds of diode laser likely to be encountered in the laboratory. The next chapter will be devoted to a description of specific semiconductor lasers and their characteristics.

## 8.2 Solid-state physics background

When we derived the Planck formula, we obtained a *distribution function* which gives the number of photons per mode as a function of the photon energy (or frequency). Although this function was derived using a model of discrete, equally spaced energy levels, it could also have been derived entirely from the *behavior of the photon wavefunction under an interchange of identical particles.* Particles with *integral spin*, such as photons, are called *bosons* and have wavefunctions that are *symmetrical upon exchange of identical particles.* One can then use *purely statistical arguments* to derive the *Bose–Einstein distribution function.* Electrons, on the other hand, are *fermions* (particles with half-integral spin), whose wavefunctions are *antisymmetrical under particle exchange.* Using similar statistical methods, one can derive the probability, $f(E)$, of finding an electron with energy $E$. This probability is given by the *Fermi–Dirac distribution function*:

$$f(E) = \frac{1}{e^{(E-E_f)/k_BT} + 1}, \tag{8.1}$$

where $E_f$ is called the *Fermi energy* and is essentially a chemical potential. The functional form of this distribution resembles that of the Bose–Einstein distribution (with a plus sign instead of a minus sign in the denominator), although there is no Fermi energy in the Bose–Einstein distribution, since the *particle number* is not conserved in a photon gas in thermal equilibrium. However, the Fermi–Dirac distribution differs in one very important respect: $f(E) \leq 1$. This is a concise statement of the *Pauli exclusion principle*, which allows *at most one electron in a quantum state at a time.* Plots of $f(E)$ at absolute zero ($T = 0$) and at a small finite temperature appear in Fig. 8.1. From the plots, we see that the Fermi energy is just the maximum energy

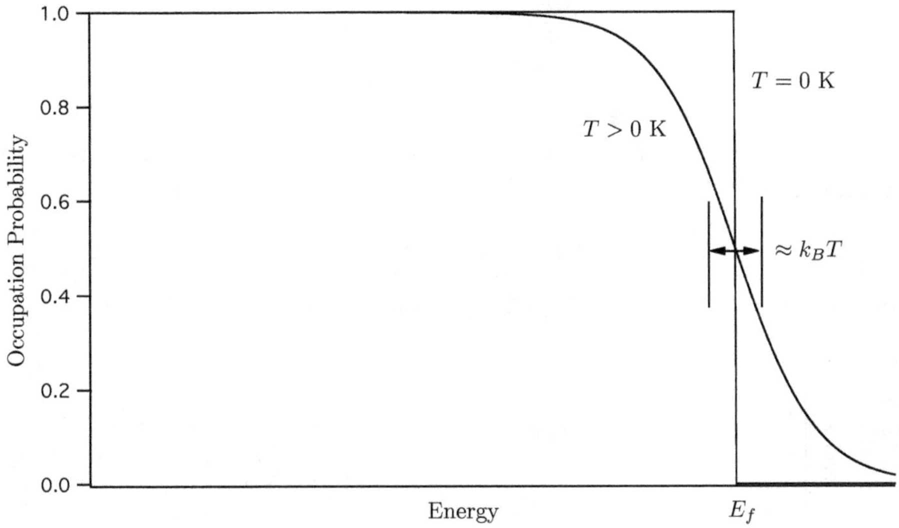


**Fig. 8.1** Fermi–Dirac distribution function.

of the electrons at absolute zero, when all of the possible levels are filled. The Fermi energy can also be defined at non-zero temperatures as *the energy for which the occupation probability is one half.* At non-zero temperature, the hard edge will have a finite breadth which is approximately equal to $k_B T$.

The *distribution of electron energies* is obtained by multiplying $f(E)$ by the *number of electron states per frequency interval*, $\rho(E)$. To derive the density of states, we will proceed in a similar fashion to our approach with photons and model the electron states as plane waves whose allowed values of $k$ are determined by the boundary conditions. A common practice in solid-state physics is to take a cubic sample of the crystal and replicate it infinitely in all six directions. We then use *periodic boundary conditions* and require that the plane wave solutions repeat themselves as one crosses each crystal boundary. If the edge length of the cube is $L$, then the possible values of $\boldsymbol{k}$ are

$$\boldsymbol{k} = \left( \frac{2\pi l}{L}, \frac{2\pi m}{L}, \frac{2\pi n}{L} \right), \qquad l, m, n \text{ integers.} \tag{8.2}$$

Each cell in $k$-space has a volume of $8\pi^3/L^3 = 8\pi^3/V$, where $V$ is the crystal volume. Thus, the number of states in a *shell* of radius $k$ and thickness $dk$ is

$$\text{Number of states} = 2\left(\frac{V}{8\pi^3}\right)(4\pi k^2 dk) = \frac{Vk^2 dk}{\pi^2}, \tag{8.3}$$

where the factor of two is due to the two possible electron *spin states*. Note that we do *not* divide by 8 as we did with photons, since changing the sign of a component of $\boldsymbol{k}$ yields a *different* state when using *periodic* boundary conditions. If we express the number of states as a function of energy, we use the dispersion relation for a *free particle*:

$$E = \frac{(\hbar k)^2}{2m} \tag{8.4}$$

$$\Rightarrow k = \sqrt{\frac{2mE}{\hbar^2}}, \quad dk = \frac{1}{2\hbar}\sqrt{\frac{2m}{E}}dE, \tag{8.5}$$

with the result that the *density of states* (number of states per energy interval per unit volume), $\rho(E)$, is

$$\rho(E) = \frac{1}{2\pi^2}\left(\frac{2m}{\hbar^2}\right)^{3/2}\sqrt{E}. \tag{8.6}$$

The *density of electron energies* is $\rho(E)f(E)$ and is plotted in Fig. 8.2. One can derive

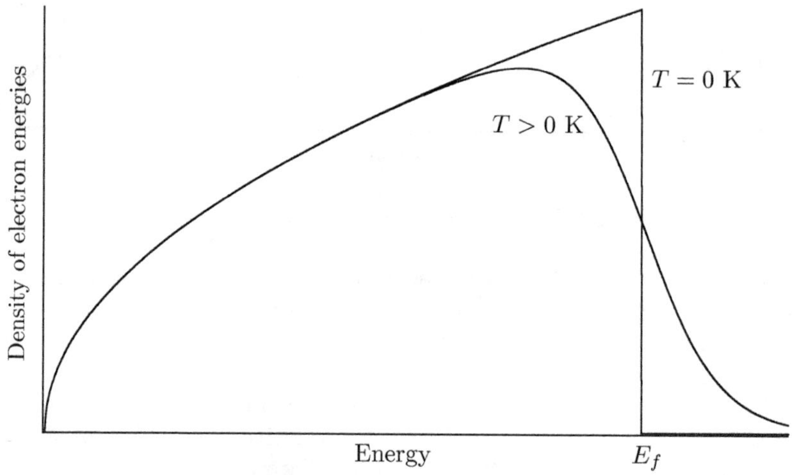


**Fig. 8.2** Density of electron energies as a function of energy.

a number of useful results (such as heat capacities of metals, etc.) solely from this model (called the Sommerfeld model), where the electrons are treated as free particles subject to the Fermi–Dirac distribution. However, all of the phenomena of interest to us will need to include the influence of a *periodic crystal lattice*.

The electronic wavefunction in a crystal is considered to be a plane wave *modulated* by a spatial function, $u_k(\boldsymbol{r})$, whose periodicity is that of the crystal lattice. The wavefunction is called a *Bloch function*:

$$\text{Bloch function:} \quad \psi(\boldsymbol{r}) = u_k(\boldsymbol{r})e^{i\boldsymbol{k}\cdot\boldsymbol{r}}. \tag{8.7}$$

The *Bloch theorem* states that this is a solution of the Schrödinger equation with a potential whose periodicity is the same as that of $u_k(\boldsymbol{r})$. Because of the lattice, the energy spectrum will be a series of discrete *bands* of allowed energies, between which are forbidden regions. We can provide some motivation for this statement by considering a one-dimensional crystal in the delta-function limit of the *Kronig–Penney model*. The time-independent Schrödinger equation is

$$\frac{d^2\psi}{dx^2} + \frac{2m}{\hbar^2}(E - V(x)) = 0, \tag{8.8}$$

where $E$ is the total energy and $V(x)$ is an infinite series of delta functions whose spacing is $a$. The delta functions are derived from a rectangular function of width $b$ and height $V_0$ in the limit of infinite height and infinitesimal width with the *area* kept constant. The *strength* of the potential is proportional to the area and is represented by the parameter, $P$,

$$P = \lim_{\substack{b\to 0\\ \beta\to\infty}} \frac{\beta^2 ab}{2}, \tag{8.9}$$

where

$$\beta = \frac{1}{\hbar}\sqrt{2m(V_0 - E)}. \tag{8.10}$$

One can show that the condition for the existence of a solution to Schrödinger's equation with this periodic potential is

$$P\frac{\sin\alpha a}{\alpha a} + \cos\alpha a = \cos ka, \tag{8.11}$$

where

$$\alpha = \frac{1}{\hbar}\sqrt{2mE}. \tag{8.12}$$

The left-hand side of eqn 8.11 is plotted in Fig. 8.3. Clearly, for this equation to have a *real* solution, the value of the left-hand side evaluated at $\alpha a$ must lie between the two dotted lines in the figure, since the range of $\cos ka$ is between $+1$ and $-1$. The energy eigenvalues are proportional to $\alpha^2$ and the spectrum consists of a series of *bands* of allowed energies with *disallowed* energies in the *gaps* between the bands. One can numerically solve the transcendental equation for $\alpha a$ as a function of $ka$ to display the band structure more clearly. The solution for $P = 3\pi/2$ (a fairly strong potential) appears in Fig. 8.4, where the electron energy ($\propto (\alpha a)^2$) is plotted as a function of $ka$ from 0 to $4\pi$.

When $P = 0$, $\alpha = k$, and we have a free electron, whose energy is

$$\text{Free electron:} \qquad E = \frac{(\hbar k)^2}{2m}. \tag{8.13}$$

We have plotted the free electron energy ($(\alpha a)^2$ in our units) as a dashed line, and one can see the resemblance between the energy of a free electron and that of an electron in a crystal lattice. A quadratic dependence of the energy near the top and bottom of the bands is strongly suggested by the plot. As $P$ becomes larger, the bands shrink until they become points whose values are

$$\text{Very strong potential:} \qquad E = \frac{n^2\hbar^2}{8ma^2}, \qquad n = 1, 2 \ldots \infty. \tag{8.14}$$

These are the eigenvalues for a particle in a box of width $a$ with impenetrable walls.

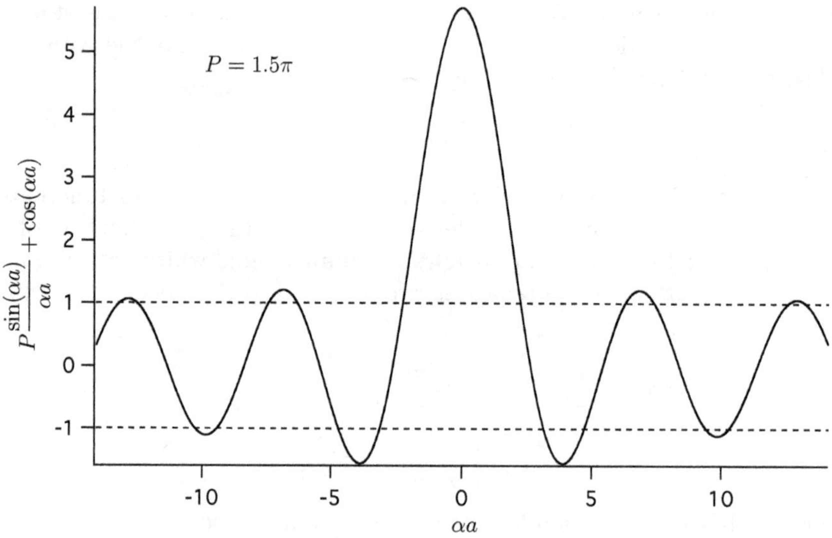


**Fig. 8.3** Condition for the existence of solution to periodic delta function potential problem. Only those values of $\alpha a$ for which the function lies between the dotted lines can yield a valid solution.

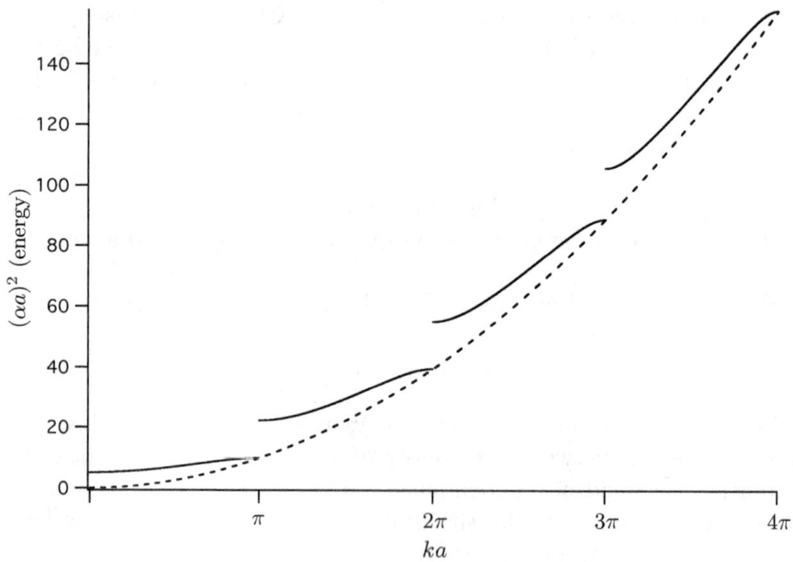


**Fig. 8.4** Energy eigenvalues for a periodic delta function potential with $P = 3\pi/2$. The band structure is clearly visible. A free-particle energy curve (dashed line) is shown for comparison.

The band theory (together with the Pauli exclusion principle) is very useful in explaining some familiar properties of materials, such as the reason that some substances are conductors and others are insulators. If there are $N$ atoms in a sample, it is easy to show that there will be $2N$ electronic states in each band. Thus, if there is one valence electron (as in the alkali metals), the band will be half full and the electrons are free to move and occupy unfilled states. These substances are therefore *conductors.* Materials composed of atoms with two valence electrons will have filled bands, since the number of electrons is equal to the number of states. If the bands are non-overlapping, the electrons will not be able to move without violating the Pauli exclusion principle, and these substances are *insulators.* Some substances with two valence electrons and *overlapping bands*, such as the alkaline earth metals, are also conductors (the two overlapping bands behave like a *single partially filled band*).

We have seen from the Bloch theorem that the electron wavefunctions are *modulated* plane waves. The electrons are actually *wave packets* made up of a large number of states having a range of $k$ values. These wave packets will have a *group velocity*,

$$\text{Group velocity:} \quad v_g = \frac{d\omega}{dk} = \frac{1}{\hbar}\frac{dE}{dk}, \tag{8.15}$$

where $E$ is the electron energy. If we apply an electric field, $E_0$, the work $dE$ done in time $dt$ is

$$dE = eE_0 v_g dt. \tag{8.16}$$

But,

$$dE = \frac{dE}{dk}dk = \hbar v_g dk \Longrightarrow dk = \frac{eE_0}{\hbar}dt \Longrightarrow \frac{dk}{dt} = \frac{eE_0}{\hbar}. \tag{8.17}$$

The acceleration is

$$\frac{dv_g}{dt} = \frac{1}{\hbar}\frac{d^2E}{dk^2}\frac{dk}{dt}. \tag{8.18}$$

Thus, we have

$$\frac{dv_g}{dt} = \frac{d^2E}{dk^2}\frac{eE_0}{\hbar^2}. \tag{8.19}$$

The classical equivalent of this is

$$\text{Classically:} \quad \frac{dv}{dt} = \frac{eE_0}{m}. \tag{8.20}$$

In order to allow one to use the classical equation when an external field is applied, we will endow the electron with an *effective mass*, $m^*$, which is due to the interaction with the lattice. With this, we have in a crystal

$$\frac{dv_g}{dt} = \frac{eE_0}{m^*}, \tag{8.21}$$

$$\text{where} \quad m^* = \hbar^2(d^2E/dk^2)^{-1}. \tag{8.22}$$

A semiconductor is a material which has a very small amount of room-temperature conductivity, even though the lowest band is filled at $T = 0$. This conductivity can be explained with the aid of Fig. 8.5, which displays the *energy curves* for the two lowest

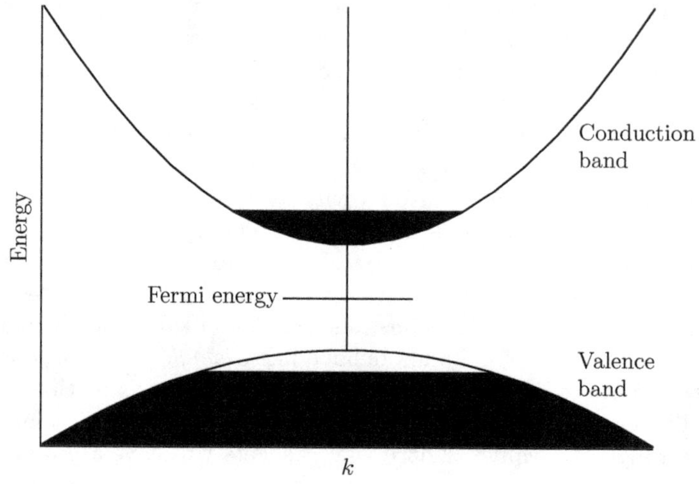


**Fig. 8.5** The energy bands in an intrinsic semiconductor

bands. We assume that the energy depends quadratically on $k$ and that the bottom of a band opens upward while the top of a band opens downward (as in the Kronig–Penney model). An idealized semiconductor which is completely free from impurities is called an *intrinsic semiconductor*. At absolute zero, the lowest band (which we call the *valence band*, since it contains electrons attached to the lattice centers) will be filled and there can be no intrinsic conductivity. At temperatures above absolute zero, there will be some *thermal* excitation of the valence electrons into the next higher band, which is empty at absolute zero. This partially filled band is called the *conduction band*, since it can support electron motion. The *band gap* (minimum energy difference between the bands) is about 1 eV in many semiconductors, though it can vary from 0.1 eV to 2 eV. Thus, the thermal excitation at room temperature is very small (room temperature corresponds to 0.026 eV). Using the band model, the classification of a substance as a conductor, semiconductor or insulator can be viewed as a measure of the *size of the band gap*. An insulator has a large band gap, a semiconductor has a modest-size gap and a conductor has either no band gap or overlapping bands. It should be kept in mind that an intrinsic semiconductor is an idealization, since the conductivity in even the purest materials will still probably be dominated by impurities, particularly at low temperatures.

The flow of current from the electrons in the conduction band is fairly easy to understand, since it is similar to conduction in a metal. What is a little unusual in a semiconductor is that the electrons in the *valence band* can *rearrange themselves* when a field is applied and will also contribute to the current flow. Since the valence band is mostly filled, we can model this contribution to the current as the motion of positively charged *vacancies*, called *holes*, in the valence band. In the following sequence, we show how this is possible by writing down an expression for the current density in the valence band:

$$\boldsymbol{J}_{valence} = \frac{1}{V} \sum_{valence} (-e)\boldsymbol{v}_i = \frac{1}{V} \sum_{all} (-e)\boldsymbol{v}_i - \frac{1}{V} \sum_{empty} (-e)\boldsymbol{v}_i = \frac{1}{V} \sum_{empty} e\boldsymbol{v}_i, \quad (8.23)$$

where $V$ is the volume and the second summation vanishes since no current is possible in a filled band. Thus we see how conduction in the valence band can be considered to be due to *positive* charge carriers (holes). One can show by equating the number of empty states in the valence band to the number of filled states in the conduction band that the Fermi energy in an intrinsic semiconductor lies approximately halfway between the bottom of the conduction band and the top of the valence band.

The number of free electrons and the equal number of holes in an intrinsic semiconductor is very small, since there is little thermal excitation into the conduction band. One can control the type and number of charge carriers by *doping* the semiconductor. Doping is the deliberate addition of a controlled amount of some impurity to a semiconductor in order to control its physical properties (mainly conductivity). The *dopant* usually has a valence which is one greater or one less than that of the semiconductor. In the former case, the dopant yields a *free electron* and the material is called *n-type*, since the *majority carriers* are electrons. When the valence is one less, the material is called *p-type* and the majority carriers are holes. The situation is illustrated in Fig. 8.6 for the case of a silicon semiconductor (valence 4) doped with arsenic (valence 5) or boron (valence 3). It has been experimentally established that the dopant atoms actually *replace* the original atoms rather than filling the interstitial spaces between them. The silicon atoms make *covalent* bonds with their four neighbors, and the substituted

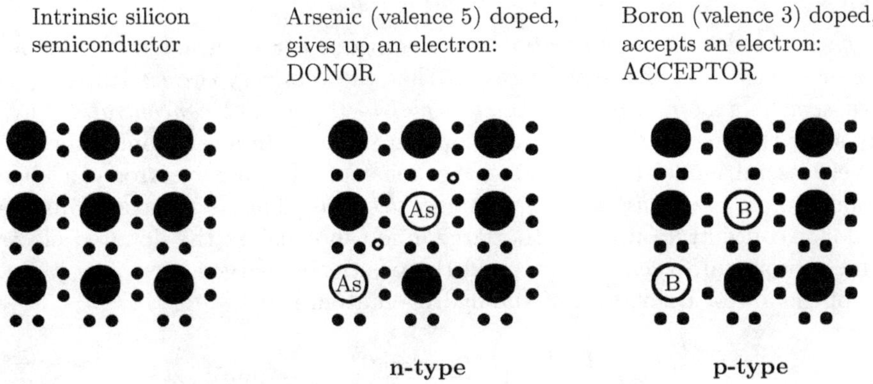


**Fig. 8.6** Doping of silicon by arsenic and boron to produce n-type and p-type semiconductors.

arsenic *donor* atom has one electron left over, which is loosely bound and can easily be *ionized* into the conduction band. In similar fashion, the boron atom has a site which can easily accept an additional electron, forming a hole in the valence band.

One can determine the density of the majority carriers using the kinetic form of the *law of mass action*, which states that the rate of a chemical reaction is proportional to the product of the concentrations of the reactants. In our case, the reactants are electrons and holes, and the two possible *reactions* are the creation of a photon when an electron *annihilates* a hole and the creation of a *hole–electron pair* when a photon is absorbed (the photons are part of the *blackbody radiation* field). If $n$ is the density of electrons and $p$ is the density of holes, the two rates are

$$e + h \longrightarrow \text{photon:} \quad \text{Rate} = B(T)np \tag{8.24}$$

$$\text{photon} \longrightarrow e + h\text{:} \quad \text{Rate} = A(T), \tag{8.25}$$

where $A(T)$ and $B(T)$ are similar to the Einstein coefficients for the two processes and $T$ is included to show the strong temperature dependence of these reactions. At equilibrium, the two rates are *equal*:

$$A(T) = B(T)np \Longrightarrow np = \frac{A(T)}{B(T)}. \tag{8.26}$$

The $A$ and $B$ coefficients are independent of the doping. Thus, for an intrinsic semiconductor, $n = p = n_i$, where $n_i$ is the intrinsic electron (or hole) density, and for a doped semiconductor, we have

$$np = n_i^2. \tag{8.27}$$

This rule breaks down in so-called *degenerate semiconductors*. These are so heavily doped that the Fermi level lies inside the conduction or valence band and the conductivity is therefore similar to that of a metal. The rule also breaks down when the system is not in *thermal equilibrium* (for example, when a current flows). For significant electron doping,

$$n \gg p \Longrightarrow n \approx \rho_n, \tag{8.28}$$

where $\rho_n$ is the *dopant concentration*. An analogous equation holds for p-type semiconductors. Therefore, the general rule is that the *majority carrier concentration in a doped semiconductor is approximately equal to the dopant concentration.* (We will shortly show that most of the dopant atoms are *ionized* in a semiconductor.)

A very useful model for analyzing the behavior of donor electrons in an n-type semiconductor is the *Bohr model for donor electrons*. This is illustrated in Fig. 8.7. If the donor concentration is not too large, one can consider the donated electron to be in an approximate *Bohr orbit* around the positively charged core. The well-known energy of the lowest Bohr orbit of the hydrogen atom is

$$E = -\left(\frac{1}{4\pi\epsilon_0}\right)^2 \frac{m_{electron}e^4}{2\hbar^2} = -13.6 \text{ eV}. \tag{8.29}$$

To account for the electrical polarization of the semiconductor material, we replace $\epsilon_0$ by $\epsilon$, the actual permittivity of the medium. The electron mass $m$ is replaced with the effective mass, $m^*$. These changes can be carried out by multiplying $-13.6$ eV by two factors:

$$E = -\left(\frac{\epsilon_0}{\epsilon}\right)^2 \left(\frac{m^*}{m_{electron}}\right)(13.6 \text{ eV}) = -0.025 \text{ eV}, \tag{8.30}$$

where we used a dielectric constant $(\epsilon/\epsilon_0)$ of 11.8 and an effective mass of $0.26 \times m_{electron}$ for silicon. Thus, room-temperature thermal energy (about 0.026 eV) is sufficient to ionize a large fraction of the donor atoms. Although the arguments are much

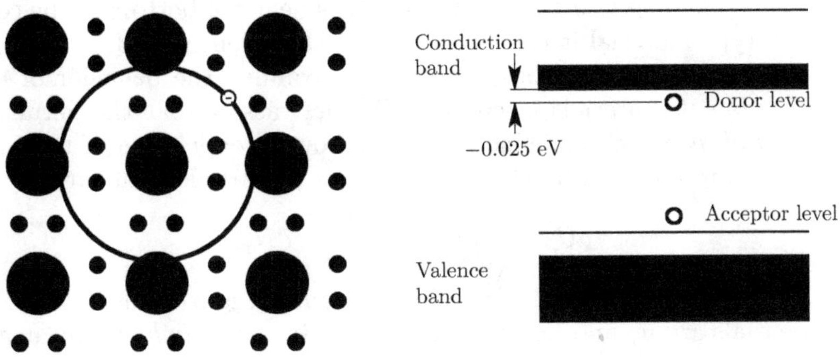


**Fig. 8.7** Illustration of Bohr model for calculating the energy levels of donor and acceptor states.

less transparent, one can analyze the acceptor sites in a similar fashion and obtain a similar result (although the acceptor states will be *slightly above* the top of the valence band).

We saw from eqn 8.22 that the effective mass is *inversely proportional to the curvature of the E versus k (dispersion) function.* To simplify our discussion, we will mostly consider semiconductors with fairly simple energy curves: the minima and maxima of the curves will be at $k = 0$. This is an example of a *direct semiconductor*, whose valence-band maximum and conduction-band minimum occur at the same value of $k$. Our $E$ versus $k$ plots will be qualitatively similar to those in the Kronig–Penney model: the conduction band plot is an upward opening parabola and the valence band plot is a downward opening parabola. One might wonder whether this implies that the effective mass of a hole is negative, since it is inversely proportional to $\partial^2 E/\partial k^2$. The negative curvature of the valence band energy plot is for *electrons* in the valence band. When one considers holes (electron *vacancies*) the mass *changes sign.* Using this model, both holes and electrons behave like free particles, but with a mass which is different from that of a free electron. The energies of the electrons and holes, $E_n$ and $E_p$, are given with respect to the *bottom of the conduction band, $E_c$, and top of the valence band, $E_v$, respectively,* and are

$$\begin{aligned} \text{electrons:} \quad E_n &= E_c + \frac{(\hbar k)^2}{2m_e^*} \\ \text{holes:} \quad E_p &= E_v - \frac{(\hbar k)^2}{2m_h^*}. \end{aligned} \tag{8.31}$$

It was stated (but not proved) that the Fermi energy of an *intrinsic* semiconductor is approximately centered in the forbidden area between the valence and conduction bands. What, then, is the location of the Fermi energy in a doped (*extrinsic*) semiconductor? The result can be obtained by integrating the density of electron energies over all energies and setting the result equal to the number of particles. As the dopant level is increased, the Fermi energy moves toward the band corresponding to the *type* of dopant: the conduction band for donors and the valence band for acceptors. Thus, for moderate to heavy doping in n-type material, $E_f$ is near the bottom of the conduction band, and in p-type material it is near the top of the valence band.

In our analysis of p–n junctions, we need to examine the behavior of the Fermi levels of two different materials in contact. We first assert that the Fermi level is a *chemical potential*, $\mu$, which is defined by the following expression of the first law of thermodynamics for a system in which the number of particles can vary:

$$dU = T\ dS - p\ dV + \mu\ dN. \tag{8.32}$$

If we consider a system composed of two phases containing $n_1$ and $n_2$ particles whose chemical potentials are $\mu_1$ and $\mu_2$, then the change in the *Helmholtz free energy* ($F = U - TS$) is

$$dF = p\ dV - S\ dT + \mu_1\ dn_1 + \mu_2\ dn_2. \tag{8.33}$$

For the two-phase system, the total number of particles is conserved:

$$dn_1 = -dn_2. \tag{8.34}$$

Therefore, at constant volume and temperature, the *equilibrium* condition ($dF = 0$) is

$$dF = \mu_2\ dn_2 - \mu_1\ dn_2 = 0, \tag{8.35}$$

which implies that

$$\mu_1 = \mu_2. \tag{8.36}$$

Thus, if there is no infinite barrier to the flow of electrons between the n and p materials, the Fermi energies will be the same. This analysis is true for *equilibrium situations*; we will introduce a slightly different approach when we discuss optical gain in a semiconductor in the presence of an electrical current.

With these preliminaries, we can discuss our final topic in this background section: the behavior of a p–n junction. Figure 8.8 shows the behavior of pieces of n-doped and p-doped material under four different conditions: before contact, immediately after contact, after contact when thermal equilibrium is established and forward biased. Note that the figures display the levels as a function of *position* ($x$) instead of $k$ and the levels are therefore *flat*. Before contact, the two elements have their own Fermi levels which are close to the edges of the appropriate bands. *Immediately after contact*, electrons will flow from the n material to the p material and holes will flow in the opposite direction, driven by the unequal chemical potentials (Fermi levels) in the two media. Two things will happen when the current flow stops: the Fermi levels will be equal (thermal equilibrium) and there will be *charged regions* on either side of the junction. The charged region is lacking in *free carriers* (since they have *recombined*) and is called the *depletion region*. Two equal and opposite currents will flow at equilibrium: the *diffusion current* due to *majority carriers* diffusing across the junction and the *drift current* due to minority carriers which are driven by the electric field from the separated positively and negatively charged regions. If an *external* emf is applied to the junction, the band structures (including the Fermi levels) of the two elements will *shift relative to each other* by the amount of the emf. A *forward bias* is shown in the figure (positive terminal of the battery attached to p material and

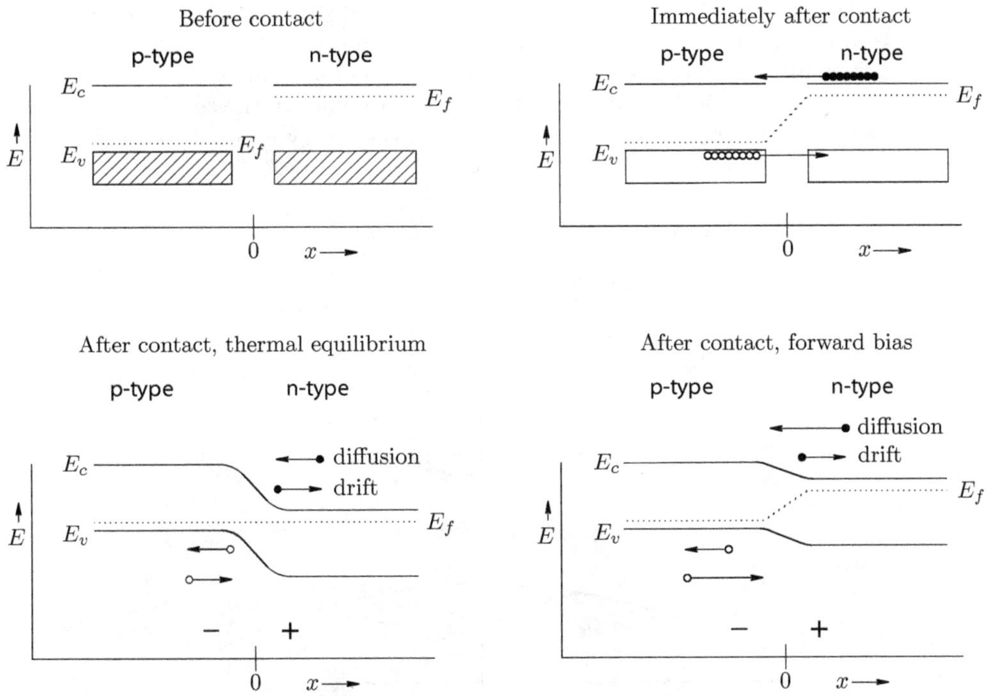


**Fig. 8.8** Behavior of p and n material before contact, after contact and forward biased. Filled circles are *electrons* and empty circles are *holes*.

negative connected to the n material). As can be seen, the forward bias reduces the barrier to the *majority carriers* (diffusion current), allowing the current to increase. The *minority carriers* are not affected, since they experience *no barrier* regardless of the emf and are always swept to the other side of the junction by the electric field. *Rectification* is due to the asymmetry of this process: reverse bias will increase the barrier to the majority carriers, reducing the current. The spatially dependent Fermi level in the composite p–n junction with an applied emf is an example of the breakdown of the utility of a single Fermi level when there is no longer thermal equilibrium.

## 8.3 Optical gain in a semiconductor

Despite the obvious differences between free atoms and bulk semiconductors, the optical processes which can take place in the latter are remarkably similar to those that occur in the former. As is the case in atoms, radiative processes in semiconductors act on *electrons*, promoting them from one level to another. The three processes in a semiconductor are

$$\text{Spontaneous emission:} \quad e + h \longrightarrow \text{photon} \tag{8.37}$$

$$\text{Absorption:} \quad \text{photon} \longrightarrow e + h \tag{8.38}$$

$$\text{Stimulated emission:} \quad \text{photon} + e + h \longrightarrow \text{photon} + \text{photon}. \tag{8.39}$$

In a semiconductor, the levels which are involved in radiative transitions reside in the valence and conduction *bands*. The hole/electron model actually slightly obscures the fact that these processes are simply *interband transitions* due to optical radiation. Thus, spontaneous emission is due to an electron dropping from the conduction band to the valence band with the emission of a photon whose energy is equal to the difference in the electron energies.

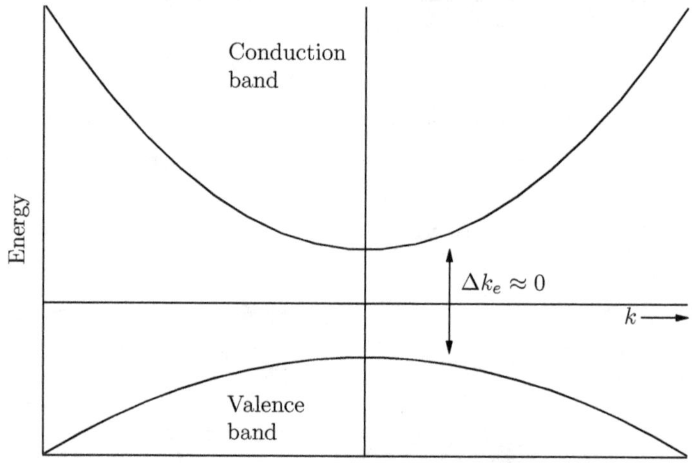


**Fig. 8.9** Energy diagram showing an optical transition in a doped semiconductor.

A diagram of the conduction and valence band energies versus $k$ is shown in Fig. 8.9. Transitions are indicated by the vertical arrow. To analyze the optical behavior, one needs to know the change in the *electron* wave vector, $k_e$, during a transition. This can be determined by the requirement of *momentum conservation* during the transition. In a *direct semiconductor*, the bottom of the conduction band is at the same value of $k$ as the top of the valence band. Since the states at room temperature are very near the band edges, there is no need for a change in the lattice vibration state (phonon change) when a transition takes place. In the absence of phonon changes, the conservation of momentum reduces to:

$$\Delta k_e = k_e' - k_e = k_{photon}, \tag{8.40}$$

where the primed quantity refers to the electron momentum after the transition. At typical diode-laser wavelengths (about 1 $\mu$m),

$$k_{photon} \approx 6 \times 10^4 \text{ cm}^{-1}. \tag{8.41}$$

In the conduction band, $k_e$ is

$$k_e = m_e^* v_e/\hbar \approx (3k_B T_e m_e^*)^{1/2}/\hbar \approx 3 \times 10^6 \text{ cm}^{-1} \tag{8.42}$$

for $T_e = 300$ K. Thus we see that

$$k_{photon} \ll k_e \tag{8.43}$$

and

$$\text{Conservation of momentum} \Longrightarrow \Delta k_e \approx 0. \tag{8.44}$$

The *selection rule* is that $k_e$ does not change during a transition: transitions are vertical in the energy versus $k_e$ diagram. A second selection rule is that the *spin state does not change* during a transition, since the field does not act on the spin. Summarizing the selection rules for a *direct* semiconductor:

$$\begin{aligned} \text{Conservation of energy:} \quad & \Delta E_e = h\nu \\ \text{Conservation of momentum:} \quad & \Delta k_e = 0 \\ \text{Conservation of spin:} \quad & \Delta S = 0, \end{aligned} \tag{8.45}$$

where $\Delta E_e$ is the change in the electron energy.

In order for a semiconductor to be useful in a laser (or light-emitting diode), the following process must occur with a reasonable probability:

$$e + h \longrightarrow h\nu, \tag{8.46}$$

where the photon energy, $h\nu$, must be greater than the band gap, $E_g$. This "reaction" easily conserves energy and momentum in a *direct* semiconductor, as we showed earlier. In an *indirect* semiconductor, the momentum conservation becomes a problem, since a much greater momentum than that carried by the photon must be supplied. This additional momentum can come from phonons (crystal vibration modes) and this process is therefore much less probable than electron–hole annihilation in a direct semiconductor. Thus, direct semiconductors, such as gallium arsenide, are invariably used in semiconductor lasers (and indirect semiconductors, such as silicon or germanium, are rarely used).

From our earlier definition of the Fermi level, a serious problem for laser transitions should be evident. In order for the three optical process discussed previously to occur, there must be holes and electrons *in the same place*, or, equivalently, there must be electrons in the conduction band and vacancies in the valence band at the same point in space. One can see that the existence of *a single* Fermi level prevents this from happening. An n-doped semiconductor will have its Fermi level near the conduction band and will therefore have almost *no vacancies* in the valence band. An analogous problem exists for p-doped material.

This problem is resolved by realizing that the Fermi level is only meaningful under conditions of *thermal equilibrium.* When a current flows, the populations *do not* obey the Fermi–Dirac distribution, since the material is *no longer in thermal equilibrium.* The conditions can undoubtedly reach a *steady state*, but this is not the same thing as being in thermal equilibrium. The solution is to define *two separate Fermi levels: one for the conduction band and one for the valence band.* These are called *quasi-Fermi levels* and are justified by the fact that the thermal equilibration time *within* the bands ($\approx 10^{-12}$ s) is much shorter than the equilibration time *between* the bands ($\approx 10^{-9}$ s). The levels in a direct semiconductor with quasi-Fermi levels ($E_{fc}$ and $E_{fv}$) are shown in Fig. 8.10. The quasi-Fermi levels can be on *either side of the band edges*, depending

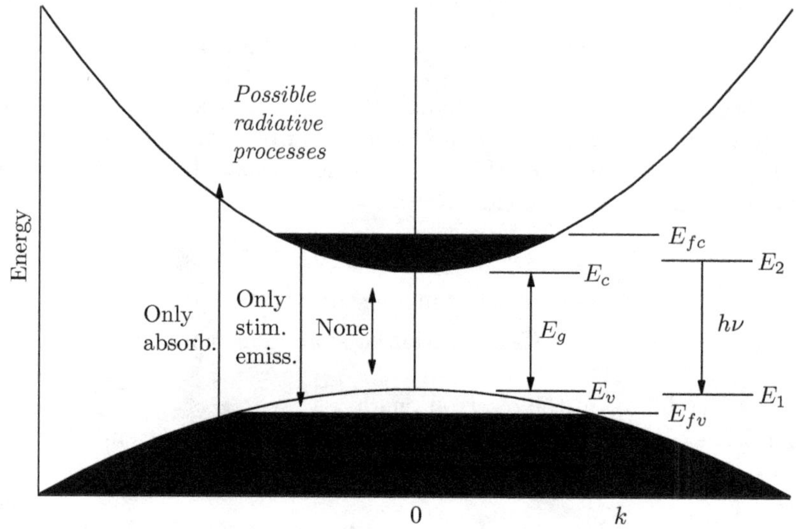


**Fig. 8.10** Band structure of a direct semiconductor showing quasi-Fermi levels. There is sufficient injected current for the $E_{fc}$ to be above the conduction edge and $E_{fv}$ to be below the valence edge in this drawing; this is the condition for the existence of optical gain.

upon the carrier concentration. If $E_2$ is an energy in the conduction band and $E_1$ is an energy in the valence band, the two distribution functions are:

$$f_c(E_2) = \frac{1}{e^{(E_2 - E_{fc})/k_B T} + 1} \tag{8.47}$$

$$f_v(E_1) = \frac{1}{e^{(E_{fv} - E_1)/k_B T} + 1}. \tag{8.48}$$

The *nature* of the optical process will be determined by the comparison between the photon energy and the difference in quasi-Fermi energies. Consulting Fig. 8.10, it should be apparent that when $h\nu < E_g$, no optical processes can take place. When $E_g < h\nu < E_{fc} - E_{fv}$, there is only *stimulated emission*, since absorption would violate the Pauli principle. This is the range in which laser behavior takes place. When $h\nu > E_{fc} - E_{fv}$, only photon absorption can occur. (These conclusions are only strictly valid at absolute zero, but are good approximations to actual behavior at non-zero temperatures.)

We will use the Einstein A and B coefficients to derive the gain for a semiconductor. The two important processes – absorption and stimulated emission – occur at the following rates in the presence of blackbody radiation whose spectral density is $\rho(\nu)$:

$$\text{Absorption:} \quad R_{12} = B_{12} \times \rho(\nu) \times \rho_j(\nu) \times [f_v(E_1)(1 - f_c(E_2)] \tag{8.49}$$

$$\text{Stimulated emission:} \quad R_{21} = B_{21} \times \rho(\nu) \times \rho_j(\nu) \times [f_c(E_2)(1 - f_v(E_1)], \tag{8.50}$$

where $E_1$ is the energy of the lower laser level and $E_2$ is the energy of the upper laser level. The first two factors are familiar from the Einstein phenomenological approach discussed in Chapter 5. The factor in brackets in the absorption expression is just the product of the probability of finding an electron at $E_1$ in the valence band and the probability of finding *an unoccupied state* in the conduction band at $E_2$. The bracketed factor in the stimulated emission expression has an analogous explanation. The factor $\rho_j(\nu)$ is the *joint density of states* for the transition at frequency $\nu$. This is equal to the number of states that begin in one band and end in the other with an energy difference $h\nu$. Only one density of states is used to determine $\rho_j(\nu)$, since each level in one band involves *a single* level in the other band. To calculate $\rho_j(\nu)$, we start with expressions for $k_c$ and $k_v$:

$$k_c = \sqrt{\frac{2m_e^*(E_2 - E_c)}{\hbar^2}} \tag{8.51}$$

$$k_v = \sqrt{\frac{2m_h^*(E_v - E_1)}{\hbar^2}}. \tag{8.52}$$

We equate these two expressions, since the wave vectors are equal according to the momentum selection rule, obtaining

$$E_2 - E_c = \frac{m_h^*}{m_e^*}(E_v - E_1), \tag{8.53}$$

from which we get the differential form:

$$dE_2 = -\frac{m_h^*}{m_e^*}dE_1. \tag{8.54}$$

Consulting the energy-level diagram, we can see that $(E_v - E_1) + (E_2 - E_c) + E_g = h\nu$, from which we obtain

$$\begin{aligned} E_2 - E_c &= \frac{m_h^*}{m_h^* + m_e^*}(h\nu - E_g) \\ E_1 - E_v &= -\frac{m_e^*}{m_h^* + m_e^*}(h\nu - E_g). \end{aligned} \tag{8.55}$$

The value of $k$ is obtained by substituting this expression for $E_2 - E_c$ into eqn 8.51:

$$k = \frac{1}{\hbar}\left\{\frac{2m_e^* m_h^*}{m_e^* + m_h^*}(h\nu - E_g)\right\}^{1/2}. \tag{8.56}$$

We arbitrarily take one of the state densities and evaluate it subject to the requirement that $E = h\nu = E_2 - E_1$ and $dE = dE_2 - dE_1$. The joint density of states is:

$$\rho_j(E)dE = \frac{1}{\pi^2}k^2 dk \Longrightarrow \rho_j(E) = \frac{k^2}{\pi^2}\left(\frac{dk}{dE}\right)_{E=E_2-E_1}. \tag{8.57}$$

The derivative can be evaluated by subtracting the two expressions in eqns 8.31 to obtain

$$dE = dE_2 - dE_1 = \hbar^2 \left( \frac{1}{m_e^*} + \frac{1}{m_h^*} \right) kdk \Longrightarrow \frac{dk}{dE} = \frac{m_r^*}{\hbar^2 k}, \tag{8.58}$$

where $m_r^*$ is the *reduced mass* defined by

$$\frac{1}{m_r^*} = \frac{1}{m_e^*} + \frac{1}{m_h^*}. \tag{8.59}$$

Using eqns 8.56, 8.57 and 8.58, the result for $\rho_j$ is

$$\rho_j(\nu) = \frac{h}{2\pi^2} \left( \frac{2m_r^*}{\hbar^2} \right)^{3/2} (h\nu - E_g)^{1/2}, \tag{8.60}$$

where $\rho_j(\nu) = h\rho_j(h\nu)$.

The remainder of the gain calculation is identical to our approach in Chapter 5 using the Einstein A and B coefficients. The rate of change of the upper laser level population is obtained by subtracting the stimulated emission rate from the absorption rate (compare eqn 5.145):

$$\begin{aligned} \frac{dN_2}{dt} &= \{B_{12}[f_v(E_1)(1 - f_c(E_2)] - B_{21}[f_c(E_2)(1 - f_v(E_1)]\} \rho_j(\nu)\rho(\nu) \\ &= B\rho_j(\nu)\rho(\nu)[f_v(E_1) - f_c(E_2)], \end{aligned} \tag{8.61}$$

where $B_{12} = B_{21} = B$, since the multiplicities are the same ($g_{1,2} = 2$, due to the electron spin). We next convert the radiation density to a monochromatic intensity and use the relationship between the $B$-coefficient and the $A$-coefficient to obtain

$$\frac{dN_2}{dt} = \int_0^\infty Ag(\nu - \nu_0) \frac{\lambda^2}{8\pi n^2} \rho_j(\nu_0)[f_v(E_1) - f_c(E_2)] \frac{I_\nu}{h\nu} d\nu_0 \tag{8.62}$$

$$= A \frac{\lambda^2}{8\pi n^2} \rho_j(\nu)[f_v(E_1) - f_c(E_2)] \frac{I_\nu}{h\nu}, \tag{8.63}$$

where the integrand is obtained from eqn 5.149 (with the addition of the $\rho_j(\nu_0)$ and $f_v(E_1) - f_c(E_2)$ factors) and is the contribution to the rate from a transition described by the lineshape function $g(\nu - \nu_0)$, whose peak is at the frequency $\nu_0$. The total rate for a monochromatic field at the frequency $\nu$ is obtained by summing (integrating) over $\nu_0$. The principal broadening in semiconductors is due to electron–phonon dephasing collisions whose collision time in GaAs is about 0.1 ps, which corresponds to a Lorentzian width of about 1.6 THz. We will shortly see that the statistical factors in the gain have a width which is about ten times larger than this. We therefore consider $g(\nu - \nu_0)$ to be a delta function (a very broad one!) when we evaluate the integral. Finally, we multiply by $h\nu$ to convert $N_2$ to the *energy stored in the upper state* and recognize the factor multiplying $I_\nu$ on the right-hand side as the gain:

$$\gamma(\nu) = A \frac{\lambda^2}{8\pi n^2} \rho_j(\nu)[f_v(E_1) - f_c(E_2)]. \tag{8.64}$$

The last factor is roughly equivalent to the *inversion*, $\Delta N$, in our earlier discussion of laser gain (Chapter 5). We note that, despite the fact that the $f$ terms depend upon

$E_1$ and $E_2$, they can be readily expressed as functions of $h\nu$ and the fixed band and Fermi energies using eqns 8.55.

In order to calculate the gain for a given electron and hole density, one needs to know the location of the quasi-Fermi levels. This requires the solution of two integral equations for $E_{fc}$ and $E_{fv}$:

$$n = \int_0^\infty \rho_c(E_2) f_c(E_2) dE_2 \tag{8.65}$$

$$p = \int_0^\infty \rho_v(E_1) f_v(E_1) dE_1, \tag{8.66}$$

where $n$ and $p$ are the electron and hole densities, while $\rho_c$ and $\rho_v$ are the densities of states for the conduction and valence bands. Both of these equations are of the form

$$F = \int_0^\infty \frac{x^{1/2}}{e^{x-\eta}+1} dx, \tag{8.67}$$

where the unknown is $\eta$. There are a number of approximate solutions to these equations; one fairly simple approach (Suhara (2004) and Joyce et al. (1977)) is to express the difference between the logarithm of the Fermi–Dirac distribution and the logarithm of the Boltzmann distribution as a power series in the carrier density. We first define $N_c$ and $N_v$:

$$N_c = 2\left(2\pi k_B \frac{T}{h^2}\right)^{3/2} (m_e^*)^{3/2} \tag{8.68}$$

$$N_v = 2\left(2\pi k_B \frac{T}{h^2}\right)^{3/2} ((m_{hh}^*)^{3/2} + (m_{lh}^*)^{3/2}), \tag{8.69}$$

where $m_{hh}^*$ and $m_{lh}^*$ are the "heavy hole" and "light hole" effective masses. Then,

$$\begin{aligned} \frac{E_{fc}-E_c}{k_B T} &= \ln\left(\frac{n}{N_c}\right) + \sum_{i=1}^{4} A_i \left(\frac{n}{N_c}\right)^i \\ \frac{E_v - E_{fv}}{k_B T} &= \ln\left(\frac{p}{N_v}\right) + \sum_{i=1}^{4} A_i \left(\frac{p}{N_v}\right)^i, \end{aligned} \tag{8.70}$$

where

$$A_1 = 3.53553 \times 10^{-1} \tag{8.71}$$

$$A_2 = -4.95009 \times 10^{-3} \tag{8.72}$$

$$A_3 = 1.48386 \times 10^{-4} \tag{8.73}$$

$$A_4 = -4.42563 \times 10^{-6}. \tag{8.74}$$

The quasi-Fermi levels for gallium arsenide are plotted in Fig. 8.11 using these formulas.

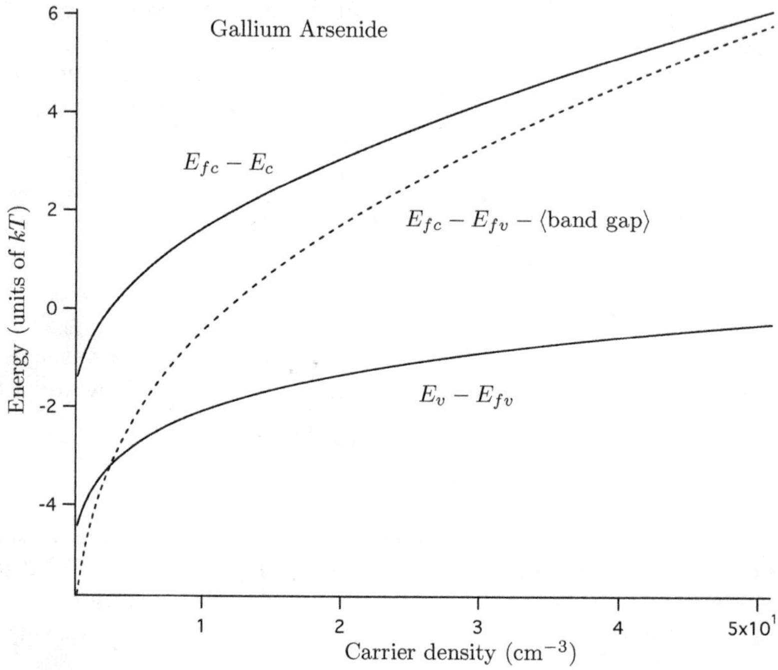


**Fig. 8.11** Quasi-Fermi levels relative to band edges in GaAs. The dashed curve is the quasi-Fermi level difference minus the band gap; the *transparency* condition occurs at the zero of this curve.

We have not discussed actual pumping mechanisms, although we implicitly assumed that, by some means, majority carriers could be injected into the bands so that the condition for gain, $E_{fc} - E_{fv} > E_g$, could be realized. As can be seen in Fig. 8.11, $E_{fc}$ can be *below* the conduction band and $E_{fv}$ can be *above* the valence band for low carrier concentrations. As the carrier concentration increases, $E_{fc}$ increases and $E_{fv}$ is reduced. The *threshold* for gain occurs when $E_{fc} - E_{fv} = E_g$. We call this the *transparency condition*, and the corresponding carrier density is $N_{tr}$. When an injection current is applied, *charge neutrality* requires that

$$n + N_A^- = p + N_D^+, \tag{8.75}$$

where $N_A^-$ and $N_D^+$ are the number densities of *ionized* acceptors and donors. If we assume that most acceptors and donors are ionized, it can be shown that $n, p \gg N_A^-, N_D^+$ at reasonably high injection levels, and one can approximate $n \approx p$. From this, we see from the figure that the transparency density for GaAs is about $1.2 \times 10^{18}$ $cm^{-3}$, which agrees with the actual value.

Although the actual expression for the gain as a function of the carrier density, $N$, is quite involved, we will assume that the relation is approximately *linear* (a good approximation for bulk semiconductors but not for quantum wells). Then, the gain can be written as

$$\gamma = \sigma(N - N_{tr}). \tag{8.76}$$

Thus, $N_{tr}$ and $\sigma$ are two of the several useful parameters which characterize a particular semiconductor for use in lasers or LEDs (the others are the electron–hole recombination time, effective masses and the band gap). The parameter $\sigma$ is sometimes called the *differential gain constant.*

A plot of the gain versus $h\nu$ appears in Fig. 8.12. It is interesting to note that the expression for $\gamma(\nu)$ is also an expression for *absorption* when $h\nu > E_{fc} - E_{fb}$. The parameters for the plot are:

$$\begin{aligned} E_g &= 1.42 \text{ eV} \\ E_{fc} - E_c &= 0.071 \text{ eV} \\ E_v - E_{fv} &= 0.039 \text{ eV} \\ m_e^* &= 0.067 \times m_{electron} \\ m_{hh}^* &= 0.46 \times m_{electron} \\ m_{lh}^* &= 0.08 \times m_{electron}. \end{aligned}$$

The quasi-Fermi levels are taken from eqns 8.70 for an electron and hole density of $1.75 \times 10^{18}$ cm$^{-3}$. Of course, when $h\nu < E_g$ there is neither gain nor absorption.

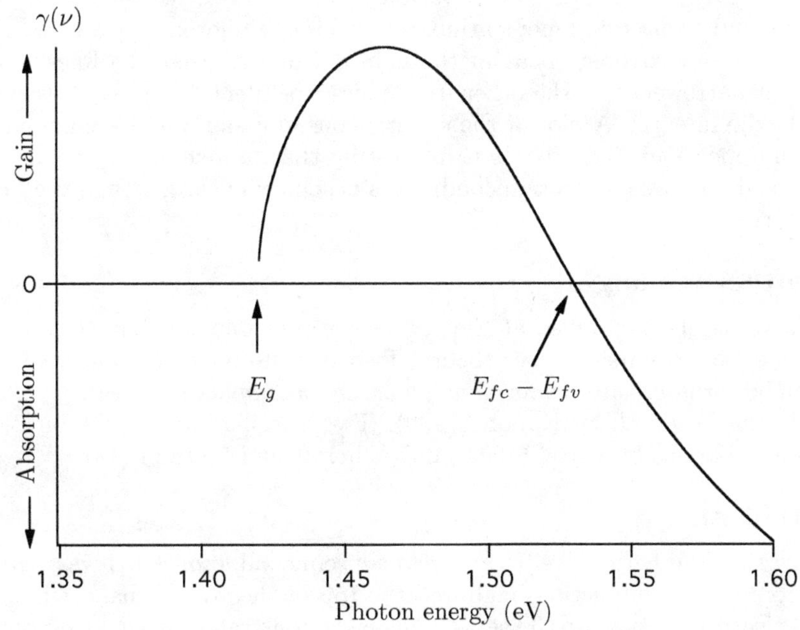


**Fig. 8.12** Gain versus $h\nu$ in GaAs.

Using very similar arguments to those involved in the gain calculation, the *spontaneous decay* rate, $R(\nu)$, can be expressed as

$$R(\nu) = Af_c(E_2)(1 - f_v(E_1))\rho_j(\nu). \tag{8.77}$$

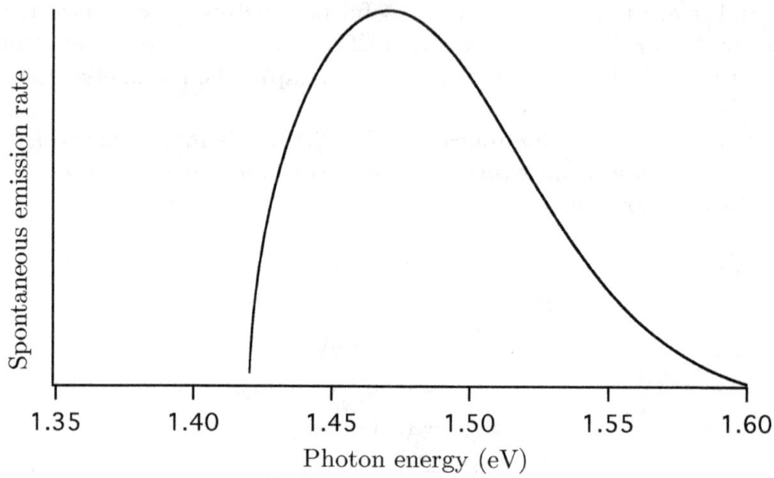


**Fig. 8.13** Spontaneous emission rate versus $h\nu$ in GaAs.

The factors after the $A$ are the probability that there is a conduction electron at $E_2$ times the probability that there is a vacancy at $E_1$ times the number of ways this can happen while conserving momentum and yielding a photon of energy $h\nu$. We have made the same approximation as in the gain calculation that the lineshape factor, $g(\nu)$, is a bit narrower than the other frequency-dependent factors and used this fact to evaluate the integral. A plot of the spontaneous emission for the same parameters as the gain appears in Fig. 8.13. It is interesting that, unlike in the atomic case, the spontaneous decay breadth in semiconductors is somewhat larger than the breadth of the gain curve.

## 8.4 Further reading

The books on lasers mentioned at the end of the previous chapter all have excellent treatments of semiconductor laser theory. Two very useful additional books devoted to semiconductor lasers are *Semiconductor Laser Fundamentals* by Suhara (2004) and *Semiconductor Lasers I* by Kapon (1999). Two excellent textbooks on solid-state physics are the books by Kittel (2004) and Ashcroft and Mermin (1976).

## 8.5 Problems

(8.1) Show that the Fermi level in an *intrinsic* semiconductor is halfway between the bottom of the conduction band and the top of the valence band when the hole and electron masses are the same. This is demonstrated by equating the number of free electrons to the number of holes, which implies that

$$\int_{E_g}^{\infty} \rho_e(E - E_g) f_e(E) dE = \int_{-\infty}^{0} \rho_h(-E) f_h(E) dE, \tag{8.78}$$

where the electron and hole distribution functions satisfy $f_e + f_h = 1$ (since conduction electrons are due to valence *vacancies*), and one can approximate the distribution functions as $f_{e,h} = e^{\pm(E_f - E)/k_B T}$ (since we anticipate that $E - E_f \gg k_B T$). The quantity $E_g$ is the band-gap energy, and it is convenient for the energy zero to be at the top of the valence band.

(8.2) Calculate the gain of GaAs at 855 nm (near the peak of the gain curve) using the parameters given in the text for Fig. 8.12. Assume that $\tau = 3 \times 10^{-9}$ s.

(8.3) Show that the *joint density of states* can be written as

$$\rho_j(h\nu) = \frac{1}{2}\left[\frac{1}{\rho_c(E_2)} - \frac{1}{\rho_v(E_1)}\right]^{-1}, \tag{8.79}$$

where $h\nu = E_2 - E_1$ and $\rho_{c,v}$ are the densities of states in the conduction and valence bands.

(8.4) The determination of the quasi-Fermi levels is difficult at finite temperatures; at absolute zero it can be accomplished by using an integral similar to that found in problem 1 with $f$ equal to a step function. Determine the location of the quasi-Fermi levels in GaAs at $T = 0$, assuming an electron density in the conduction band of $5{\times}10^{18}$ cm$^{-3}$ and an equal hole density in the valence band.

# 9 Semiconductor diode lasers

## 9.1 Introduction

The previous chapter provided the theoretical background for semiconductor lasers; this chapter will describe some specific systems and discuss their characteristics. We will proceed in historical order, starting with the homojunction laser and introducing each new development up to the strained junction quantum well laser. We will describe longitudinal mode control with distributed feedback and Bragg reflector lasers and the more experimentally ubiquitous external cavity diode laser. The properties of all of these lasers will be discussed in some detail. We will end with a description of semiconductor laser amplifiers and injection locked diode lasers. These latter techniques together with the external cavity diode laser provide cost-effective and efficient solid-state competitors for dye and Ti–sapphire (CW) lasers, and should eventually help to render them obsolete.

## 9.2 The homojunction semiconductor laser

The earliest semiconductor lasers were made from a p–n junction in a direct semiconductor (usually gallium arsenide). Since there is only one junction between materials of the same type (but differently doped), these lasers are called *homojunction lasers*. They have serious shortcomings and are no longer in use; we describe them both for historical reasons and also because they have a very simple structure.

**Table 9.1** Parameters for gallium arsenide.

| | GaAs |
|---|---|
| $\lambda$ (nm) | 850 |
| $E_g$ (eV) | 1.424 |
| $m_e/m_{electron}$ | 0.067 |
| $m_{hh}/m_{electron}$ | 0.46 |
| $m_{lh}/m_{electron}$ | 0.08 |
| $N_{tr}$ ($10^{18}$ cm$^{-3}$) | 1.2 |
| $\sigma$ ($10^{-16}$ cm$^2$) | 1.5 |
| $\tau$ (ns) | 3 |

The important parameters for GaAs are tabulated in Table 9.1. Gallium arsenide is a *binary* III–V semiconductor: it is composed of equal amounts of a group III element (gallium) and a group V element (arsenic). The valence bands in GaAs are derived from atomic p-states, which have three possible levels ($l = 1, \quad 2l + 1 = 3$), excluding electron spin. In the crystalline form, there are three possible valence bands, with different curvatures of $E_v(k)$ (i.e., different effective masses). Since the density of states is proportional to $m^{*3/2}$, the most important valence band is the one with the least curvature (and greatest effective mass): the so-called *heavy hole* whose mass ($m_{hh}$) is given in the table.

A schematic of a homojunction laser is shown in Fig. 9.1 and a flat band diagram of the levels (forward biased) appears in Fig. 9.2. The reason for using a p–n junc-

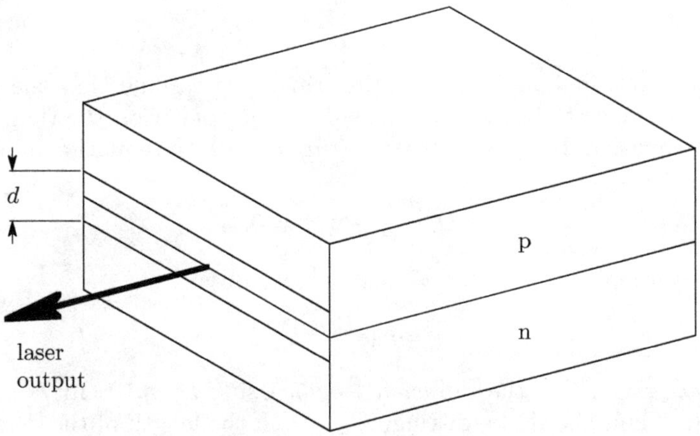


**Fig. 9.1** Schematic of homojunction laser.

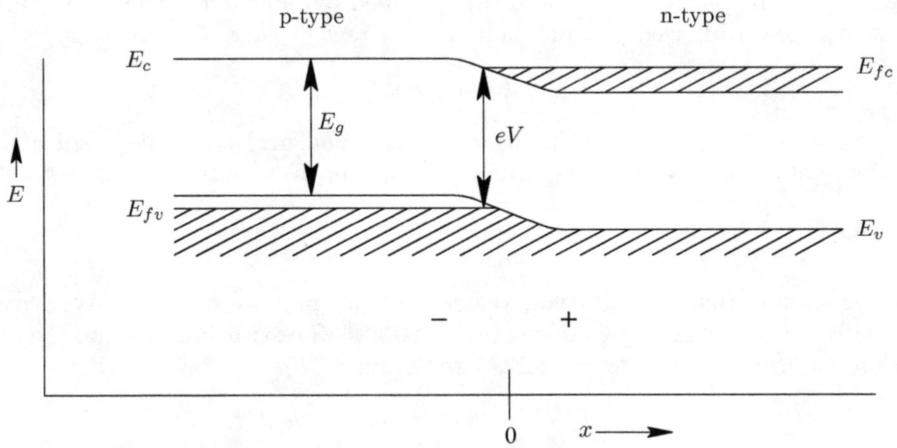


**Fig. 9.2** Energy bands in a forward-biased (with voltage $V$) homojunction laser.

tion is that it is a convenient place for greatly enhanced electron–hole recombination (necessary for laser action) which is confined to a fairly small region. The size of the region can be determined from the *diffusion equation.* The electron diffusion constant, $D_e$, is defined by:

$$n\boldsymbol{v}_e = -D_e \nabla n, \tag{9.1}$$

where $n$ is the electron density and $\boldsymbol{v}_e$ is its diffusion velocity. From the *continuity equation,*

$$\nabla \cdot (n\boldsymbol{v}_e) + \frac{\partial n}{\partial t} = -\langle \text{recombination rate per unit volume} \rangle = -n/\tau, \tag{9.2}$$

one obtains the diffusion equation:

$$-D_e \nabla^2 n + \frac{\partial n}{\partial t} = -n/\tau, \tag{9.3}$$

where $\tau$ is the electron–hole recombination time. We assume that the electrons are *generated elsewhere,* so the generation rate does not appear on the right-hand side of the continuity equation. In the steady state ($\partial n/\partial t = 0$), the equation in one dimension is

$$D_e \frac{d^2 n}{dx^2} - n/\tau = 0. \tag{9.4}$$

Using a trial solution of the form $n = n_0 e^{-x/L}$, we obtain

$$n = n_0 e^{-x/\sqrt{D_e \tau}}, \tag{9.5}$$

where $L = \sqrt{D_e \tau}$ is called the *diffusion length.* Using $D_e \approx 10$ cm$^2$/s and $\tau = 3$ ns, $L = d \approx 1$ $\mu$m. (This should be distinguished from the length of the *depletion region,* which is calculated in a completely different way and is about ten times smaller in GaAs.) The size of the active region is therefore about 1 $\mu$m in a GaAs homojunction laser.

Semiconductor lasers are almost always pumped by *injection current*: the electron current in a forward biased p–n junction. The current density, $\boldsymbol{J}$, is given by

$$\boldsymbol{J} = n_0 e \boldsymbol{v}_e. \tag{9.6}$$

Assuming the current is flowing in the $x$-direction and neglecting the small $y$ and $z$ dependence of $\boldsymbol{J}$, the continuity equation (eqn 9.2) in the steady state ($\partial n/\partial t = 0$) is

$$v_e \frac{dn}{dx} = -n/\tau, \tag{9.7}$$

where we assume that the diffusion velocity, $v_e$, is spatially constant. We have already derived the spatial dependence of $n = n_0 e^{-x/d}$. Substituting this into the above equation and using the expression for $\boldsymbol{J}$, we obtain

$$v_e n_0 = J_x/e = n_0 d/\tau \Longrightarrow J_x = e n_0 d/\tau. \tag{9.8}$$

One can estimate the current density needed for laser operation by starting with the *transparency density,* which is $1.2 \times 10^{18}$ cm$^{-3}$ in GaAs. This electron density is generated by the injected current. Using $\tau = 3$ ns and $d = 1$ $\mu$m, the needed current density is

$$J_x = \frac{n_e e d}{\tau} \approx 6.4 \times 10^3 \text{ A/cm}^2. \tag{9.9}$$

This is the minimum current density for *laser gain*; the *threshold* current density will be a bit larger. We see that a fairly significant current density is needed in a homojunction laser and that to reduce this current density we need to reduce $d$.

A second problem with the homojunction design is due to the cavity mode in the p–n junction. As can be seen from the drawing (Fig. 9.1), the end facets of the laser material are used as the cavity mirrors. One can coat the facets to increase the reflectivity, but the high gain in the semiconductor medium allows the laser to work adequately with uncoated facets. The reflectivity of the facets is given by the Fresnel formula for normal incidence:

$$\text{Power reflectivity} = \left(\frac{n-1}{n+1}\right)^2 = 0.32, \tag{9.10}$$

where the index of refraction, $n$, is about 3.6 for GaAs. The size of the cavity mode can be estimated from our work with standing-wave cavities. The waist radius in a *confocal* cavity of length $L$ is $\sqrt{\lambda L/2\pi n}$ which is about 4.7 $\mu$m in a 500 $\mu$m long cavity when $\lambda = 1$ $\mu$m. The actual waist size will be larger since the facets are closer to being flat than concave. The mode size is much larger than the *active region* ($d$) and the field outside the gain region will be attenuated. Some method for confining the cavity mode is needed to reduce the physical extent of the field.

As a result of these two deficiencies, a very substantial current is needed to obtain useful radiation from a homojunction laser, and these lasers are normally operated either in pulse mode at room temperature or CW at 77 K (liquid nitrogen temperature) due to the heating from this large current.

## 9.3 The double heterostructure laser

It should be clear from the discussion of the homojunction laser that the threshold current would be greatly decreased if one could reduce the size of both the active region and the cavity mode. The first successful attempt to reduce these quantities employed the *heterostructure* configuration in which there are one or more junctions involving materials with *different compositions* and dopings (unlike the GaAs homojunction or silicon or germanium p–n junctions which consist of a *single* material with different dopings). Semiconductor lasers constructed in this way always have *two junctions* and are called *double heterostructure lasers*. The importance of this technology was acknowledged by the 2000 Nobel prize in physics, which was awarded to three scientists for "developing semiconductor heterostructures".

When we discussed the homojunction laser, we introduced the material gallium arsenide, which differs from the previously discussed *single-element* materials (silicon and germanium) in being a *compound*. Certain families of crystalline compounds have a much greater probability of being *direct-bandgap* semiconductors, in which radiative interactions occur with high probability. In addition to *binary* compounds, such as GaAs, direct semiconductors suitable for laser construction include *ternary* compounds, such as $Al_xGa_{1-x}As$, and *quaternary* compounds, such as $In_{1-x}Ga_xAs_yP_{1-y}$, where $x$ and $y$ are the *molar fractions* of the constituents. From the appearance of the subscripts, it should be clear that the ternary and quaternary compounds have similar structures to GaAs, since the total number of atoms from group III and group V are independent of the values of $x$ or $y$. For example, one can consider $Al_xGa_{1-x}As$ to be GaAs with $x$ fraction of Ga atoms replaced with Al atoms. Changing $x$ and $y$ changes the band gap, the index of refraction and the *physical size* of the crystal cells. As $x$ is increased in $Al_xGa_{1-x}As$, the band gap increases and the refractive index decreases.

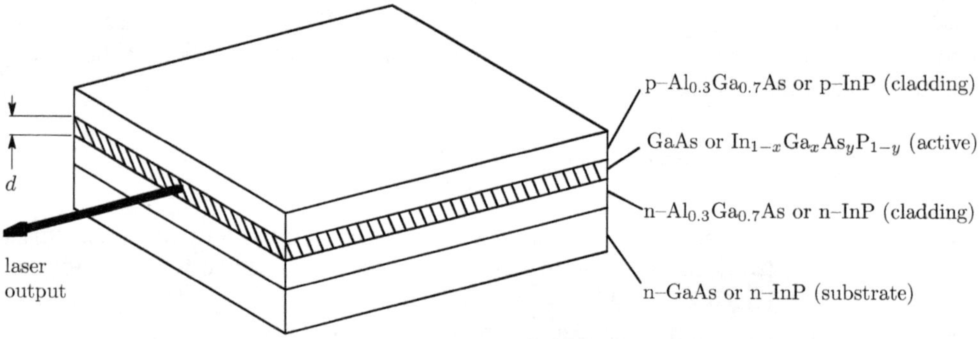


**Fig. 9.3** Schematic of a double heterostructure laser.

The only constraint on the materials on either side of the junctions is that they have essentially the same *lattice constants* (dimensions of lattice) – a condition called *lattice matching*. Most semiconductor lasers are composed of atoms from group III and group V (so-called III–V compounds) of the periodic table and all of these compounds form crystals with *cubic* geometry. If the lattice spacings of the materials differ by more that 0.1%, there can be *strains* which will often have adverse effects on the interface (though strains are deliberately used in some *quantum-well* lasers). The use of ternary and quaternary compounds increases the number of degrees of freedom so that lattice matching can be obtained without adversely affecting other desirable properties of the materials. A diagram of a common double heterojunction laser using two possible sets of compounds appears in Fig. 9.3.

The use of the double heterostructure confines both the carriers (electrons and holes) and the cavity mode to the *active region*, of width $d$. Since $d$ can be ten times smaller than its value in the homojunction laser, the threshold is reduced by more than an order of magnitude. An aid to understanding why this can occur appears in Fig. 9.4, where a band diagram of a forward-biased double heterostructure laser is shown together with a plot of the refractive index as a function of position.

As can be seen from the band diagram, with a sufficient forward bias there is a *well* in the active GaAs region which is energetically favorable for the confinement of both electrons and holes (until they recombine). The electrons are injected from the left and the holes from the right, and both carriers are exposed to forces (from the band energy gradients) which confine them in the active region where laser action can take

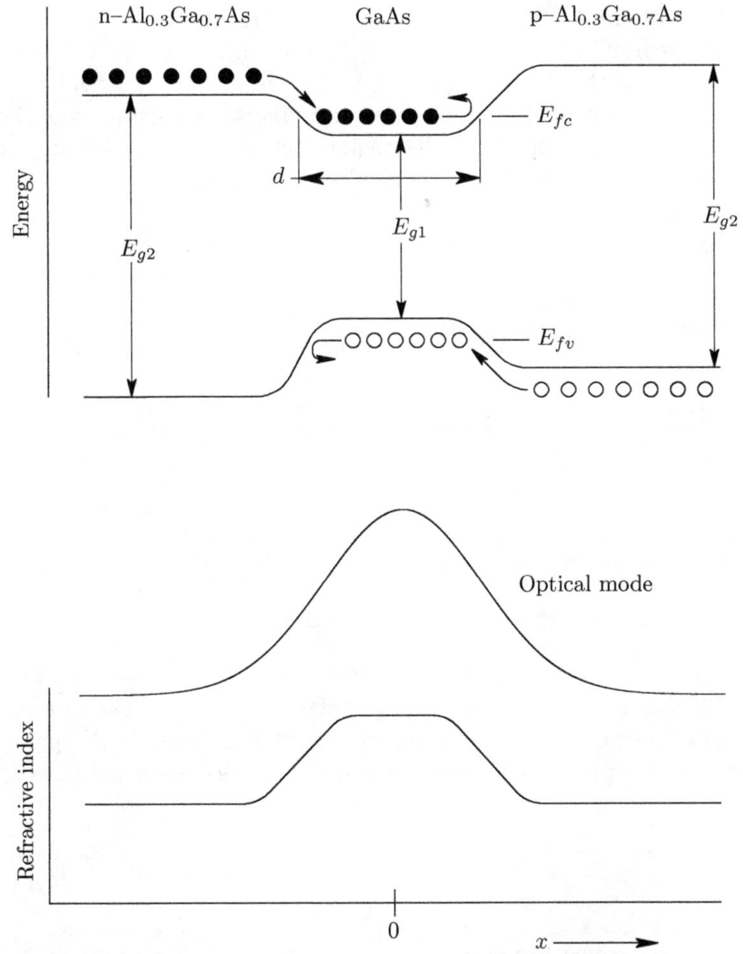


**Fig. 9.4** Plots of band energies and refractive indices in double heterostructure.

place. Thus, unlike the case of the homojunction laser where we relied on uncontrolled diffusion and recombination to define the active region, we can now "engineer" the active region to be in a much smaller layer of material whose size is under our control (within limits).

Photon confinement is due to the *lower* index of refraction on either side of the active region. The heterostructure forms an *optical waveguide* which confines the field, much like an optical fiber (see Chapter 10 for an analytical treatment of these waveguides). The lower index material is called the *cladding*. The *tail* of the field distribution, which propagates in the cladding, has very little absorption since the photon energy (slightly above $E_{g1}$) likely satisfies the condition for transparency in the cladding ($h\nu < E_{g2}$). The simultaneous occurrence of both carrier and field confinement is due to the fortuitous *inverse relationship* between the band gap and refractive index in

III–V crystals: as the band gap increases, the refractive index decreases. Thus, the smaller band gap of the GaAs active region, which results in carrier confinement, is also a region of greater index of refraction, which results in field confinement.

If we assume that the active region is well-defined and the field distribution is known (from optical waveguide theory), we can make some estimates of the threshold gain and injection current density. The region between $x = -d/2$ and $x = d/2$ is characterized by a threshold gain, $\gamma_t$, and absorption, $\alpha_a$, and the regions in the n-type and p-type material are described by absorptions $\alpha_n$ and $\alpha_p$. The gain and losses are weighted by the intensity distributions using the $\Gamma$ quantities:

$$\Gamma_a = \frac{\int_{-d/2}^{d/2} |E|^2 dx}{\int_{-\infty}^{\infty} |E|^2 dx} \tag{9.11}$$

$$\Gamma_n = \frac{\int_{-\infty}^{-d/2} |E|^2 dx}{\int_{-\infty}^{\infty} |E|^2 dx} \tag{9.12}$$

$$\Gamma_p = \frac{\int_{d/2}^{\infty} |E|^2 dx}{\int_{-\infty}^{\infty} |E|^2 dx}, \tag{9.13}$$

where

$$\Gamma_a + \Gamma_n + \Gamma_p = 1, \tag{9.14}$$

and the subscripts $a, n$ and $p$ refer to the active region, the $n$-region and the $p$-region, respectively. These parameters allow one to take into account the imperfect *overlap* between the internal field and the active region of the laser. If the gain and losses are fairly uniform within each of these three regions, the weighted gain and loss are approximately

$$\text{gain} = \gamma \Gamma_a \tag{9.15}$$

$$\text{loss} = \alpha_a \Gamma_a + \alpha_n \Gamma_n + \alpha_p \Gamma_p. \tag{9.16}$$

In order to calculate the threshold gain, we need the *laser oscillation condition* and the *semiconductor gain expression*, which we will repeat here (at threshold):

$$\text{Laser oscillation condition:} \quad \gamma_t = \alpha - \frac{2}{l} \ln(r_1 r_2) \tag{9.17}$$

$$\text{Semiconductor gain:} \quad \gamma_t = \sigma(N_t - N_{tr}). \tag{9.18}$$

From the gain formula,

$$N_t = N_{tr} + \frac{\gamma_t}{\Gamma_a \sigma}, \tag{9.19}$$

where we multiplied the right-hand side of the gain expression by the *weight function*, $\Gamma_a$. The threshold gain is obtained from the laser oscillation condition (replacing $\alpha$ by the total weighted loss):

$$\gamma_t = -\frac{2}{l} \ln(r_1 r_2) + \alpha_a \Gamma + \alpha_n (1 - \Gamma), \tag{9.20}$$

where we let $\alpha_n = \alpha_p$ and $\Gamma_n = \Gamma_p$ and dropped the subscript from $\Gamma_a = \Gamma$ (the quantity $\Gamma$ is called the *beam confinement factor*). Finally, we can use $J = eNd/\tau$ to obtain the threshold current density:

$$J_t = \left(\frac{ed}{\tau}\right)\left[N_{tr} + \frac{\gamma_t}{\Gamma\sigma}\right]. \tag{9.21}$$

We should note that for very small values of $d$, the confinement factor $\Gamma$ is proportional to $d^2$ (this is derived in Chapter 10). Thus, the threshold current is proportional to $d$ until the active region becomes very thin, at which point the second term in brackets dominates, and the threshold current will *increase* with further reductions in $d$. When the two bracketed terms are equal, half of the injected current is used merely to establish quasi-Fermi levels appropriate for laser gain. The minimum threshold current in GaAs occurs when $d \approx 0.1$ $\mu$m.

We can estimate the threshold current density using the following parameters: $\Gamma \approx 1$, $r_1^2 = 0.32$ (GaAs), $r_2 = 1$ and $l = 500$ $\mu$m. The mirror loss is 23 cm$^{-1}$, and if we let $\alpha_a = 10$ cm$^{-1}$ the threshold gain is $\gamma_t = 33$ cm$^{-1}$. The values of $\tau$, $N_{tr}$ and $\sigma$ are taken from Table 9.1. Assuming $d = 0.1$ $\mu$m, the threshold current density is about 800 A/cm$^2$, which is about a factor of 10 less than the threshold in a homojunction laser.

For the laser shown in Fig. 9.3, the confinement of the field along the plane of the junction is not nearly as good as the confinement perpendicular to the plane. This laser will therefore not, in general, operate in the lowest *transverse* mode. To improve the transverse-mode behavior, the *stripe geometry* shown in Fig. 9.5 is used. Two variants are in the figure: a *gain-guided* laser and an *index-guided* laser. The

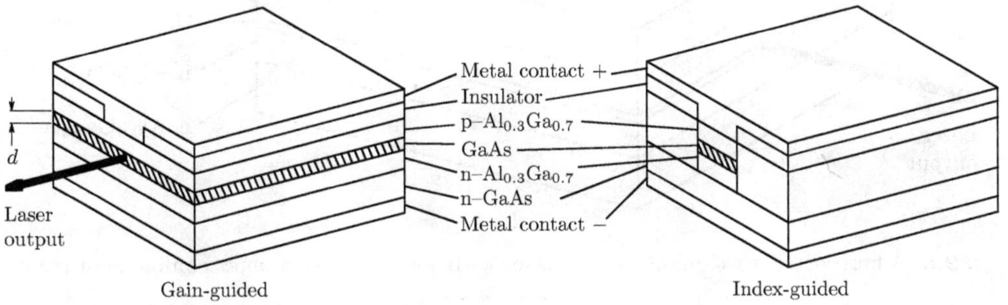


**Fig. 9.5** A *gain-guided* (left) and *index-guided* (right) double heterostructure stripe laser.

gain-guided configuration uses a very narrow electrode to inject the carriers into a narrow region of the active area of the laser, thereby ensuring that laser gain only occurs there. The index-guided approach uses in addition a very narrow stripe of active GaAs, surrounded on all sides by media with a lower index of refraction. The field is confined in a two-dimensional optical waveguide formed by the active medium and the lower index cladding. This is occasionally called a *buried heterostructure laser* and is much better at confining the field than the gain-guided approach.

## 9.4 Quantum-well lasers

In the pursuit of lower thresholds by reducing the thickness of the active layer, we saw that the confinement factor suffers when $d$ is reduced to less than about 0.1 $\mu$m. However, there are significant advantages to be obtained when $d$ is on the order of the de Broglie wavelength of the carriers (about 100 Å for electrons). Quantum effects then become prominent and useful changes occur to the density of states and other parameters. Lasers with these very small active regions are called *quantum well lasers* and are the dominant type of semiconductor laser in use today (2009), due to their extremely low thresholds and high efficiency. The very thin active region provides little or no field confinement and these lasers therefore usually employ a separate field confinement structure which is similar to the one used in a double heterostructure laser. The size of the optical confinement region is on the order of the *optical wavelength*, and the size of the active gain region is on the order of the *electron de Broglie wavelength.* Such lasers are called *separate confinement heterostructure quantum-well lasers.*

An AlGaAs-GaAs double heterostructure single quantum-well laser with a separated confinement region is shown in Fig. 9.6 (with apologies to the reader for the jargon, which can get fairly dense at times; we will not burden the reader with the numerous and confusing acronyms which proliferate in this field). The active region is

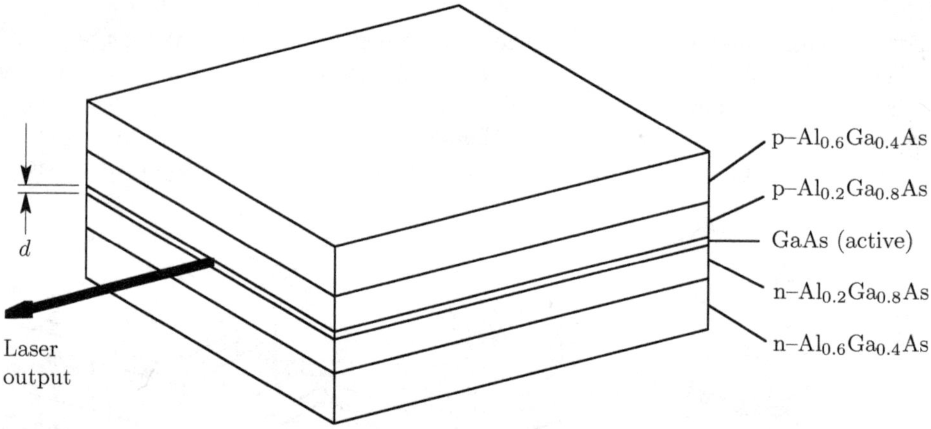


**Fig. 9.6** A heterostructure quantum-well laser with a separate *step index* confinement region.

a 100-Å wide layer of GaAs surrounded by 0.1 $\mu$m layers of $Al_{0.2}Ga_{0.8}As$. The cladding for the field confinement consists of layers of $Al_{0.6}Ga_{0.4}As$, whose refractive index is smaller than that of $Al_{0.2}Ga_{0.8}As$.

The band structure of a quantum-well laser is shown in Fig. 9.7. Instead of having the abrupt *step index* shown in Fig. 9.6, the layer on either side of the active GaAs region has a *smooth transition* between $Al_{0.6}Ga_{0.4}As$ and $Al_{0.2}Ga_{0.8}As$, resulting in the linear change in band energy shown in the figure. (These lasers are called GRINSCH lasers: GRaded INdex Separate Confinement Heterostructure). This arrangement does a better job than a step index in confining the carriers to the extremely thin active

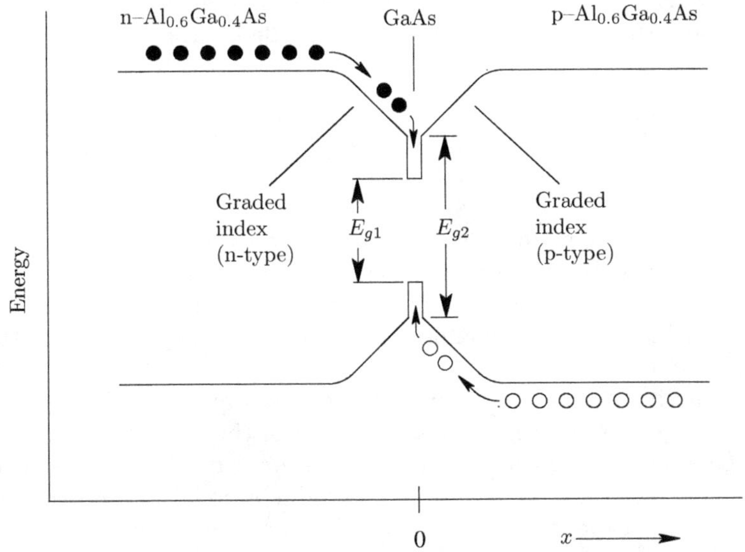


**Fig. 9.7** Band structure of a quantum-well laser using a graded index.

GaAs region. The field confinement is due to the optical waveguide structure formed by the graded index. The thinness of the active region requires a different approach to the electron wavefunction and density of states.

When we discussed the electron wavefunction in *bulk semiconductors* (those with dimensions large compared to the carrier de Broglie wavelength), we assumed that the wavefunction was a modulated plane wave with a large number of possible values for the wave vector, $\boldsymbol{k}$. When one of the dimensions of the active material becomes comparable to the carrier de Broglie wavelength, one must treat the eigenfunction corresponding to that coordinate differently from the others. In the limit of an extremely thin layer, that wave vector component will have *a single value*. The Hamiltonian is written with the special coordinate separated from the others:

$$H = T + V = -\frac{\hbar^2}{2m^*}\nabla_\perp^2 - \frac{\hbar^2}{2m^*}\frac{\partial^2}{\partial x^2} + V, \tag{9.22}$$

where $\nabla_\perp$ is the transverse Laplacian, which operates on the coordinates perpendicular to the junction normal (i.e., parallel to the junction surface). Separating variables, the wavefunction is a product of a two-dimensional Bloch wavefunction (a function of the coordinates, $\boldsymbol{r}_\perp$) and a function of $x$, $\phi(x)$,

$$\psi(\boldsymbol{r}) = u_{\boldsymbol{k}_\perp}(\boldsymbol{r}_\perp)e^{i\boldsymbol{k}_\perp \cdot \boldsymbol{r}_\perp}\phi(x). \tag{9.23}$$

It is very straightforward to solve Schrödinger's equation for an *infinite barrier* in the $x$-direction:

$$\phi(x) = \sin(l\pi x/L_x), \qquad l = 1, 2, \cdots \infty \tag{9.24}$$

$$E_l = \frac{l^2\hbar^2\pi^2}{2m^*L_x^2} = l^2E_1 \qquad E_1 = \frac{\hbar^2\pi^2}{2m^*L_x^2}, \tag{9.25}$$

where, to be consistent with the literature, we call the dimensions of the active region $L_x, L_y$ and $L_z$ and the origin ($x = 0$) is at the left edge of the active region. Thus, the wavefunctions and energies are

$$\psi(\boldsymbol{r}) = u_{k_\perp}(\boldsymbol{r}_\perp)e^{i\boldsymbol{k}_\perp \cdot \boldsymbol{r}_\perp} \sin(l\pi x/L_x) \tag{9.26}$$

$$E = \frac{\hbar^2 k_\perp^2}{2m^*} + l^2 E_1. \tag{9.27}$$

The energy expression is applicable to either electrons or holes, where one uses the effective mass ($m^*$) appropriate to the type of carrier being considered. While an infinite barrier is not very realistic, it demonstrates all of the *qualitative* features of a quantum well without excessive complications (it overestimates some of the energies by a factor of about 2).

It is worth noting that a quantum well is essentially a *two-dimensional* physical system and all of its useful properties are a result of this reduction in the number of dimensions. One can reduce the dimensionality even further and construct *quantum wires* (one-dimensional) and *quantum dots* (zero-dimensional). Lasers have been made from both of these latter structures and they have some advantages over quantum wells. As the dimensionality is reduced, the states become more clustered around the laser oscillation energy, and the density of states and gain become larger (though the gain bandwidth will be reduced). The extreme is a zero-dimension quantum dot, which is essentially an artificial atom with only discrete energy levels. All of these structures are made possible by the recent advances in the ability to grow very thin layers of semiconductor materials using techniques such as molecular beam epitaxy. Remarkably, the transitions between the different layers occur on the scale of *a few atomic diameters.*

We will save some steps in the calculation of the gain by recognizing that the only quantities which are different from the bulk semiconductor case are the *density of states* and the *joint density of states.* Thus, we will use eqn 8.64 for the gain and replace $\rho_j(\nu)$ with the quantity appropriate for a quantum well. We start with the density of states.

The density of states in two dimensions is calculated in exactly the same way as it is in three dimensions. We use *periodic boundary conditions*, for which the allowed values of $\boldsymbol{k}_\perp$ are

$$\boldsymbol{k}_\perp = \left(\frac{2\pi m}{L_y}, \frac{2\pi n}{L_z}\right), \qquad m, n \text{ integers.} \tag{9.28}$$

The area of each cell is $4\pi^2/(L_y L_z) = 4\pi^2/A$, where $A = L_y L_z$ is the perpendicular area of the active region. Thus, the number of states between $k_\perp$ and $k_\perp + dk_\perp$ is

$$\text{Number of states} = 2\left(\frac{A}{4\pi^2}\right)(2\pi k_\perp dk_\perp) = \frac{A k_\perp}{\pi} dk_\perp, \tag{9.29}$$

where the factor of 2 is due to the number of spin states. From the equation for the energy due to the $y$ and $z$ motion (eqn 9.27), the number of states per unit area for $l = 1$ is

$$\rho(E)dE = \frac{m^*}{\pi\hbar^2} dE. \tag{9.30}$$

When $l = 2$, the carrier can be found in either the $l = 1$ or $l = 2$ band and the density doubles. Generalizing to any value of $l$,

$$\rho(E) = \sum_{l=1}^{\infty} \frac{m^*}{\pi\hbar^2} H(E - E_l), \tag{9.31}$$

where

$$H(x) = \begin{cases} 0 & : \quad x < 0 \\ 1 & : \quad x \geq 0 \,. \end{cases} \tag{9.32}$$

Graphs of $\rho(E)$ for a quantum well and a bulk semiconductor are shown in Fig. 9.8. Since the separation between the various values of $E_l$ is proportional to $1/L_x^2$, the

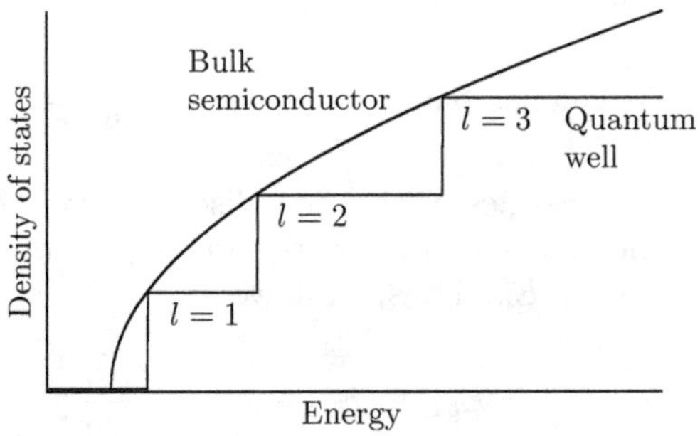


**Fig. 9.8** Density of states for quantum well and bulk semiconductor (for reference).

quantum-well steps rapidly approach the continuous bulk semiconductor curve as the size of the well is increased.

In an earlier section we derived selection rules for momentum, energy and spin in a bulk semiconductor. Quantum wells have an additional quantum number, $l$. The selection rule for $l$ can be determined by writing down the dipole matrix element for a transition between the conduction band and the valence band using the quantum-well wavefunctions. For simplicity, we will assume that the wave is polarized in the $y$-direction, so the operator responsible for optical transitions can be written as

$$\text{Interaction operator } (y\text{-polarization}): \quad e\boldsymbol{E}(\boldsymbol{r}_\perp) \cdot \boldsymbol{r}_\perp e^{i\boldsymbol{k}_{opt}\cdot\boldsymbol{r}_\perp}, \tag{9.33}$$

where $\boldsymbol{k}_{opt}$ is the wave vector for the radiation. Although the field is $E\hat{\boldsymbol{y}}$, we write the dot product in terms of $\boldsymbol{r}_\perp$ to facilitate the interpretation of the first factor of the matrix element. The transition matrix element can be factored into the product of two integrals: one operating on the transverse coordinates ($y$ and $z$) and one acting on the $x$ coordinate. Due to the direction of polarization, the latter has no coordinate dependence. The matrix element is then

$$\langle\psi_c(\boldsymbol{r})|H_I|\psi_v(\boldsymbol{r})\rangle = \int u_{kc}^*(\boldsymbol{r}_\perp)e^{-i\boldsymbol{k}\cdot\boldsymbol{r}_\perp}(e\boldsymbol{E}\cdot\boldsymbol{r}_\perp e^{i\boldsymbol{k}_{opt}\cdot\boldsymbol{r}_\perp})u_{kv}(\boldsymbol{r}_\perp)e^{i\boldsymbol{k}\cdot\boldsymbol{r}_\perp}dydz \\ \times \int \phi_c^*(x)\phi_v(x)dx. \tag{9.34}$$

The first integral yields the $\boldsymbol{k}$ selection rule already considered. From the orthogonality of the $\phi$ functions (sines) for different values of $l$, the selection rule for $l$ is

$$\text{Quantum-well selection rule on } l\text{:}\quad \Delta l = 0. \tag{9.35}$$

Although we used a convenient polarization in deriving this rule, it can be shown that the rule is valid for any polarization.

One very useful feature of quantum wells is the fact that one can *change the effective band gap* by changing $L_x$. If we measure the energies from the bottom of the conduction and top of the valence band of the *bulk material*, the conduction-band energy, $E'_c$, and the valence-band energy, $E'_v$, are given by (for $l = 1$)

$$\text{Conduction band:}\quad E'_c = E_{1c} + \frac{\hbar^2 k_\perp^2}{2m_c^*} \tag{9.36}$$

$$\text{Valence band:}\quad E'_v = E_{1v} + \frac{\hbar^2 k_\perp^2}{2m_v^*}, \tag{9.37}$$

where $E_{1c}$ and $E_{1v}$ are the energies of the first discrete level ($E_1$) in the conduction and valence bands and the valence-band energy increases *downward*. The transition energy, $E' = h\nu$, is $E_{g1} + E'_c + E'_v$. Thus, we have

$$h\nu = E' = E_{g1} + (E_{1c} + E_{1v}) + \frac{\hbar^2 k_\perp^2}{2m_r^*}. \tag{9.38}$$

We see from this that the minimum transition energy is not the band gap, $E_{g1}$, as it would be in a bulk semiconductor, but instead it is the band gap plus the quantity $E_{1c} + E_{1v}$. These terms are proportional to $1/L_x^2$ and we therefore have the ability to change the effective band gap by changing $L_x$.

The joint density of states is particularly simple to derive since $\rho(E)$ is independent of $E$ for a given $l$. We will begin by assuming that the transitions occur between states with $l = 1$, which is certainly the case at low injection current levels. We then need to express the density of states as a function of the transition frequency, $\nu$. This is done in two steps. First, we use the relationship

$$\rho(k_\perp)dk_\perp = \rho(E')dE', \tag{9.39}$$

where, again, $E'$ is the *transition energy*. We saw above that $\rho(k_\perp) = k_\perp/\pi$. From this, we obtain $\rho(E')$

$$\rho(E') = \rho(k_\perp)\frac{dk_\perp}{dE'} = \left(\frac{k_\perp}{\pi}\right)\frac{m_r^*}{\hbar^2 k_\perp} = \frac{m_r^*}{\pi\hbar^2}, \tag{9.40}$$

where the derivative is obtained by differentiating eqn 9.38. Note that this differs from the $\rho(E)$ derived earlier, since it is a function of the *transition energy*, $E'$. The second step uses $E' = h\nu$ and $\rho(E')dE' = \rho_j(\nu)d\nu$:

$$\rho_j(\nu) = \rho(E')\frac{dE'}{d\nu} = \left(\frac{m_r^*}{\pi\hbar^2}\right)h = \frac{2m_r^*}{\hbar}. \tag{9.41}$$

When we substitute $\rho_j(\nu)$ into the gain equation, we need a *volumetric* density. To convert the $\rho_j(\nu)$ that we have derived into a volumetric density, we simply divide by $L_x$:

$$\rho_j(\nu) \longrightarrow \frac{\rho_j(\nu)}{L_x}. \tag{9.42}$$

Substituting the joint density into the gain equation, eqn (8.64), we obtain

$$\gamma(\nu) = A\frac{\lambda^2}{8\pi n^2}\left(\frac{2m_r^*}{\hbar L_x}\right)[f_v(E_1) - f_c(E_2)]. \tag{9.43}$$

The actual gain will need to take into account the fairly small overlap between the active region and the field, given by the quantity $\Gamma$. If the lateral extent of the field is $W$, the beam confinement factor is

$$\Gamma = \frac{\int_{-L_x/2}^{L_x/2} E^2 dx}{\int_{-\infty}^{\infty} E^2 dx} \approx \frac{L_x}{W}. \tag{9.44}$$

Note that this is a different result from that which was claimed in connection with the double heterostructure laser, where it was stated that $\Gamma$ is proportional to $d^2$ when $d$ is very small. The difference is due to the presence of a *separate confinement structure* in the quantum-well laser, so the field mode width is *independent* of $L_x$. Making the substitution $\gamma(\nu) \longrightarrow \Gamma\gamma(\nu)$, we obtain

$$\text{Quantum-well gain:} \quad \gamma(\nu) = A\frac{\lambda^2}{8\pi n^2}\left(\frac{2m_r^*}{\hbar W}\right)[f_v(E_1) - f_c(E_2)]. \tag{9.45}$$

The gain can be estimated using $A = 3 \times 10^8$ s$^{-1}$ (3 ns carrier lifetime), $\lambda = 1$ $\mu$m, $m_r^* = .058 \times m_{electron}$, $n = 3.6$ and $W = 0.1$ $\mu$m (the same as the $d$ in the double heterostructure laser). Assuming that the second bracketed factor is unity, the result is

$$\text{Quantum-well gain:} \quad \gamma \approx 102 \text{ cm}^{-1}. \tag{9.46}$$

This is fairly respectable (a total gain of $\approx 3$ in typical diodes) and is an upper bound achieved with a very large injection current, which will cause $f_v(E_1) - f_c(E_2)$ to approach unity. The gain can be increased even further by the use of *multiple quantum wells.* The improvement is less than proportional to the number of wells, particularly at lower injection currents.

It can be shown that the *transparency density* in a quantum well is roughly the same as that in a bulk semiconductor. The threshold current (eqn 9.21) relative to a double heterostructure laser is therefore reduced by the ratio of the active region widths ($d$). The quantum-well laser is the final step in a sequence of lasers (homojunction$\rightarrow$double heterostructure$\rightarrow$quantum well) in which the threshold current is reduced by reducing the size of the active region.

We conclude this section by mentioning *strained quantum-well lasers*, where lattice matching is deliberately violated to introduce strains in the interface between the layers. An example is a laser constructed from an InGaAs active region surrounded by AlGaAs. By varying the In/Ga molar ratio, one can change the band gap by a factor of 1.5, allowing great flexibility in the peak wavelength (0.84 $\mu$m to 1.3 $\mu$m). The lattice constants can differ by more than 3%, which causes considerable strain at the interface. This changes the effective mass of the holes (they are affected by the lattice parameters to a greater extent than the electrons). A factor of 2 reduction in the heavy hole mass will reduce the density of states and the transparency density by a factor of 2 and increase the differential gain ($\sigma = d\gamma/dN$) by the same factor (the explanations for these changes require a more involved treatment then we gave above).

In summary, the quantum-well laser can provide useful laser gain with a threshold current density which is roughly ten times smaller than that of a bulk semiconductor double heterostructure laser. The laser wavelength can to some extent be tailored by changing the well size. Further reductions in the threshold current and increases in the gain can be obtained using the strained quantum-well configuration.

## 9.5 Distributed feedback lasers

In the section on heterostructure lasers, it was stated that *transverse modes* could be controlled by confining the field with a two-dimensional structure in a *stripe geometry laser*. The *index-guided* configuration is particularly well suited to ensuring oscillation in the *lowest transverse mode*. In pursuit of a *spectrally narrow* source of radiation (needed for both telecommunication and atomic physics experiments), it is also important to control the *longitudinal modes*, since semiconductor lasers are inhomogeneously broadened (mainly from spatial hole burning) and will tend to oscillate in a number of longitudinal modes, as shown in Fig. 9.9.

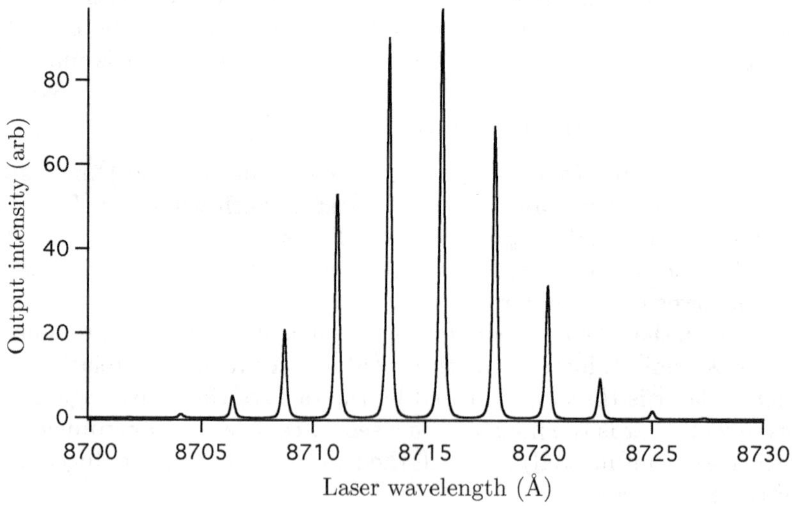


**Fig. 9.9** Computer-generated output spectrum of a 380-$\mu$m-long Fabry–Pérot laser displaying multiple longitudinal modes.

Longitudinal mode control in a monolithic semiconductor laser requires an integral frequency-selective element. This is accomplished by either using frequency-selective end mirrors or by arranging that the entire active region provides the necessary feedback only at certain selected frequencies. The latter type of laser is called a *distributed feedback laser* (DFB laser) and the former is called a *distributed Bragg reflection laser* (DBR laser). The advantage of these techniques is that they can be employed in a monolithic device, with all of the attendant advantages of small size, simplicity, economy, etc. In a later section we will describe *external cavity diode lasers*, which use an *extended cavity* to both force single-mode operation and reduce the Schawlow–Townes linewidth, which can be large due to the very small size (and consequently large linewidth) of the semiconductor laser cavities.

The *distributed feedback laser* dispenses with the reflecting facets for feedback and uses instead a spatially periodic medium for this purpose (the facets can still be reflecting; for simplicity we will assume that they are antireflection coated). The *index-coupled* variant employs a periodic *index of refraction* and the *gain-coupled* approach uses a periodic *gain*. We will discuss the index-coupled double heterostructure laser, which is displayed in Fig. 9.10.

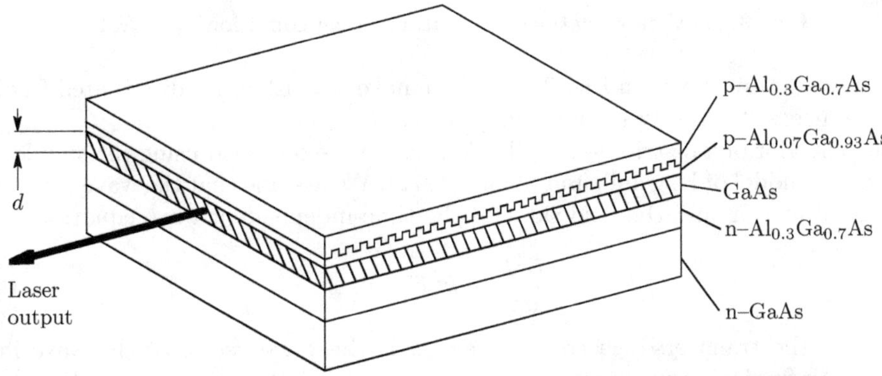


**Fig. 9.10** Distributed feedback double heterostructure index-coupled laser. The index is square-wave modulated since the $Al_{0.3}Ga_{0.7}As$ layer and the $Al_{0.07}Ga_{0.93}As$ layer have different indices of refraction and the mode extends into both layers.

The basic principle is that the reflected wave is generated by *Bragg reflection* from the periodic index of refraction, which is produced by the structure shown in the figure. The optical wave samples both sides of the corrugated interface between two different media and thus sees a spatially modulated index of refraction. The periodic index forms a one-dimensional *grating*, defined by the vector $\boldsymbol{K}$ of magnitude $K = 2\pi/\Lambda$, where $\Lambda$ is the spatial period. One can consider $K$ to be the *spatial frequency of the grating.* If the wave vectors of the forward and reflected optical waves are $\boldsymbol{k}_f$ and $\boldsymbol{k}_r$, the grating can be considered to apply a *spatial phase modulation* to the forward wave, whose spatial frequency is $k_f$. The modulation will generate *sidebands* at integral multiples of the spatial frequency of the grating, and one of these sidebands will be at the wave vector ($\boldsymbol{k}_r$) of the reflected wave:

$$\boldsymbol{k}_r = \boldsymbol{k}_f - q\boldsymbol{K}, \qquad q = 1, 2, 3, \cdots \infty, \tag{9.47}$$

where we ignore the "upper sidebands". The magnitudes of $\boldsymbol{k}_f$ and $\boldsymbol{k}_r$ are the same and the directions are *opposite*; therefore

$$k_f = k_r = \frac{q\pi}{\Lambda} \Longrightarrow \lambda = 2n\Lambda/q, \tag{9.48}$$

where $\lambda$ is the optical wavelength and $n$ is the *average* index of refraction seen by the wave. This is called the *Bragg condition.* The vector relations (for $q = 1$) are shown in Fig. 9.11. Detailed analysis will show that the Bragg condition is not exactly satisfied

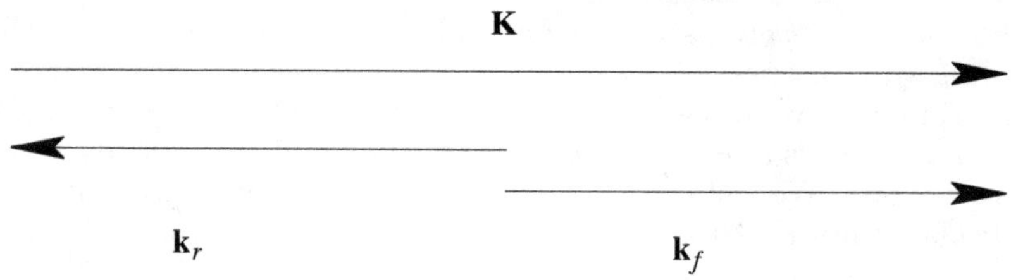


**Fig. 9.11** Vector relations illustrating Bragg condition $k_f = K/2$.

using the structure shown in Fig. 9.10 and to make a satisfactory distributed feedback laser, one needs to slightly modify the geometry.

The behavior of a distributed feedback laser can be obtained using the simple, one-dimensional model of Kogelnik and Shank (1972). We assume that the wave propagates in the $\pm z$ direction and the field satisfies the time-independent wave equation,

$$\frac{\partial^2 E}{\partial z^2} + k^2 E = 0, \tag{9.49}$$

where $E$ is the transverse electric field and $k$ is the wave vector of the wave in the medium. We further assume that the medium has a periodic gain, $\gamma(z)$, and a periodic index of refraction, $n(z)$, which have a simple sinusoidal dependence on $z$:

$$n(z) = n_0 + n_1 \cos 2\beta_0 z \tag{9.50}$$

$$\gamma(z) = \gamma_0 + \gamma_1 \cos 2\beta_0 z \qquad \beta_0 = \frac{\pi}{\Lambda}. \tag{9.51}$$

An index-coupled grating has $\gamma_1 = 0$ and a gain-coupled grating has $n_1 = 0$. The *average* values of the index of refraction and gain are $n_0$ and $\gamma_0$. For consistency with the above reference, the gain is defined for the *field* (not the intensity) and is *positive* for absorption. For this grating, the Bragg condition is

$$\text{Bragg condition:} \quad \lambda = \frac{2n\pi}{\beta_0} \quad \text{or} \quad k_f = \beta_0, \tag{9.52}$$

where $k_f$ is the wave vector of the radiation in the absence of the grating.

The modulations of $n(z)$ and $\gamma(z)$ are considered to be small ($n_1 \ll n_0$, $\gamma_1 \ll \gamma_0$) and the field variation due to the gain is fractionally small over one period of the grating ($\gamma_0 \ll \beta_0$). The square of the field's wave vector ($k^2$) in the presence of the grating is approximately

$$k^2 = k_f^2 + 2i\gamma_0 k_f + 4\kappa k_f \cos 2\beta_0 z, \tag{9.53}$$

where $\kappa$ is the *coupling constant* between the forward and reflected wave due to the grating and is given by

$$\text{Coupling constant:} \quad \kappa = \frac{\pi n_1}{\lambda} + \frac{i\gamma_1}{2} \tag{9.54}$$

and we have dropped terms of second order in small quantities (including $\gamma_0$). Note that the index coupling and gain coupling are in quadrature – this is a very important difference with significant consequences. The coupling, $\kappa$, is equal to the amount of feedback per unit length provided by the grating ($\kappa L$ is the total coupling, equivalent to the mirror reflectivity in a Fabry–Pérot laser).

If the field, $E(z)$, *approximately* satisfies the Bragg condition, it can be expressed as the sum of a forward wave with amplitude, $A_f(z)$, and a reflected wave with amplitude, $A_r(z)$,

$$E(z) = A_f(z)e^{-i\beta_0 z} + A_r(z)e^{i\beta_0 z}, \tag{9.55}$$

where the amplitudes are slowly varying. The field expression can be substituted into the wave equation (including eqn 9.53), yielding two coupled first-order differential equations when the coefficients of different exponentials are equated to zero:

$$\begin{aligned} i\kappa A_r &= -\frac{dA_f}{dz} + (\gamma_0 - i\delta)A_f \\ i\kappa A_f &= \frac{dA_r}{dz} + (\gamma_0 - i\delta)A_r. \end{aligned} \tag{9.56}$$

The second derivatives of the amplitudes were ignored, and $\delta$ is a measure of the difference between the laser frequency, $\omega$, and the Bragg frequency, $\omega_0 = \beta_0 c/n_0$,

$$\delta = \frac{k^2 - \beta_0^2}{2\beta_0} \approx \frac{n_0(\omega - \omega_0)}{c}. \tag{9.57}$$

The coupled differential equations can be solved using a trial solution of the form,

$$A_f = f_1 e^{\gamma' z} + f_2 e^{-\gamma' z} \tag{9.58}$$

$$A_r = r_1 e^{\gamma' z} + r_2 e^{-\gamma' z}, \tag{9.59}$$

where $\gamma'$ is

$$\gamma'^2 = \kappa^2 + (\gamma_0 - i\delta)^2. \tag{9.60}$$

If we take the origin of the coordinate system at the midpoint of the medium, the *boundary conditions* at each end are

$$\text{Boundary conditions:} \quad A_f(-L/2) = A_r(L/2) = 0, \tag{9.61}$$

since the forward field grows from zero as the wave propagates to the right and the reflected field behaves similarly starting from the right side of the medium. From the coupled differential equations (eqns 9.56), one observes that the equations are symmetric under $z \to -z$ and $A_f \leftrightarrow A_r$. We therefore expect *symmetric* ($E(z) = E(-z)$)and *antisymmetric* ($E(z) = -E(-z)$) solutions. Additional relations among the coefficients are obtained using the boundary conditions at each end. Thus, one has

$$f_1 = \pm r_2 \tag{9.62}$$

$$f_2 = \pm r_1 \tag{9.63}$$

$$\frac{f_1}{f_2} = \frac{r_2}{r_1} = -e^{\gamma' L}, \tag{9.64}$$

and the solutions are

$$A_f = \sinh \gamma'(z + L/2) \tag{9.65}$$

$$A_r = \pm \sinh \gamma'(z - L/2), \tag{9.66}$$

where an overall factor which is the same for $A_f$ and $A_r$ has been dropped.

The *eigenvalues* ($\gamma'$) can be obtained by substituting the above solutions into the coupled equations (eqns 9.56). After a bit of algebra, one obtains

$$\begin{aligned} \gamma' + (\gamma_0 - i\delta) &= \pm i\kappa e^{\gamma' L} \\ \gamma' - (\gamma_0 - i\delta) &= \mp i\kappa e^{-\gamma' L}. \end{aligned} \tag{9.67}$$

In the limit of *weak coupling* ($\kappa \ll \gamma_0$), one can use eqn 9.60 to obtain

$$\gamma' \approx \pm(\gamma_0 - i\delta). \tag{9.68}$$

Substituting this into eqns 9.67,

$$2(\gamma_0 - i\delta) \approx \pm i\kappa e^{(\gamma_0 - i\delta)L}, \tag{9.69}$$

whose square modulus gives us the threshold condition for the laser in the *high-gain* limit:

$$\text{Threshold condition } (\kappa \ll \gamma_0): \qquad 4(\gamma_0^2 + \delta^2) \approx |\kappa|^2 e^{2\gamma_0 L}. \tag{9.70}$$

(Note that in the high-gain limit, we still require that $\gamma_0 \ll \beta_0$). The spectral properties of the grating can be obtained from the left-hand side of this equation: when $\delta = \gamma_0$, the coupling squared ($|\kappa|^2$) will need to be increased by a factor of 2 from its value at resonance ($\delta = 0$) to maintain threshold. If we consider $\gamma_0$ to be the *spectral width* of the grating, the wavelength and frequency resolution is

$$\frac{\Delta\lambda}{\lambda} = -\frac{\Delta\nu}{\nu} \approx \frac{\gamma_0}{k_f}. \tag{9.71}$$

We thus obtain the familiar result that the width is inversely proportional to the *length of the grating* (or the number of grating lines) since the round-trip gain is proportional to $\gamma_0 L$ and this is fixed for a laser with a fixed loss.

The location of the laser *grating* resonances can be obtained from eqn 9.69 by comparing the real and imaginary parts of both sides. We will use a *polar* representation of $\kappa$:

$$\kappa = |\kappa| e^{i\phi}. \tag{9.72}$$

Near the Bragg frequency ($\delta \ll \gamma_0$),

$$\delta L = (q + \tfrac{1}{2})\pi + \phi \qquad \text{(integral } q\text{)}. \tag{9.73}$$

Expressing this as a ratio of the deviation from the Bragg frequency ($c\delta/2\pi n_0$) to the Fabry–Pérot free spectral range ($c/2n_0 L$):

$$\text{Number of free spectral ranges from Bragg frequency} = q + \tfrac{1}{2} + \frac{\phi}{\pi}. \tag{9.74}$$

This demonstrates that the modes have approximately the same separation as in a Fabry–Pérot laser. A purely *gain-coupled* laser ($n_1 = 0$) will have $\phi = \pi/2$ and will therefore have a resonance at the exact Bragg frequency. An *index-coupled* laser ($\gamma_1 = 0$) has a *real* $\kappa$ and therefore *will not have any resonances at the Bragg frequency.* Furthermore, an index-coupled laser will have two low-threshold modes, separated by approximately one free spectral range. Since these modes have an equal gain, the laser can oscillate on *either mode*, a not very desirable state of affairs. There are several remedies for this; the most commonly used cure is to place an abrupt $\pi/2$ *phase shift* in the grating index modulation at the exact center. This is shown in Fig. 9.12. A laser

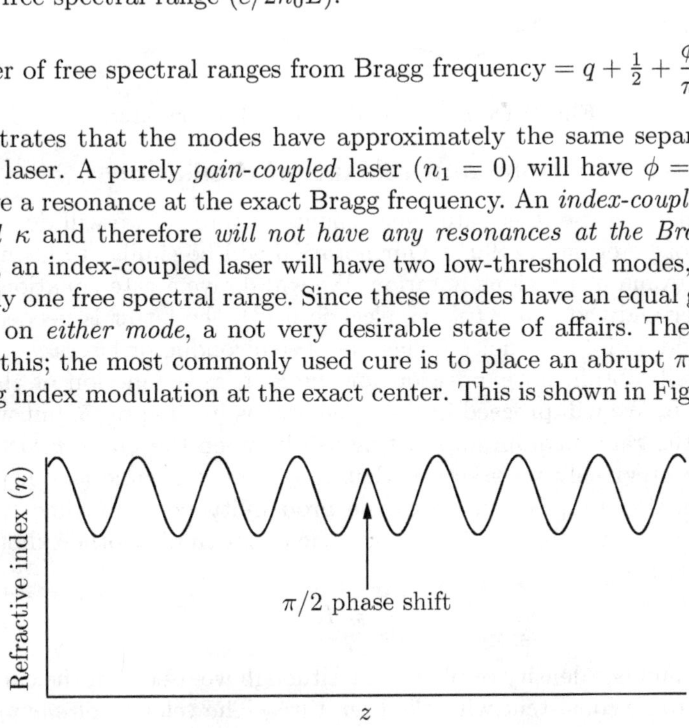


**Fig. 9.12** Sinusoidal index modulation with $\pi/2$ phase shift at the center.

with this kind of grating will have the lowest threshold mode at the Bragg frequency.

Another approach to longitudinal mode control is to use Bragg reflection gratings at each end of the laser in place of the reflecting facets which are typical in a Fabry–Pérot laser. A laser of this kind is called a distributed Bragg reflection (DBR) laser and is shown schematically in Fig. 9.13. These lasers are slightly easier to fabricate since the gain region is physically separate from the gratings. They have the disadvantage of using shorter gratings which have poorer frequency selectivity than in the DFB laser. This is compensated a bit by the shorter active cavity, whose modes are further apart. DFB lasers seem to be the preferred single-longitudinal-mode monolithic semiconductor laser choice at present.

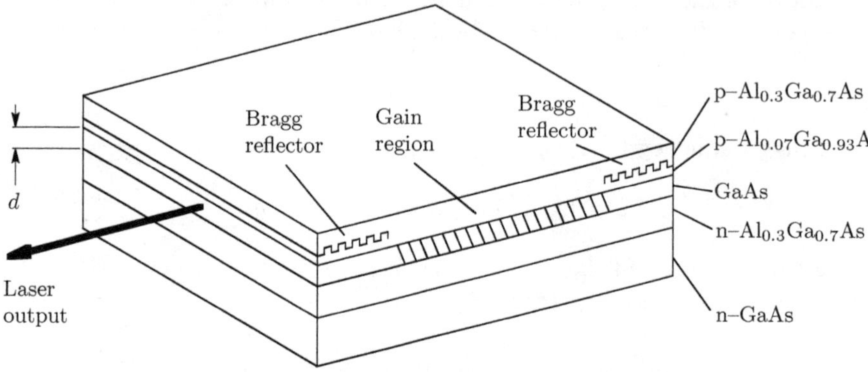


**Fig. 9.13** Distributed Bragg reflection laser.

## 9.6 The rate equations and relaxation oscillations

Although we derived the laser rate equations in Chapter 5, we will do so again using the approach of Petermann (1991). Our notation will be similar to his and will differ in some cases from our previous notation. We will derive a rate equation for both the photon number and the *phase* (of the electric field); the latter is needed to properly account for modulation and noise behaviors of semiconductor lasers.

First, we will obtain an expression for the gain as a function of the stimulated emission rate, $R$. We will proceed in the same way as in Chapter 5, but will anticipate dispersion in the gain medium and distinguish between the *group velocity* and *phase velocity* where previously we assumed that they were the same (no dispersion). The stimulated emission rate is defined as the probability per unit time that a photon incident on the gain medium will stimulate the emission of another photon. The rate $R$ therefore satisfies

$$\frac{dS}{dt} = RS, \tag{9.75}$$

where $S$ is the number density of photons. (Although we used $n$ in the earlier chapters, we use $S$ here to be consistent with the literature.) The velocity of *energy transport* is the *group velocity*, $v_g$, defined as

$$v_g = \frac{d\omega}{dk}, \tag{9.76}$$

where we assume that the medium is weakly dispersive and that the index of refraction increases with frequency (*normal dispersion*). The magnitude of the photon flux is $Sv_g$ and this quantity satisfies a *continuity equation* in one dimension:

$$\frac{\partial(Sv_g)}{\partial z} + \frac{\partial S}{\partial t} = \langle\text{source rate}\rangle, \tag{9.77}$$

where the *source rate* is the rate of increase in the photon number density due to some external source – in this case the stimulated emission. In the steady state the time derivative is zero and we obtain

$$\frac{d(Sv_g)}{dz} = RS. \tag{9.78}$$

If we multiply both sides by $h\nu$, we obtain an equation for the growth of the *intensity*, $I = h\nu S v_g$

$$\frac{dI}{dz} = \frac{R}{v_g} I. \tag{9.79}$$

From the definition of the *laser gain* due to stimulated emission, $g_{st}$,

$$g_{st} = \frac{R}{v_g}. \tag{9.80}$$

The actual gain, $g$, is obtained by multiplying $g_{st}$ by the *mode confinement factor*, $\Gamma$:

$$\text{Gain:} \quad g = \Gamma g_{st} = \Gamma R / v_g. \tag{9.81}$$

In a semiconductor gain medium of length, $L$, with absorption coefficient, $a$, the electric field of a wave propagating in the $z$-direction will be

$$E(z) = E_0 e^{-i\beta z + (g-a)z/2}, \tag{9.82}$$

where $\beta$ is the *propagation constant* of the wave (we use $\beta$ instead of $k$ since the wave is in a *guiding structure* rather than in free space). From this equation and the requirement that the field replicate itself in one round trip, the laser mode frequencies are given by

$$\nu = \frac{qc}{2L\mu}, \tag{9.83}$$

where $q$ is an integer and $\mu$ is the refractive index. We will now consider a laser slightly above threshold, where the threshold carrier density is $N_{th}$ and the threshold oscillation frequency is $\nu_{th}$. We will assume that the refractive index depends only upon the frequency (due to dispersion) and carrier density. Expanding the refractive index around the threshold values of frequency and carrier density,

$$\mu = \mu_0 + \frac{\partial \mu}{\partial \nu}(\nu - \nu_{th}) + \frac{\partial \mu}{\partial N}(N - N_{th}). \tag{9.84}$$

If we assume that the last two terms are much smaller than the first, we can substitute this equation into eqn 9.83 and obtain

$$\left\{1 + \frac{\nu_{th}}{\mu_0}\frac{\partial \mu}{\partial \nu}\right\}(\nu - \nu_{th}) = -\frac{\nu_{th}}{\mu_0}\frac{\partial \mu}{\partial N}(N - N_{th}), \tag{9.85}$$

where $\nu_{th} = qc/2L\mu_0$. It is easy to show that the factor in curly brackets is $\mu_g/\mu_0$, where $\mu_g$ is the *group index* defined by

$$\mu_g \equiv \frac{c}{v_g}. \tag{9.86}$$

(Group velocity is discussed in some detail in Chapter 11). From this, we obtain the equation governing semiconductor laser frequency modulation:

$$\nu - \nu_{th} = -\frac{\nu_{th}}{\mu_g}\frac{\partial \mu}{\partial N}(N - N_{th}). \tag{9.87}$$

After one round trip, one can easily see from eqn 9.82 that the field is multiplied by the *round-trip gain*, $G_r$

$$G_r = r_1 r_2 e^{-2i\beta L + (g-a)L}, \tag{9.88}$$

where $r_{1,2}$ are the field reflectivities of the end facets. Since the conditions are slightly different from those at threshold, we use the expansion of the refractive index (based upon eqn 9.84) to determine $\beta$:

$$\beta = \frac{\omega}{c}\mu = \frac{\omega_{th}}{c}\left\{\mu_0 + \frac{\partial \mu}{\partial N}(N - N_{th}) + \frac{\mu_g}{\omega_{th}}(\omega - \omega_{th})\right\}, \tag{9.89}$$

where $\omega_{th}$ and $\omega$ are angular frequencies at threshold and slightly above threshold. The gain can be written as the product of a frequency independent factor, $G_1$, and a frequency dependent factor, $G_2$:

$$G_r = G_1 G_2. \tag{9.90}$$

From eqns 9.88 and 9.89, the two gain factors are

$$G_1 = r_1 r_2 e^{(g-a)L} e^{-i\phi_G} \qquad \text{where } \phi_G = \frac{2\omega_{th} L}{c}\frac{\partial \mu}{\partial N}(N - N_{th}) \tag{9.91}$$

$$G_2 = e^{-it_{rt}(\omega - \omega_{th})}. \tag{9.92}$$

In the second equation, $t_{rt}$ is the round trip time and is equal to $2L\mu_g/c$. In deriving these equations, we used the fact that $2\omega_{th}L\mu_0/c$ is an integral multiple of $2\pi$.

In eqn 9.82, we gave the electric field as a function of $z$. Since we seek a *rate equation* for the electric field, we now consider the time dependence of the field, which we write as the product of a factor oscillating at $\omega_{th}$ and a slowly varying part, $E_s(t)$:

$$E(t) = E_s(t)e^{i\omega_{th}t}. \tag{9.93}$$

We will invoke the usual self-consistency argument and require that after one round trip the field replicates itself:

$$E(t) = G_1 G_2 E(t) = G_1 e^{it_{rt}\omega_{th}}\left\{e^{-i\omega t_{rt}}E(t)\right\}. \tag{9.94}$$

The bracketed quantity on the right can be shown to be equal to $E(t - t_{rt})$. This equality is demonstrated by expanding $E(t - t_{rt})$ in a Taylor series and evaluating the various time derivatives using $d/dt \to i\omega$ (since $E(t) \propto e^{i\omega t}$). An expansion of the factor in brackets will yield the same result. Thus, we obtain:

$$E_s(t) = G_1 E_s(t - t_{rt}). \tag{9.95}$$

Since $E_s$ is *slowly varying*, we can expand it to first order in $t_{rt}$:

$$E_s(t) = G_1 E_s(t - t_{rt}) = G_1 \left( E_s(t) - t_{rt} \frac{dE_s(t)}{dt} \right). \tag{9.96}$$

Placing the derivative on the left-hand side yields a differential equation for the field:

$$\frac{dE_s}{dt} = \frac{1}{t_{rt}}(1 - 1/G_1)E_s. \tag{9.97}$$

Since $1/G_1$ is close to unity, one can expand it:

$$\frac{1}{G_1} = e^{-\ln(r_1 r_2) - (g-a)L + i\phi_G} \approx 1 - \ln(r_1 r_2) - (g - a)L + i\phi_G. \tag{9.98}$$

The final result is obtained by substituting this into the differential equation, eqn 9.97:

$$\text{Equation for field:} \quad \frac{dE_s}{dt} = \left[ i(\omega - \omega_{th}) + \tfrac{1}{2}(\Gamma R - 1/\tau_p) \right] E_s. \tag{9.99}$$

The term involving $R$ comes from eqn 9.81:

$$g \frac{c}{\mu_g} = \Gamma R. \tag{9.100}$$

The $\omega - \omega_{th}$ term is from eqns 9.87 and 9.91, and the *photon lifetime*, $\tau_p$, is

$$\frac{1}{\tau_p} = v_g \left( a - \frac{1}{L} \ln(r_1 r_2) \right). \tag{9.101}$$

The rate equation for the electric field (eqn 9.99) has a very simple interpretation. The equation states that $E_s(t)$ has an oscillatory component at frequency $\omega - \omega_{th}$ (and thus $E(t)$ oscillates at $\omega$, as expected) and grows exponentially at a rate equal to the difference between $\Gamma R/2$ and $1/(2\tau_p)$. Indeed, a non-rigorous derivation could have been made from these considerations.

The field is normalized so that the photon number density, $S$, is $|E_s|^2$ and the phase, $\phi$, is just the phase of $E_s$. With these assignments, it is easy to show that the rate equations for the photon number density and electric field phase can be obtained from

$$\frac{dS}{dt} = 2\text{Re}\left\{ E_s^* \frac{dE_s}{dt} \right\} \tag{9.102}$$

$$\frac{d\phi}{dt} = \frac{1}{S}\text{Im}\left\{ E_s^* \frac{dE_s}{dt} \right\}. \tag{9.103}$$

The rate equations are:

$$\text{Number density:} \quad \frac{dS}{dt} = (\Gamma R - 1/\tau_p)S \tag{9.104}$$

$$\text{Phase:} \quad \frac{d\phi}{dt} = \omega - \omega_{th}. \tag{9.105}$$

We have neglected spontaneous emission in deriving these equations.

The phase rate equation can be reformulated to make it more useful for linewidth calculations, which will be carried out later in this chapter. From eqns 9.87 and 9.105,

$$\frac{d\phi}{dt} = \omega - \omega_{th} = -\frac{\omega_{th}}{\mu_g}\frac{\partial\mu}{\partial N}(N - N_{th}) = -\frac{\omega_{th}}{\mu_g}\frac{\partial\mu'}{\partial\mu''}\frac{\partial\mu''}{\partial N}(N - N_{th}). \tag{9.106}$$

The quantities $\mu'$ and $\mu''$ are respectively the real and imaginary parts of the refractive index, which is given by

$$\mu = \mu' - i\mu''. \tag{9.107}$$

The imaginary part is proportional to the gain, $g$, which is

$$g = -\frac{4\pi\mu''}{\lambda}, \tag{9.108}$$

while the real part is the usual index of refraction. (We did something similar using a complex susceptibility in Chapter 5.) The derivative of the real part with respect to the imaginary part is very important in semiconductor theory and is called the *linewidth enhancement factor*, $\alpha$:

$$\text{Linewidth enhancement factor:} \qquad \alpha = \frac{\partial\mu'}{\partial\mu''} = \frac{\partial\mu'/\partial N}{\partial\mu''/\partial N}. \tag{9.109}$$

This quantity is responsible for a substantial increase in the Shawlow–Townes linewidth in diode lasers and has a numerical value typically between 3 and 10. It describes the coupling between the gain and index of refraction in semiconductors. Since $g_{st}$ increases with $N$ while the index of refraction, $\mu'$, decreases, $\alpha$ is always positive (due to the minus sign used for the imaginary part of $\mu$). As a result of the inverse relation between $\mu'$ and $N$, the index of refraction in the *wings* of the wave will be increased relative to the center and the confinement of the field in the laser diode waveguide structure will be harmed. The parameter $\alpha$ is a measure of this (adverse) tendency and it is therefore often called the *antiguiding parameter*. The reason that $\alpha \neq 0$ in semiconductor lasers is due to the somewhat *unsymmetric* gain curve in semiconductors (Chapter 8). Many laser gain curves (e.g., Lorentzians and Gaussians) are symmetric about their peak and thus the dispersion relation of the index of refraction has a zero at the peak of the gain curve. A medium with such a dispersion relation will have a zero $\alpha$.

Using the $\alpha$ parameter and the relation between the gain and $\mu''$, we obtain

$$\frac{d\phi}{dt} = \tfrac{1}{2}\alpha v_g \frac{\partial g}{\partial N}(N - N_{th}). \tag{9.110}$$

A final transformation of this equation is based upon the relation between $g$ and $R$ and the fact that the value of $\Gamma R$ at threshold is just $1/\tau_p$ (from the photon number density rate equation):

$$\frac{d\phi}{dt} = \tfrac{1}{2}\alpha(\Gamma R - 1/\tau_p), \tag{9.111}$$

where the product of the derivative and $N - N_{th}$ is just $\Delta g$, the gain increment above threshold.

Spontaneous emission is accounted for by adding the term involving $R_{sp}$ to the right side of the photon number equation

$$\frac{dS}{dt} = (v_g g - 1/\tau_p)S + \frac{R_{sp}}{\tau_r}N, \tag{9.112}$$

where $\tau_r$ is the spontaneous emission lifetime (lifetime for electron–hole recombination). The added term, which is small compared to the other terms, is the rate of spontaneous photons which are emitted into the laser mode.

A third (and final) rate equation is that for the carriers and will be presented without a formal derivation, since the terms are fairly self-explanatory:

$$\text{Carriers:} \quad \frac{dN}{dt} = \frac{I}{eV} - \frac{N}{\tau_r} - \Gamma RS, \tag{9.113}$$

where $I$ is the injected current and $V$ is the volume of the active region. The first term on the right is the electron injection rate, the second is the recombination rate and the third is the rate of stimulated emission (which causes recombination). In summary, the three rate equations are

$$\text{Photon density:} \quad \frac{dS}{dt} = (v_g g - 1/\tau_p)S + \frac{R_{sp}}{\tau_r}N \tag{9.114}$$

$$\text{Electric field phase:} \quad \frac{d\phi}{dt} = \tfrac{1}{2}\alpha(v_g g - 1/\tau_p) \tag{9.115}$$

$$\text{Carrier density:} \quad \frac{dN}{dt} = \frac{I}{eV} - \frac{N}{\tau_r} - v_g g S. \tag{9.116}$$

We will derive the small signal behavior by considering *small, time-dependent displacements* of $I, N$ and $S$ from their steady-state values $I_0, N_0$ and $S_0$:

$$\begin{aligned} I(t) &= I_0 + I_1(t) \\ N(t) &= N_0 + N_1(t) \\ S(t) &= S_0 + S_1(t). \end{aligned} \tag{9.117}$$

Since the gain depends upon the carrier density and photon density, we will also expand it around the operating point defined by $I_0$, $N_0$, and $S_0$:

$$g = g_0 + g_N N_1 + g_S S_1, \tag{9.118}$$

where $g_N$ and $g_S$ are derivatives of the gain with respect to $N$ and $S$. The parameter $g_S$ is a measure of nonlinear *gain compression* due to saturation and is less than zero. Substituting into the rate equations and dropping second-order quantities in the small displacements we obtain

$$\begin{aligned} \frac{dN_1}{dt} &= \frac{I_1}{eV} - \left(\frac{1}{\tau_r} + v_g g_N S_0\right) N_1 - v_g(g_0 + g_S S_0)S_1 \\ \frac{dS_1}{dt} &= \left(v_g g_N S_0 + \frac{R_{sp}}{\tau_r}\right) N_1 - \left(\frac{R_{sp}N_0}{\tau_r S_0} - v_g g_S S_0\right) S_1, \end{aligned} \tag{9.119}$$

where we also used the relations among the steady-state quantities obtained by setting the time derivatives in the original rate equations equal to zero. Differentiating the second equation with respect to time and using the first equation for $dN_1/dt$ and the second equation for $N_1$, we obtain

$$\frac{d^2S_1}{dt^2} + A\frac{dS_1}{dt} + BS_1 = \frac{v_g g_N S_0 + R_{sp}/\tau_r}{eV} I_1, \tag{9.120}$$

where

$$A \approx \frac{1}{\tau_r} + v_g(g_N - g_S)S_0 \tag{9.121}$$

$$B \approx v_g\left(\frac{g_N}{\tau_p} - \frac{g_S}{\tau_r}\right)S_0. \tag{9.122}$$

In the above, we have ignored the spontaneous emission terms ($R_{sp} \to 0$) and have used the approximate expression $v_g g_0 \approx 1/\tau_p$ (gain equals loss), which is valid near threshold.

A formally identical differential equation will be encountered in a chapter on control systems and is solved for a step change in the driving term ($I_1$) using Laplace transforms. The driving term is

$$I_1(t) = I_{10}u(t), \tag{9.123}$$

where

$$u(t) = \begin{cases} 0 & : \ t \leq 0 \\ 1 & : \ t > 0\,. \end{cases} \tag{9.124}$$

We will write the equation in a more standard form:

$$\frac{d^2S_1}{dt^2} + 2\zeta\omega_n\frac{dS_1}{dt} + \omega_n^2 S_1 = \frac{v_g g_N S_0}{eV} I_{10}u(t), \tag{9.125}$$

where we have ignored spontaneous emission in the right-hand driving term. In most cases, the gain saturation term, $g_S$, is much smaller than the gain carrier dependence, $g_N$. Ignoring $g_S$, the parameters $\omega_n$ and $\zeta$ are

$$\omega_n^2 = \frac{v_g g_N S_0}{\tau_p} \tag{9.126}$$

$$\zeta = \frac{1}{2\omega_n}\left(\frac{1}{\tau_r} + v_g g_N S_0\right) = \frac{1}{2}\sqrt{\frac{\tau_p}{v_g g_N S_0}}\left(\frac{1}{\tau_r} + v_g g_N S_0\right). \tag{9.127}$$

The quantities $v_g g_N S_0$ and $1/\tau_r$ are both approximately $10^9$ s$^{-1}$. The photon lifetime, $\tau_p$, is given by eqn 9.101 and, for a typical laser, $\tau_p \approx 10^{-12}$ s. Thus, $\omega_n \approx 3 \times 10^{10}$ s$^{-1}$ (about 5 GHz) and $\zeta \approx 0.03$ (extreme *underdamping*). The solution for $S_1(t)$ is

$$S_1(t) = \frac{v_g g_N S_0}{eV}(1 - \frac{1}{\beta}e^{-\zeta\omega_n t}\sin(\omega_n\beta t + \theta)) \tag{9.128}$$

$$\text{where } \beta = \sqrt{1-\zeta^2}$$

$$\theta = \tan^{-1}\beta/\zeta.$$

The damping time is $1/\zeta\omega_n \approx 10^{-9}$ s. A plot of the photon number (which is proportional to the laser power) as a function of time for a step change in injection current

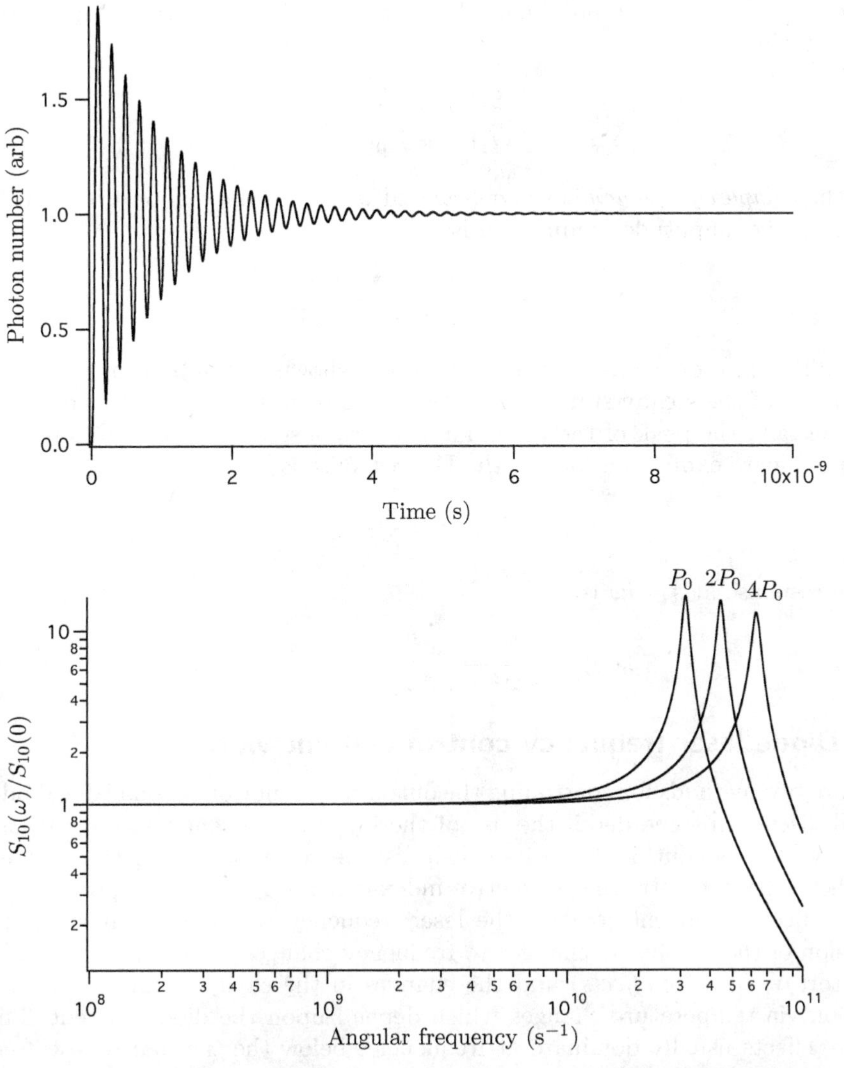


**Fig. 9.14** The transient response to a step in injection current (top) and the modulation response as a function of angular frequency (bottom). Three laser powers differing by factors of 2 are plotted, showing the power dependence of the modulation bandwidth ($P_0$ is an arbitrary starting point).

at $t = 0$ appears at the top of Fig. 9.14. The damped oscillation is called *relaxation oscillation* and occurs to some extent in all lasers. The oscillation frequency is very high (several GHz) in semiconductor lasers and is proportional to the square root of the laser power. A small amount of relaxation oscillation is always present and is driven by technical noise and by spontaneous emission, which continually generates very small power *steps* which ring as shown in the plot.

The response to a *sinusoidal* modulation of the injection current can be obtained by substituting

$$S_1(t) = S_{10}e^{st} \tag{9.129}$$

$$I_1(t) = I_{10}e^{st}, \tag{9.130}$$

where the *complex frequency* is $s = \sigma + i\omega$ and we ignore the real part in what follows. The size of the amplitude modulation is

$$S_{10} = \frac{1}{s^2 + 2\zeta\omega_n s + \omega_n^2} \frac{v_g g_N S_0}{eV} I_{10}. \tag{9.131}$$

A normalized plot of the modulation response is shown at the bottom of Fig. 9.14 for three values of the steady-state power. As can be seen, the bandwidth increases with the power, and the peak of the underdamped response is at approximately $\omega_n$. From the approximate expression for $\omega_n$, the DC response is

$$S_{10}(0) = \frac{\tau_p}{eV} I_{10} \tag{9.132}$$

and the response at $s = i\omega$ is

$$S_{10}(\omega) = \frac{\omega_n^2}{-\omega^2 + 2i\zeta\omega_n\omega + \omega_n^2} \frac{\tau_p}{eV} I_{10}(\omega). \tag{9.133}$$

## 9.7 Diode laser frequency control and linewidth

There are two methods for controlling the output frequency of a monolithic diode laser. We will discuss in some detail the use of the injection current for this purpose. The frequency of a monolithic diode laser can also be controlled using the temperature, since the cavity geometry and refractive index depend upon the temperature.

The injection current controls the laser frequency through two mechanisms: the conversion of the amplitude changes to frequency changes via the linewidth broadening factor, $\alpha$, (*carrier effects*) and the changes in the cavity geometry and index of refraction via temperature changes which depend upon the diode current. The temperature effects usually dominate at frequencies below the thermal roll-off frequency of about 1–10 MHz. The $\alpha$-dependent carrier effect rolls off near the frequency of relaxation oscillations (several GHz). We will discuss the temperature-dependent effects first.

The frequency deviation, $\Delta\nu$, due to a change in temperature, $\Delta T$, is given by

$$\frac{\Delta\nu}{\nu} = -(\alpha_L + \alpha_\mu)\Delta T. \tag{9.134}$$

The quantity $\alpha_L$ is the thermal expansion coefficient and is about $0.6\times10^{-5}$/degree for GaAs, while $\alpha_\mu$ is the thermal refractive index coefficient ($(1/\mu)d\mu/dT$) and is about $10^{-4}$/degree for GaAs. Thus, for a 850 nm laser, the *static* (DC) temperature tuning sensitivity of a GaAs laser is about 37 GHz/degree. It is important to distinguish this frequency shift, which is due to change in the laser cavity resonance frequency, from the frequency shift due to the *dependence of the peak of the gain curve on temperature.* The latter value of $d\nu/dT$ is about four times as large as the former and is used for gross tuning of laser diodes. The shift of the gain peak with temperature will mainly result in *mode hops* as the temperature is changed. The fact that there are two temperature tuning mechanisms with different temperature sensitivities can cause there to be frequency *gaps* in the tuning of the lasers; these can only be removed with an eternal cavity or by injection locking.

The time dependence of the diode temperature can be obtained from the thermal equation in the active medium:

$$C_t \frac{\partial T}{\partial t} - k_t \nabla^2 T = W_i, \tag{9.135}$$

where $C_t$ is the heat capacity, $k_t$ is the thermal conductivity to the environment and $W_i$ is the heat source due to the injection current. The exact solution is rather complicated and is obtained by solving this equation in each layer of the diode. From simple considerations, one can show that the *transfer function* for the temperature is given by

$$\frac{T(\omega)}{I(\omega)} = \frac{C_{th}}{1 + i(\omega/\omega_{th})}, \tag{9.136}$$

where $I(\omega), T(\omega)$ are the Fourier transforms of the current and temperature and $\omega_{th}$ is the *thermal cut-off frequency.*

Treatment of the carrier effect begins with eqn 9.87, which we will reproduce here:

$$\nu - \nu_{th} = -\frac{\nu_{th}}{\mu_g} \frac{\partial \mu}{\partial N}(N - N_{th}).$$

Using the definition of $\alpha$ and $\mu''$, this equation is equivalent to

$$\nu - \nu_{th} = \frac{v_g \alpha}{4\pi} \frac{\partial g}{\partial N}(N - N_{th}). \tag{9.137}$$

One would think that obtaining the derivative of $N$ with respect to $I$ would complete the calculation. Unfortunately, this would significantly underestimate the size of the frequency shift. For frequencies below about 1 GHz, the main source of the frequency shift is actually the *gain compression,* parametrized with $g_S$. This behavior is similar to that of an anharmonic oscillator, whose resonant frequency depends upon the oscillation amplitude. To study this analytically, we will first rewrite the photon number rate equation as

$$\frac{dS}{dt} = v_g \left[ \frac{\partial g}{\partial N}(N - N_{th}) + g_S S \right] S + \frac{R_{sp} N}{\tau_r}. \tag{9.138}$$

We have simply expanded the gain around the threshold carrier density ($N_{th}$) and used the facts that $g_N = \partial g/\partial N$ and $v_g g_0 = 1/\tau_p$. Eliminating the term involving $\partial g/\partial N$ in the above two equations, we obtain

$$\nu - \nu_{th} = \frac{\alpha}{4\pi} \left[ \frac{1}{S}\frac{dS}{dt} - v_g g_S S - \frac{R_{sp} N}{S \tau_r} \right]. \tag{9.139}$$

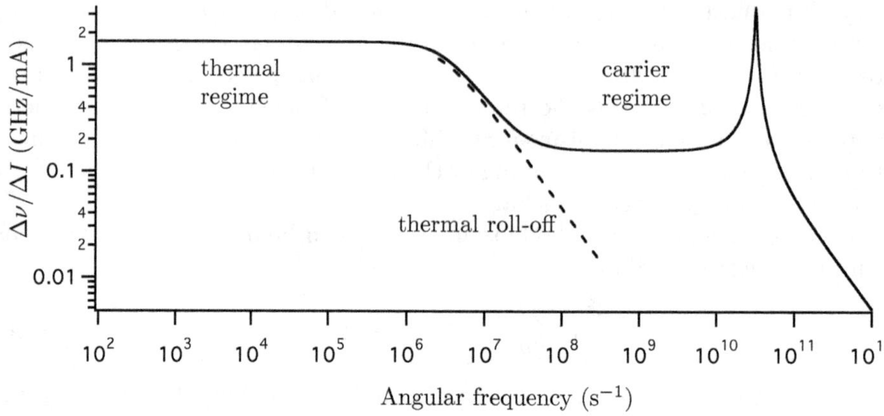


**Fig. 9.15** Theoretical plot of $\Delta\nu/\Delta I$.

Well above threshold, we can ignore the third term in brackets (the spontaneous emission term). If we consider a sinusoidal variation at $\omega_m$ of both $\nu$ and $S$ about their average values, $\bar{\nu}, \bar{S}$,

$$\begin{aligned} \nu(t) &= \bar{\nu} + \Delta\nu e^{i\omega_m t} \\ S(t) &= \bar{S} + \Delta S e^{i\omega_m t}, \end{aligned} \tag{9.140}$$

then the ratio of $\Delta\nu$ to $\Delta S$ is

$$\frac{\Delta\nu}{\Delta S} = \frac{\alpha}{4\pi\bar{S}}(i\omega_m - \omega_g), \tag{9.141}$$

where

$$\omega_g = v_g g_S \bar{S} \tag{9.142}$$

and we assume that the changes in the parameters are much smaller than their average values. (Note that the $\omega_g$ contribution is positive since $g_S$ is *negative*.) The magnitude of $\omega_g$ is in the GHz region for diode lasers and therefore the frequency response is fairly flat between about 10 MHz and 1 GHz. The dependence of the laser frequency on the injection current is obtained by multiplying the derivatives:

$$\frac{d\nu}{dI} = \frac{d\nu}{dS}\frac{dS}{dI}. \tag{9.143}$$

The second factor is eqn 9.133. Including the temperature effects (eqn 9.136),

$$\frac{d\nu}{dI} = \left(\frac{\alpha\tau_p}{4\pi eV\bar{S}}\right)\frac{\omega_n^2(i\omega_m - \omega_g)}{-\omega_m^2 + 2i\zeta\omega_n\omega_m + \omega_n^2} - \frac{\nu(\alpha_L + \alpha_\mu)C_{th}}{1 + i(\omega_m/\omega_{th})}. \tag{9.144}$$

This function is plotted in Fig. 9.15 for $\omega_n = 3 \times 10^{10}$ s$^{-1}$, $|\omega_g| = \omega_n$, $\zeta = 0.03$, a DC thermal sensitivity of 1.6 GHz/mA and a DC carrier sensitivity of about 0.1 GHz/mA.

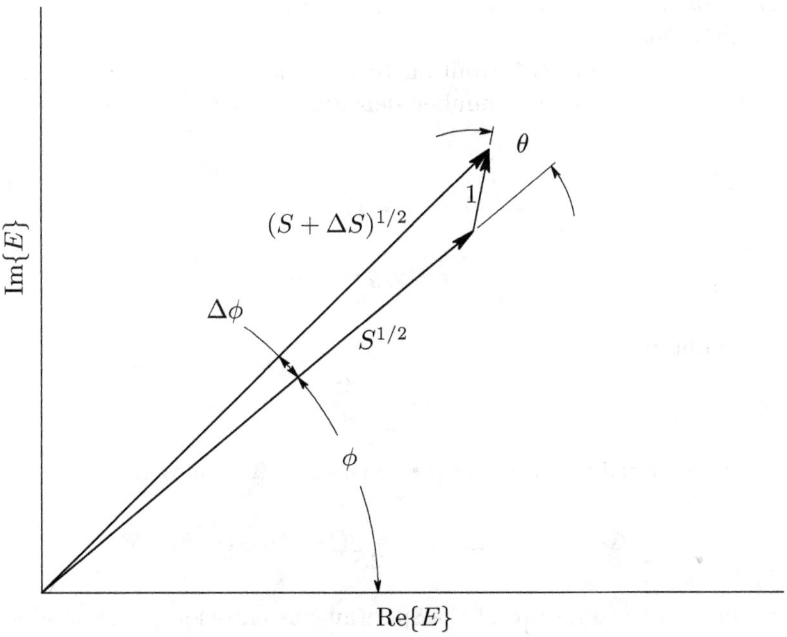


**Fig. 9.16** Phasor plot showing change in photon number density, $S$, and electric field phase, $\phi$, due to the spontaneous emission of one photon.

We will conclude this section by justifying the designation of $\alpha$ as the linewidth enhancement factor: we will derive the expression for the linewidth of a bare diode laser, using the approach of Henry (1982). As in the preceding, we assume that the square modulus of the electric field, $E$, represents the number density of photons, $S$, and that the phase of the field is $\phi$:

$$E = \sqrt{S}e^{i\phi}. \tag{9.145}$$

We are interested in the changes in $\phi$ and $S$ due to a single spontaneous emission. A phasor plot of the situation appears in Fig. 9.16. The change in the field, $\Delta E$, is of unit magnitude (since it represents a single photon) and is given by

$$\Delta E = e^{i(\phi+\theta)}, \tag{9.146}$$

where $\theta$ is random. There will be two contributions to the change in the photon phase: one due to geometry and indicated in the figure and the other due to the coupling of $\phi$ and $S$ via the rate equations. The geometrical change in phase, $\Delta\phi_1$, due to the single emission is

$$\Delta\phi_1 = \frac{\sin\theta}{\sqrt{S}} \qquad (\theta \text{ random}). \tag{9.147}$$

Using the law of cosines, the change in $S$ is

$$\Delta S = 1 + 2\sqrt{S}\cos\theta. \tag{9.148}$$

The average value of $\Delta S$ is one, as expected, since the spontaneous event should add one photon to the field.

To determine the second contribution to the phase change, we will use the rate equations for phase and photon number density, without the spontaneous emission term:

$$\frac{dS}{dt} = (v_g g - 1/\tau_p)S \tag{9.149}$$

$$\frac{d\phi}{dt} = \tfrac{1}{2}\alpha(v_g g - 1/\tau_p). \tag{9.150}$$

From these, we obtain

$$\frac{d\phi}{dt} = \frac{\alpha}{2S}\frac{dS}{dt}. \tag{9.151}$$

The rate equation contribution to the phase change, $\Delta\phi_2$, is therefore

$$\Delta\phi_2 = -\frac{\alpha}{2S}\Delta S = -\frac{\alpha}{2S}(1 + 2\sqrt{S}\cos\theta). \tag{9.152}$$

The minus sign is due the nature of the spontaneous emission process. When a spontaneous photon is emitted, there is a *step change* in the field intensity which causes damped relaxation oscillations, after which the intensity returns to its steady state. Thus, $S$ changes from $S+\Delta S$ (immediately after emission) to the steady-state value $S$ and the change is negative. This latter process – the restoration of the steady state – is the source for this component of the line broadening, not the spontaneous emission (whose contribution is the $\Delta\phi_1$ term). The total phase change is

$$\Delta\phi = \Delta\phi_1 + \Delta\phi_2 = -\frac{\alpha}{2S} + \frac{1}{\sqrt{S}}[\sin\theta - \alpha\cos\theta]. \tag{9.153}$$

The first term is constant and it causes the phase to increase at rate

$$\Delta\omega = \frac{d}{dt}\Delta\phi = -\frac{\alpha R_{spont}}{2S}, \tag{9.154}$$

where $R_{spont}$ is the spontaneous emission rate. This is just a constant frequency shift ($\Delta\omega$) due to a reduction in the stimulated rate by spontaneous emission. The second term results in phase *fluctuations*, whose accumulated square value is

$$\langle\Delta\phi^2\rangle = \left\langle\left(\sum_{i=1}^{N} S^{-1/2}[\sin\theta_i - \alpha\cos\theta_i]\right)^2\right\rangle, \tag{9.155}$$

where $N$ is the (large) number of observed emissions and the $\langle\rangle$ denotes an *ensemble average*. This can easily be summed and averaged since the cross terms all average to zero ($\theta_i$ is random) and the $\sin^2$ and $\cos^2$ average to 1/2. The sum can be performed by multiplying each (identical, after averaging) term by $R_{spont}t$, where $t$ is the observation time:

$$\langle \Delta\phi^2 \rangle = \frac{R_{spont}(1+\alpha^2)t}{2S}. \tag{9.156}$$

The resultant electric field executes a *random walk* after each emission event. From the theory of random walks, the autocorrelation function of the field is

$$\langle E^*(t)E(0)\rangle = |E(0)|^2 e^{-|t|/t_{coh}}, \tag{9.157}$$

where the coherence time, $t_{coh}$, is given by

$$1/t_{coh} = \frac{\langle \Delta\phi^2 \rangle}{2t} \tag{9.158}$$

and the power spectrum is the Fourier transform of the autocorrelation function. The power spectrum is a Lorentzian whose full width at half maximum is

$$\Delta\nu = \frac{1}{\pi t_{coh}} = \frac{R_{spont}}{4\pi S}(1+\alpha^2). \tag{9.159}$$

This last result is the important one. Carrying the calculation to completion will yield the standard Shawlow–Townes linewidth multiplied by $1+\alpha^2$. Thus, diode lasers will have linewidths which are ten to a hundred times as large as expected from an *ideal* laser with the same power and cavity configuration ($\alpha$ varies from 3 to 10). When this is combined with the small cavity length, we obtain linewidths of tens of MHz or greater. The external cavity, described in the next section, allows one to greatly reduce this large linewidth.

## 9.8 External cavity diode lasers (ECDLs)

We have seen that monolithic distributed feedback lasers can be operated in both a single transverse and single longitudinal mode. However, the spectral breadth can be fairly large since the Shawlow–Townes line breadth is proportional to the square of the cavity width, and this is very large in the extremely short feedback structures found in monolithic DFB lasers. Moreover, the linewidth enhancement factor further increases the spectral breadth by the factor $1 + \alpha^2$. A significant electronic tuning range is difficult to achieve in a DFB laser since, as was shown in the previous section, the tuning range due to changes in the injection current is fairly small and one is forced to use the fairly slow mechanism of diode temperature changes to tune the diode over a large range. Both of these limitations can be overcome by using the diode as a gain medium in a much larger cavity composed of mirrors, gratings and other possible frequency-selection elements. Such a laser is called an external cavity diode laser (ECDL). A longer cavity reduces the Shawlow–Townes width by the inverse of the square of the cavity length and eliminates the effects of the $\alpha$ parameter since the active medium length is a very small fraction of the overall cavity length. With the use of a coarse frequency selecting element (usually a grating) together with cavity length changes, a large tuning range can be obtained without changing the diode current or temperature.

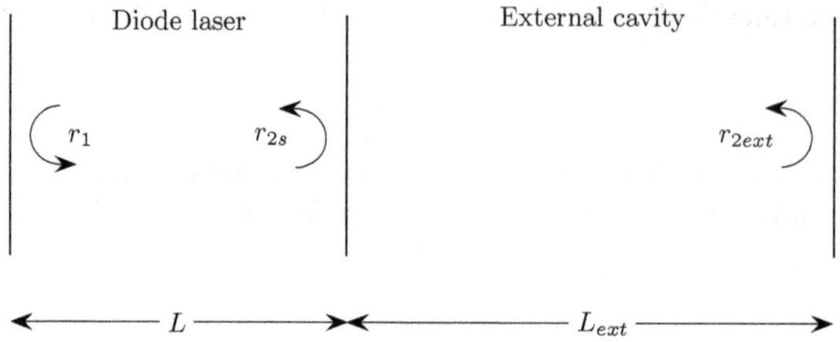


**Fig. 9.17** Schematic of laser cavity reflectors together with external reflector.

Although commercial ECDLs are available, many atomic physics experimentalists manufacture their own units using commercial monolithic diodes and readily available optical components. If one chooses this route, one is immediately confronted with the problem of the existing coatings (if any) on the diode. The best results are obtained if one of the diode facets has zero reflectivity and the other is highly reflecting. Then, the diode is truly just a gain element in a larger cavity. On the other hand, if the facets are not antireflection coated, one is left with a much more complicated system since the laser resonator will consist of two coupled cavities: one due to the monolithic laser diode and the other due to the external cavity. In the following we will analyze the coupled cavities in an external cavity laser and describe some geometries which facilitate tuning over a relatively wide frequency range.

Our analysis is due to Petermann (1991) and can be used to treat small to moderate amounts of feedback. Thus, we can describe an external cavity laser without anti-reflection coatings and analyze the adverse effects of small amounts of feedback on a solitary diode laser. We will assume that there is a reflective element with field reflectivity $r_{2ext}$ placed a distance $L_{ext}$ from the output facet of a laser diode whose facets have field reflectivities $r_1$ and $r_{2s}$ and whose length is $L$ (Fig. 9.17). This analysis assumes that the external cavity supports only a single reflection, which is a fairly realistic assumption since $r_{2s}$ due to an uncoated facet is about 0.57 (in GaAs). The effective reflectivity, $r_2$, of the output facet is now frequency-dependent and is given by

$$r_2(\omega) = r_{2s} + r_{2ext}(1 - |r_{2s}|^2)e^{-i\omega\tau_{ext}}, \tag{9.160}$$

where $\tau_{ext} = 2L_{ext}/c$ is the round trip time in the external cavity and the bracketed expression describes *two* transmissions through the output facet. The two laser oscillation conditions are

$$r_1|r_2|e^{(g'_{th}-a)L} = 1 \tag{9.161}$$

$$2\beta L + \phi_r = 2\pi m \qquad \text{(integral } m\text{)}, \tag{9.162}$$

where $\phi_r$ is the phase of the complex number, $r_2$, $g'_{th} - a$ is the threshold gain minus loss of the *compound* cavity and $\beta$ is the propagation constant in the guiding structure of the diode. We can define a *coupling parameter*, $\kappa_{ext}$,

$$\kappa_{ext} = \frac{r_{2ext}}{r_{2s}}(1 - |r_{2s}|^2). \tag{9.163}$$

When $\kappa_{ext} \ll 1$, $|r_2|$ and $\phi_r$ are given by

$$|r_2| = r_{2s}(1 + \kappa_{ext}\cos\omega\tau_{ext}) \tag{9.164}$$
$$\phi_r = -\kappa_{ext}\sin\omega\tau_{ext}. \tag{9.165}$$

We seek the change in threshold gain and phase ($\phi_r$) due to the presence of the feedback. From the gain condition (eqn 9.161) and eqn 9.164, the gain change is

$$g'_{th} - g_{th} = -\frac{\kappa_{ext}}{L}\cos\omega\tau_{ext}, \tag{9.166}$$

where $g_{th}$ is the threshold gain in the *absence* of the external cavity. The phase equations with and without feedback are

$$\text{No feedback:} \quad \frac{2\mu L\omega_{th}}{c} = 2\pi m \tag{9.167}$$
$$\text{With feedback:} \quad \frac{2L}{c}(\mu + \Delta\mu)(\omega_{th} + \Delta\omega) + \phi_r = 2\pi m + \Delta\phi, \tag{9.168}$$

where we used $\mu\omega/c$ for $\beta$ and $\Delta\omega = \omega - \omega_{th}$. The quantity $\Delta\phi$ is the difference between the round-trip phase change and $2\pi m$; in the presence of feedback, we will require that $\Delta\phi$ be zero (or a multiple of $2\pi$) for laser oscillation to take place. Expanding the above equation and keeping only terms to first order in small quantities,

$$\Delta\phi = \frac{2L}{c}(\mu\Delta\omega + \omega_{th}\Delta\mu) + \phi_r, \tag{9.169}$$

where we used the first equation to eliminate zero-order quantities. We will make our by now familiar expansion of $\mu$ around the threshold values of the carrier density and frequency:

$$\Delta\mu = \frac{\partial\mu}{\partial N}(N - N_{th}) + \frac{\partial\mu}{\partial\omega}(\omega - \omega_{th}), \tag{9.170}$$

where $N_{th}$ is the carrier density without feedback. Substituting this into the equation for $\Delta\phi$ yields

$$\Delta\phi = \frac{2L}{c}\left\{\Delta\omega\left[\mu + \omega_{th}\frac{\partial\mu}{\partial\omega}\right] + \omega_{th}\frac{\partial\mu}{\partial N}(N - N_{th})\right\} + \phi_r. \tag{9.171}$$

The expression in the left-hand inner brackets is just $\mu_g$, the group index of refraction. The term on the right can be expressed in terms of $\alpha$:

$$\frac{\partial\mu}{\partial N}(N - N_{th}) = -\frac{\alpha c}{2\omega_{th}}\frac{\partial g}{\partial N}(N - N_{th}) = -\frac{\alpha c}{2\omega_{th}}(g'_{th} - g_{th}). \tag{9.172}$$

Including this and the expression for the change in gain (eqn 9.166) and $\phi_r$ in the $\Delta\phi$ equation results in

$$\Delta\phi = \frac{2\mu_g L}{c}(\omega - \omega_{th}) + \kappa_{ext}[-\sin\omega\tau_{ext} + \alpha\cos\omega\tau_{ext}]. \tag{9.173}$$

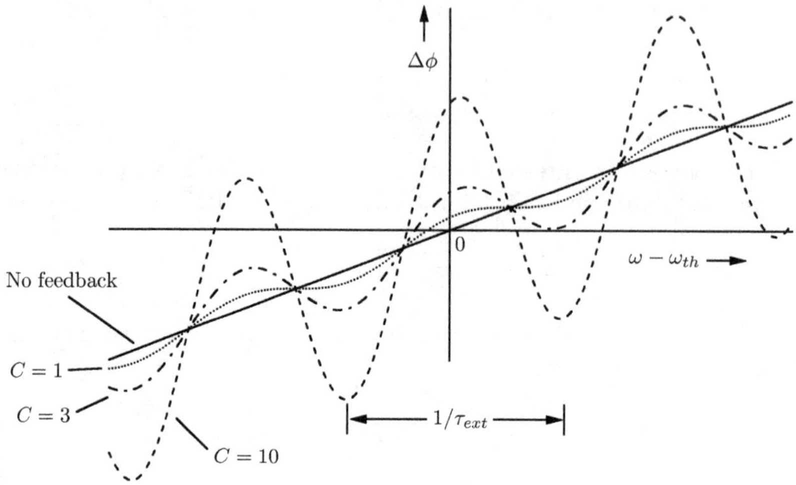


**Fig. 9.18** Plots of $\Delta\phi$ versus $\omega - \omega_{th}$ for a variety of values of the coupling parameter, $C$.

Writing this in terms of the round-trip time in the diode ($\tau_L = 2\mu_g L/c$) and combining the trigonometric functions:

$$\Delta\phi = \tau_L(\omega - \omega_{th}) + \kappa_{ext}\sqrt{1+\alpha^2}\sin(\omega\tau_{ext} - \arctan\alpha). \tag{9.174}$$

A slightly more useful coupling parameter than $\kappa_{ext}$ is the parameter $C$ defined as

$$C \equiv \frac{\tau_{ext}}{\tau_L}\kappa_{ext}\sqrt{1+\alpha^2}. \tag{9.175}$$

The final version of our equation for $\Delta\phi$ is

$$\Delta\phi = \tau_L(\omega - \omega_{th}) + C\frac{\tau_L}{\tau_{ext}}\sin(\omega\tau_{ext} - \arctan\alpha). \tag{9.176}$$

Although the $\Delta\phi$ versus $\Delta\omega$ function might seem a bit obscure, it actually provides quite a bit of insight into the behavior of a diode laser under varying amounts of feedback. A family of plots of this function for a variety of values of $C$ appears in Fig. 9.18. The diode laser with external feedback can only oscillate when $\Delta\phi$ is zero or a multiple of $2\pi$. Without feedback, the function is a straight line through the origin, confirming that the laser operates only at $\omega_{th}$ ($\Delta\phi = 0$). The parameter $C$ is defined so that the function is strictly monotonic when $C < 1$. Thus, with weak feedback ($C < 1$), the laser still oscillates at a single frequency, though the frequency will be shifted. This frequency shift is a sensitive function of $\tau_{ext}$, so even a small change of the position of the external reflector (on the order of a wavelength of light) will have a large effect on the laser frequency. One can immediately see from eqn 9.174 that the maximum frequency shift when $C < 1$ is

$$\Delta\omega_{max} = \frac{\kappa_{ext}}{\tau_L}\sqrt{1+\alpha^2}. \tag{9.177}$$

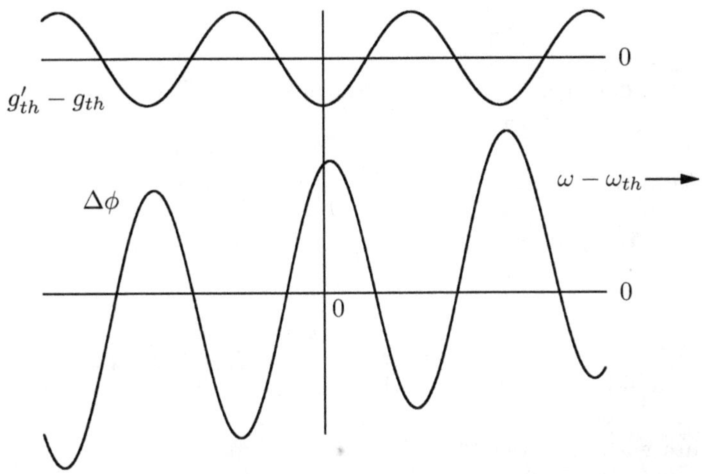


**Fig. 9.19** Plots of $\Delta\phi$ and $g'_{th} - g_{th}$ versus $\omega - \omega_{th}$ for a Littrow configuration external cavity diode laser ($C = 28$).

For example, for a small external field reflectivity of $10^{-4}$ and a laser cavity length of 300 $\mu$m, the shift is about 18 MHz ($\alpha = 5$) and will occur when $L_{ext}$ changes by $\lambda/4$. Thus optical isolators are necessary if frequency stability is important.

An external cavity diode laser uses for feedback either the first-order light back-reflected from a grating (Littrow configuration) or the light reflected from the combination of a grating and mirror (Littman configuration). In the Littrow configuration, about 25% of the light can be returned to the diode ($r_{2ext} = 0.5$) and the coupling parameter, $\kappa_{ext}$, for an uncoated diode ($r_{2s} = 0.56$) is about 0.6, which is on the edge of the range assumed in the above calculations. For a diode length of 300 $\mu$m and external cavity length of 1 cm, the parameter $C$ is about 28 ($\alpha = 5$). Curves of $\Delta\phi$ and threshold gain versus frequency are shown in Fig. 9.19. There are a number of possible laser frequencies, separated by about one half the free spectral range of the external cavity. The likely laser frequencies will be those with the *smallest* threshold gain, which eliminates alternate zero crossings of $\Delta\phi$ and results in the expected mode separation.

According to the Shawlow–Townes linewidth formula, the diode laser spectral linewidth is proportional to the *square* of the laser cavity linewidth, which is proportional to $1/\tau_{eff}^2$, where $\tau_{eff}$ is the *effective round-trip delay*. The latter is the round-trip delay which yields the correct *photon lifetime* and can be given in general by

$$\tau_{eff} = \frac{d\Delta\phi}{d\omega} = \tau_L F, \tag{9.178}$$

where

$$F = 1 + C\cos(\omega\tau_{ext} - \arctan\alpha). \tag{9.179}$$

Thus, the modification to the linewidth due to the external cavity is

$$\Delta\nu = \frac{\Delta\nu_0}{F^2}, \tag{9.180}$$

where $\Delta\nu_0$ is the linewidth without the external cavity. For a large $C$, the oscillation frequencies are those for which

$$\text{Oscillation frequencies:} \quad \sin(\omega\tau_{ext} - \arctan\alpha) \approx 0 \qquad (C \gg 1). \tag{9.181}$$

Thus, $F \approx 1 + C$ at the laser emission frequencies and

$$\Delta\nu = \frac{\Delta\nu_0}{(1+C)^2}. \tag{9.182}$$

For the above example, the external cavity reduces the linewidth by nearly a factor of 1000.

We will state without proof that the frequency modulation sensitivity will be *reduced* by the factor $F$, which approaches $1 + C$ in the strong feedback regime:

$$\left(\frac{\partial\nu}{\partial I}\right)_{feedback} = \frac{1}{F}\left(\frac{\partial\nu}{\partial I}\right)_{nofeedback} \Longrightarrow \frac{1}{1+C}\left(\frac{\partial\nu}{\partial I}\right)_{nofeedback}. \tag{9.183}$$

One can make the following plausibility argument for this result. In an external cavity laser with an antireflection coated output facet, the ratio of the modulation sensitivity to that of a bare diode should be the inverse of the ratio of the cavity lengths. From eqn 9.178, this ratio is $F$, since it is the ratio of $\tau_{eff}$ to $\tau_L$ and $\tau_{eff}$ is the *round-trip time* for the extended cavity laser when the output facet has no reflections.

We will complete our theoretical examination of external cavity diode lasers with a discussion of potential instabilities which can occur for intermediate levels of feedback. There are two regimes in which we expect unconditional stability: the low-feedback regime, with $C < 1$, where there can be adverse effects on the laser frequency due to changes in the feedback phase but in which the laser will otherwise behave normally, and the high-feedback regime, where the output facet is antireflection coated and the feedback is very strong. In the latter case, the laser will have a short active region in a larger optical cavity and will behave like the lasers discussed in earlier chapters and therefore warrants little additional analysis.

Researchers Tkach and Chraplyvy (1986) analyzed the effect of feedback on DFB lasers and established five distinct regimes. Distributed feedback lasers behave similarly to Fabry–Pérot lasers in this regard, though $\tau_L$ will be a little shorter than expected for a laser of a given length and $r_{2s}$ will depend upon the DFB coupling parameter ($\kappa$) and is a bit smaller than the facet reflectivity in an uncoated Fabry–Pérot laser. Their conclusions are graphically portrayed in Fig. 9.20. In region I, $C < 1$ and the laser is unconditionally stable as discussed above. In region II, the possibility of multimode emission exists and the laser will often jump between modes, causing the linewidth to greatly increase. As the feedback is increased, one will enter another region of stability, region III. There the laser will oscillate on the line with the smallest linewidth rather than the line with the lowest threshold gain. Region IV is the regime of *coherence collapse*, where the linewidth becomes extremely great (up to 100 GHz)

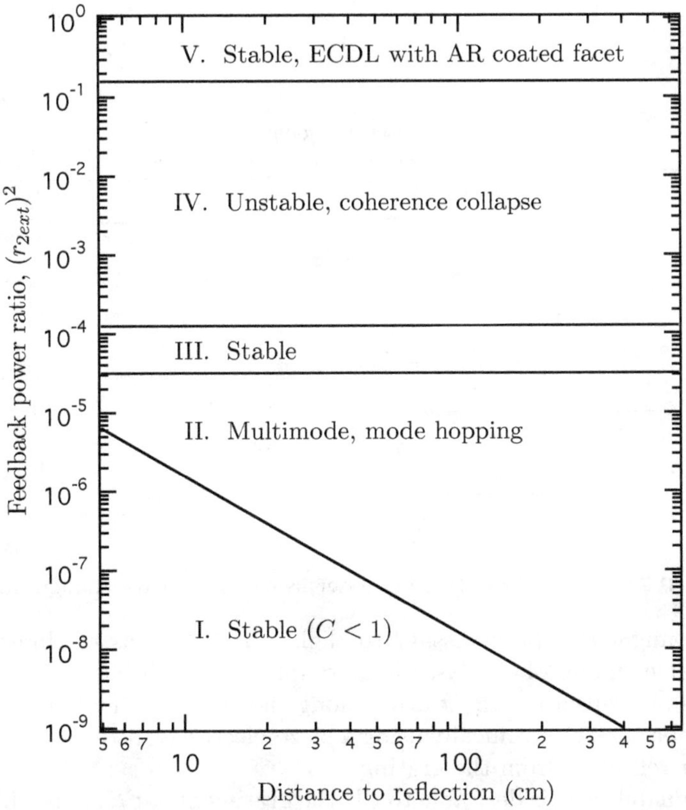


**Fig. 9.20** Stability regions for 1.5 $\mu$m DFB laser (from Tkach (1986)).

and the relaxation oscillations (at several GHz) are continuously excited. It is speculated (Henry (1986)) that this instability occurs when the coupling coefficient ($\kappa_{ext}/\tau_L$ in Henry's theory) becomes comparable to the relaxation frequency. This would occur when $\kappa_{ext} \approx 2 \times 10^{-2}$. The transition to coherence collapse should be independent of the length of the external cavity. The final region, Region V, is the desirable place to operate an external cavity diode laser and is in effect when the laser output facet is antireflection coated and the feedback ratio is fairly large. Unfortunately, an external cavity diode laser with uncoated facets can operate close to region IV and possibly experience the adverse effects of this regime.

We will complete this section with a discussion of some of the practical aspects of external cavity design and describe the *Littrow* and *Littman* feedback approaches. The Littrow feedback scheme is shown schematically in Fig. 9.21. The feedback element is a grating arranged in the *Littrow configuration*, where the diffracted light is retroreflected from the grating back to the laser along the same path as the incident light. The laser emission is expanded with a collimating lens to illuminate as many lines of the grating as possible in order to increase its frequency selectivity. Due to the extreme feedback sensitivity of the laser diode, the lens should be very well antireflection coated;

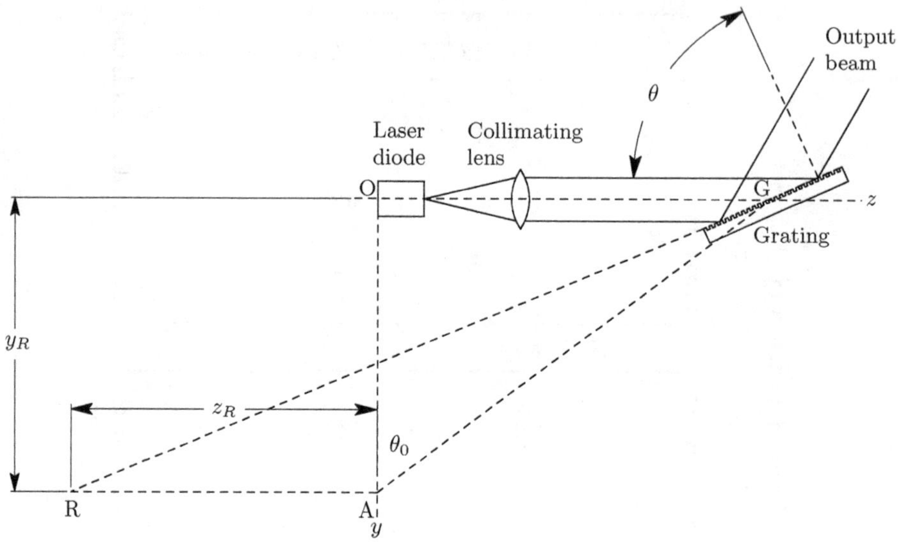


**Fig. 9.21** External cavity diode laser using the Littrow configuration.

in practice, it might still be necessary to slightly tilt the lens to eliminate feedback problems. Coarse tuning of the laser is accomplished by tilting the grating and fine tuning is done by translating the grating along the $z$-axis without changing its angle. The latter can be done electronically using a piezo-electric transducer. The laser output is the specular reflection from the grating.

For wide tunability, it is desirable to tilt the grating about an axis which maintains *tracking* between the grating center frequency and the cavity resonance. Otherwise, mode hops can occur as the laser is tuned with the grating. We will now determine the location of the pivot point which accomplishes this (following the treatment of Labachelerie (1993)). Referring to the figure, the grating wavelength, $\lambda_r$, is

$$\lambda_r = 2a \sin\theta, \tag{9.184}$$

where $a$ is the grating period. If $L$ is distance $OG$ in the figure (the cavity length), cavity resonances occur when

$$L = m\lambda_m/2, \tag{9.185}$$

where $m$ is an integer (we will ignore the diode refractive index at present) and $\lambda_m$ is the cavity resonance wavelength. We will first adjust the geometry so that, at $\theta_0$ and $L_0$, the two wavelengths are the same:

$$\lambda_m(L_0) = \lambda_r(\theta_0). \tag{9.186}$$

We seek a location of the pivot point (whose coordinates are $y_R$ and $z_R$) which preserves this equality over the largest possible range of wavelengths. The wavelength difference is $F(\theta)$:

$$F(\theta) = \lambda_m(\theta) - \lambda_r(\theta). \tag{9.187}$$

In attempting to minimize $F(\theta)$, we will find the location of the pivot point for which the first two derivatives with respect to $\theta$ are zero.

There is a fairly subtle point which, if overlooked, will give a less than optimum result. It turns out that there is a *phase shift* when the grating is translated in its plane by amount $t_0$; this shift advances the phase by $-2\pi$ when $t_0$ increases by the grating period, $a$. Thus the round-trip phase shift is

$$\Delta\phi = \frac{4\pi}{\lambda}L - \frac{2\pi t_0}{a}. \tag{9.188}$$

The function $F(\theta)$ is now

$$F(\theta) = \frac{2L(\theta)}{m + t_0(\theta)/a} - 2a\sin\theta. \tag{9.189}$$

If $t(\theta_0) = 0$, the expression for $F(\theta)$ is

$$F(\theta) = 2a\left[\frac{L(\theta)\sin\theta_0}{L_0 + t_0(\theta)\sin\theta_0} - \sin\theta\right]. \tag{9.190}$$

From simple geometry, $L(\theta)$ is given by

$$L(\theta) = -y_R\left[\frac{\sin\theta_0}{\cos\theta} - \tan\theta\right] - z_R\left[\frac{\cos\theta_0}{\cos\theta} - 1\right] + L_0\frac{\cos\theta_0}{\cos\theta}. \tag{9.191}$$

This is an equation for $L(\theta)$ in terms of $L_0$ and the corresponding angle $\theta_0$, which satisfies the equation $F(\theta_0) = 0$. The first derivative of $F(\theta)$ vanishes when

$$y_R = \frac{L_0}{\tan\theta_0}, \tag{9.192}$$

where we have used for $t_0(\theta)$, to first order:

$$t_0(\theta) = \left[\frac{L_0 - z_R}{\cos\theta_0}\right](\theta - \theta_0). \tag{9.193}$$

The pivot point is thus constrained to a horizontal line which passes through point $A$ in the figure. We can refine the pivot point location by obtaining the second derivative of $F(\theta)$ along the above horizontal line. We find that $z_R = -L_0$: the pivot point is behind the rear facet of the diode by $L_0$.

The advantages of the Littrow approach are simplicity and efficiency (it introduces less loss than the Littman configuration). Its principal disadvantage is that the output beam angle changes with wavelength tuning. This angle problem can be eliminated by rigidly attaching to the grating assembly a mirror which is parallel to the grating (Fig. 9.22). This preserves the angle of the output beam, although the beam will be displaced parallel to itself by an amount whose upper bound is $\approx 2L\Delta\theta$ for an angular change of $\Delta\theta$, where $L$ is the distance between the grating and the mirror.

The Littman approach (Liu (1981)) uses a grating at near grazing incidence for gross frequency tuning and reflects the first-order diffracted beam back using a mirror

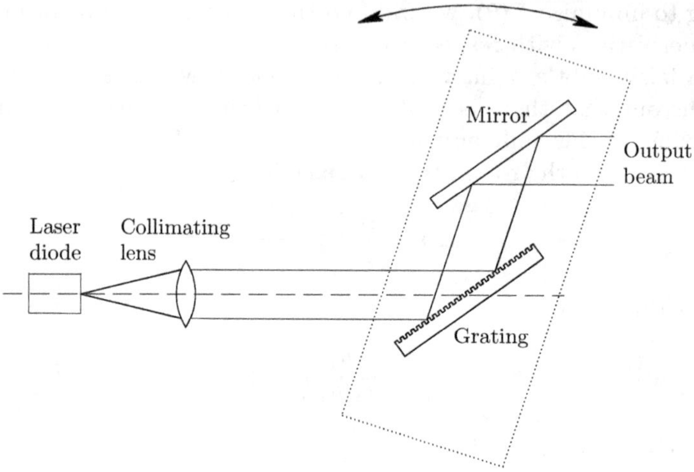


**Fig. 9.22** Arrangement which eliminates angle changes with wavelength tuning.

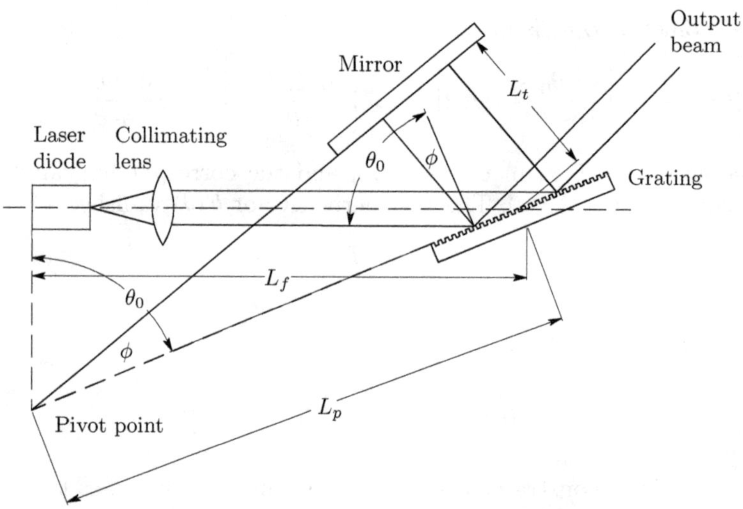


**Fig. 9.23** Geometry used in Littman approach with self-tracking.

which can be rotated. As with the Littrow approach, the output beam is the specular reflection from the grating, whose orientation is fixed. Tuning is accomplished by rotating the mirror (which also changes the cavity length). This scheme was originally used in dye lasers, and the pivot point of the mirror was chosen to achieve tracking between cavity modes and the grating frequency. The tracking for the Littman configuration is *exact* over the entire tuning range of the grating. Figure 9.23 illustrates the geometry and defines the distances and angles used in the following analysis. As with the Littrow calculation, we will assume that the optical length of the diode is equal to its physical length.

For tracking, the following two wavelengths need to be the same

$$\text{Cavity mode:} \quad \lambda_m = \frac{2}{m}L(\phi) = \frac{2}{m}[L_f + L_t(\phi)] \tag{9.194}$$

$$\text{Grating resonance:} \quad \lambda_r = \frac{a}{p}(\sin\theta_0 + \sin\phi), \tag{9.195}$$

where $a$ is the grating period, $p$ is the grating order, $m$ is the cavity axial order, $L(\phi)$ is the optical path length between the end facet of the diode and the mirror and $\phi$ is the angle between the grating and mirror normals. The first equation is the resonance condition for the cavity and the second equation is the standard grating master relationship. From the figure,

$$L(\phi) = L_f + L_p \sin\phi. \tag{9.196}$$

Thus, we have

$$\frac{2}{m}(L_f + L_p \sin\phi) = \frac{a}{p}(\sin\theta_0 + \sin\phi). \tag{9.197}$$

This single equation implies that tracking will occur if the following two equations are satsified:

$$\frac{2}{m}L_f = \frac{a}{p}\sin\theta_0 \tag{9.198}$$

$$\frac{2}{m}L_p = \frac{a}{p}. \tag{9.199}$$

Thus, $L_f/L_p = \sin\theta_0$, which implies, as the figure indicates, that the pivot point is at the intersection of the mirror and grating planes and directly below the rear facet of the diode.

## 9.9 Semiconductor laser amplifiers and injection locking

Despite the efforts of many semiconductor laser researchers, it has proven to be very difficult to obtain high power (1 W or more) from a diode laser together with a narrow-frequency spectrum and good spatial mode. An output power of more than ≈100 mW results in optical damage to the output facet of the small-area diode lasers described in this chapter. Increasing the width of the active medium from several $\mu$m to 100 $\mu$m or more to allow higher-power operation severely degrades the transverse mode of the laser. These wide-stripe lasers can produce more than 1 W and are useful for pumping other lasers but are very poor sources of narrowband radiation with a well-behaved transverse mode. The best solution at present is to use a semiconductor *traveling-wave amplifier* to increase the power of a narrowband diode to useful levels.

Traveling-wave amplifiers are ordinary diode lasers with antireflection coatings at each end. They suffer from the same problems as diode lasers: optical damage unless the cross-sectional area is made fairly large. While a broad-area amplifier will probably not have a problem with transverse mode quality, it is fairly inefficient and has relatively low gain. The best solution is to combine the efficiency of a narrow-area device with the robustness of a broad-area diode and make the active region *flared*. If the flare is

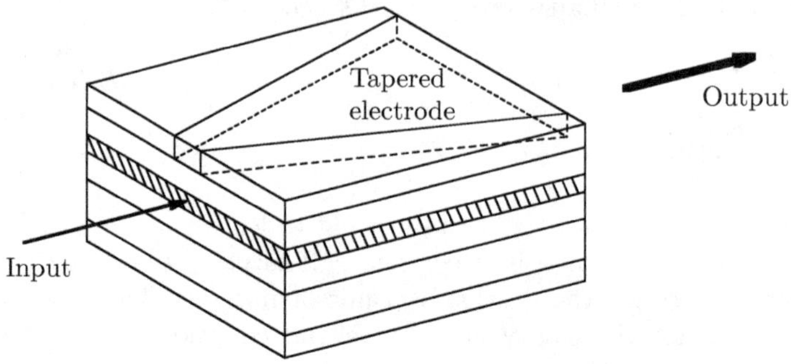


**Fig. 9.24** Gain-guided tapered amplifier. The tapered optical waveguide is defined by the tapered injection current distribution.

smooth and gradual, the mode will adiabatically fill the area as the wave progresses and the intensity (and therefore electric field) at the output facet will be below the damage threshold even for output powers in the 1 W range. A schematic of a gain-guided flared traveling-wave amplifier appears in Fig. 9.24.

The tapered (flared) amplifier will also solve another problem: the power limitation due to saturation of the active medium in a diode with a narrow channel. To see why this is so, we will analyze a general traveling-wave amplifier.

The behavior of the intensity as a function of position in a traveling-wave amplifier with uniform cross section is described by the equation

$$\frac{dI}{dz} = \left(\frac{g_0}{1 + I/I_s} - a\right) I, \tag{9.200}$$

where $g_0$ is the unsaturated gain, $I_s$ is the saturation intensity and $a$ is the absorption coefficient due to loss in the active medium. The first term in brackets is just the saturated gain (eqn 5.170). This equation does not have a closed solution. We will examine certain limiting cases by first ignoring the absorption term, $a$. One can *formally* integrate the equation, obtaining

$$\ln \frac{I_{out}}{I_{in}} + \frac{I_{out} - I_{in}}{I_s} = g_0 L, \tag{9.201}$$

where $L$ is the length of the amplifier. The power gain of the amplifier, $G$, is then

$$G = \frac{I_{out}}{I_{in}} = G_0 \exp\left(-\frac{I_{out} - I_{in}}{I_s}\right) = G_0 \exp\left(-\frac{(G-1)I_{out}}{GI_s}\right), \tag{9.202}$$

where $G_0$ is the small-signal gain (in the absence of saturation)

$$G_0 = e^{g_0 L}. \tag{9.203}$$

If $G_0 \gg 1$, the value of $I_{out}$ which reduces the gain to one-half its small-signal value is

$$I_{out} \approx (\ln 2) I_s. \tag{9.204}$$

Thus, we see that the gain rapidly diminishes when the intensity approaches the saturation intensity. When $I \gg I_s$ at the input, the gain is unity and the amplifier is essentially *transparent*. This interesting result can be obtained by integrating eqn 9.200 in the limit $I \gg I_s$ and $I \gg g_0 I_s L$. Typically, saturating *powers* in laser diodes are 100 mW or less.

We turn next to the efficiency of the amplifier. The extracted intensity is defined by $I_{out} - I_{in}$ and is given by

$$I_{extr} \equiv I_{out} - I_{in} = I_s \ln \frac{G_0}{G}. \tag{9.205}$$

The *maximum extracted intensity*, $I_{avail}$ (*available intensity*) is given by

$$I_{avail} = I_s \ln G_0, \tag{9.206}$$

since $G = 1$ at high power. If we define the *efficiency*, $\eta$, as $I_{extr}/I_{avail}$,

$$\eta = 1 - \frac{\ln G}{\ln G_0}. \tag{9.207}$$

The last term is just the ratio of the gains expressed in dB units. Thus, to extract half of the power, we lose half of the gain, expressed in dB. We therefore have the conflicting requirements that to obtain the maximum output intensity and efficiency, we need the output to be near the saturating level, but not too near, since the gain suffers when the output begins to saturate the gain medium.

If we include the absorption term in eqn 9.200, we can obtain another implicit formula for $I_{out}/I_{in}$ by integration:

$$\ln \frac{I_{out}}{I_{in}} = (g_0 - a)L + \frac{g_0}{a} \ln \left( \frac{g_0 - a(1 + I_{out}/I_s)}{g_0 - a(1 + I_{in}/I_s)} \right). \tag{9.208}$$

From eqn 9.200, one can see that the intensity approaches an asymptote, $I_{max}$, when the expression in brackets approaches zero:

$$I_{max} = (g_0/a - 1)I_s. \tag{9.209}$$

The value of $z$ where the intensity is near maximum is given by $L_{max}$:

$$L_{max} \approx \frac{1}{g_0 - a} \ln \left[ \frac{g_0 - a}{a} \frac{I_s}{I_{in}} \right]. \tag{9.210}$$

Thus, there is no point in making the amplifier longer than $L_{max}$. Beyond this length, all of the stimulated photons are absorbed and the traveling-wave intensity does not increase.

The above expressions are all written in terms of the *intensity*, which is simply proportional to the power in an amplifier with uniform cross section. The saturation intensity is *fixed* and is given approximately by

$$I_s = \frac{h\nu}{\sigma\tau}, \tag{9.211}$$

for a transition whose cross section is $\sigma$ and whose decay time is $\tau$. The saturating *power* can, however, be changed by changing the cross-sectional area of the active

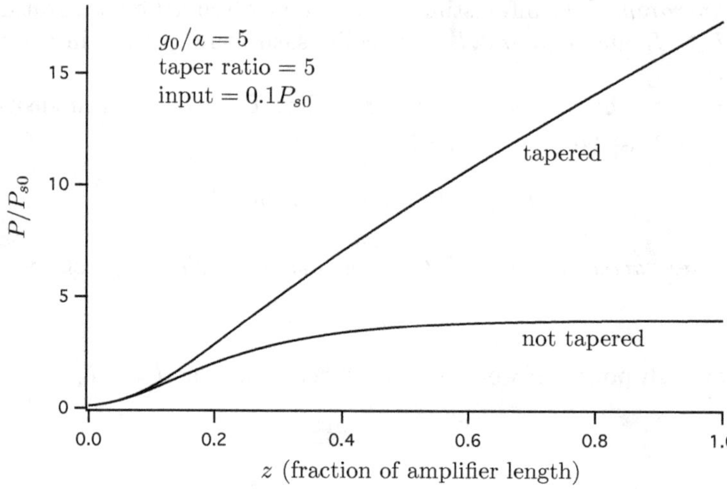


**Fig. 9.25** Plots of power as a function of position in tapered and uniform amplifier (the differential equations were easily integrated using the free computer program, *Octave*). $P_{s0}$ is the saturating power at the amplifier *input.*

medium. If the width of the active medium is linearly tapered, the saturating power is given by

$$P_s = P_{s0}(1 + k_t z), \tag{9.212}$$

where $P_{s0}$ is the saturating power at the input end ($z = 0$) and the taper is described by the constant $k_t$. For example, if the width increases by a factor of 10, $k_t$ would be $9/L$. (It is implicitly assumed that the field fills the active area of the medium as its width increases.) The equation for the traveling-wave amplifier, written in terms of power, is now

$$\frac{dP}{dz} = \left( \frac{g_0}{1 + \frac{P}{P_{s0}(1+k_t z)}} - a \right) P. \tag{9.213}$$

It should come as no surprise that the taper removes many of the limitations on the amplifier gain due to saturation. Plots of the growth of the power of the traveling wave for both a tapered and a uniform amplifier appear in Fig. 9.25. The input power is $0.1 \times P_s$ and both amplifiers show the expected exponential growth until the power approaches $P_s$. Then, the uniform amplifier power ceases to grow while the tapered amplifier power grows in a linear fashion allowed by the linear taper in the amplifier width.

A traveling-wave amplifier operates *below threshold* for laser oscillation, since the facets are antireflection coated. It is also possible to use an *above-threshold* laser oscillator to coherently increase the power of a low-power laser. The latter approach is called *injection locking.*

Injection locking works by sending a relatively weak laser beam, the *master beam,* into an already oscillating laser which is emitting the *slave beam.* The frequency difference between the two beams is normally fairly small and the master beam will be amplified by the gain medium in the slave laser. When the amplified master intensity is above the slave-laser intensity, the master beam can *take over* the slave laser and the latter's emission will be at the exact master frequency. This phenomenon is called *injection locking*, and the *locking bandwidth* is the maximum frequency difference between the two lasers which will result in injection locking; it depends upon the powers of the two lasers and the mode separation in the slave laser. We will derive the locking bandwidth using the rate equations for the laser field and phase.

We will first write down the time dependence of the two fields:

$$\text{Slave laser:} \quad E_{slave}(t) = E_s(t)e^{i\omega_s t+\phi(t)} \tag{9.214}$$
$$\text{Master laser:} \quad E_{master}(t) = E_i e^{i\omega_m t}. \tag{9.215}$$

The rate equations, including the injected field, are derived from eqns 9.99 and 9.105:

$$\frac{dE_s}{dt} - \tfrac{1}{2}(\Gamma R - 1/\tau_p)E_s = \eta f_d E_i \cos(\Delta(t)) \tag{9.216}$$
$$\frac{d\phi}{dt} - (\omega_s - \omega_{th}) = \eta f_d (E_i/E_s)\sin(\Delta(t)). \tag{9.217}$$

Note that the first equation is lacking the $\omega_s - \omega_{th}$ term, which is present in the rate equation derived earlier in the chapter. This is due to the definition of $E_s(t)$ in eqn 9.214: the rate equations derived earlier defined $E_s$ by factoring out the time dependence $e^{i\omega_{th}t}$ from the electric field. The above definition factors out $e^{i\omega_s t}$; thus, the $\omega_s - \omega_{th}$ term is not needed.

The right-hand terms have the following rationale. The *geometrical* relations are illustrated in Fig. 9.26. First, the injected field adds *in phase* to $E_{slave}$ and *in quadrature* to $\phi$. Thus, the injected field increases the internal *field* by $E_i \cos(\Delta(t))$ and the *phase* by $(E_i/E_s)\sin(\Delta(t))$, where $\Delta(t)$ is the instantaneous phase angle between $E_{master}$ and $E_{slave}$ and is given by

$$\Delta(t) = \Delta\omega t - \phi(t). \tag{9.218}$$

The quantity $\Delta\omega = \omega_m - \omega_s$ is the difference between the two laser frequencies. The *coupling constant* is $\eta f_d$, where $f_d = 1/t_{rt}$ is the mode spacing ($t_{rt}$ is the *round trip* time) and $\eta$ is the *coupling efficiency* into the slave laser, including mode matching and transmission efficiencies. The internal field, $E_{slave}$, bounces back and forth at the rate $f_d$, and therefore the *rate of change* of $E_s$ and $\phi$ due to the injected field is equal to the rate at which the internal field presents itself to the injected field ($f_d$) times the increment in the field and phase due to the injection.

If $\Delta N$ is the increment of the carrier density over threshold, then

$$\omega_s = \omega_{th} + \tfrac{1}{2}\alpha v_g g_N \Delta N \tag{9.219}$$
$$v_g g(N) = v_g(g_{th} + g_N\Delta N) = 1/\tau_p + v_g g_N \Delta N = \Gamma R, \tag{9.220}$$

where we used eqn 9.110, and the rate equations are now

$$\frac{dE_s}{dt} = \tfrac{1}{2}v_g g_N \Delta N E_s + \eta f_d E_i \cos\Delta(t) \tag{9.221}$$
$$\frac{d\phi}{dt} = \tfrac{1}{2}\alpha v_g g_N \Delta N + \eta f_d (E_i/E_s)\sin\Delta(t), \tag{9.222}$$

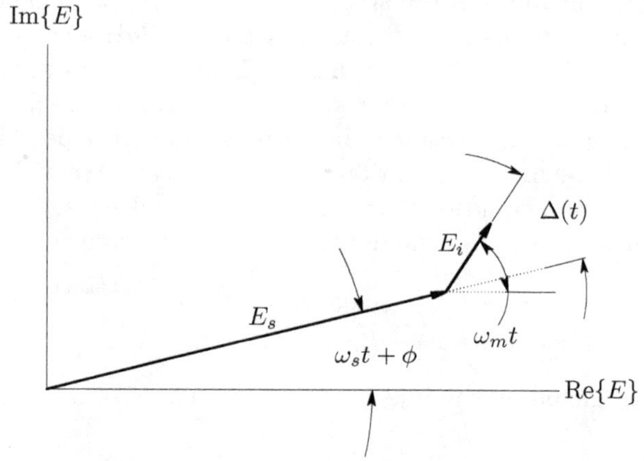


**Fig. 9.26** Phasors of slave-laser field envelope ($E_s$) and injected field amplitude ($E_i$) justifying $E_i \cos\Delta$ for the field increment and $(E_i/E_s)\sin\Delta$ for the phase increment. These increments become *rates* after multiplication by $f_d$.

If $\phi_L$ is the (constant) phase difference between the lasers when the slave is *locked* to the master, then

$$\phi(t) = \Delta\omega t + \phi_L. \tag{9.223}$$

There is a subtle difference between this $\Delta\omega$ and the previous one. This one is the frequency change of the slave laser from its free-running (no injection) value to its locked frequency. Since the two $\Delta\omega$ values are *numerically* the same, from eqn 9.218 we obtain

$$\Delta(t) = -\phi_L. \tag{9.224}$$

The steady-state solution to the first rate equation is obtained by setting $dE_s/dt = 0$, from which we can obtain

$$\Delta N = -\frac{2\eta f_d}{v_g g_N}\frac{E_i}{E_s}\cos\phi_L. \tag{9.225}$$

From the second rate equation and eqns 9.223 and 9.225, we obtain

$$\Delta\omega = -\eta f_d \frac{E_i}{E_s}[\sin\phi_L + \alpha\cos\phi_L]. \tag{9.226}$$

Setting $\eta = 1$ and combining the trigonometric functions,

$$\Delta\omega = -f_d \frac{E_i}{E_s}\sqrt{1+\alpha^2}\sin(\phi_L + \arctan\alpha). \tag{9.227}$$

Since the sine function is constrained to be in the interval $[-1, 1]$, the *locking half bandwidth* is

$$\Delta\omega_{locking} = \frac{c}{2L\mu}\frac{E_i}{E_s}\sqrt{1+\alpha^2}. \tag{9.228}$$

We notice the characteristic $\sqrt{1+\alpha^2}$ factor in the expression for the bandwidth. This factor would not have appeared if we had used the general result for the injection locking bandwidth found in a number of books and papers on the subject. On the other hand, part of this enhanced bandwidth might be unstable, as discussed in a number of papers (Mogensen (1985), for example).

## 9.10 Miscellaneous characteristics of semiconductor lasers

**Table 9.2** Wavelengths and materials for currently available diode lasers.

| Material | Wavelength range | Notes |
|---|---|---|
| AlGaN | 350–400 nm | Developmental |
| GaInN | 375–440 nm | Mostly at $\approx$ 400 nm |
| AlGaInP/GaAs | 620–680 nm | |
| $Ga_{0.5}In_{0.5}P$/GaAs | 670–680 nm | |
| GaAlAs/GaAs | 750–900 nm | |
| GaAs/GaAs | 904 nm | |
| InGaAs/GaAs | 915–1050 nm | Strained layer |
| InGaAsP/InP | 1100–1650 nm | |

We will conclude this chapter with a discussion of some of the laser diode properties which have not been mentioned earlier in this and the preceding chapter. First, the laser emission wavelength is, of course, material dependent. We briefly mentioned some binary, ternary and quaternary compounds which are used in diode lasers. Table 9.2 lists the wavelengths and materials of some currently available semiconductor lasers.

We have discussed the spectral characteristics of semiconductor lasers, but have not mentioned the far-field pattern, which is usually elliptical. If the elliptical spot at the output of an *index-guided* laser has semi-major axes of length $w_x$ and $w_y$, the angular divergences in the $x$ and $y$ planes are

$$\theta_{x,y} = \sin^{-1}\left(\frac{\lambda}{w_{x,y}\pi}\right). \tag{9.229}$$

We specified an index-guided laser since the output facet of such a laser can be considered as a source of a spherical wave whose center is the center of the facet but whose divergence angles are different in the two orthogonal planes. On the other hand, the source of the spherical wave for the $x$-plane is displaced from that for the $y$-plane in a *gain-guided* laser and therefore the beam is astigmatic.

In order to obtain a *circular beam* from a diode laser, one must use a corrective optical element which has a different magnification in the $x$- and $y$-planes: this is called an *anamorphic* optical system. There are two commonly used approaches. One can use one or more cylindrical lenses to individually adjust the magnification in the two planes. A system with two cylindrical lenses appears in Fig. 9.27. If the focal points of *both* lenses are placed at the output facet (for an index-guided laser) and

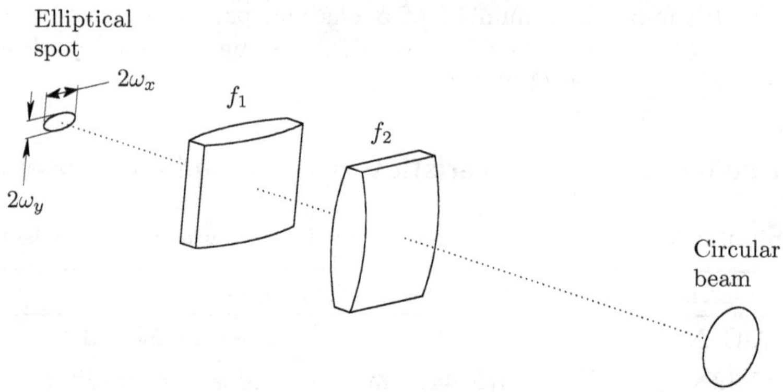


**Fig. 9.27** Use of a pair of cylindrical lenses to make a circular beam from an elliptical spot generated by a diode laser.

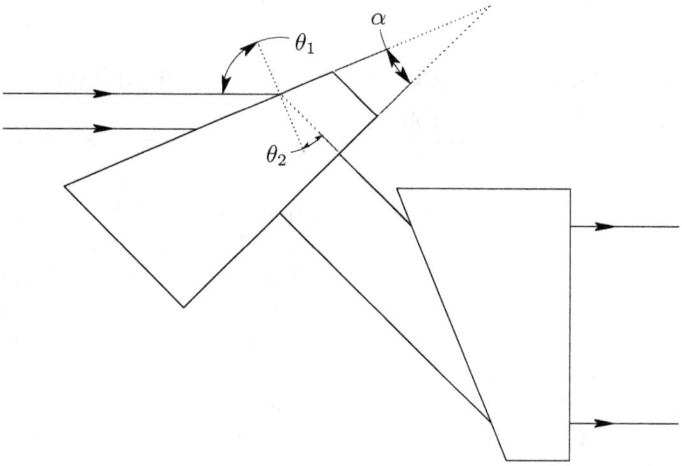


**Fig. 9.28** Use of a pair of anamorphic prisms to magnify a beam in one plane.

the focal lengths are $f_1$ and $f_2$, the different magnifications will generate a circular collimated beam when

$$\frac{f_1}{f_2} = \frac{w_y}{w_x}. \tag{9.230}$$

The advantage of this approach is that the beam axis is not changed, and small changes can be made by changes in the separation between the lenses. A possible disadvantage is due to the greater difficulty in manufacturing cylindrical lenses with $\lambda/4$ surface figures.

One can also use a pair of prisms to expand the beam along one plane. The common approach is to use *anamorphic prisms* in the configuration shown in Fig. 9.28. The two prisms are identical and are cut so that the beam enters each prism at Brewster's angle and exits normal to the output surface. If $L$ is the distance along the prism surface between the two input rays, then it is easy to see that

$$w_1 = L\cos\theta_1 \tag{9.231}$$

$$w_2 = L\cos\theta_2, \tag{9.232}$$

where $w_1$ and $w_2$ are the incident and transmitted beam widths for each prism. Dividing these equations, the magnification $M$ is

$$M = \frac{w_2}{w_1} = \frac{\cos\theta_2}{\cos\theta_1}. \tag{9.233}$$

At Brewster's angle, $\theta_1 = \tan^{-1} n$, where $n$ is the index of refraction of the prism material. Using this expression in the above equation, one can easily show that

$$\text{One prism at Brewster's angle:} \quad M = n. \tag{9.234}$$

From the requirement that the beam exit normal to the prism surface, it is easy to see that $\alpha = \theta_2$ and

$$\alpha = \tan^{-1}\frac{1}{n}. \tag{9.235}$$

The magnification for two prisms is $M^2 = n^2$. One can *tune* the magnification to some extent by rotating both prisms, though one will sacrifice either the Brewster incident angle or the normal exit beam. This system displaces the beam and only works well for one polarization: the orthogonal polarization suffers great Fresnel loss at the near grazing incidence to the prism input surfaces.

## 9.11 Further reading

The three types of semiconductor laser in order of development (homojunction, double heterostructure and quantum well) are well described in the books by Yariv (1989) Svelto (2004), Milonni (1988) (except for the quantum well), Verdeyen (1995) and Suhara (2004). The distributed feedback laser is discussed in the classic paper by Kogelnik (1972) and in the books by Yariv, Verdeyen and Svelto. One of the best and most complete discussions of diode laser modulation and noise is in the book by Petermann (1991). Our discussion of the linewidth of semiconductor lasers is based upon the classic treatment by Henry (1982). Petermann's book also contains a good treatment of the effect of optical feedback and an early discussion of external cavity diode lasers. Our discussion of the various feedback regions is based upon the paper by Tkach and Chraplyvy (1986). The treatment of *tracking* using the Littrow configuration is in the paper by Labachelerie (1993) and the analogous treatment using the Littman configuration is in the paper by Lui (1981). A complete discussion of flared amplifiers is in the paper by Lang (1993), and our analysis of injection locked diode lasers is based upon the paper by Mogensen et al. (1985). The book by Siegman (1986) also contains some useful material on both laser amplifiers and injection locking.

## 9.12 Problems

(9.1) A diode laser which uses GaAs as the active medium has a cavity length of 300 $\mu$m (uncoated facets) and emits 1 mW of radiation at 830 nm. Calculate the mode spacing and the Schawlow–Townes linewidth assuming an antiguiding factor equal to 5.

(9.2) A double heterostructure diode laser has an active layer "normalized" thickness, $D$, given by

$$D = 2\pi(n_{act}^2 - n_{clad}^2)^{1/2}\frac{d}{\lambda} \qquad \begin{aligned} n_{act} &= \text{active index} \\ n_{clad} &= \text{cladding index} \end{aligned} \tag{9.236}$$

and a confinement parameter, $\Gamma$, which is approximately

$$\Gamma \approx \frac{D^2}{2 + D^2}, \tag{9.237}$$

where $d$ is the *physical* thickness of the active layer. Using these expressions (and others) determine the approximate value of $d$ which minimizes the threshold in GaAs at 850 nm. (Assume that $D \ll 2$, the diode is 0.05 cm long, the facets are uncoated, bulk absorption can be ignored and the cladding index is 3.4.)

(9.3) Calculate the output power of a diode laser using the following approach. If $\eta$ is the *quantum efficiency* (ratio of number of internal stimulated photons to number of injected electrons), first calculate the number of stimulated photons as a function of $\eta$, the injected current, $I$, and the *threshold* current, $I_{th}$. Next, convert this to an *internal* power (using the frequency, $\nu$) and finally determine the output power by multiplying by the ratio of the output coupling to the total internal plus external loss. Using this expression, determine the *slope efficiency* ($dP/dI$).

(9.4) Assume that a laser amplifier is based upon the quantum-well structure described in the text. If the bulk absorption is 10 cm$^{-1}$, what is the longest *practical* length of an amplifier before the overall gain is limited by saturation? Assume the input power is one tenth of the saturating power.

# 10
# Guided-wave devices and fiber lasers

## 10.1 Introduction

Although the demonstration of laser gain (and oscillation) in an optical fiber is nearly as old as the laser itself, only in recent years have fiber lasers matured sufficiently to be useful in an atomic physics laboratory. An optical fiber is in many ways an ideal structure for a laser. A single-mode fiber confines both the laser mode and pumping radiation to a small cross-sectional area with a relatively long interaction length and nearly perfect overlap between the laser mode and the pumping radiation; this allows efficient pumping and provides good beam quality of the emitted radiation. The mechanical stability of a fiber together with the small Schawlow–Townes linewidth in a fiber resonator makes it relatively straightforward to construct a laser with a very narrow linewidth. Finally, the simplicity and ruggedness of the mechanical components of a fiber laser allow very reliable lasers to be built at low cost.

In this chapter, we will provide a theoretical and practical survey of the state of the fiber laser art. We will begin with an analysis of optical fibers and will introduce this subject by studying a simple "slab" optical waveguide, whose treatment is considerably more tractable than that of a fiber. Several practical aspects of fiber technology, such as fiber dispersion and coupling, will be mentioned and a number of fiber optic components will be described. A necessarily qualitative discussion of the atomic physics of rare earth ions in a glass medium will be provided and the level structures of several specific rare-earth-doped fibers will be discussed. Finally, we will describe some current continuous-wave fiber lasers (leaving mode-locked fiber lasers for a later chapter).

## 10.2 Slab waveguide: preliminary analysis

We will begin our study of guided-wave structures with the simplest possible system: a two-dimensional symmetric planar (slab) waveguide, displayed in Fig. 10.1. As shown in the figure, the structure consist of three sheets of transparent material in contact with one another, with the center section (core) having an index of refraction, $n_2$, which is greater than that of the media (cladding) on either side, whose indices are the same. Since there is an abrupt change in the index at the interface, these waveguides are called *step-index waveguides.* (We will not study waveguides with a smoothly varying index of refraction – so-called *graded-index waveguides.*) The beam propagates along the $z$-direction and the waveguide is uniform in the $y$-direction. In choosing this simple configuration for our introduction to guided-wave optics, we will avoid some of the

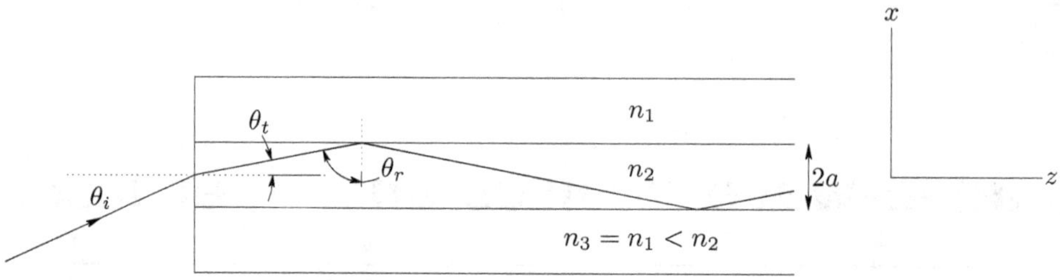


**Fig. 10.1** Side view of two-dimensional planar waveguide showing indices of refraction of the three layers needed for beam confinement.

geometrical complexities of a cylindrically symmetric fiber while describing most of the important characteristics of all guided-wave devices.

Our analysis will be based initially upon a ray approach and will proceed in several steps to a fully wave-based treatment. Assume that we have a ray incident on the input surface of the waveguide making an angle $\theta_i$ with the waveguide axis. Upon entering the central region of the waveguide, the refracted ray makes an angle $\theta_t$ with the axis. The condition for the confinement of the wave in the central layer is that the *internal* ray be *totally internally reflected* at the boundary between the core (index $n_2$) and the cladding (index $n_3 = n_1$). This will happen when the following relation is satisfied:

$$\sin\theta_r > \frac{n_1}{n_2}, \tag{10.1}$$

or, in terms of the angle $\theta_t$,

$$\cos\theta_t > \frac{n_1}{n_2}. \tag{10.2}$$

Using Snell's law at the input surface,

$$\sin\theta_i = n_2 \sin\theta_t, \tag{10.3}$$

where the input ray is assumed to be in a medium with unity index. Eliminating $\theta_t$ in the above equations, we obtain the *numerical aperture*, $NA$, of the waveguide:

$$NA \equiv \langle\text{index}\rangle \times \sin(\text{acceptance half angle}) = \sin\theta_{i,m} = \sqrt{n_2^2 - n_1^2}, \tag{10.4}$$

where $\theta_{i,m}$ is the *maximum* $\theta_i$ of a ray which is totally internally reflected. Thus, in principle, all of the incident light within an angle $\theta_{i,m}$ of the $z$-axis will be coupled into the waveguide and light outside this angle will be largely lost (by leaking into the cladding). A measure of the fractional index difference is the parameter, $\Delta$, defined by

$$\Delta \equiv \frac{n_2^2 - n_1^2}{2n_2^2} \approx 1 - \frac{n_1}{n_2}, \tag{10.5}$$

where the approximation assumes that the fractional index difference is small. Typically, $\Delta < 0.01$, which implies that $\theta_{i,m} \approx n_2\sqrt{2\Delta} < 12°$ for $n_2 = 1.5$. For a waveguide with $\Delta = 0.01$ and $n_2 = 1.5$, the numerical aperture is about 0.21.

Some insight into the wave behavior of the guided radiation can be obtained by observing that the planar waveguide resembles a solid Fabry-Pérot etalon with very off-axis light (this treatment closely follows that in the well-known text on electromagnetism by Jackson (1999)). In this model, the wave progresses down the waveguide in a zigzag fashion through multiple reflections from the core-cladding interface. Appreciable progress down the waveguide requires that the wave interferes *constructively* with itself at each reflection. In order for this to occur, the *transverse* component of the phase accumulated by the wave after two internal reflections must be an integral multiple ($q$) of $2\pi$. If the phase shift for each internal reflection is $\phi$, then

$$\frac{4n_2 a\omega}{c}\sin\theta_t + 2\phi = q(2\pi). \tag{10.6}$$

For an electric field perpendicular to the plane of incidence, the phase shift, $\phi$, is obtained from Fresnel's formula (Jackson (1999), eqns 7.39 and 7.42):

$$\phi_{TE} = -2\tan^{-1}\sqrt{\frac{2\Delta}{\sin^2\theta_t} - 1}. \tag{10.7}$$

The waveguide modes for this polarization are called *transverse electric* (TE) since the electric field is always perpendicular to the $z$-axis, while the magnetic field has some component along the $z$-axis. Combining these two equations, the allowed values of $\theta_t$ are obtained from solutions to the transcendental equation:

$$\tan\left(V\xi - \frac{q\pi}{2}\right) = \sqrt{\frac{1}{\xi^2} - 1}, \tag{10.8}$$

where

$$\begin{aligned} \xi &\equiv \frac{\sin\theta_t}{\sqrt{2\Delta}} \quad \text{and} \\ V &\equiv \frac{n_2\omega a}{c}\sqrt{2\Delta}. \end{aligned} \tag{10.9}$$

The case where the magnetic field is perpendicular to the plane of incidence is called *transverse magnetic* (TM) and is described by a slightly different Fresnel phase shift upon reflection:

$$\phi_{TM} = -2\tan^{-1}\left(\frac{1}{1-2\Delta}\sqrt{\frac{2\Delta}{\sin^2\theta_t} - 1}\right). \tag{10.10}$$

The transcendental equation for both TE and TM modes can be written as:

$$\tan\left(V\xi - \frac{q\pi}{2}\right) = f\sqrt{\frac{1}{\xi^2} - 1}, \tag{10.11}$$

where

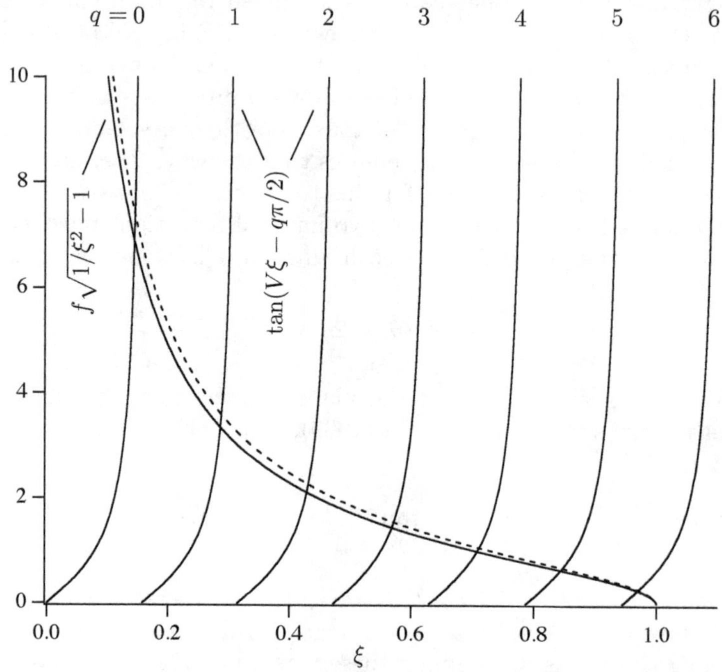


**Fig. 10.2** Graphical solution of eqn 10.11 for $V = 10$. The periodic curves are for the left side of the equation and the two non-periodic curves display the right side of the equation. The solid line of the latter is for TE modes and the dashed line is for TM. A fairly large value of $\Delta$ (0.04) was used for readability. The fourteen allowed values of $\xi$ are at the intersections.

$$\begin{aligned} \text{TE:} \quad & f = 1 \\ \text{TM:} \quad & f = \frac{1}{1-2\Delta}. \end{aligned} \tag{10.12}$$

Since the periodicity of the tangent function is already accounted for by including the $q\pi/2$ term in the argument, we restrict the argument of the tan to be in the interval $[-\pi, \pi]$. Then, we can assign values of $q$ to each curve in Fig. 10.2.

The quantity $V$, which is proportional to the optical frequency, is often called the *normalized frequency parameter* of the waveguide. In Fig. 10.2, each side of eqn 10.11 is plotted; the allowed values of $\xi$ are at the intersections between the two sets of curves. Since the separation, $\delta\xi$, between modes is roughly $\pi/(2V)$, the number of modes between $\xi = 0$ an $\xi = 1$ (its maximum value) is about $2V/\pi$ for each polarization and twice this $(4V/\pi)$ for both TE and TM modes. The fourteen intersections in the figure roughly agrees with this value. We therefore conclude that single-mode operation of the waveguide occurs when $V < 1$. This criterion will be stated more precisely later on.

The above approach is useful in correctly obtaining the parameters for the modes supported by the waveguide. One can even determine the fields in the *core* using this method. However, the calculation of the cladding fields requires a complete solution of Maxwell's equations in all three regions of the waveguide and an application of the appropriate boundary conditions. This will be done in the next section.

## 10.3 Wave propagation in a slab waveguide

Continuing with our slab waveguide model, we will next investigate the behavior of the electric and magnetic fields using Maxwell's equations. In the absence of sources, the field equations (Jackson (1999), eqns 8.125) in a medium with a spatially dependent index of refraction are:

$$\begin{aligned} \nabla^2 \boldsymbol{E} + n^2 k_0^2 \boldsymbol{E} &= -\nabla \left[ \frac{1}{n^2} (\nabla n^2) \cdot \boldsymbol{E} \right] \\ \nabla^2 \boldsymbol{H} + n^2 k_0^2 \boldsymbol{H} &= i\omega\epsilon_0 (\nabla n^2) \times \boldsymbol{E}, \end{aligned} \tag{10.13}$$

where $k_0 = \omega/c$ is the magnitude of the vacuum wave vector. These equations serve as the basis for all of our waveguide theory using a wave-based approach. In a *homogeneous medium*, the right-hand sides of these equations are zero and the equations are the familiar homogeneous wave equations that can be solved *separately* for the electric and magnetic fields. In an inhomogeneous medium, the electric and magnetic fields are coupled together by the non-zero right-hand terms, complicating the analysis. There are two ways of coping with this. One technique is to treat the waveguide as a single medium with an index $(n(x, y, z))$ which depends upon the spatial coordinates and solve the equations subject to boundary conditions at infinity. This might be a suitable approach for *graded-index* waveguides where the index is a smooth function of position. A more tractable approach for the *step-index* waveguide we are studying is to separately solve the equations (with zero right-hand sides) in each homogeneous region and match the solutions at the boundaries using the well-known boundary conditions for the electric and magnetic fields. We will use the latter approach in all that follows. It should be kept in mind that, even if one uses the second approach and solves the wave equation in each homogeneous region, the presence of the $\nabla n^2$ terms in eqns 10.13 will still couple the $\boldsymbol{E}$ and $\boldsymbol{H}$ fields in certain geometries (the coupling will be due to the boundary conditions in these geometries).

The time-independent wave equations for the three layers of the slab waveguide are:

$$\begin{aligned} \text{Layers 1 and 3:} \quad & \nabla^2 u + n_1^2 k_0^2 u = 0 \\ \text{Layer 2:} \quad & \nabla^2 u + n_2^2 k_0^2 u = 0, \end{aligned} \tag{10.14}$$

where $u$ is a component of the electric or magnetic field. The solutions will be written as the product of a function, $u_0(x)$, of the transverse coordinate, $x$, and a traveling-wave function depending only on $z$ and $t$ and having the form of a plane wave with propagation constant $\beta_z$:

$$u(\boldsymbol{r}, t) = u_0(x) e^{i(\beta_z z - \omega t)}. \tag{10.15}$$

The wave equations in each layer are now:

$$\begin{aligned} \text{Layers 1 and 3:} \quad & \nabla_t^2 u_0 + (n_1^2 k_0^2 - \beta_z^2) u_0 = 0 \\ \text{Layer 2:} \quad & \nabla_t^2 u_0 + (n_2^2 k_0^2 - \beta_z^2) u_0 = 0, \end{aligned} \tag{10.16}$$

where $\nabla_t = \frac{\partial^2}{\partial x^2}$ is the *transverse Laplacian* (all derivatives with respect to $y$ are zero since the waveguide is uniform in that direction). Note that the value of $\beta_z$ *is the same in all three regions.* The nature of the propagating wave in each region depends upon the sign of the bracketed quantity: if it is positive, the solution will have an oscillatory transverse dependence and if it is negative, the transverse dependence will be exponentially damped:

$$\begin{aligned} n_{1,2}^2 k_0^2 - \beta_z^2 &> 0 \quad \text{Oscillatory} \\ n_{1,2}^2 k_0^2 - \beta_z^2 &< 0 \quad \text{Damped.} \end{aligned} \tag{10.17}$$

It will be useful to examine the behavior of the wave for various values of $\beta_z$:

1. $\beta_z > n_1 k_0, n_2 k_0$. The transverse field dependence is exponentially damped in all three regions. It is impossible to satisfy the boundary conditions at the region interfaces with a finite field and therefore there is no propagating wave.
2. $n_2 k_0 > \beta_z > n_1 k_0$. The transverse dependence is oscillatory in the central region (core) and exponentially damped in the cladding. This is the condition which results in propagating waves in the core without leakage into the cladding and is the desired range of values.
3. $n_1 k_0, n_2 k_0 > \beta_z$. The transverse dependence is oscillatory in all three regions and the wave will leak into the cladding, with little actual propagation in the core.

Additional insight into the wave behavior can be obtained by solving the wave equation for $u_0(x)$ in the *core* under condition 2. It is easy to see that the $x$-dependence is $e^{\pm i\beta_x x}$, where

$$\beta_x = \sqrt{n_2^2 k_0^2 - \beta_z^2}. \tag{10.18}$$

Therefore, the full coordinate and time dependence of the wave for condition 2 is:

$$u(x, z, t) \propto e^{i(\pm\beta_x x + \beta_z z - \omega t)}. \tag{10.19}$$

From this equation, it is evident that $u(x, z, t)$ can be considered to be a wave propagating (internally) in the direction $\theta_t$ relative to the waveguide axis ($z$-axis), where

$$\tan\theta_t = \frac{\beta_x}{\beta_z}. \tag{10.20}$$

The minimum allowed $\beta_z$ for confinement occurs when $\beta_z = n_1 k_0$ (condition 2). The cosine of $\theta_t$ corresponding to the minimum $\beta_z$ is:

$$\text{Minimum } \beta_z\text{:} \quad \cos\theta_t = \frac{1}{\sqrt{1+\tan^2\theta_t}} = \frac{n_1}{n_2}. \tag{10.21}$$

Since $\beta_z$ decreases as $\cos\theta_t$ decreases, the minimum $\beta_z$ corresponds to the minimum $\cos\theta_t$ and we have the condition for the confinement of the wave in region 2:

$$\text{Confinement condition:} \quad \cos\theta_t > \frac{n_1}{n_2}. \tag{10.22}$$

This is precisely the same as the condition for total internal reflection obtained in the previous section using a ray model (eqn 10.2).

In the remainder of this section we will obtain a complete solution of the *vector* wave equation in the slab waveguide. In order to do so we will assume that the electric and magnetic fields have longitudinal as well as transverse components. Indeed, it can be shown that the fields in a guiding wave structure with a *simply-connected* cross-section must have at least one longitudinal component (Jackson (1999), section 8.2). A simply connected waveguide has a uniform region completely surrounded by a dielectric or metallic surface. *Multiply connected* geometries, such as coaxial cables and bifilar (two-wire) structures, can have *transverse electromagnetic* (TEM) modes in which *both* $\boldsymbol{E}$ and $\boldsymbol{H}$ are perpendicular to the waveguide axis.

The electric and magnetic fields will have the $z$-dependence of a monochromatic plane wave:

$$\begin{aligned} \boldsymbol{E}(x,y,z) &= \boldsymbol{E}(x,y)e^{i(\beta_z z-\omega t)} \\ \boldsymbol{H}(x,y,z) &= \boldsymbol{H}(x,y)e^{i(\beta_z z-\omega t)} \end{aligned} \tag{10.23}$$

and each of the six components of $\boldsymbol{E}(x,y)$ or $\boldsymbol{H}(x,y)$ will satisfy the wave equation:

$$\nabla_t^2 u + (n^2 k_0^2 - \beta_z^2)u = 0. \tag{10.24}$$

It will be useful to separate out the *longitudinal* and *transverse* components of the fields:

$$\begin{aligned} \boldsymbol{E} &= \boldsymbol{E}_t + E_z\hat{z} \\ \boldsymbol{H} &= \boldsymbol{H}_t + H_z\hat{z} \end{aligned} \tag{10.25}$$

and also to write the $\nabla$ differential operator as the sum of transverse and longitudinal terms:

$$\nabla = \nabla_t + \hat{z}\frac{\partial}{\partial z}, \tag{10.26}$$

where

$$\nabla_t = \hat{x}\frac{\partial}{\partial x} + \hat{y}\frac{\partial}{\partial y}. \tag{10.27}$$

Wave propagation in the absence of sources is described by Faraday's and Ampere's laws:

$$\begin{aligned} \nabla \times \boldsymbol{E} &= i\omega\mu_0\boldsymbol{H} \\ \nabla \times \boldsymbol{H} &= -i\omega\epsilon\boldsymbol{E}, \end{aligned} \tag{10.28}$$

where we assume there is no magnetic material ($\mu = \mu_0$). Substituting eqns 10.23 into the above equations and considering only the transverse components, we obtain:

$$\begin{aligned} \nabla_t \times (\hat{z}E_z) &= i\beta_z\hat{z} \times \boldsymbol{E}_t + i\omega\mu_0\boldsymbol{H}_t \\ \nabla_t \times (\hat{z}H_z) &= -i\omega\epsilon\boldsymbol{E}_t + i\beta_z\hat{z} \times \boldsymbol{H}_t. \end{aligned} \tag{10.29}$$

(In the above equations and in every equation from now on we are using the fields with the $z$-dependence factored out.) These equations can be solved for the transverse fields, yielding:

$$\begin{aligned}\boldsymbol{E}_t &= \frac{i}{\beta_z^2 - (\omega n/c)^2}\left[\beta_z \nabla_t E_z - \omega\mu_0 \hat{z} \times \nabla_t H_z\right] \\ \boldsymbol{H}_t &= \frac{i}{\beta_z^2 - (\omega n/c)^2}\left[\beta_z \nabla_t H_z + n^2\epsilon_0\omega \hat{z} \times \nabla_t E_z\right],\end{aligned} \tag{10.30}$$

where we have used $\epsilon = n^2\epsilon_0$. One observes that the transverse fields can be obtained from the *gradients* of the $z$-components of the fields, which behave as *potentials*. The $z$-components in a homogeneous medium are solutions of the wave equation:

$$\nabla_t^2 \begin{Bmatrix} E_z \\ H_z \end{Bmatrix} + (n^2k_0^2 - \beta_z^2)\begin{Bmatrix} E_z \\ H_z \end{Bmatrix} = 0. \tag{10.31}$$

Equations 10.30 are the same as those used in the analysis of rectangular, metallic waveguides (for transport of *microwave* radiation). When analyzing such a waveguide, the modes fall naturally into two different categories: those with $H_z \neq 0, E_z = 0$, called *transverse electric* (TE) modes, and those with $H_z = 0, E_z \neq 0$, called transverse magnetic (TM) modes. In a dielectric waveguide with rectangular geometry (such as the slab waveguide we are studying), a similar separation of modes exists (eqns 10.13 for the $z$-component of the fields are not coupled). When we study waveguides with cylindrical geometry (fibers), this neat separation is no longer possible in most cases and one can have *hybrid modes* (HE and EH) with both non-zero $E_z$ and non-zero $H_z$.

The procedure will be to solve eqn 10.31 in each of the three regions of the waveguide and obtain the transverse fields using eqns 10.30. We will then match the fields at the interface by imposing the appropriate *boundary conditions*. In an ideal metallic waveguide, in which there can be *surface charges and currents*, the boundary conditions are the continuity of the *normal* $\boldsymbol{B}$ components and the *tangential* $\boldsymbol{E}$ components. A dielectric waveguide does not support charges or currents at the interface between the media and therefore has two additional boundary conditions: the continuity of the *tangential* $\boldsymbol{H}$ and *normal* $\boldsymbol{D}$ fields. The important boundary conditions for a dielectric waveguide are:

$$\text{Boundary conditions: continuity of} \begin{Bmatrix} \text{tangential } \boldsymbol{E} \\ \text{tangential } \boldsymbol{H} \end{Bmatrix} \text{at the interfaces.} \tag{10.32}$$

The TE modes are those for which $E_z = 0$ and $H_z \neq 0$. The "potential" is $H_z$, which satisfies the equation:

$$\text{TE:} \quad \frac{\partial^2 H_z}{\partial x^2} + \beta_x^2 H_z = 0, \quad \text{where } \beta_x = \sqrt{n^2k_0^2 - \beta_z^2} \tag{10.33}$$

in each of the three regions. In region 1 and region 3, $\beta_x$ is pure imaginary. To avoid an imaginary $\beta_x$, in these regions we will use $\alpha_x$ defined as:

$$\text{Regions 1 and 3:} \quad \alpha_x \equiv \sqrt{\beta_z^2 - n_1^2k_0^2}. \tag{10.34}$$

Recalling our requirement that the fields be oscillatory in the core and have an exponentially damped $x$-dependence in the cladding, the solutions are

$$\text{TE:}\quad \begin{aligned} H_z^{clad1} &= Ae^{-\alpha_x x} \\ H_z^{core} &= B\sin\beta_x x + C\cos\beta_x x \\ H_z^{clad3} &= De^{\alpha_z x}, \end{aligned} \tag{10.35}$$

where the superscript refers to the core or the cladding in region 1 or region 3. Due to the symmetry of the waveguide about the plane $x = 0$, the solutions must be either symmetric or antisymmetric. We will treat the symmetric case first:

$$\text{TE (symmetric):}\quad \begin{aligned} H_z^{clad1} &= Ae^{-\alpha_x x} \\ H_z^{core} &= B\sin\beta_x x \\ H_z^{clad3} &= -Ae^{\alpha_x x}. \end{aligned} \tag{10.36}$$

The only transverse field components are $H_x$ and $E_y$ (since $\partial/\partial y = 0$ in a uniform slab) and they are are obtained from eqns 10.30:

$$\begin{aligned} E_y^{clad1} &= A\frac{i\omega\mu_0}{\alpha_x}e^{-\alpha_x x} \\ E_y^{clad3} &= A\frac{i\omega\mu_0}{\alpha_x}e^{\alpha_x x} \\ E_y^{core} &= B\frac{i\omega\mu_0}{\beta_x}\cos\beta_x x \\ H_x^{clad1} &= -A\frac{i\beta_z}{\alpha_x}e^{-\alpha_x x} \\ H_x^{clad3} &= -A\frac{i\beta_z}{\alpha_x}e^{\alpha_x x} \\ H_x^{core} &= -B\frac{i\beta_z}{\beta_x}\cos\beta_x x. \end{aligned} \tag{10.37}$$

Note that the division into "symmetric" and "antisymmetric" modes is determined by the *transverse fields* and not by the $z$-components, and that the latter have *opposite* symmetry from former. (Unfortunately, the literature is not always consistent in this classification.) The most straightforward way to invoke the boundary conditions at $x = a$ is to use the following relation:

$$\text{TE boundary conditions} \rightarrow \left(\frac{E_y^{clad1}}{H_z^{clad1}}\right)_{x=a} = \left(\frac{E_y^{core}}{H_z^{core}}\right)_{x=a}, \tag{10.38}$$

which combines both boundary conditions into one expression. Substituting the fields evaluated at $x = a$, one obtains

$$\text{TE (symmetric):}\quad \tan\beta_x a = \frac{\alpha_x}{\beta_x}. \tag{10.39}$$

Writing this equation in terms of $\beta_z$, we obtain:

$$\tan\left(a\sqrt{n_2^2k_0^2-\beta_z^2}\right)=\sqrt{\frac{(n_2^2-n_1^2)k_0^2}{n_2^2k_0^2-\beta_z^2}-1}. \tag{10.40}$$

This is a transcendental equation for $\beta_z$. The tan function is periodic with a period of $\pi$. We are therefore allowed to include an integral multiple of $\pi$ in the argument of the tangent. We can actually do a bit more than this. By analyzing the *antisymmetric* solutions in the same way, we obtain

$$\text{TE (antisymmetric):} \quad -\cot\beta_x a=\frac{\alpha_x}{\beta_x}. \tag{10.41}$$

The minus cotangent function is displaced by $\pi/2$ from the tangent; we therefore include all possible TE modes by solving the transcendental equation:

$$\tan\left(a\sqrt{n_2^2k_0^2-\beta_z^2}-q\pi/2\right)=\sqrt{\frac{(n_2^2-n_1^2)k_0^2}{n_2^2k_0^2-\beta_z^2}-1}, \tag{10.42}$$

where $q$ is an integer. If we make the sensible definition,

$$\tan\theta_t=\frac{\beta_x}{\beta_z}, \tag{10.43}$$

it is straightforward to show that our result is exactly the same as we obtained earlier (eqn 10.8) when we required that the wave in the core be totally internally reflected at the interface with the cladding. The quantities $\xi$, $\sin\theta_t$ and $V$ in eqn 10.9 from the previous section are

$$\begin{aligned}\xi&=\frac{1}{k_0}\sqrt{\frac{n_2^2k_0^2-\beta_z^2}{n_2^2-n_1^2}}\\ \sin\theta_t&=\frac{1}{n_2k_0}\sqrt{n_2^2k_0^2-\beta_z^2}\\ V&=k_0a\sqrt{n_2^2-n_1^2}.\end{aligned} \tag{10.44}$$

From the above, the following useful expressions for $\beta_x$ and $\alpha_x$ can be obtained:

$$\begin{aligned}\beta_x&=\frac{V\xi}{a}\\ \alpha_x&=\frac{V}{a}\sqrt{1-\xi^2}.\end{aligned} \tag{10.45}$$

In the literature, the parameters $U$ and $W$ are frequently used instead of $\beta_x$ and $\alpha_x$:

$$\begin{aligned}U&\equiv a\beta_x\\ W&\equiv a\alpha_x.\end{aligned} \tag{10.46}$$

These are related to the waveguide frequency parameter, $V$, by

$$V^2=U^2+W^2. \tag{10.47}$$

The quantities $U$ and $W$ depend upon the mode while $V$ does not.

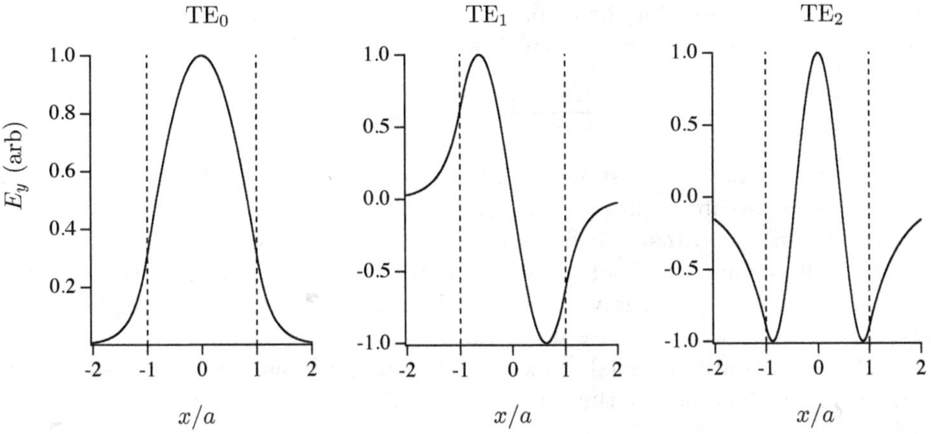


**Fig. 10.3** Electric fields for the three TE modes in a slab waveguide with $V = 4$. The dashed lines are at the interfaces between the core and the cladding.

Since the fields are functions of $\beta_x x$ and $\alpha_x x$, their $x$-dependence is a function of $U$, $W$ and $x/a$. We can write all of the fields for the lowest (symmetric) TE mode in terms of these parameters,

$$
\begin{aligned}
H_z^{core} &= \sin(Ux/a) \\
H_z^{clad1} &= \sin(U)e^{-W(x/a-1)} \\
H_z^{clad3} &= -\sin(U)e^{W(x/a+1)} \\
E_y^{core} &= \frac{i\omega\mu_0 a}{U}\cos(Ux/a) \\
E_y^{clad1} &= \cos(U)\frac{i\omega\mu_0 a}{U}e^{-W(x/a-1)} \\
E_y^{clad3} &= \cos(U)\frac{i\omega\mu_0 a}{U}e^{W(x/a+1)} \\
H_x^{core} &= -\frac{i\beta_z a}{U}\cos(Ux/a) \\
H_x^{clad1} &= -\cos(U)\frac{i\beta_z a}{U}e^{-W(x/a-1)} \\
H_x^{clad3} &= -\cos(U)\frac{i\beta_z a}{U}e^{W(x/a+1)},
\end{aligned}
\qquad (10.48)
$$

where we have used the boundary conditions to obtain a relation between $A$ and $B$ and set $B = 1$. The electric field for the lowest three TE modes is displayed in Fig. 10.3. Each mode is defined by a solution of eqn 10.2 for $\xi$, from which one can obtain $\beta_x$ and $\alpha_x$ using eqns 10.45. Each solution is characterized by a value of $q$, as seen in Fig. 10.2. Since $V = 4$, these are the only TE modes for this waveguide. In general, the value of $q$ in $\text{TE}_q$ modes is equal to the number of zero-crossings of the transverse field (neglecting the "crossings" at $\pm\infty$).

From the expressions for the fields for the lowest TE mode, we can determine the ratio between the $H_z^{core}$ to $H_x^{core}$, which we anticipate will be small:

$$\frac{|H_z^{core}|}{|H_x^{core}|} = \frac{\beta_x}{\beta_z} = \sqrt{2\Delta}. \tag{10.49}$$

Since $\Delta$ is very small in most waveguides, one can often employ the *weak guiding principle*, which assumes that $n_2 \approx n_1$. We then ignore the $z$-components of the field and obtain the transverse fields as solutions to the wave equation subject to some simple boundary conditions. Although this approximation was not necessary in the simple slab waveguide, it will be used in our study of fiber waveguides, whose treatment can otherwise be quite involved.

The TM modes are obtained in exactly the same way as the TE modes except for the equation which embodies the boundary conditions:

$$\text{TM boundary conditions} \rightarrow \left(\frac{H_y^{clad1}}{E_z^{clad1}}\right)_{x=a} = \left(\frac{H_y^{core}}{E_z^{core}}\right)_{x=a}. \tag{10.50}$$

The application of this equation yields

$$\begin{aligned} \text{TM (symmetric):} \quad \tan\beta_x a &= \frac{n_2^2}{n_1^2}\frac{\alpha_x}{\beta_x} \\ \text{TM (antisymmetric):} \quad \cot\beta_x a &= -\frac{n_2^2}{n_1^2}\frac{\alpha_x}{\beta_x}. \end{aligned} \tag{10.51}$$

These equations differ from the equations for TE modes by the factor $n_2^2/n_1^2$ on the right-hand side. It is easy to show that this factor is just $f = 1/(1-2\Delta)$, which completes the agreement between the results obtained by solution of the wave equation and the results obtained using the total internal reflection model.

Since we are ultimately interested in single-mode waveguides, it will be useful to mention an alternative graphical solution based upon a modified version of eqn 10.11. If we multiply both sides by $\xi$ we obtain

$$\xi\tan(V\xi - q\pi/2) = f\sqrt{1-\xi^2}, \tag{10.52}$$

where $f = 1$ for TE modes and $f = 1/(1-2\Delta)$ for TM modes. The right-hand side is immediately identified as the equation for a circle of unity radius for TE modes and an ellipse for TM modes. A plot of both sides of the equation appears in Fig. 10.4. As was done earlier, we restrict the argument of the tan function to be in the range $[-\pi, \pi]$ so that we can identify each curve with a $q$ value. From the equation, it is evident that the second crossing ($q = 1$) of the $\xi$-axis by the periodic function (left-hand side) will occur when $V\xi - \pi/2 = 0$ or $V\xi = \pi/2$. The criterion for single-mode operation is that this crossing occurs when $\xi > 1$ (so that it does not intersect the circle with unit radius). Therefore the single-mode criterion is:

$$\text{Single mode:} \quad V < \pi/2. \tag{10.53}$$

Since the symmetric modes correspond to even values of $q$ and the antisymmetric modes correspond to odd values, this mode is a symmetric TE or TM mode.

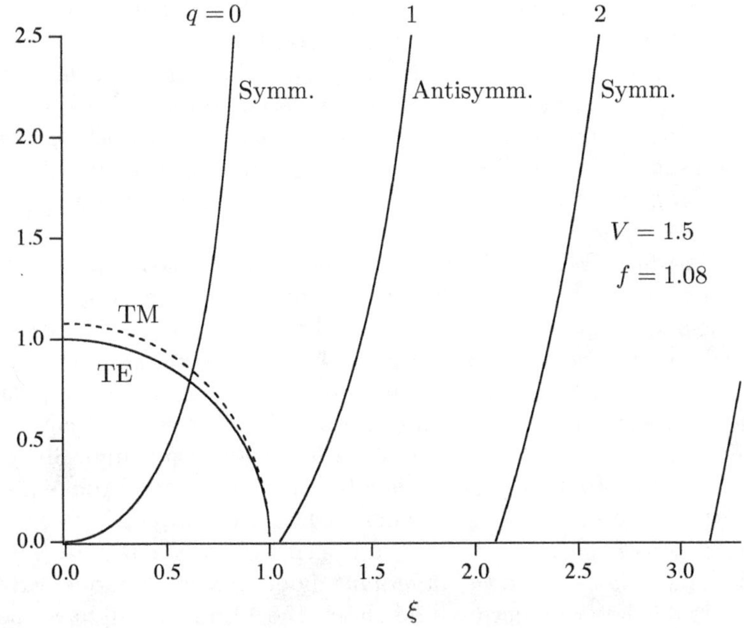


**Fig. 10.4** Plot of both sides of eqn 10.52 for $V = 1.5$. There is only one TE and TM mode; the character of the mode depends upon the light polarization. A slightly larger value of $V$ would allow additional modes.

We can rewrite eqn 10.52 in terms of the parameters $U, V$ and $W$ by multiplying both sides by $V$ and using eqns 10.45 and 10.46:

$$U \tan(U - q\pi/2) = fW. \tag{10.54}$$

We have already established that the minimum value of $\beta_z$ for the mode to be confined is $\beta_z = n_1 k_0$. If $\beta_z$ is less than this, the $x$-dependence of the mode will be oscillatory in *both* the core and cladding and the wave will leak out through the cladding. From this, we conclude that the lowest possible value for $W$ is $W = 0$. Since the frequency is a monotonic function of $W$, the *cutoff frequency is determined from*:

$$\text{Cutoff criterion:} \quad W = 0. \tag{10.55}$$

This cutoff is different for each mode and is a *lower bound* on the frequency for which the mode will propagate in the waveguide. The cutoff frequency, $\omega_{c,q}$, for mode $q$ of a slab waveguide is determined by the zeros of the tan function. Thus we obtain:

$$\text{Cutoff frequencies (slab waveguide): solutions of } \; U = V = q\pi/2$$
$$\Longrightarrow \omega_{c,q} = \frac{q\pi c}{2an_2\sqrt{2\Delta}}. \tag{10.56}$$

What happens when $q = 0$? We immediately obtain $\omega_{c0} = 0$, which neatly shows that the $TE_0$ and $TM_0$ modes have a zero-frequency cutoff ($TE_q$ and $TM_q$ modes have the same cutoff frequency). The deceptively simple eqn 10.54 allows us to determine the cutoff frequencies for each mode and also to establish the possibility that a mode has a zero frequency cutoff and is therefore the only mode in a single-mode waveguide. We will make use of equations similar to this when we study optical fibers. Equation 10.54 (and others like it) is often called in the literature an *eigenvalue equation.* Such a designation is a slightly unusual use of the phrase and should be understood to mean that the equation allows one to determine the parameters governing the propagation of the *eigenmodes* in the waveguide.

Whether a mode is TE or TM depends upon the polarization of the light. The mode will be a pure TE or TM mode if the input electric field polarization is along the $y$-axis or $x$-axis, respectively. For other polarizations, both TE and TM modes will be excited. As can be seen from Fig. 10.4, $\xi$ and therefore $\beta_z$ will be different for TE and TM modes. The wave will therefore acquire some elliptical polarization as it propagates in the waveguide, since the two polarizations will propagate at slightly different speeds. The waveguide behaves like a *birefringent* medium. Since $f > 1$, the value of $\xi$ for TM modes will be greater than that for TE modes. From eqns 10.44, this implies that $\alpha_x$ will be larger for TE modes. A larger $\alpha_x$ implies tighter confinement and a larger portion of the wave confined in the core of the waveguide. Therefore, one can consider the TE mode to be the *dominant* mode in a single-mode waveguide. For example, in a double heterostructure diode laser, the TE mode will have the lower loss (due to its better confinement) and will tend to be the one with the lowest threshold.

Before proceeding to cylindrical waveguides (fibers), we will do one more calculation: since we have expressions for the electric field in both the core and the cladding, we will calculate the *confinement parameter*, $\Gamma$, mentioned during our discussion of double heterostructure diode lasers. Recall that $\Gamma$ is defined as

$$\Gamma \equiv \frac{\int_{-a}^{a} E^2 dx}{\int_{-\infty}^{\infty} E^2 dx} = \frac{\int_{0}^{a} E^2 dx}{\int_{0}^{\infty} E^2 dx}, \tag{10.57}$$

where the last equality is due to the symmetry or antisymmetry of the fields in a symmetric waveguide. From eqns 10.48, the electric field for the lowest symmetric TE mode varies with $x$ as

$$\begin{aligned} E_y &= E_0 \cos(U) e^{-W(x/a-1)} & x &> a \\ E_y &= E_0 \cos(Ux/a) & -a &< x < a \\ E_y &= E_0 \cos(U) e^{W(x/a+1)} & x &< -a \end{aligned} \tag{10.58}$$

where various constants are lumped into $E_0$ and the $\cos(U)$ factor in the first and third equations ensures that the fields match at $x = \pm a$. Performing the integrations, one obtains

$$\Gamma = \frac{\frac{\sin 2U}{2U} + 1}{\frac{\sin 2U}{2U} + 1 + \frac{\cos^2 U}{W}}. \tag{10.59}$$

It will be useful to write this expression in terms of $\Gamma'$, the asymptotic value of $\Gamma$ as $a \to 0$. As $a$ approaches zero, $U$ also approaches zero and the two sinc terms $(\sin 2U/2U)$ together with the $\cos^2 U$ term approach unity. Therefore,

$$\Gamma' \approx \frac{2}{2+\frac{1}{W}} \approx 2W \qquad (W \ll 1). \tag{10.60}$$

For small $a$ (and $q = 0$),

$$U \tan U = W \Longrightarrow W \approx U^2 \tag{10.61}$$

and therefore

$$W/U \to 0 \quad \text{as } a \to 0. \tag{10.62}$$

Since $U^2 + W^2 = V^2$, we have $V^2 \approx U^2 \approx W$ for small $a$ and

$$\Gamma' = 2V^2. \tag{10.63}$$

Using this in the equation for $\Gamma$, we obtain the exact formula:

$$\begin{aligned} &\text{Exact formula:} \quad \Gamma = \frac{\Gamma'}{\Gamma' + B(U)} \\ &\text{where } B(U) = \frac{2U}{\tan U} \frac{1}{\frac{\sin 2U}{2U} + 1} \qquad 0 < U < \pi/2. \end{aligned} \tag{10.64}$$

Following Botez (1972), we will assert that we can replace $B(U)$ by unity and only suffer at most a 1.5% error in calculating $\Gamma$:

$$\begin{aligned} &\text{Approximate formula:} \quad \Gamma = \frac{\Gamma'}{1+\Gamma'} \\ &\text{where } \Gamma' = 2V^2 = \frac{8\pi^2 a^2}{\lambda^2}(n_2^2 - n_1^2). \end{aligned} \tag{10.65}$$

The argument for this approximation is more or less as follows. The function $B(U)$ varies from unity to 0.96 for the first half of the argument range ($0 < U < \pi/4$) and therefore its replacement by unity is a good approximation in this region. For the upper half of the argument range, $B(U)$ drops smoothly from 0.96 to zero. However, for that range, $\Gamma'$ rapidly increases from $(\pi/2)^2$ to $\infty$ (eqns 10.61 and 10.63), greatly reducing the influence of a poor approximation to $B$ on the final result. The greatest error (about 1.5%) occurs when $U \approx 1$.

This expression for $\Gamma$ has confirmed the $a^2$ dependence of $\Gamma$ for small $a$ which was asserted in our discussion of double heterostructure lasers. This result also appears in one of the problems given at the end of the Chapter 9.

We will summarize our analysis of slab waveguides with the following points:

- A slab waveguide consists of a sheet of material with index $n_2$ sandwiched between two sheets of index $n_1 < n_2$.
- Only light whose propagation vectors make an angle less than $\sqrt{n_2^2 - n_1^2}$ with the waveguide axis will be accepted by the waveguide and propagate in it. The *numerical aperture* (NA) of the waveguide is

$$\text{NA} = \sqrt{n_2^2 - n_1^2}. \tag{10.66}$$

- The modal capacity of the waveguide is determined by the parameter $V$, defined by:

$$V \equiv k_0 a \sqrt{n_2^2 - n_1^2} \tag{10.67}$$

where $2a$ is the thickness of the core of the waveguide. A waveguide will only transport a single mode if $V < \pi/2$ and the number of modes in a multimode slab waveguide is approximately $\pi V/4$.
- The possible modes in the waveguide are identified by the integer $q$ as $\mathrm{TE}_q$ and $\mathrm{TM}_q$. The quantity $q$ takes on all the positive integers starting at zero and is equal to the number of zero-crossings of the transverse fields.
- TE modes have a completely transverse electric field but allow a magnetic field with a longitudinal component. Similarly, TM modes have a transverse magnetic field and a small longitudinal electric field.
- Whether a mode is TE or TM depends upon the polarization of the light. For polarization in the plane of the waveguide the mode will be TE, and for polarization perpendicular to the plane, the mode will be TM. For any other polarization a combination of TE and TM modes will propagate and the waveguide will exhibit *birefringence.*
- All modes but those with $q = 0$ have a lower bound (cutoff) on the frequency: waves with a lower frequency than the cutoff will not propagate in the specified mode. The $\mathrm{TE}_0$ and $\mathrm{TM}_0$ modes have no such lower bound.
- The size of the longitudinal component of the field relative to the transverse component is proportional to $\sqrt{2\Delta}$, where $\Delta$ is the fractional index difference between the core and the cladding. For typical waveguides with very small $\Delta$, one can ignore the longitudinal field and assume that the modes are approximately *transverse electromagnetic* (the *weak guiding principle*).
- Since the waveguide is symmetrical about its mid-plane, the transverse fields will be either symmetrical or antisymmetrical about this plane. The lowest TE and TM modes have symmetrical fields. The symmetry of the fields then alternates with increasing mode number

## 10.4 Wave propagation in a fiber – ray theory

An optical fiber consists of a cylindrical glass core of radius $a$ and refractive index $n_2$ surrounded by a cylindrical cladding with index $n_1 < n_2$ (Fig. 10.5). As might be expected, the ray theory for a fiber is a bit more complicated than for a slab waveguide, since the cross-sectional geometry of a fiber is two-dimensional and that of a slab is one-dimensional. It is convenient to divide the rays in a fiber into two distinct categories: *meridional rays*, which propagate in a plane which contains the fiber axis and *skew rays*, which never cross the axis of the core. The two types of ray are illustrated in Fig. 10.5.

The treatment of meridional rays in a fiber is exactly the same as the treatment of all rays in a slab waveguide, since the geometry is the same for both situations. Thus, the *numerical aperture* is

$$\mathrm{NA} = \sqrt{n_2^2 - n_1^2} = n_2\sqrt{2\Delta} \tag{10.68}$$

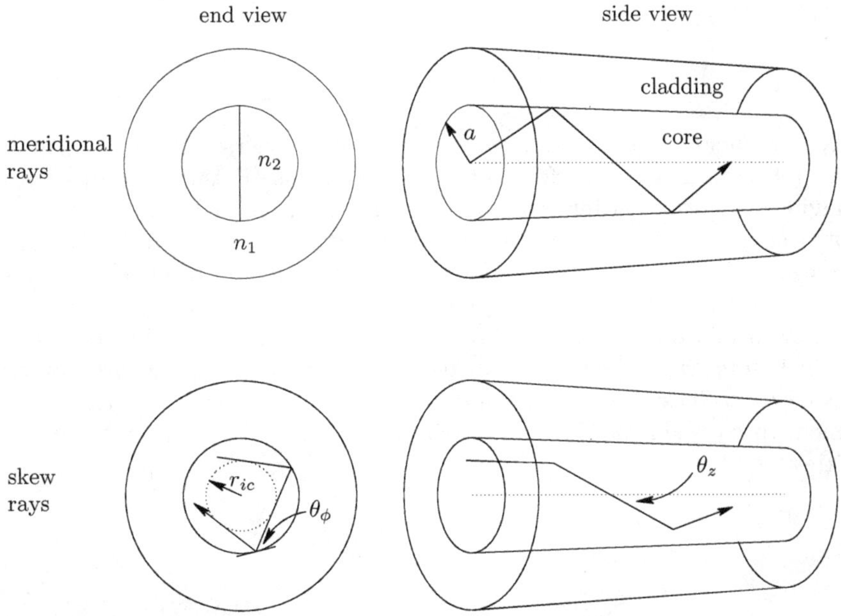


**Fig. 10.5** Side view and end view of optical fiber showing the paths of meridional and skew rays.

and the $V$-number is

$$V = k_0 a\sqrt{n_2^2 - n_1^2} = k_0 a n_2 \sqrt{2\Delta} \qquad k_0 = \frac{\omega}{c}. \tag{10.69}$$

The geometry of skew rays is a bit more complicated than that for meridional rays. The ray direction is described by two angles: $\theta_z$ is the angle the ray makes with the fiber axis ($z$-axis) and $\theta_\phi$ is the angle between the projection of the ray onto a plane perpendicular to the fiber axis and a tangent to the core (Fig. 10.5). According to the law of reflection, these angles are constant as the ray proceeds along the fiber. The bundle of skew rays is completely contained between two concentric cylindrical *caustic* surfaces: an outer one at the surface of the core and an inner one at radius $r_{ic}$, which is given by

$$r_{ic} = a\cos\theta_\phi. \tag{10.70}$$

For meridional rays, $\theta_\phi = \pi/2$ and $r_{ic} = 0$. If $\alpha$ is the angle that a ray makes with the *normal* to the core, then

$$\cos\alpha = \sin\theta_\phi \sin\theta_z. \tag{10.71}$$

This expression can be viewed as the projection of a unit vector, which points in the direction of a skew ray, onto a normal to the core. We first project the ray vector onto a plane perpendicular to the fiber axis and obtain a factor of $\sin\theta_z$ (not cosine, from the way we defined $\theta_z$). The second factor of $\sin\theta_\phi$ is obtained by projecting onto the surface normal and recognizing that the angle $\theta_\phi$ is the complement to the angle between the original projected ray and the normal to the core surface.

We define the *critical* angle, $\alpha_c$, to be

$$\sin\alpha_c = \frac{n_1}{n_2}. \tag{10.72}$$

If the ray were incident on a *planar surface*, it would be refracted if $\alpha < \alpha_c$ and totally internally reflected if $\alpha > \alpha_c$. However, the expressions (Fresnel formulae) governing the behavior of light at an interface are different for a curved surface. Therefore, $\alpha_c$ does not describe the onset of total internal reflection for skew rays (for meridional rays, for which the intersection of the plane of incidence and the surface is a straight line, the results are the same as for a planar surface). It turns out that there is a third class of skew rays which are neither refracted nor totally internally reflected. These are so-called *tunneling rays*, which are not bound in the core and are not refracted in the usual way. These rays are often called *leaky rays*. If we define the $\theta_c$ to be the complement to $\alpha_c$, then the three categories of skew rays are defined by the following inequalities:

$$\begin{aligned} \text{Bound rays:} &\quad 0 < \theta_z < \theta_c \\ \text{Refracted rays:} &\quad \theta_c < \theta_z < \pi/2 \\ \text{Leaky rays:} &\quad \theta_c < \theta_z < \pi/2; \quad \alpha_c < \alpha < \pi/2. \end{aligned} \tag{10.73}$$

The condition for bound rays is the same as in a planar waveguide. However, rays with $\theta_z > \theta_c$ *and* $\alpha_c < \alpha$ (leaky rays) seem to satisfy the criterion for total internal reflection from a planar surface ($\alpha_c < \alpha$) at the same time as the criterion for refraction ($\theta_c < \theta_z$). This class of ray can propagate in the fiber, but with greater loss than for bound rays. Therefore, the acceptance angle for a fiber will be a bit larger than that for a slab waveguide with the same $\Delta$. A complete treatment of leaky rays can be found in the references (e.g., Snyder (1983))

Before leaving the topic of the ray analysis, we would like to make three observations. First, the *travel time of the ray in the fiber is independent of $\theta_\phi$ and depends only on $\theta_z$*. This can be demonstrated as follows. If $L$ is the distance a ray travels between reflections from the surface of the core, then the *axial* distance, $\Delta z$, between bounces is:

$$\Delta z = L\cos\theta_z. \tag{10.74}$$

The optical path length is $n_2 L$ between bounces and therefore the time, $t$, taken, is

$$t = \frac{n_2 L}{c} = \frac{n_2 \Delta z}{c\cos\theta_z} \tag{10.75}$$

and the time per unit axial distance is

$$\frac{t}{\Delta z} = \frac{n_2}{c\cos\theta_z}, \tag{10.76}$$

which is independent of $\theta_\phi$. Thus, the presence of skew rays does not cause pulse broadening. The second observation is that the meridional rays correspond to TE and TM modes and are independent of the azimuthal angle in both the ray and wave theory. The existence of TE/TM modes is due to the *decoupling* of $\boldsymbol{E}$ and $\boldsymbol{H}$ in the vector wave equations (eqns 10.13) for the $z$-components of the fields when the fields are independent of the azimuthal angle. It turns out that these modes do have a low-frequency cutoff, unlike in a slab waveguide. Finally, as one might expect, the skew rays, which spiral around the core of the fiber in a helical path, have both a non-zero $H_z$ and $E_z$; i.e., they correspond to *hybrid modes*, symbolized by EH and HE. The only mode which lacks a low-frequency cutoff is a hybrid mode (the $\text{HE}_{11}$ mode) and is therefore the mode which propagates in a single-mode fiber.

## 10.5 Wave propagation in a fiber – wave theory

There are two approaches to the wave analysis of cylindrical fibers. One can use the "rigorous theory", in which one solves the wave equation for the $z$-components of the electric and magnetic fields and substitutes the resulting "potentials" into eqns 10.30 to determine the transverse components. The fields and propagation parameters ("eigenvalues") are then obtained by applying the appropriate boundary conditions at the core/cladding interface. This method is somewhat involved when treating cylindrical waveguides, since one has six field components to contend with and the general solution requires *two potentials*: both the electric and magnetic fields can have longitudinal components. A much simpler approach is to recognize that a fiber with very small $\Delta$ (fractional index difference) can be analyzed using the "weak guiding approximation", in which the longitudinal components are small enough to be ignored (eqn 10.49). We will use the second approach in all that follows.

Both approaches begin with the solution to the time-independent wave equation in cylindrical coordinates. The coordinate system is shown in Fig. 10.6. The three differential operators in cylindrical coordinates are:

$$\begin{aligned}
\nabla\phi &= \hat{r}\frac{\partial\phi}{\partial r} + \hat{\theta}\frac{1}{r}\frac{\partial\phi}{\partial\theta} + \hat{z}\frac{\partial\phi}{\partial z} \\
\nabla\cdot\boldsymbol{A} &= \frac{1}{r}\frac{\partial}{\partial r}(rA_r) + \frac{1}{r}\frac{\partial A_\theta}{\partial\theta} + \frac{\partial A_z}{\partial z} \\
\nabla^2\psi &= \frac{1}{r}\frac{\partial}{\partial r}\left(r\frac{\partial\psi}{\partial r}\right) + \frac{1}{r^2}\frac{\partial^2\psi}{\partial\theta^2} + \frac{\partial^2\psi}{\partial z^2}
\end{aligned} \tag{10.77}$$

where $\phi$ and $\psi$ are scalar functions and $\boldsymbol{A}$ is a vector field. Substituting the expression for the Laplacian into eqn 10.31 (without the $z$-term, since we are interested in the *transverse* Laplacian):

$$\frac{\partial^2\psi}{\partial r^2} + \frac{1}{r}\frac{\partial\psi}{\partial r} + \frac{1}{r^2}\frac{\partial^2\psi}{\partial\theta^2} + (n^2k_0^2 - \beta_z^2)\psi = 0 \tag{10.78}$$

where $\psi$ is any of the four transverse field components in the weak guiding approximation (and the $z$-components of the fields in the rigorous approach).

The wave equation is solved using the *separation of variables* method. One assumes a product solution,

$$\psi(r,\theta) = R(r)\Theta(\theta) \tag{10.79}$$

and obtains:

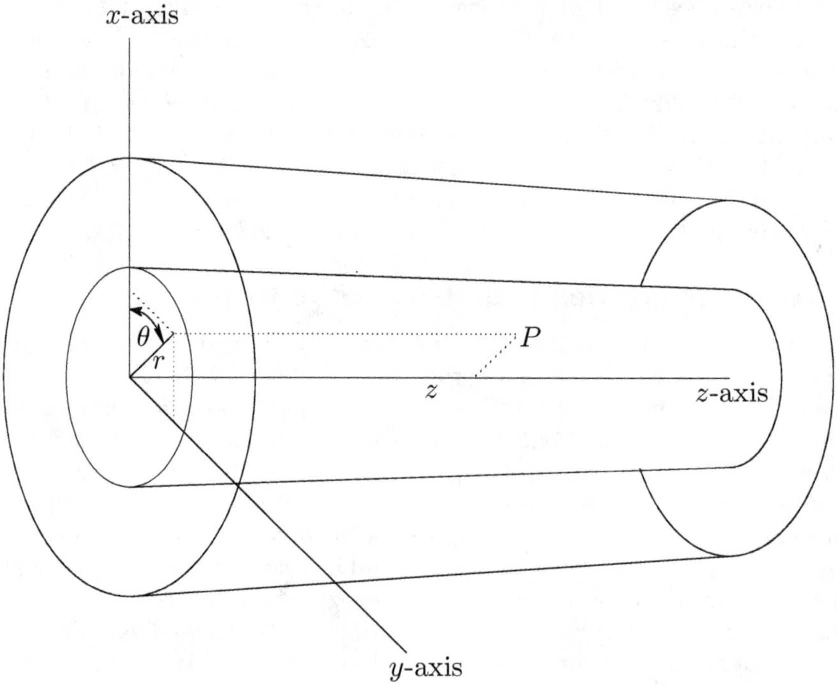


**Fig. 10.6** The cylindrical coordinate system used in the wave analysis of an optical fiber. The coordinates of the point $P$ are $r, \theta$ and $z$.

$$\frac{r^2}{R(r)}\frac{\partial^2 R(r)}{\partial r^2} + \frac{r}{R(r)}\frac{\partial R(r)}{\partial r} + r^2(n^2k_0^2 - \beta_z^2) = -\frac{1}{\Theta(\theta)}\frac{\partial^2\Theta(\theta)}{\partial\theta^2}. \qquad (10.80)$$

Since the left side depends only on $r$ and the right only on $\theta$, they both must be equal to a constant, which we will call $\nu^2$. The two resulting ordinary differential equations are:

$$\frac{d^2\Theta}{d\theta^2} + \nu^2\Theta = 0$$
$$\frac{d^2R}{dr^2} + \frac{1}{r}\frac{dR}{dr} + \left(n^2k_0^2 - \beta_z^2 - \frac{\nu^2}{r^2}\right)R = 0. \qquad (10.81)$$

The solution to the first equation is

$$\Theta(\theta) = A\sin\nu\theta + B\cos\nu\theta \qquad A, B \text{ constants} \qquad (10.82)$$

where $\nu$ is necessarily integral in order that $\Theta(\theta) = \Theta(\theta + 2\pi)$: a rotation by 360° must leave the function unchanged. The second equation is *Bessel's equation* and has two forms – one for the core and the other for the cladding:

$$\text{Core:} \quad \frac{d^2R}{dr^2} + \frac{1}{r}\frac{dR}{dr} + \left(\beta_r^2 - \frac{\nu^2}{r^2}\right) R = 0$$

$$\text{Cladding:} \quad \frac{d^2R}{dr^2} + \frac{1}{r}\frac{dR}{dr} + \left(-\alpha_r^2 - \frac{\nu^2}{r^2}\right) R = 0 \qquad (10.83)$$

$$\text{where } \beta_r = \sqrt{n_2^2 k_0^2 - \beta_z^2}$$

$$\text{and } \alpha_r = \sqrt{\beta_z^2 - n_1^2 k_0^2}.$$

Note that the first equation is transformed into the second when $r \to ir$. The solutions to these equations are called *Bessel functions* of order $\nu$. For each value of $\nu$, there are two linearly independent solutions for each equation, as required for a second-order differential equation. The solutions in the core (first equation) are called $J_\nu(\beta_r r)$ and $Y_\nu(\beta_r r)$. (The $Y_\nu$ Bessel functions are occasionally called $N_\nu$.) The two solutions in the cladding (second equation) are $I_\nu(\alpha_r r)$ and $K_\nu(\alpha_r r)$. Plots of the four types of Bessel functions for the three lowest orders appear in Fig. 10.7. If one is unfamiliar with Bessel functions, it is helpful to consider the analogy between Bessel functions and the much more familiar *circular* functions (sines and cosines), which are solutions to the azimuthal equation (and the wave equation in Cartesian coordinates). For a given Bessel function order, there are two linearly independent, oscillatory functions ($J_\nu$ and $Y_\nu$) of a real argument and two linearly independent damped functions ($I_\nu$ and $K_\nu$), which are linear combinations of $J_\nu$ and $Y_\nu$ with an imaginary argument. The relations between these two sets of solutions are:

$$I_\nu(x) = i^{-\nu} J_\nu(ix)$$

$$K_\nu(x) = \frac{\pi}{2} i^{\nu+1} \left[J_\nu(ix) + iY_\nu(ix)\right]. \qquad (10.84)$$

In the circular case, the analogous functions are sine and cosine for a real argument and sinh and cosh for an imaginary argument. Unlike the circular functions, Bessel functions have an infinite number of integral orders and are *damped*, oscillatory functions when the argument is real. Furthermore, one of the oscillatory functions ($Y_\nu$) tends to $-\infty$ as $r \to 0$, unlike the case with circular functions.

For a cylindrical waveguide, we will define the following quantities, which have exactly the same meaning as their rectangular counterparts:

$$U = \beta_r a$$

$$W = \alpha_r a \qquad \text{where } V^2 = U^2 + W^2. \qquad (10.85)$$

From the Bessel function behavior when $r = 0$ and when $r \to \infty$, the only physically acceptable radial solutions corresponding to propagating waves ($n_1 k_0 < \beta_z < n_2 k_0$, i.e., real $\beta_r$ and real $\alpha_r$) are

$$R(r) = J_\nu(\beta_r r) \qquad r < a$$

$$R(r) = K_\nu(\alpha_r r) \qquad r > a. \qquad (10.86)$$

The other functions tend to $-\infty$ at $r = 0$ and to $\infty$ as $r \to \infty$ and are therefore rejected. The complete solutions are

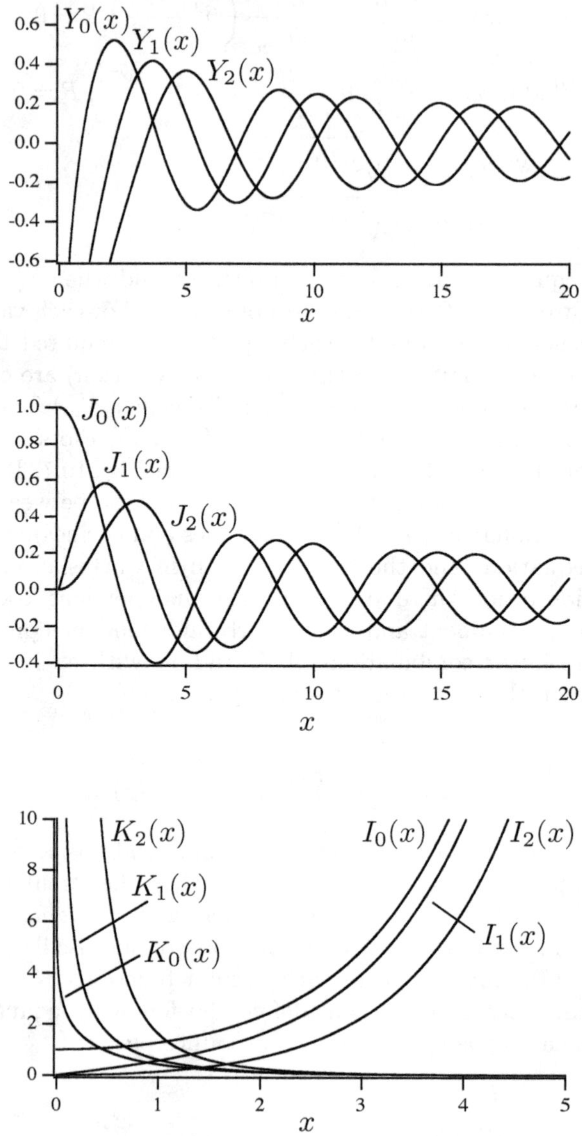


**Fig. 10.7** The four types of Bessel functions.

$$\begin{aligned} \psi(r,\theta,z) &= J_\nu(\beta_r r)(A\sin\nu\theta + B\cos\nu\theta)e^{i\beta_z z} \qquad r < a \\ \psi(r,\theta,z) &= K_\nu(\alpha_r r)(A\sin\nu\theta + B\cos\nu\theta)e^{i\beta_z z} \qquad r > a. \end{aligned} \tag{10.87}$$

One should note that the radial order ($\nu$) is the same as the azimuthal order. This gives rise to the familiar patterns which are displayed in most texts which treat laser modes using cylindrical coordinates.

We will now invoke the weak guiding approximation. In addition to assuming negligible longitudinal fields, this approximation uses boundary conditions which ignore the vectorial nature of the fields. All four *Cartesian* field components and their radial derivatives are required to be continuous at the interface between the core and cladding. The only connection to the vectorial nature of the fields is the assumption that they are *linearly polarized*. For this reason the resulting modes are called LP (linearly polarized) modes.

$$\begin{aligned}\text{Weak guiding approximation:}\quad & \psi = E_x, E_y, H_x, H_y \\ & \psi \text{ is solution of scalar wave equation} \\ & \psi \text{ is continuous at core/cladding boundary} \\ & \frac{\partial\psi}{\partial r} \text{ is continuous at core/cladding boundary.}\end{aligned} \tag{10.88}$$

Thus, the approach treats the four field components as solutions of the *scalar wave equation*, which requires that the scalar solution ($\psi$) and its spatial derivative ($\partial\psi/\partial r$) be continuous everywhere. For this reason, the weak binding approximation is often called the *scalar approximation*. Clearly, the boundary conditions are only rigorously valid when $n_1 = n_2$ and the medium is completely homogeneous. This technique is, however, useful in describing the modes (particularly the lowest) in actual fibers with a finite (but small) $\Delta$.

The determination of the modal parameters is done in much the same way as with a slab waveguide. The transverse dependence of $\psi$ is given by

$$\begin{aligned}\psi(r,\theta) &= \frac{1}{J_\nu(U)} J_\nu(Ur/a)(A\sin\nu\theta + B\cos\nu\theta) \qquad r < a \\ \psi(r,\theta) &= \frac{1}{K_\nu(W)} K_\nu(Wr/a)(A\sin\nu\theta + B\cos\nu\theta) \qquad r > a,\end{aligned} \tag{10.89}$$

where the leading factors ensure continuity of $\psi$ at the core/cladding interface. Invoking the continuity of the radial derivative, we obtain

$$\frac{1}{J_\nu(U)}\left(\frac{\partial J_\nu(Ur/a)}{\partial r}\right)_{r=a} = \frac{1}{K_\nu(W)}\left(\frac{\partial K_\nu(Wr/a)}{\partial r}\right)_{r=a}. \tag{10.90}$$

In order to evaluate the Bessel function derivatives, we use the following properties of $J_\nu(x)$ and $K_\nu(x)$:

$$\begin{aligned}\frac{dJ_\nu(x)}{dx} &= \tfrac{1}{2}(J_{\nu-1}(x) - J_{\nu+1}(x)) \\ \frac{dK_\nu(x)}{dx} &= -\tfrac{1}{2}(K_{\nu-1}(x) + K_{\nu+1}(x)) \\ J_\nu(x) &= \frac{x}{2\nu}(J_{\nu-1}(x) + J_{\nu+1}(x)).\end{aligned} \tag{10.91}$$

After a little algebra, we obtain the *eigenvalue equation* for an optical fiber:

$$\text{Eigenvalue equation:}\quad \frac{UJ_{\nu-1}(U)}{J_\nu(U)} + \frac{WK_{\nu-1}(W)}{K_\nu(W)} = 0. \tag{10.92}$$

In a later section, when we discuss *fiber dispersion*, we will provide an *approximate* solution to this equation. As in a slab waveguide, the cutoff frequencies are determined by setting $W = 0$. From the following limiting forms of $K_\nu(x)$ as $x \to 0$,

$$\begin{aligned}\lim_{x\to 0} K_0(x) &= -\left[\ln\left(\frac{x}{2}\right) + 0.5772\cdots\right] \\ \lim_{x\to 0} K_\nu(x) &= \frac{(\nu-1)!}{2}\left(\frac{2}{x}\right)^\nu \quad \text{for } \nu > 0,\end{aligned} \tag{10.93}$$

we observe that the second term on the left side of eqn 10.92 approaches zero for all $\nu$ as $W \to 0$. Thus the cutoff frequencies are the solutions to the equations

$$\begin{aligned}\nu = 0: &\quad J_1(V) = 0 \\ \nu = 1: &\quad J_0(V) = 0 \\ \nu > 1: &\quad J_{\nu-1}(V) = 0\end{aligned} \tag{10.94}$$

where we have used the relations

$$\begin{aligned}J_{-\nu}(U) &= (-1)^\nu J_\nu(U) \\ K_{-\nu}(W) &= K_\nu(W)\end{aligned} \tag{10.95}$$

and the fact that $U = V$ when $W = 0$. Since $J_1(0) = 0$ and $J_0(0) = 1$, the cutoff value of $V$ for $\nu = 0$ is zero. Thus, the zero-order mode has a zero-frequency cutoff. For $\nu \geq 1$,

$$\lim_{V\to 0} \frac{V J_{\nu-1}(V)}{J_\nu(V)} \neq 0 \tag{10.96}$$

since the leading terms of $J_\nu(V)$ for $\nu > 0$ are proportional to $V^\nu$ (and $J_0(0) = 1$). Thus, the only zero-frequency-cutoff mode is the one for which $\nu = 0$. The LP mode labeling follows these rules:

$$\begin{aligned}&\text{LP (linearly polarized modes): } \mathrm{LP}_{\nu q} \\ &\quad \nu = \text{order of Bessel function and azimuthal function} \\ &\quad q = \text{index of eigenvalue obtained from eqn 10.92 starting with } q = 1.\end{aligned} \tag{10.97}$$

The lowest mode is the $LP_{01}$ mode. From Fig. 10.7, one observes that the smallest Bessel function zero is the first zero of $J_0(x)$. This zero is the cutoff value of $V$ for the $LP_{11}$ mode, which is therefore the next higher mode after the $LP_{01}$ mode. This particular number is familiar to anyone who studies frequency modulation (it is the modulation index which results in carrier suppression) and is equal to 2.4048. Thus, the single-mode criterion is:

$$\text{Single mode: } 0 < V < 2.4048. \tag{10.98}$$

One can construct curves of both terms of eqn 10.92 and determine the eigenvalues ($U$ or $\beta_r$) at the intersections. Unlike for the slab waveguide, there are many sets of curves since each index ($\nu$) has its own set. Curves for $\nu = 0$ and $\nu = 5$ appear in Fig. 10.8 for

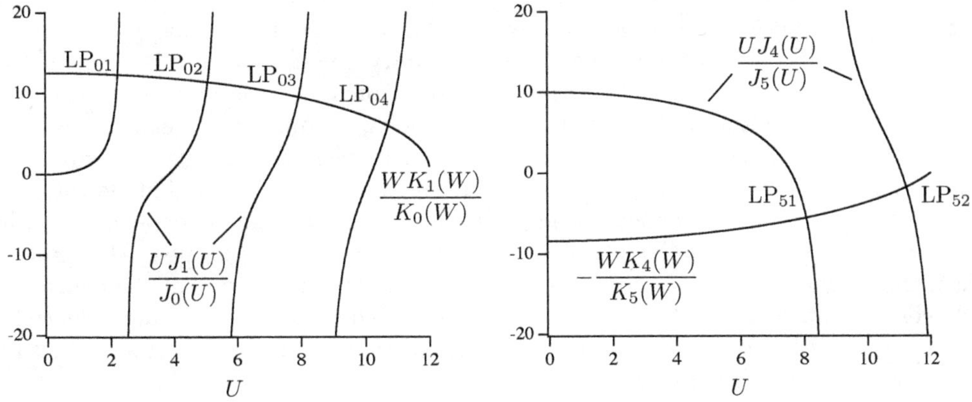


**Fig. 10.8** Graphical solution of eigenvalue equation for fiber with $V = 12$. The parameter $W$ is equal to $\sqrt{V^2 - U^2}$.

a waveguide whose frequency parameter ($V$) is 12. As is seen in the curves, we label the $q$-values in order of increasing $U$. We will state without proof that the number of modes supported by a multimode fiber when $V \gg 1$ is given by:

$$\text{Number of modes in multimode fiber} \approx V^2/2. \tag{10.99}$$

In summary, when $V < 2.4048$, a step-index optical fiber will support a single mode with no lower bound on the frequency and the linearly polarized field will be azimuthally invariant and have a radial dependence given by:

$$\begin{aligned} E(r) &= \frac{E_0}{J_0(U)} J_0(Ur/a) \qquad r < a \\ E(r) &= \frac{E_0}{K_0(W)} K_0(Wr/a) \quad r > a \end{aligned} \tag{10.100}$$

where the value of $U$ corresponding to a particular $V$ is obtained by the solution to the transcendental equation:

$$\frac{UJ_1(U)}{J_0(U)} = \frac{WK_1(W)}{K_0(W)} \qquad W = \sqrt{V^2 - U^2}. \tag{10.101}$$

When $V < 1$, we find that $U \approx V$ with the result that $W$ is very small. Thus, although single-mode fibers do not have a low-frequency cutoff, low-frequency fields are poorly bound.

Before leaving the topic of fiber modes, we would like to try to make some sense of the somewhat confusing mode designations for hybrid modes and try to relate them to the modes we have been studying. Hybrid modes are analyzed using the rigorous approach described above (and used in the analysis of the slab waveguide). The "potentials" in the core are:

$$\text{Hybrid modes (EH}_{\nu q}\text{ and HE}_{\nu q}\text{):} \quad \begin{aligned} E_z &\propto J_\nu(Ur/a) \cos \nu\theta \\ H_z &\propto J_\nu(Ur/a) \sin \nu\theta. \end{aligned} \tag{10.102}$$

Note that the first index ($\nu$) in the mode designation is the index of the Bessel function of the *longitudinal fields*. The fact that one of the components is a sine and the other a cosine is an indirect consequence of the boundary conditions (and is explained in the paper by Snitzer (1961)). The *cylindrical components* of the transverse fields are calculated by taking the transverse gradients of the longitudinal fields (eqn 10.30). This has no effect on the azimuthal index but raises and lowers the transverse field Bessel function index by one unit. Thus, the transverse fields of the $\nu$th hybrid mode will have a $J_{\nu+1}$ or $J_{\nu-1}$ radial dependence and a $\sin\nu\theta$ or $\cos\nu\theta$ azimuthal dependence. The EH and HE modes are distinguished by the dominant longitudinal field: the mode is HE if the magnetic field dominates and EH otherwise. It turns out that the transverse field components will have a $J_{\nu-1}$ dependence for HE fields and a $J_{\nu+1}$ dependence for EH fields. A rather complicated *eigenvalue equation* is obtained from the transverse fields by applying the appropriate boundary conditions. The second index ($q$) in the mode designation identifies the solution to the eigenvalue equation; by convention it starts with $q = 1$.

As an example, consider the $HE_{11}$ mode. For this mode, the dominant longitudinal field is $H_z$:

$$\text{HE}_{11}\ \text{mode:}\quad H_z \propto J_1(Ur/a)\sin\theta. \tag{10.103}$$

Since this is an HE mode, the transverse field looks like $J_0(Ur/a)\sin\theta$. The $\sin\theta$ factor can convert the cylindrical component to a Cartesian component and we are left with the functional form of the $LP_{01}$ mode. From the eigenvalue solution for this mode, it can be shown that it has no low-frequency cutoff. Thus, the $HE_{11}$ hybrid mode is *approximately* the same as the $LP_{01}$ mode.

As mentioned in the section on ray analysis, the TE and TM modes in a fiber are azimuthally invariant. They therefore can be considered to be "hybrid" modes with $\nu = 0$. From eqns 10.102, we immediately see that $H_z = 0$ when $\nu = 0$, so these equations describe the longitudinal fields for TM modes. Using eqns 10.30, the only non-zero transverse field components in the core are

$$\text{TM modes:}\quad \begin{aligned} H_\theta &\propto J_1(Ur/a) \\ E_r &\propto J_1(Ur/a) \end{aligned} \tag{10.104}$$

where we used the fact that $dJ_0(x)/dx = -J_1(x)$. Since the waveguide is rotationally invariant, we could have rotated the coordinate system by 90° about the $z$-axis and exchanged the sine and cosine field dependences. Then, we would have obtained TE mode fields, whose transverse components are

$$\text{TE modes:}\quad \begin{aligned} E_\theta &\propto J_1(Ur/a) \\ H_r &\propto J_1(Ur/a). \end{aligned} \tag{10.105}$$

These are "donut" modes: cylindrically symmetric field distributions which are zero at the center. From the eigenvalue equation, it can be shown that TE/TM modes in a fiber all have a non-zero cutoff. As with the slab waveguide, the $TE_{\nu q}$ and $TM_{\nu q}$ modes have different eigenvalues.

Table 10.1 shows the approximate equivalences between hybrid and linearly polarized modes.

**Table 10.1** Approximate equivalences between LP, HE and EH modes in a fiber.

| LP mode | Hybrid mode |
|---|---|
| $LP_{0m}$ | $HE_{1m}$ |
| $LP_{1m}$ | $HE_{2m}$ |
| $LP_{lm}$ $l \geq 2$ | $HE_{l+1,m}$, $EH_{l-1,m}$ |

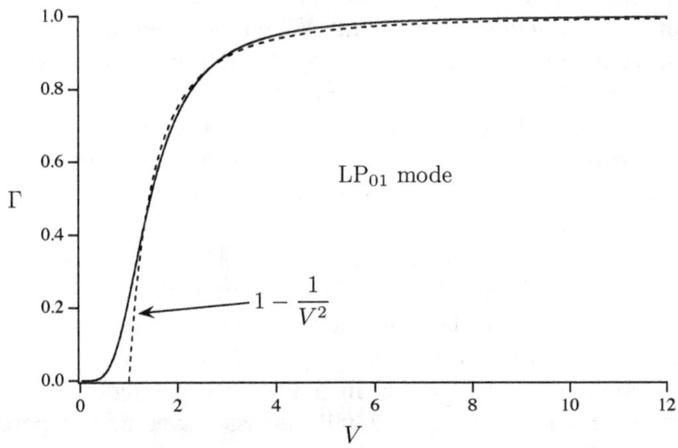


**Fig. 10.9** Plot of confinement factor (Γ) versus normalized frequency, $V$, for the $LP_{01}$ mode of a step index fiber. The solid line is based upon the exact formula and the dashed line is the approximate relation shown on the plot.

As was done at the end of our discussion of slab waveguides, we will calculate the *confinement factor*, Γ, for the lowest ($LP_{01}$) mode of an optical fiber. This quantity is defined as:

$$\Gamma = \frac{\int_0^a \int_0^{2\pi} E^2(r,\theta) r dr d\theta}{\int_0^\infty \int_0^{2\pi} E^2(r,\theta) r dr d\theta}. \tag{10.106}$$

Using the field equation (eqn 10.89) and the eigenvalue equation (eqn 10.92) together with several Bessel function integral relations, we obtain:

$$\Gamma = \frac{W^2}{V^2} + \frac{U^2}{V^2} \frac{K_\nu^2(W)}{K_{\nu+1}(W) K_{\nu-1}(W)} \quad (\text{LP}_{\nu q} \text{ mode}) \tag{10.107}$$

for a general $LP_{\nu q}$ mode. For a given $V$, the values of $W$ and $U$ are obtained from the eigenvalue equation. Figure 10.9 displays Γ versus $V$ for the $LP_{01}$ mode using the above formula and the approximate relation between $V$ and $U$ given in the section on *dispersion* (eqn 10.111). For values of $V > 1.2$, a good simple approximation for Γ is

$$\text{Approximation for } V > 1.2\text{:}\ \ \Gamma \approx 1 - \frac{1}{V^2} \quad (\text{LP}_{01} \text{ mode}). \tag{10.108}$$

## 10.6 Dispersion in fibers and waveguides

In optics, *dispersion* refers to the dependence of the index of refraction of a medium (or the *phase velocity* of a wave) on the frequency of a wave. An understanding of fiber dispersion is extremely important when fibers are used for long-distance communication, since the amount of dispersion determines how much a propagating pulse will spread and it therefore sets a limit on the ultimate bandwidth (and information carrying capacity) of a fiber. When fibers are used in lasers, dispersion can influence the mode pulling in the laser and will be important in determining the minimum pulse duration in a *mode-locked* fiber laser, as is discussed in Chapter 11.

In a fiber, dispersion is a measure of the dependence of the axial propagation constant, $\beta_z$, on the frequency of the light. When we used the ray model, we found that the internal propagation angle of the ray, $\theta_t$, depended on the frequency of the light. This dependence leads to a dependence of the axial speed of the wave on the frequency, since this speed depends upon the degree of folding of the zigzagging ray as it makes its way down the fiber. With the wave theory, we discovered that we needed to solve a transcendental equation to determine the dependence of $U$ on $V$, which provides the frequency dependence of $\beta_z$.

There are three kinds of dispersion in a fiber. *Material dispersion* is due to the dependence of the index of refraction of the fiber material on the wavelength of the light. *Modal dispersion* refers to the differing propagation constants for different *modes* and is, of course, absent in a single-mode fiber. Finally, *waveguide dispersion* is a measure of the dependence of $U$ on $V$ in a fiber for a single mode and neglecting material dispersion. In the following, we will discuss waveguide dispersion.

To determine the frequency dependence of $\beta_z$, we first write it as a function of $U$:

$$\begin{aligned}\beta_z &= \sqrt{n^2k_0^2 - \left(\frac{U}{a}\right)^2} \qquad k_0 = \omega/c \\ &= n_2k_0\sqrt{1 - \left(\frac{U}{n_2k_0a}\right)^2}.\end{aligned} \tag{10.109}$$

Since $V = k_0an_2\sqrt{2\Delta}$, we can eliminate $n_2k_0a$ and obtain

$$\begin{aligned}\beta_z &= n_2k_0\sqrt{1 - \frac{2U^2}{V^2}\Delta} \\ &\approx n_2k_0\left(1 - \frac{U^2}{V^2}\Delta\right)\end{aligned} \tag{10.110}$$

when $\Delta \ll 1$. (Some treatments include an additional $\Delta$ in the brackets due to a slight difference in the definition of $\Delta$ which becomes insignificant when $\Delta \ll 1$.) This provides the frequency dependence of $\beta_z$, since $U$ depends upon $V$ and both $V$ and $k_0$ depend upon $\omega$. Thus, to determine the dispersion relation of $\beta_z$ one first needs a relation between $U$ and $V$. An approximate solution to the *transcendental eigenvalue equation* (eqn 10.92) with an error of less than 2% for all modes was provided by Gloge (1971):

$$\begin{aligned}&\mathrm{LP}_{01}: \quad U = \frac{(1+\sqrt{2})V}{1+(4+V^4)^{1/4}} \\ &\mathrm{LP}_{\nu q}: \quad U = U_c \exp\left\{\frac{1}{s}\sin^{-1}\left(\frac{s}{U_c}\right) - \frac{1}{s}\sin^{-1}\left(\frac{s}{V}\right)\right\} \qquad \nu \neq 0.\end{aligned} \tag{10.111}$$

The quantity $s$ is given by

$$s = \sqrt{U_c^2 - \nu^2 - 1} \tag{10.112}$$

and the parameters $U_c$ are determined by the following rules:

$$\begin{aligned} \mathrm{LP}_{0q}: &\ (q \neq 1)\ U_c \text{ is } q\text{th root of } J_1(U) \text{ beginning with zeroth root} = 0 \\ \mathrm{LP}_{\nu q}: &\ (\nu \neq 0)\ U_c \text{ is } q\text{th root of } J_{\nu-1}(U). \end{aligned} \tag{10.113}$$

In Fig. 10.10 we have plotted $\beta_z$ versus $\omega$ for the first few modes of a weakly guiding fiber. One should take notice of the fact that the curves lie between the line which

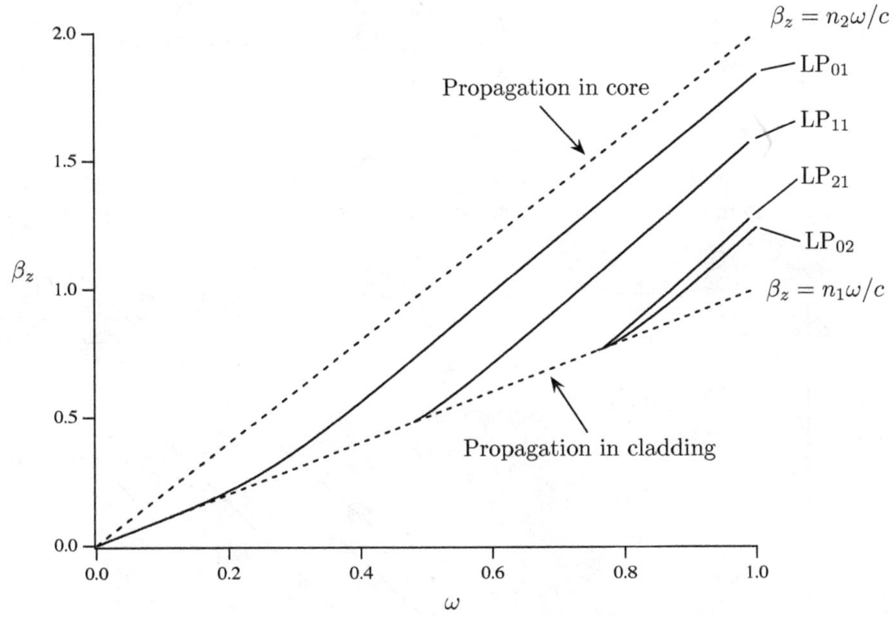


**Fig. 10.10** Plots of the dispersion curves for some of the lowest modes of a weakly guiding fiber. The frequency varies from zero to 1, which corresponds to $V = 8$. The slope difference between the core and cladding lines is highly exaggerated.

describes propagation in the cladding and that which describes propagation in the core. For high frequencies (and short wavelengths), the properties approach those of the core and at the cutoff frequency the propagation constant is that of the cladding. The $\mathrm{LP}_{01}$ curve extends all the way to $\omega = 0$, but at lower values (approximately where $V = 1$) the curve is similar to that of the cladding (which means that $W \ll U$ and the wave is poorly confined).

The treatment of dispersion in a planar slab waveguide is similar to the treatment in a fiber. We have already noted that one needs the functional form of $U$ as a function of $V$. There are undoubtedly approximations which are similar to those for a fiber, but in the absence of such an approximation one can obtain dispersion curves in the following way. We first define the *effective index of refraction*, $n_{eff}$, from the equation:

$$\beta_z = \frac{n_{eff}\omega}{c}. \tag{10.114}$$

Thus, $n_{eff}$ is the ratio of the speed of light in a vacuum to the axial speed of the wave in the waveguide and it varies between $n_1$ and $n_2$. We then substitute $\beta_z$ into the eigenvalue equation,

$$a\sqrt{n_2^2 k_0^2 - \beta_z^2}\tan\left(a\sqrt{n_2^2 k_0^2 - \beta_z^2} - q\pi/2\right) = af\sqrt{\beta_z^2 - n_1^2 k_0^2} \tag{10.115}$$

to obtain two *parametric equations* for $\omega$ and $\beta_z$ as a function of $n_{eff}$:

$$\frac{\omega}{\omega_{c1}} = \sqrt{\frac{n_2^2 - n_1^2}{n_2^2 - n_{eff}^2}}\left(q + \frac{2}{\pi}\tan^{-1} f\sqrt{\frac{n_{eff}^2 - n_1^2}{n_2^2 - n_{eff}^2}}\right)$$
$$\beta_z = \frac{n_{eff}\omega}{c}, \tag{10.116}$$

where we used the expression for the $TE_1$ cutoff, $\omega_{c1}$, from eqn 10.56. Curves can be generated by allowing $n_{eff}$ to vary in some subset of the interval $[n_1, n_2]$. Plots of the

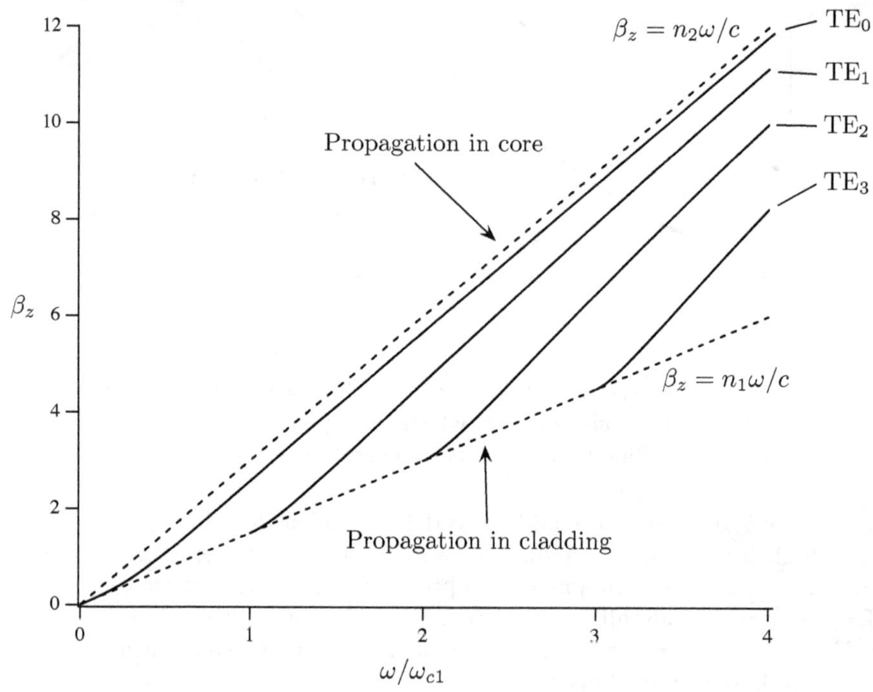


**Fig. 10.11** Plots of the dispersion curves for the four lowest modes of a slab waveguide. The frequency is expressed as a multiple of the cutoff frequency for the $TE_1$ mode ($\omega_{c1}$). The slope difference between the core and cladding lines is highly exaggerated. The range of the parameter, $n_{eff}$, was adjusted to obtain the same frequency range in all four plots.

four lowest TE modes ($f = 1$) are displayed in Fig. 10.11. The similarity between the slab dispersion curves and those for a fiber should be noted.

## 10.7 Coupling into optical fibers

In this section we will investigate the coupling of a monochromatic Gaussian beam into a single mode optical fiber. In addition to determining the parameters for the optimum coupling we will discuss the losses which occur when there are various misalignments between fibers which are "spliced" together. Some of the mathematical details of the calculations will be left out. Much of the content of this section appears in a paper by Marcuse (1977).

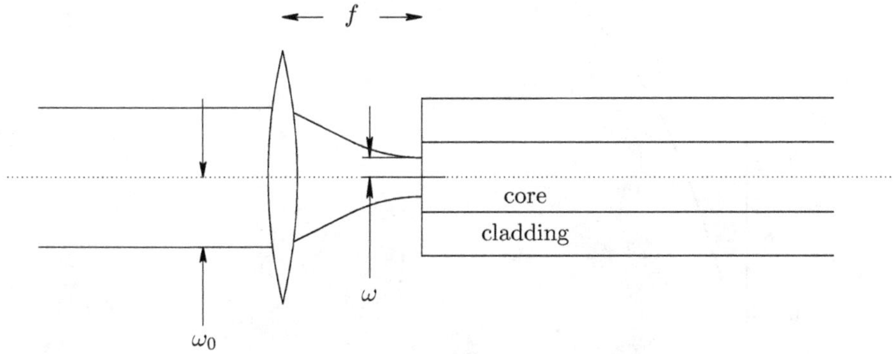


**Fig. 10.12** Schematic of lens used to couple a Gaussian beam into a single mode fiber.

We would first like to determine the optimum way to couple a Gaussian beam into a single-mode fiber. Figure 10.12 displays a Gaussian beam with radius $\omega_0$ being focused to a waist (radius $\omega$) at the input end of a fiber. If the beam has power $P$, the $y$-component of the Gaussian electric field amplitude is given by:

$$\text{Gaussian beam:} \quad E_y = \sqrt{\frac{4Z_0 P}{\pi n_1 \omega^2}} \exp\left(-\frac{r^2}{\omega^2}\right) \qquad Z_0 = \sqrt{\frac{\mu_0}{\epsilon_0}}, \tag{10.117}$$

where $Z_0 = 377\ \Omega$ is the optical "impedance" of free space and $n_1$ is the refractive index of the cladding. The $x$-component of the magnetic field of the $LP_{01}$ fiber mode with the same power is:

$$\text{LP}_{01}\text{ mode:} \quad H_x = -\sqrt{\frac{2n_1 P}{\pi Z_0}} \left(\frac{W}{aVJ_1(U)}\right) \begin{cases} J_0(Ur/a) & r < a \\ \frac{J_0(U)}{K_0(W)} K_0(Wr/a) & r > a. \end{cases} \tag{10.118}$$

The fraction, $T$, of the power coupled into the $LP_{01}$ mode is the square modulus of $c_0$, which is the *overlap integral* between the Gaussian beam and the fiber mode:

$$\begin{aligned} c_0 &= \frac{1}{2P} \int_0^{2\pi} d\theta \int_0^{\infty} E_y H_x r dr \\ T &= |c_0|^2. \end{aligned} \tag{10.119}$$

The optimum beam waist, $\omega$, depends upon the fiber $V$-parameter; the dependence is well approximated by the following empirical relation:

$$\frac{\omega}{a} = 0.65 + \frac{1.619}{V^{3/2}} + \frac{2.879}{V^6}. \tag{10.120}$$

The optimum coupling, $T$, as a function of $V$ was calculated on a computer by Marcuse (1977) and is plotted in Fig. 10.13. At each point in the plot, the waist radius was optimized using eqn 10.120. Equation 10.120 is plotted in Fig. 10.14 for the same

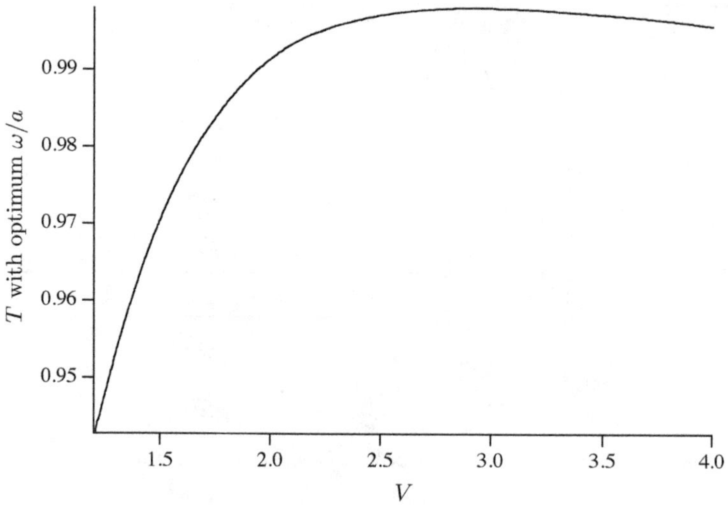


**Fig. 10.13** Fraction of power coupled from a Gaussian beam into the lowest fiber mode as a function of frequency parameter, $V$. The optimum waist size was used at each point. (from Marcuse (1977))

range of values of $V$ as the transmission plot. It should be observed that for a greater than threefold change of $V$, the transmission remains above 94%. The very broad peak is for a $V$ of about 2.8, which allows a second mode to propagate in the fiber. At the maximum $V$ for single-mode operation ($V = 2.4$), the transmission is still above 99%. Later in this section we will provide an expression for the transmission loss if eqn 10.120 is not satisfied.

It would be interesting to compare these results with the "common wisdom" (for multimode fibers) that the optimum coupling into a fiber occurs when one matches the numerical aperture of the coupling lens to that of the fiber. Using the dimensions from Fig. 10.12, the waist size is approximately

$$\omega \approx \frac{\lambda f}{\pi \omega_0} = \frac{\lambda}{\pi \langle \mathrm{NA} \rangle}, \tag{10.121}$$

where the numerical aperture (NA) of the lens is taken as $\omega_0/f$. If we match the numerical apertures, then

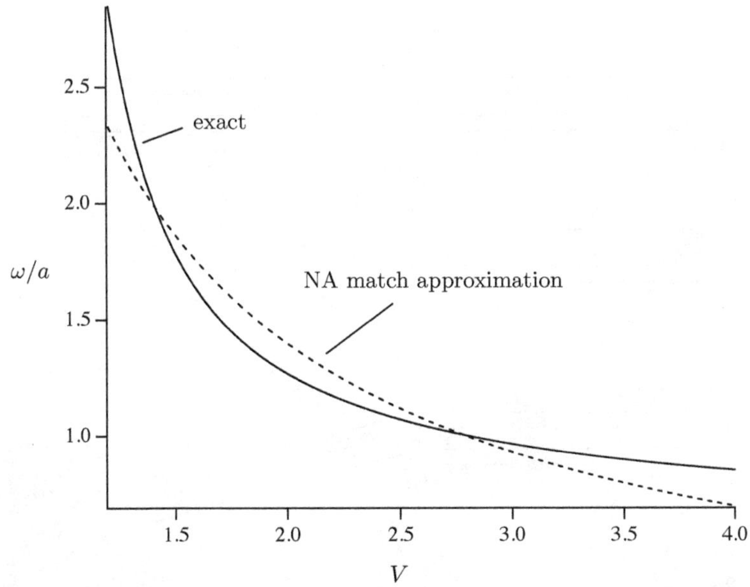


**Fig. 10.14** Relation between $\omega/a$ and $V$ for optimum coupling into a single-mode fiber. The solid line is the "exact" relation and the dashed line is the "improved NA match" approximation, $\omega/a = 2.8/V$.

$$\langle \mathrm{NA} \rangle = V/k_0 a. \tag{10.122}$$

Substituting into the equation for $\omega$, we obtain:

$$\text{NA matching criterion:} \quad \omega/a = \frac{2}{V}. \tag{10.123}$$

It is tempting to slightly modify this matching criterion to make it resemble the exact relation; an "improved" matching criterion obtained by a least square fit of a hyperbola to eqn 10.120 is:

$$\text{“Improved” NA matching criterion:} \quad \omega/a = \frac{2.8}{V}. \tag{10.124}$$

This is plotted in Fig. 10.14 as a dashed curve. Stated in words, the approximation requires that the NA of the lens be about 29% smaller than that of the fiber.

The coupling together of two fibers is called a "splice", and it would be useful to examine the possible losses that can occur when the alignment between the coupled fibers is not perfect. The four mismatches are illustrated in Fig. 10.15. The analysis of the loss uses the Gaussian approximation: the fiber field is assumed to be a Gaussian beam whose waist size is obtained from eqn 10.120. The calculations are fairly straightforward using the Gaussian beam theory already worked out in earlier chapters. To simplify matters, we will treat each of the four mismatches individually and assume the fibers are identical in all other ways. The waist radius will be $\omega$ for all of the cases except the last, where the waist radii for the two fibers are $\omega_1$ and $\omega_2$.

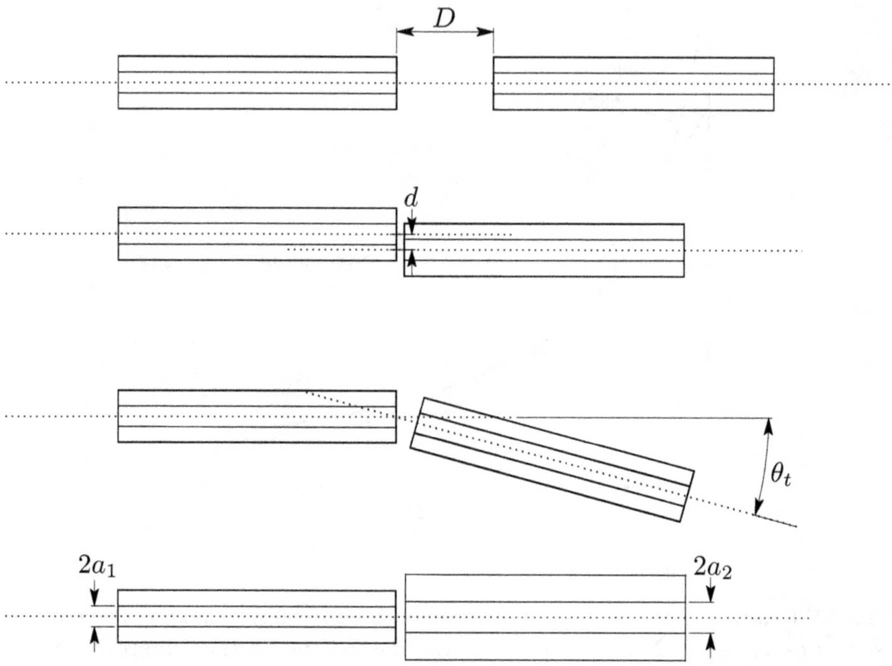


**Fig. 10.15** Four possible mismatches between fibers when spliced together.

The longitudinal fiber separation case begins with the definition of the normalized separation distance, $Z$:

$$Z \equiv \frac{D}{n_1 k_0 \omega^2}. \tag{10.125}$$

The transmission under this separation is

$$T = \frac{4Z^2 + 1}{(2Z^2 + 1)^2 + Z^2}. \tag{10.126}$$

This approaches unity as $Z \to 0$. For $Z \gg 1$, the transmission is

$$T \to \frac{1}{Z^2} = \left(\frac{n_1 k_0 \omega^2}{D}\right)^2 \quad \text{as } Z \to \infty. \tag{10.127}$$

When one of the fibers is tilted by angle $\theta_t$, the transmission is

$$T = \exp\left(-\left[\frac{n_1 \pi \omega \theta_t}{\lambda}\right]^2\right). \tag{10.128}$$

As an example, a 1° tilt in a fiber with $V = 2.4$, $\Delta = 0.002$ and $\lambda = 1\ \mu$m ($n_1 = 1.457$ yielding $a = 4.15\ \mu$m) reduces the transmission from unity to 95%.

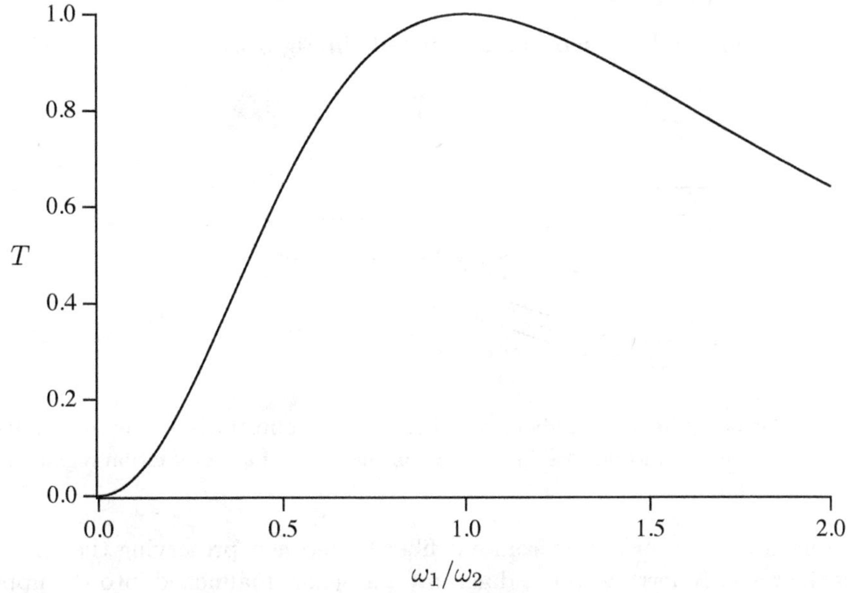


**Fig. 10.16** Transmission when two fibers with different core radii are spliced. This also provides a measure of the loss when the fiber coupling is not optimized.

The case of an offset by distance $d$ is

$$T = \exp\left(-\frac{d^2}{\omega^2}\right). \tag{10.129}$$

Finally, when there is a mismatch in the spot sizes (due to a mismatch in the core radii), the transmission is

$$T = \left(\frac{2\omega_1\omega_2}{\omega_1^2 + \omega_2^2}\right)^2. \tag{10.130}$$

In addition to providing the transmission when the fibers have different core radii, this expression also allows one to estimate the loss when the coupling from a lens is different from the optimum given in eqn 10.120. A plot of this formula appears in Fig. 10.16. The transmission reduction if fairly small for small deviations from the optimum spot size. For example, a 10% error (in either direction) will reduce $T = 1.0$ by 1%, a 20% error will reduce it by 3.3% and a 30% error will reduce it by 6.6%.

## 10.8 Fiber-optic components

In the previous section we discussed the coupling of laser light into an optical fiber using a simple lens. In this section we will survey the additional fiber-optical components likely to be found in a fiber laser. The two most important devices in this category are *fiber directional couplers* and *Bragg reflectors*. We will provide simplified descriptions of these two components together with several other devices in the following.

### 10.8.1 Directional coupler

A fiber directional coupler is shown schematically in Fig. 10.17. This is a device which

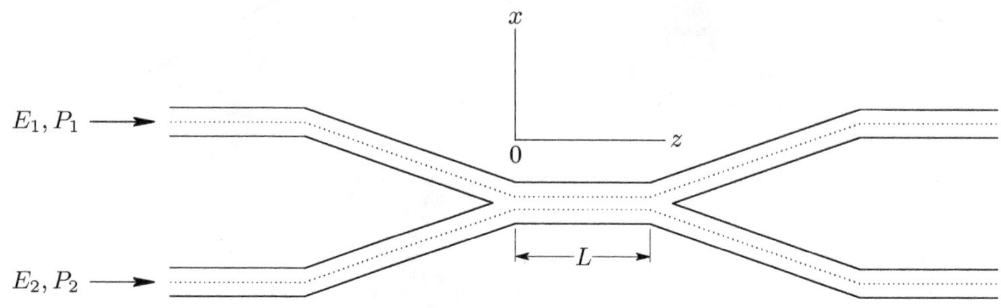


**Fig. 10.17** Schematic diagram of fiber-optical directional coupler displaying two inputs (with electric fields $E_1$ and $E_2$ and powers $P_1$ and $P_2$) on the left and an interaction region of length $L$.

couples a fixed fraction of power from one fiber to another, preserving the direction of travel of the wave. Referring to the diagram, if a beam is launched into the upper left input, with no beam in the lower left, the power will be split between the two output ports on the right and no power will appear at the lower left input.

A simplified analysis follows that given by Marcuse (1971) and assumes that the two fibers are identical and lossless. The two fibers are placed in close proximity to one another for a distance $L$, over which the claddings overlap and light can pass from one fiber to the other (via the *evanescent* wave in the cladding). We are interested only in the $z$-dependence of the electric fields, which are written as:

$$\begin{aligned} E_1(z,t) &= E_1(z)e^{i\omega t} \\ E_2(z,t) &= E_2(z)e^{i\omega t} \end{aligned} \tag{10.131}$$

Then, $E_{1,2}(z)$ are described by two coupled differential equations:

$$\begin{aligned} \frac{dE_1}{dz} &= -i\beta_z E_1 + \kappa_1 E_2 \\ \frac{dE_2}{dz} &= -i\beta_z E_2 + \kappa_2 E_1, \end{aligned} \tag{10.132}$$

where we take $z = 0$ to be at the beginning of the interaction region and the coupling between the fibers is described by the complex constants $\kappa_1$ and $\kappa_2$. These equations can be derived using *coupled mode theory* and can be justified by the following rationale. If there were no coupling, $\kappa_1 = \kappa_2 = 0$ and the solutions for the two waves would be simple traveling waves having a $e^{-i\beta_z z}$ $z$-dependence. One can consider this to be a form of *degeneracy*: there would be two modes with the same propagation constant. As a result of the coupling, the spatial growth of each wave will depend slightly (for weak coupling) on the presence of the other wave. The degeneracy will be broken by the coupling and the modes will have slightly different propagation constants.

Before proceeding to a solution, we can immediately obtain a relation between $\kappa_1$ and $\kappa_2$ based upon the conservation of power in the lossless fiber:

$$\frac{d}{dz}(P_1 + P_2) = 0. \tag{10.133}$$

From this (and the coupled equations), it is straightforward to show that $\kappa_1$ and $\kappa_2$ satisfy

$$\kappa_1 = -\kappa_2^*. \tag{10.134}$$

The equations can be solved by differentiating the first and substituting the second equation, yielding:

$$\frac{d^2 E_1}{dz} - 2i\beta_z E_1 - (\kappa_1\kappa_2 + \beta_z^2)E_1 = 0. \tag{10.135}$$

This equation is solved by substituting $e^{sz}$ for $E_1$ and solving the resulting quadratic equation:

$$s = -i\beta_z \pm \sqrt{\kappa_1\kappa_2}. \tag{10.136}$$

The general solution is:

$$\begin{aligned} E_1(z) &= f_1 e^{-i(\beta_z+\Delta\beta_z)z} + f_2 e^{-i(\beta_z-\Delta\beta_z)z} \\ E_2(z) &= f_3 e^{-i(\beta_z+\Delta\beta_z)z} + f_4 e^{-i(\beta_z-\Delta\beta_z)z} \end{aligned} \tag{10.137}$$

where the $f_i$ are constants which depend upon the boundary conditions and $\Delta\beta_z$ is

$$i\Delta\beta_z = \sqrt{\kappa_1\kappa_2} = i|\kappa_1|. \tag{10.138}$$

The boundary conditions are:

$$\begin{aligned} E_1(0) &= E_{1,0} \\ E_2(0) &= 0 \end{aligned} \tag{10.139}$$

and the solutions satisfying these boundary conditions are:

$$\begin{aligned} E_1(z) &= E_{1,0}\cos(\Delta\beta_z z)e^{-i\beta_z z} \\ E_2(z) &= -i\left(\frac{\kappa_2}{\kappa_1}\right)^{1/2} E_{1,0}\sin(\Delta\beta_z z)e^{-i\beta_z z}. \end{aligned} \tag{10.140}$$

The solutions describe fields that are periodically being exchanged between the two fibers, with a half-period, $D$, given by:

$$\text{Power exchange distance:}\quad D = \frac{\pi}{2\Delta\beta_z}. \tag{10.141}$$

At the end of the interaction region ($z = L$), the *powers* in each fiber, for a given $P_1(0)$ (and assuming that $P_2(0) = 0$) are

$$\begin{aligned} P_1(L) &= P_1(0)\cos^2\Delta\beta_z L \\ P_2(L) &= P_1(0)\sin^2\Delta\beta_z L. \end{aligned} \tag{10.142}$$

Thus, by choosing $\Delta\beta_z L$, one can have any desired coupling. For reference, the coupling between two lossless fibers can be shown to be (Marcuse (1971))

$$\Delta\beta_z = \pm\frac{\omega\epsilon_0}{4P_0}\int_{-\infty}^{\infty}\int_{-\infty}^{\infty}(n(x,y)^2 - n_{cladd}^2)\boldsymbol{E}_1^* \cdot \boldsymbol{E}_2 dxdy \tag{10.143}$$

where $n(x,y)$ is the index of refraction of the composite system, $\boldsymbol{E}_{1,2}$ are the electric fields in the two fibers and $P_0$ is the *total* power. Since the index difference is zero in the cladding, the integrals only need to be evaluated in the fiber cores. An approximation (Digonnet (1982)) for single-mode fibers is

$$\Delta\beta_z \approx \frac{\lambda}{2\pi n_2}\frac{U^2}{a^2V^2}\frac{K_0[Wh/a]}{K_1^2(W)} \tag{10.144}$$

where $\lambda$ is the vacuum wavelength of the radiation, $h$ is the core separation distance between the two fibers and the other symbols are the familiar fiber parameters.

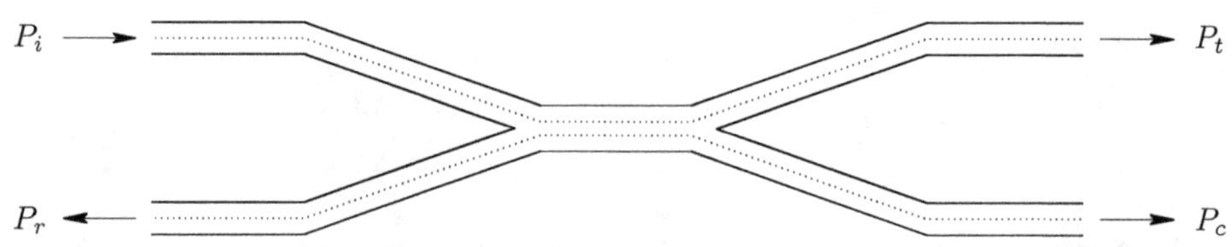


**Fig. 10.18** Diagram of fiber optical directional coupler showing the four ports and their associated powers: input ($P_i$), transmitted ($P_t$), reflected ($P_r$) and coupled ($P_c$).

The following additional comments about fiber-optic couplers are worth noting (refer to Fig. 10.18 for the port designations):

- The *coupling ratio*, $R$, in dB, is

$$R = 10\log_{10}\left[\frac{P_c}{P_c + P_t}\right] \tag{10.145}$$

- The *excess loss*, $L_i$, is

$$L_i = 10\log_{10}\left[\frac{P_i}{P_t + P_c}\right] \tag{10.146}$$

- The *insertion loss*, $I$, is

$$I = 10\log_{10}\left[\frac{P_i}{P_c}\right] \tag{10.147}$$

- The *directivity*, $D$, is

$$D = 10\log_{10}\left[\frac{P_r}{P_i}\right] \tag{10.148}$$

  In an ideal coupler, the directivity is $-\infty$ dB: $P_r = 0$ for any $P_i$.
- When $\Delta\beta_z L = \pi/4$, the input power, $P_i$, is split *evenly* between the output ports, $P_c$ and $P_t$. The coupler is then called a *3 dB coupler*.

- All of the power will be transferred from one fiber to the other when $\Delta\beta_z L = \pi/2$.
- Since $\Delta\beta_z$ depends upon wavelength, it is possible to choose $L$ so that two wavelengths appearing at the $P_i$ input are sent to separate output ports. The coupler is then called a *wavelength division demultiplexer.* The inverse process – combining different wavelengths at the $P_i$ and $P_r$ inputs into a single output port – is called *wavelength division multiplexing* (WDM). The latter process can be used to couple the *pump radiation* into a fiber amplifier.
- The couplers being described tend to be somewhat narrowband, due to the strong variation of the $\cos^2$ factor with its argument near the 3-dB point. By using *non-identical* fibers, the dependence of the coupling with wavelength can be considerably reduced.

### 10.8.2 The loop reflector

An important application of the fiber coupler is a *loop reflector*, shown schematically in Fig. 10.19. It allows one to construct an *all-fiber* resonator for use in a fiber laser. If

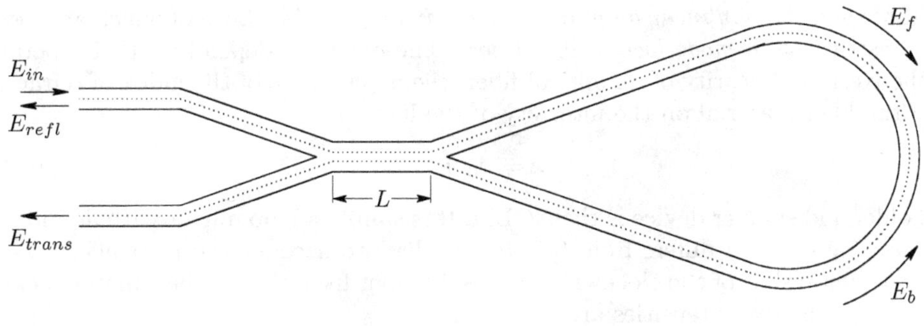


**Fig. 10.19** Diagram of loop mirror which uses a fiber coupler and loop to provide up to 100% reflection. The length of the loop is $L'$.

the two output ports of the coupler are connected together, the reflected field, $E_{refl}$, is the sum of two fields, both of which encounter the coupler twice. The forward field, $E_f$, first passes through the coupler, then enters the loop and finally returns to the coupler, crossing over to the other fiber. The backward field, $E_b$, crosses over first, then encounters the loop and finally passes through the coupler.

In order to properly analyze this device, one needs to know the phases of the coupling constants, $\kappa_1$ and $\kappa_2$; this has not been a concern until now, since we were mainly interested in how the *power* was split between the two fibers in a coupler. When the two fibers have identical propagation constants, it is shown in a paper by Snyder (1972) that the coupling constants are *pure imaginary.* Therefore, since $\kappa_1 = -\kappa_2^*$, the two coupling constants are equal and eqns 10.140 can be simplified:

$$\begin{aligned} E_1(z) &= E_{1,0}\cos(\Delta\beta_z z)e^{-i\beta_z z} \\ E_2(z) &= -iE_{1,0}\sin(\Delta\beta_z z)e^{-i\beta_z z} \end{aligned} \tag{10.149}$$

The field which crosses over the coupler suffers a 90° phase lag compared to the straight-through field.

Inserting the various factors and adding up the fields, one has

$$\begin{aligned} E_{refl} =& \left[E_{1,0}\cos(\Delta\beta_z L)e^{-i\beta_z L}\right]\left[e^{-i\delta}\right]\left[-i\sin(\Delta\beta_z L)e^{-i\beta_z L}\right] \\ &+\left[-iE_{0,1}\sin(\beta_z L)e^{-\beta_z L}\right]\left[e^{-i\delta}\right]\left[\cos(\beta_z L)e^{-i\beta_z L}\right] \\ =& -2iE_{1,0}\cos(\Delta\beta_z L)\sin(\Delta\beta_z L)e^{-i(2\beta_z L+\delta)} \\ =& -iE_{1,0}\sin(2\Delta\beta_z L)e^{-i(2\beta_z L+\delta)} \end{aligned} \tag{10.150}$$

where $\delta$ is the phase change over the length of the loop (and is the same for both directions in *single-mode* fibers). From this, the ratio of the reflected to incident power is

$$\frac{P_{refl}}{P_{in}} = \sin^2(2\Delta\beta_z L) \tag{10.151}$$

Neglecting various insertion losses and possible fiber mismatch, one can have any reflection from 0% to 100% by varying $\Delta\beta_z L$.

Two important variants of the loop mirror are the *nonlinear–optical loop mirror* (NOLM) and the *nonlinear amplifying loop mirror* (NALM), both of which are used in the construction of *mode-locked* fiber lasers. These devices depend for their operation on the Kerr nonlinearity of an optical fiber: the dependence of the index of refraction, $n$, of the fiber material on the intensity of the light, $I$:

$$n = n_0 + n_2 I. \tag{10.152}$$

The older and simpler device is the NOLM. It is simply a loop mirror with an intensity coupling ratio which differs from 0.5. If we take into account the Kerr effect, we see that the phase shift of the clockwise wave is different from that of the counterclockwise wave, since the two intensities are different.

To simplify the calculation of the NOLM transmission, we will normalize the intensity so that $I = |E|^2$. Then, we will define two nonlinear phase shifts: one for the clockwise wave and one for the counterclockwise wave.

$$\begin{aligned} \delta_{cw} &= \frac{2\pi n_2 |E_f|^2 L'}{\lambda} \\ \delta_{ccw} &= \frac{2\pi n_2 |E_b|^2 L'}{\lambda}, \end{aligned} \tag{10.153}$$

where $L'$ is the length of the loop. If we let the power transmission across the coupler be $\alpha = \sin^2(\Delta\beta_z L)$, it is straightforward to show that

$$\frac{P_{trans}}{P_{in}} = 1 - 2\alpha(1-\alpha)\left\{1 + \cos(\delta_{cw} - \delta_{ccw})\right\} \tag{10.154}$$

where

$$\delta_{cw} - \delta_{ccw} = (1-2\alpha)\frac{2\pi |E_{in}|^2 L'}{\lambda}. \tag{10.155}$$

From this last, we note that there is no power-dependent effect when $\alpha = 0.5$. When $\delta_{cw} - \delta_{ccw}$ is an *odd* multiple of $\pi$ radians, the transmission is 100%. The minimum transmission occurs when the above difference is an even multiple of $\pi$ radians and the transmission is then

$$\text{Minimum transmission:} \quad \frac{P_{trans}}{P_{in}} = 1 - 4\alpha(1-\alpha) \tag{10.156}$$

Thus, the reflectivity at the upper port is never 100% for any input power, but can approach 100% when $\alpha$ is nearly 0.5 at the expense of a great deal of power needed to switch between the two states. This state of affairs can be remedied by placing an *amplifier* close to the coupler in the loop and setting $\alpha = 0.5$. This device is then called a nonlinear amplifying loop mirror (NALM).

A schematic of a NALM appears in Fig. 10.20. Since the amplifier is next to the

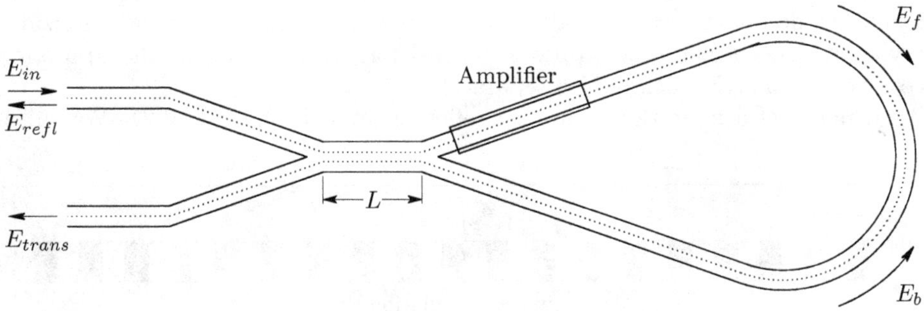


**Fig. 10.20** Schematic of nonlinear amplifying loop mirror (NALM). The pumping radiation for the amplifier is usually coupled in using a WDM (not shown) in the loop. The length of the loop is $L'$.

coupler, the loop phase shift from the counterclockwise wave is not affected by the amplifier, while the other wave's phase shift is increased by the amplifier gain. If the *power* gain of the amplifier is $g$, the phase difference between the two waves in the loop is seen to be

$$\text{NALM:} \quad \delta_{cw} - \delta_{ccw} = \frac{\pi n_2 g |E_{in}|^2 L'}{\lambda} - \frac{\pi n_2 |E_{in}|^2 L'}{\lambda} = \frac{\pi n_2 (g-1)|E_{in}|^2 L'}{\lambda} \tag{10.157}$$

where we have set $\alpha = 0.5$. The transmission of the NALM is

$$\frac{P_{trans}}{P_{in}} = g \sin^2\left(\frac{\delta_{cw} - \delta_{ccw}}{2}\right). \tag{10.158}$$

The maximum transmission occurs when $\delta_{cw} - \delta_{ccw} = \pi$ and the intensity needed to switch the NALM into this state is

$$I_{100\%} = |E_{in}|^2 = \frac{\lambda}{n_2(g-1)L'}. \tag{10.159}$$

Of course, at very low powers the device behaves like a conventional loop mirror and the transmission is zero.

### 10.8.3 Fiber Bragg gratings

It is fairly common for fiber lasers to employ *fiber Bragg grating* (FBG) mirrors at one or both ends of the fiber in order to provide the feedback needed for laser oscillation. Bragg reflectors are often more convenient than separate conventional mirrors since they can be easily welded to the ends of the fiber. In favorable cases they can be "written" directly onto the ends of the fiber providing laser gain. Furthermore, by their very nature they are narrowband devices and can help to ensure single-frequency laser oscillation. We have already discussed Bragg gratings in our discussion of distributed feedback lasers, where the entire laser used a spatially periodic gain or refractive index for laser feedback. Here we will use a similar formalism to describe a passive discrete Bragg grating reflector. Our approach is an application of *coupled mode theory*, which is actually used five times in this book (in our descriptions of a distributed feedback laser, acousto-optic modulator, directional coupler, a fiber Bragg grating and a polarization-preserving fiber). Rather than apply a general formalism to all of the appropriate devices, we will analyze each device separately.

A schematic of a fiber Bragg grating reflector appears in Fig. 10.21. We will use

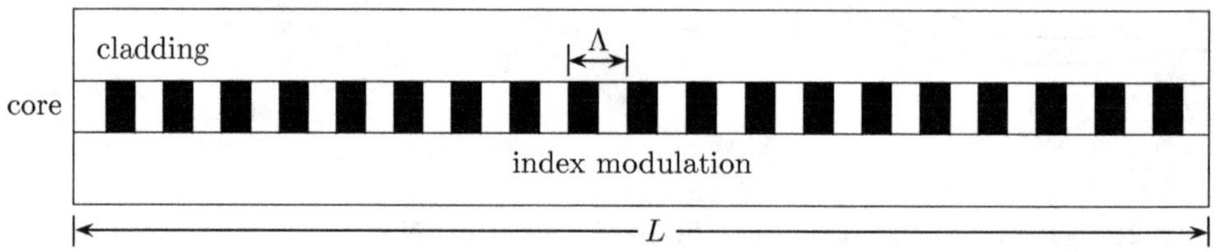


**Fig. 10.21** Schematic of fiber Bragg reflector, which consists of short length of fiber whose core has a periodic modulation of its index of refraction. The modulation spatial period is Λ.

the results from our treatment of a distributed feedback (DFB) laser but set the gain equal to zero and assume that there is only a periodic index of refraction described by:

$$n(z) = n_0 + n_1 \cos 2\beta_0 z \qquad n_0 = \text{ average index} \tag{10.160}$$

where the period, Λ, of the index modulation is $\Lambda = \pi/\beta_0$ and we assume that the modulation is very *weak* ($n_1 \ll n_0$). We seek solutions to the wave equation:

$$\frac{\partial^2 E}{\partial z^2} + k^2 E = 0 \tag{10.161}$$

where we assume that the wave vector is close to $\beta_0$,

$$E(z) = A_f(z)e^{-i\beta_0 z} + A_r(z)e^{i\beta_0 z}. \tag{10.162}$$

In the presence of the spatial index modulation, the value of $k^2$ in the wave equation is obtained from eqn 10.160, after dropping a very small term proportional to $n_1^2$:

$$k^2 = \beta^2 + 4\kappa\beta \cos 2\beta_0 z \tag{10.163}$$

where $\beta = n_0\omega/c$ is the wave vector of the wave in the fiber and $\kappa$ is the *coupling constant*

$$\kappa = \frac{\pi n_1}{\lambda}. \tag{10.164}$$

We are implicitly assuming that the fiber mode is contained entirely in the fiber core; if this is not the case, a more accurate value of $\kappa$ would be obtained by multiplying it by $\Gamma$, the fraction of power in the core. The quantity $A_r$ is the reflected wave amplitude and $A_f$ is the forward wave amplitude. Substituting $E(z)$ and $k$ into the wave equation yields two coupled equations for $A_r$ and $A_f$:

$$\begin{aligned} i\kappa A_r &= -\frac{dA_f}{dz} - i\delta A_f \\ i\kappa A_f &= \frac{dA_r}{dz} - i\delta A_r \end{aligned} \tag{10.165}$$

where $\delta$ is the displacement of $\beta$ from the Bragg wave vector, $\beta_0$,

$$\delta = \beta - \beta_0 \tag{10.166}$$

and we have dropped second derivatives of the amplitudes, since the latter are assumed to vary slowly with $z$. Although these equations can be solved in this form, it might be a little easier (and also consistent with most other coupled mode formalisms) to make the following substitutions:

$$\begin{aligned} A'_f &= A_f e^{i\delta z} \\ A'_r &= A_r e^{-i\delta z}. \end{aligned} \tag{10.167}$$

The coupled equations are now:

$$\begin{aligned} \frac{dA'_f}{dz} &= -i\kappa A'_r e^{2i\delta z} \\ \frac{dA'_r}{dz} &= i\kappa A'_f e^{-2i\delta z}. \end{aligned} \tag{10.168}$$

By differentiating the first equation and substituting the second, we obtain a second-order differential equation:

$$\frac{d^2 A'_f}{dz^2} - 2i\delta\frac{dA'_f}{dz} - \kappa^2 = 0. \tag{10.169}$$

This equation is solved in the usual way (by substituting $A'_f = e^{sz}$ and obtaining the roots of the resulting quadratic equation in $s$). The solution for $A'_f$ is

$$A'_f = e^{i\delta z}[f_1 e^{\gamma z} + f_2 e^{-\gamma z}] \tag{10.170}$$

where $f_1$ and $f_2$ are determined by the boundary conditions and

$$\gamma = \sqrt{\kappa^2 - \delta^2}. \tag{10.171}$$

The boundary conditions are:

$$\begin{aligned} A'_f(z=0) &= 1 \\ A'_r(z=L) &= 0 \end{aligned} \qquad (10.172)$$

where the length of the Bragg fiber is $L$, and we are merely stating that the incident wave is normalized to unity at the entrance to the fiber grating ($z = 0$) and the reflected wave must grow from zero at the end of the fiber grating. We obtain the complete solution by substituting the expression for $A'_f$ into the first of the coupled equations and applying the boundary conditions. We are ultimately interested in the *power reflectivity*, $R$, of the grating, which is given by

$$R = |A_r(z=0)|^2 \qquad (10.173)$$

since the forward wave is normalized to unity at the entrance to the fiber ($z = 0$). The final result is, after a bit of algebra:

$$R = \frac{\kappa^2 \sinh^2 \gamma L}{\delta^2 \sinh^2 \gamma L + \gamma^2 \cosh^2 \gamma L}. \qquad (10.174)$$

The peak reflectivity occurs when $\delta = 0$:

$$\text{Peak reflectivity:} \quad R = \tanh^2 \kappa L \qquad (\delta = 0). \qquad (10.175)$$

For small $\kappa$, $R \approx (\kappa L)^2$ at $\delta = 0$, justifying our interpretation of $\kappa$ as the field reflectivity per unit length. The *width* of the reflectivity function can be considered to be the wavelength difference between the zeros of $R$ that are on each side of $\delta = 0$. The zeros occur when $\gamma L = \pm\pi$ (and not when $\gamma = 0$ since both the denominator and numerator of the $R$ formula would then be zero). For arbitrary $\kappa L$, the width in units of $\beta$ is $2\kappa\sqrt{1 + (\pi/\kappa L)^2}$ and in wavelength units is:

$$\text{FBG width:} \quad \Delta\lambda = \frac{\kappa\lambda^2}{n_0 \pi}\sqrt{1 + (\pi/\kappa L)^2} = \frac{\lambda n_1}{n_0}\sqrt{1 + (\pi/\kappa L)^2} \qquad (10.176)$$

where we have substituted the expression for $\kappa$ from eqn 10.164. For a *strong* grating ($\kappa L \gg 1$), the square root would be approximately unity. Thus, the fractional wavelength width is roughly equal to the fractional index modulation.

The design of a fiber grating is fairly straightforward. The center wavelength determines $\beta_0$ and therefore $\Lambda$. The width determines $\kappa$ and the desired reflectivity determines $\kappa L$ and therefore $L$. For example, a typical grating for an erbium-doped fiber laser has a central wavelength of 1550 nm, a 90% reflectivity and a width of 0.2 nm. The reflectivity requires that $\kappa L = 1.82$. To meet the width requirement, $n_1/n_0 = 6.5 \times 10^{-5}$. For a silica fiber at 1550 nm, $n_0 = 1.44$, the coupling constant $\kappa = 1.88\ \text{cm}^{-1}$ and $L = 9.7$ mm. Figure 10.22 displays the reflectivity of such a fiber grating as a function of wavelength.

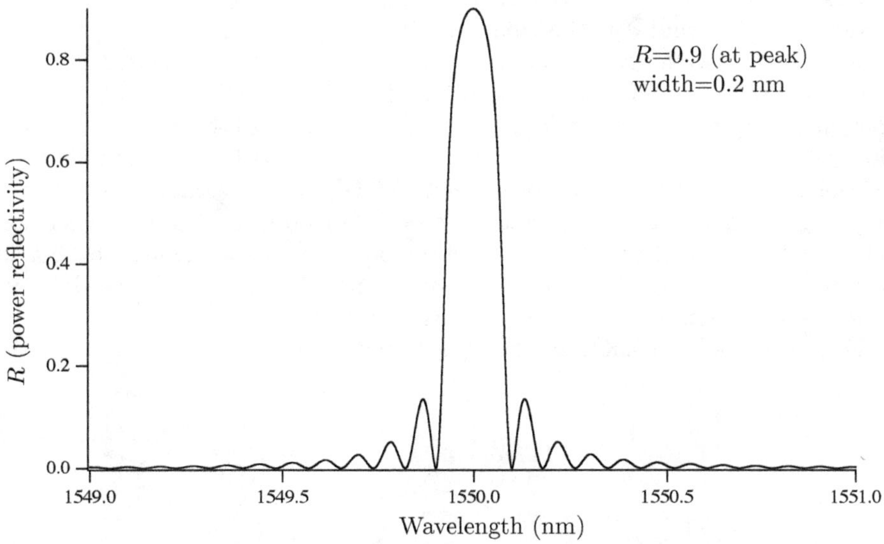


**Fig. 10.22** Reflectivity of fiber Bragg grating with $n_1/n_0 = 6.5 \times 10^{-5}$ and $L = 9.7$ mm.

### 10.8.4 Optical isolators and circulators

In our discussion of external cavity semiconductor lasers, we found that *bare lasers* are extremely sensitive to very small amounts of optical feedback from an external reflector. Fiber lasers (as do most other types of laser) share this sensitivity to optical feedback. To prevent the adverse effects of undesired optical feedback, the usual practice is to use a *Faraday isolator* immediately after the laser. As a practical matter, these devices can have an *insertion loss* of less than 1 dB and an *isolation* (ratio of forward to reverse powers) of 33 dB (a factor of 2000). This is usually enough to prevent problems due to reflections; if necessary, a second isolator can be used to double (in dB) the isolation. Although one can isolate a laser using other devices (for example, an acousto-optical modulator), we will discuss the use of Faraday isolators for this purpose.

The operation of an isolator is based upon the *Faraday effect*: the rotation of the plane of polarization of light in a medium exposed to a magnetic field which has a component along the wave vector of the light. The key feature of this effect is that *the sense of rotation of the polarization is independent of the propagation direction of the light.* In this way it differs from *optical activity*, in which the sense of the polarization rotation reverses when the direction of the light reverses. Thus, if a beam of light passes through a medium displaying a Faraday effect and is reflected in the opposite direction back through the medium, the net rotation angle will be *twice* the original rotation angle. In a medium exhibiting optical activity, the same reflected beam would emerge from the medium with no net polarization rotation.

The rotation angle, $\beta$, is determined by the empirical formula:

$$\beta = \mathcal{V}Bd \tag{10.177}$$

where $\mathcal{V}$ is the *Verdet constant*, which is characteristic of a given material, $B$ is the size of the magnetic field and $d$ is the thickness of the medium. The Verdet constant, $\mathcal{V}$, depends upon the wavelength of the light and the temperature of the medium. It can take on a large range of values; for solid or liquid media, $\mathcal{V}$ is often roughly $10^{-4}$ degree per gauss per cm (or 1 degree/Tesla-cm; 1 T=$10^4$ G). This is too small to provide a 45° rotation with a reasonable magnetic field. For wavelengths from the visible to 1 $\mu$m, optical isolators use terbium–gallium–garnet (TGG), whose Verdet constant is about $3 \times 10^{-3}$deg/cm-gauss. The large magnetic field (15000 G or 1.5 T) needed for $\pi/4$ rotation can be provided by neodymium–iron–boron permanent magnets. For longer wavelengths (1.3 $\mu$m to 1.5 $\mu$m), yttrium–iron–garnet (YIG) is usually used; its Verdet constant is several hundred times as large as that of TGG.

A discrete Faraday isolator is shown schematically in Fig. 10.23. It consists of

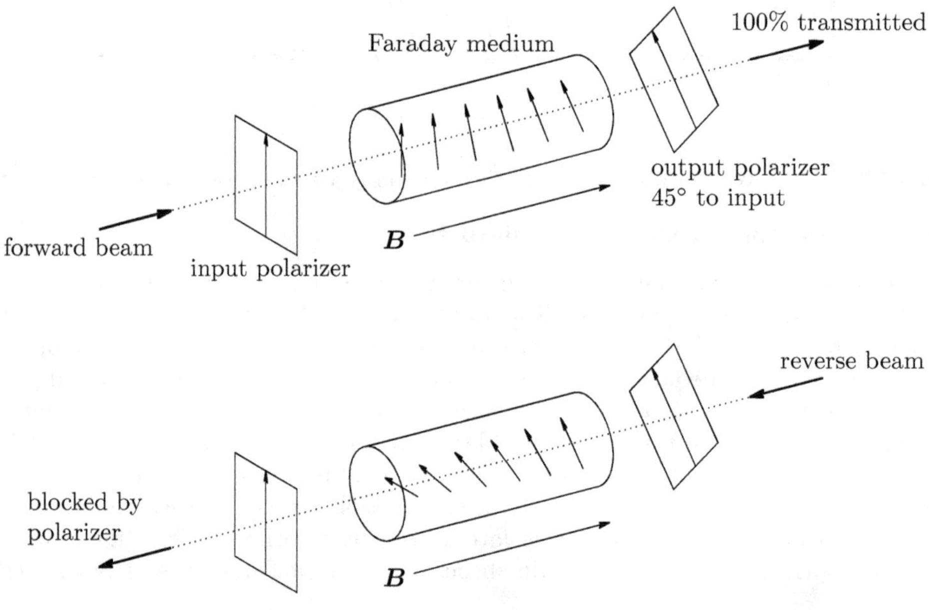


**Fig. 10.23** Schematic of Faraday isolator, which allows light propagation in only one direction. The upper diagram shows forward light whose polarization rotation, due to the Faraday effect, is just enough to allow the light to pass through the output polarizer with no attenuation. In the lower diagram, the polarization rotation is in *the same sense* as for the forward light and the light is therefore blocked by the input polarizer.

an input polarizer, a medium exhibiting a Faraday effect, a strong magnet and an output polarizer rotated so that its polarization axis is tilted by 45° from that of the input polarizer. The magnetic field is strong enough to produce a 45° rotation of the polarization for the desired wavelength. As one can see from the figure, any light propagating in the reverse direction will be rotated in the same sense as the forward light and will be blocked by the input polarizer. For use in a fiber-optical system, current practice is to use the discrete isolator described above and provide it with *fiber pigtails* to allow it to be easily incorporated into the fiber system. Optical fibers with large Verdet constants can also be used in a completely *fiber-based* isolator, though these devices do not currently provide as good performance as the bulk isolators.

*Polarization independent* isolation can be obtained by first separating the two orthogonal polarizations, then sending them each to a conventional isolator and finally recombining them. These devices are often used with non-polarization-maintaining fibers, since the polarization in these fibers can be somewhat unpredictable.

A generalization of an optical isolator is the *circulator*, shown schematically in Fig. 10.24. The usual realization of a circulator has three ports and allows light to

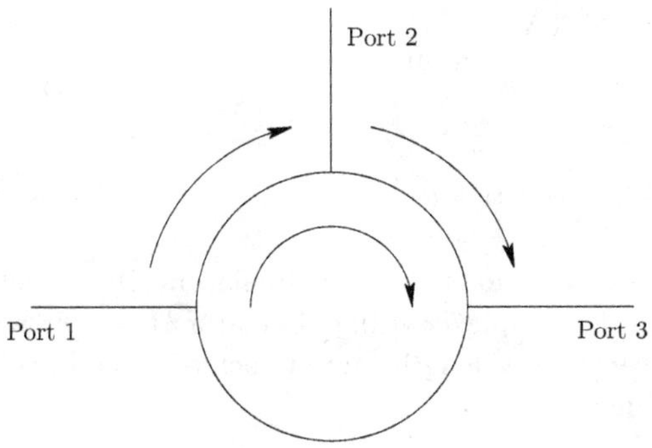


**Fig. 10.24** Schematic of optical circulator, which allows light to go only from port 1 to port 2 or port 2 to port 3.

travel only in the direction shown by the arrows. Thus, light can go from port 1 to port 2 or from port 2 to port 3, but not from port 3 to port 1 (in most devices). Due to the directivity of the device, port 1 is an input port only and port 3 is only an output port. Port 2 is both an input and output port. The construction of a circulator consists of two Faraday media together with a variety of polarizers and polarizing beam-splitters needed to provide the desired characteristics. As with the isolator, a circulator is usually constructed from bulk components and is provided with fiber pigtails to allow it to be connected to a fiber-optic system.

Figure 10.25 shows two applications for one or more optical circulators. One application is an alternative realization of a wavelength division demultiplexer, where an input beam containing three wavelengths is separated into three individual beams, each containing a single wavelength, using a combination of circulators and fiber Bragg grating reflectors. The second is a two-pass fiber amplifier, where the circulator conveniently separates the input beam from the output beam.

### 10.8.5 Amplitude and phase modulation

In Chapter 15 we provide detailed descriptions of discrete acousto-optical and electro-optical modulators; here we will focus on modulators which are likely to be found in fiber-optic systems.

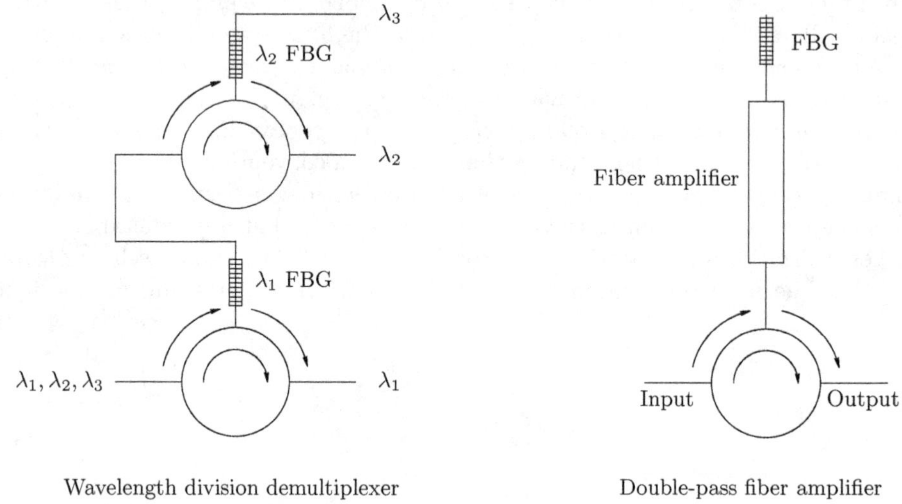


**Fig. 10.25** Two possible applications for fiber circulators. On the left is another realization for a wavelength division demultiplexer using circulators together with fiber Bragg grating (FBG) reflectors whose center wavelengths are at each of three wavelengths. On the right is a double-pass fiber amplifier.

Electro-optical phase modulators used in fiber systems are usually traveling-wave modulators provided with fiber pigtails. The electro-optical material is usually lithium niobate ($LiNbO_3$). Operationally, these modulators provide very wide modulation bandwidth (often DC to several GHz) with modest radio-frequency power requirements (a *half-wave voltage* – the voltage needed for $\pi$ radians of phase modulation – of 10 V or less). Such a modulator is shown in Fig. 10.26. The principal of operation of these devices is described in Chapter 15.

Amplitude modulation in a fiber-optic system is usually provided by a Mach–Zehnder interferometer constructed from optical waveguides on a single substrate (Fig. 10.26). One arm of the Mach–Zehnder interferometer is an electro-optical modulator. If the net phase difference between the two arms of the interferometer is $\pi$ radians, there will be no output.

To analyze the modulator, we assume that a wave that enters the modulator has the $z$-dependence,

$$\text{Input wave:}\quad E(z) = E_0 \cos(\beta_z z) \tag{10.178}$$

The wave is split into two parts and the phase difference between the two arms due to the electro-optical modulator is $\delta$ radians. The two waves are recombined and the wave which emerges is given by:

$$E(z) = \frac{E_0}{2}\left[\cos(\beta_z z) + \cos(\beta_z z + \delta)\right] = E_0 \cos(\beta_z z + \delta/2)\cos(\delta/2) \tag{10.179}$$

As a result of the electro-optical effect, $\delta$ is proportional to the RF voltage, $V_{rf}$. We therefore have an amplitude modulation of the wave through the factor $\cos(\delta/2)$. This

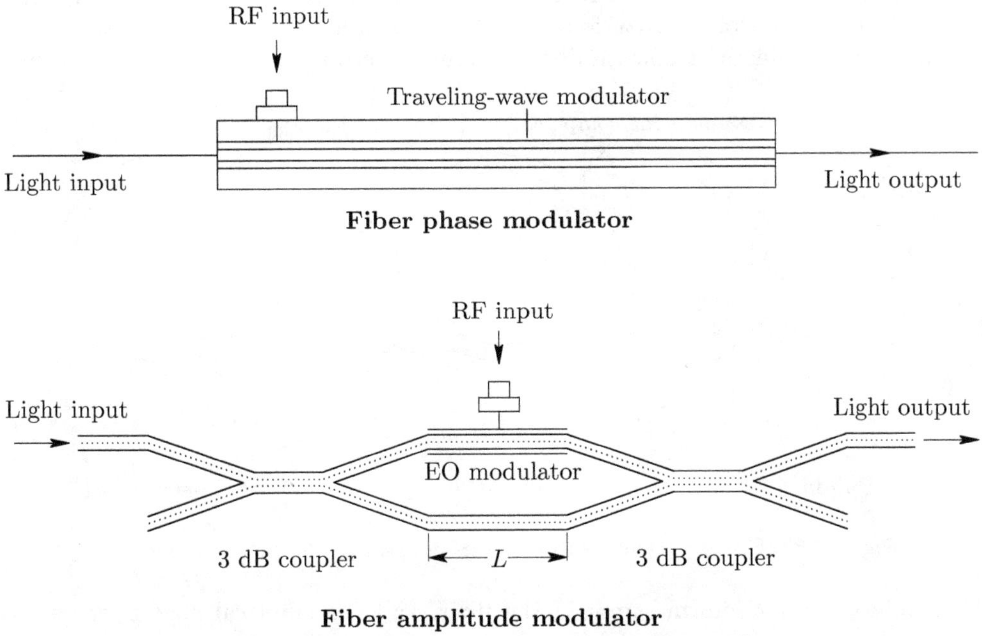


**Fig. 10.26** A schematic of a traveling-wave phase modulator with fiber pigtails (top) and a Mach–Zehnder-based amplitude modulator (bottom).

modulation is unfortunately not linear in the applied RF voltage; furthermore, there is a simultaneous phase modulation, which might not be desirable. The latter can be removed by using a second identical electro-optical modulator in the lower arm and driving it 180° out of phase with the upper modulator.

### 10.8.6 Polarization-preserving fibers

Although not strictly speaking a separate fiber component, a fiber which preserves polarization can be a useful tool in pursuit of narrow-bandwidth single-frequency operation of a fiber laser. At present, most single-frequency fiber lasers do not use polarization-preserving fibers to control polarization, but it is expected that their use will become more frequent in the future.

An ideal fiber is an isotropic medium for wave propagation: the phase velocity of the wave is independent of the polarization and a wave which is launched at one end will maintain its polarization over the entire length of the fiber. Obviously this is an unrealistic idealization and actual fibers normally do not preserve the polarization of the propagating wave, as will be shown shortly. In order to ensure that the fiber will preserve its polarization, it is necessary to make its optical properties somewhat anisotropic, with the result that the propagation constant (and phase velocity) of the wave will depend upon its polarization. The fiber is made *birefringent*: it has two different indices of refraction for two orthogonal polarizations.

The large birefringence is produced either by stress or an anisotropic geometry. Three common types of polarization-preserving fibers are shown in cross section in Fig. 10.27. The panda and bow-tie fibers produce birefringence by using components

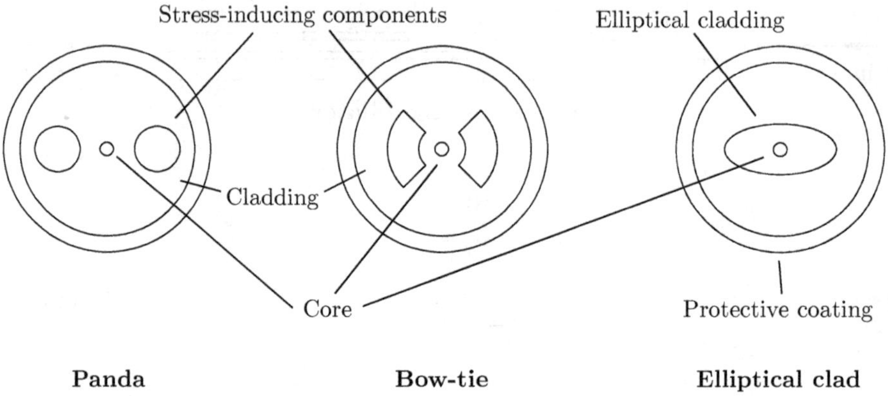


**Fig. 10.27** Cross sections of three types of polarization-preserving fiber.

which induce a non-uniform stress in the fiber, and the elliptical clad fiber uses a non-cylindrical geometry to make the fiber anisotropic. The two *principal axes* are horizontal and vertical: a wave which is polarized along either of these axes will preserve its polarization. For any other incident polarization, the polarization of the propagating wave will in general be *elliptical* and will vary as the wave progresses along the fiber.

We will use the *coupled mode* formalism to analyze these fibers. The two modes being coupled are the $x$- and $y$-components of the electric field and, as a result of the birefringence, the two propagation constants, $\beta_z^x$ and $\beta_z^y$, will be different. The coupled equations are the same as were used in analyzing the directional coupler, except for the birefringence:

$$\begin{aligned} \frac{dE_x}{dz} &= -i\beta_z^x E_x + \kappa_x E_y \\ \frac{dE_y}{dz} &= -i\beta_z^y E_y + \kappa_y E_x, \end{aligned} \tag{10.180}$$

where the coupling constants, $\kappa_{x,y}$, are due to a perturbation which mixes the polarizations. For exactly the same reason (conservation of power) as in the directional coupler case, the relation between the coupling constants is:

$$\kappa_x = -\kappa_y^*. \tag{10.181}$$

We will first solve the equations and then will give a rationale for the presence of the coupling constants. The equations are solved in exactly the same way as they were in the directional coupler analysis: by differentiating the first, substituting the second and solving the resulting second-order differential equation. The result for $E_x$ is:

$$E_x = e^{-i\bar{\beta}z}[f_1 e^{i\Delta\beta_z z} + f_2 e^{-i\Delta\beta_z z}] \tag{10.182}$$

where

$$\bar{\beta} = \tfrac{1}{2}(\beta_z^x + \beta_z^y)$$
$$i\Delta\beta_z = \sqrt{\kappa_x\kappa_y + (\beta_z^x - \beta_z^y)^2/4} \tag{10.183}$$

and the constants $f_{1,2}$ are obtained from the boundary conditions:

$$E_x(0) = E_0$$
$$E_y(0) = 0. \tag{10.184}$$

We are launching a wave polarized in the $x$-direction (along a principal axis) and are interested in the power in the $y$-polarization due to the coupling. After a bit of algebra, we find that the ratio of the power in the $y$-polarization to the initial power ($x$-polarized) is:

$$\frac{P_y(z)}{P_x(0)} = F\sin^2\Delta\beta_z z \tag{10.185}$$

where

$$F = \frac{1}{1 + \left(\frac{\beta_z^x - \beta_z^y}{2|\kappa_x|}\right)^2} \tag{10.186}$$

We therefore find that when $\beta_z^x - \beta_z^y \gg 2|\kappa_x|$, there is very little power coupled into the $y$-polarization.

We will now try to relate the coupling constants to the sort of fiber perturbation that might be expected to mix the polarizations. We would expect that the worst possible fiber perturbation is one which is applied at 45° to the polarization: this can cause a 90° rotation of the polarization when the optical path difference between the polarizations is one half-wave (as in a *half-wave plate*). We will analyze this by assuming first that a perturbation induces a birefringence whose principal axes are along the $x$- and $y$-axes. The two field components satisfy:

$$\frac{dE_x}{dz} + i\gamma_z^x E_x = 0$$
$$\frac{dE_y}{dz} + i\gamma_z^y E_y = 0, \tag{10.187}$$

where the two (different) propagation constants are $\gamma_z^{x,y}$. We now rotate the coordinate system by 45° about the $z$-axis to generate our 45° perturbation:

$$E'_x = \tfrac{1}{\sqrt{2}}(E_x + E_y)$$
$$E'_y = \tfrac{1}{\sqrt{2}}(-E_x + E_y). \tag{10.188}$$

Substituting the primed fields into the equations and adding and subtracting the resulting equations, one obtains

$$\frac{dE'_x}{dz} = -i\bar{\gamma}E'_x + \tfrac{i}{2}\Delta\gamma E'_y$$
$$\frac{dE'_y}{dz} = -i\bar{\gamma}E'_y + \tfrac{i}{2}\Delta\gamma E'_x \tag{10.189}$$

where

$$\bar{\gamma} = \tfrac{1}{2}(\gamma_z^x + \gamma_z^y)$$
$$\Delta\gamma = \gamma_z^x - \gamma_z^y. \tag{10.190}$$

We note that we have generated two coupled equations whose coupling constants are both $\frac{i}{2}\Delta\gamma$ (they are the same for both equations since $\kappa_x = \kappa_y$ when the constants are *imaginary*). We will assume that we can use these as the coupling constants in the *original* coupled equations and that including them completely describes the perturbation which is applied at 45° to the $x$-axis. We saw earlier that the polarization would be preserved when the difference in the propagation constants was greater than twice the coupling. Therefore, the criterion for polarization preservation against this kind of perturbation is:

$$\text{Polarization preservation:} \quad \beta_z^x - \beta_z^y \gg \Delta\gamma. \tag{10.191}$$

It is customary to use the *beat distance* to describe the polarization properties of a fiber. This is defined as the propagation distance which results in a full-wave optical path difference between the polarizations:

$$\text{Beat distance} = \frac{2\pi}{\beta_z^x - \beta_z^y}. \tag{10.192}$$

The beat distance is the distance an initially linearly polarized wave will travel before its linear polarization is restored. If we consider the perturbation to have an analogous "beat distance", then the criterion for a fiber to preserve its polarization against a perturbation is that *the beat distance of the perturbation be much larger than the beat distance of the fiber.* Thus, a short fiber beat distance is highly desirable in a polarization-preserving fiber. We finally note the analogy between the reduction of the coupling between polarizations obtained by increased birefringence and the quantum-mechanical situation where, in first-order perturbation theory, the mixing of wavefunctions is greatly reduced when the energy separation between the two mixed states is much larger than the coupling matrix element.

The most common fiber perturbation is the birefringence induced in a fiber by bending. From the equation (eqn 10.200) for bending birefringence appearing later in this chapter, a 90° bend of a silica fiber with a standard cladding diameter of 125 μm at 633 nm will produce a propagation constant difference of

$$90^\circ \text{ bend:} \quad \gamma_z^x - \gamma_z^y = 0.47\frac{1}{R} \text{ rad/cm} \tag{10.193}$$

where $R$ is the radius of the bend in cm. Thus, the "beat distance" of this perturbation is about $13R$. The conclusion is that a fiber with an intrinsic beat distance which is somewhat less than the radius of any likely bend will remain comfortably polarization-preserving.

Before leaving this topic, it would be useful to determine the *penalty* for a failure to align the input polarization exactly along one of the principal axes of the fiber. We will assume that the polarization is misaligned by $\delta\theta$ from the principal axis. In the absence of any other perturbation, the coupled equations are the original ones without the coupling terms:

$$\begin{aligned}\frac{dE_x}{dz} &= -i\beta_z^x E_x \\ \frac{dE_y}{dz} &= -i\beta_z^y E_y.\end{aligned} \tag{10.194}$$

To include the misalignment, we will rotate the coordinate system by $\delta\theta \ll 1$:

$$\begin{aligned}E'_x &= E_x + \delta\theta E_y \\ E'_y &= -\delta\theta E_x + E_y\end{aligned} \tag{10.195}$$

The equations in the primed system are (neglecting terms quadratic in $\delta\theta$):

$$\begin{aligned}\frac{dE'_x}{dz} &= -i\beta_z^x E'_x + i\delta\theta\delta\beta_z E'_y \\ \frac{dE'_y}{dz} &= -i\beta_z^y E'_y + i\delta\theta\delta\beta_z E'_x \qquad \delta\beta_z = \beta_z^y - \beta_z^x\end{aligned} \tag{10.196}$$

As a result of the misalignment, we have introduced a coupling constant $i\delta\theta\delta\beta_z$ into the equations. We conclude from eqn 10.186 that the maximum power in the *wrong* polarization is

$$\text{Maximum power in } y\text{-polarization:} \quad P_y = 4P_0(\delta\theta)^2 \tag{10.197}$$

Therefore, even a 5° alignment error will only add 4% of the incident power to a $y$-polarized wave.

### 10.8.7 Polarization controller

Although one approach to polarization control in a fiber laser is to use a polarization-preserving (high birefringence) fiber, most fiber lasers actually use low birefringence fibers and control the polarization state with specialized fiber elements called *polarization controllers.* These devices transform an arbitrary polarization state into another completely arbitrary polarization state.

A general polarization state can be described using an ellipse, which is the curve traced out over time by the electric field vector at a fixed point in space. The electric field components for a wave propagating along the $z$-axis are:

$$\begin{aligned}E_x &= E_{0x}\cos(\tau) \\ E_y &= E_{0y}\cos(\tau + \epsilon) \\ &\text{where } \tau = kz - \omega t.\end{aligned} \tag{10.198}$$

These equations can be considered to be parametric equations for an ellipse which is inscribed in a box with dimensions $E_{0x}$ and $E_{0y}$ with the major axis tilted by an angle $\psi$ with respect to the $x$-axis, where $\psi$ is given by:

$$\tan 2\psi = \frac{2E_{0x}E_{0y}}{E_{0x}^2 - E_{0y}^2}\cos\epsilon. \tag{10.199}$$

The ellipse is plotted in Fig. 10.28. The polarization can be completely described by two

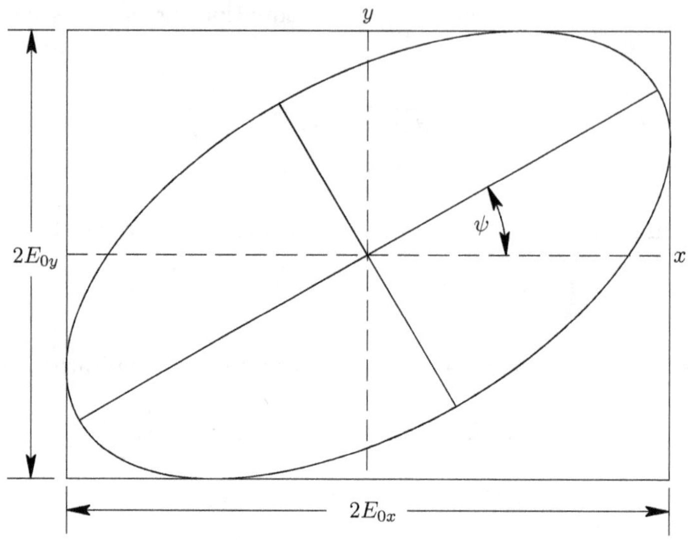


**Fig. 10.28** Locus traced by $\boldsymbol{E}$ for arbitrary polarization.

parameters: $\psi$ and $E_{0x}/E_{0y}$. When the ellipse is aligned with either of the coordinate axes ($\psi = 0$ or $\pi/2$), eqn 10.199 indicates that $\epsilon$ is $\pi/2 \pm m\pi$, for integral $m$. Thus, a quarter-wave plate, either of whose axes (fast or slow) makes an angle $\psi$ with the $x$-axis, will convert the elliptical polarization into linear polarization since the $\lambda/4$ plate simply changes the relative phase between the field components by $\pm\pi/2$. If one adds a half-wave plate, which rotates the resulting linear polarization by an arbitrary angle and a second rotatable quarter-wave plate, it is easy to see that one can convert an arbitrary polarization into another arbitrary polarization. One needs three degrees of freedom to describe this process: $E_{0x}/E_{0y}$ for each wave and the difference in the tilt angles ($\psi$) of the two ellipses. These degrees of freedom are provided by the rotation angles of the three plates.

A common realization of a polarization controller is called a "Lefevre" (or "bat-ear") controller and uses three fiber loops, each of which can be rotated about a common axis (Fig. 10.29). The bending of the fiber induces birefringence due almost entirely to stress (the geometrical component of the birefringence is several orders of magnitude smaller than that due to stress). This phenomenon was studied by Ulrich et al. (1980), and the formula for the phase difference between the two polarization components is

$$\beta_b = -7.7 \times 10^7 \left(\frac{r}{R}\right)^2 \ \text{deg/m} \tag{10.200}$$

where $r$ is the outer radius of the fiber and $R$ is the radius of the loop. This formula is for a fused silica fiber at 633 nm but should hold for all weakly guiding silica fibers, regardless of the core diameter or index profile.

There are a number of other approaches to the control of polarization in a fiber. One popular method uses a short piece of fiber which can be squeezed with a variable

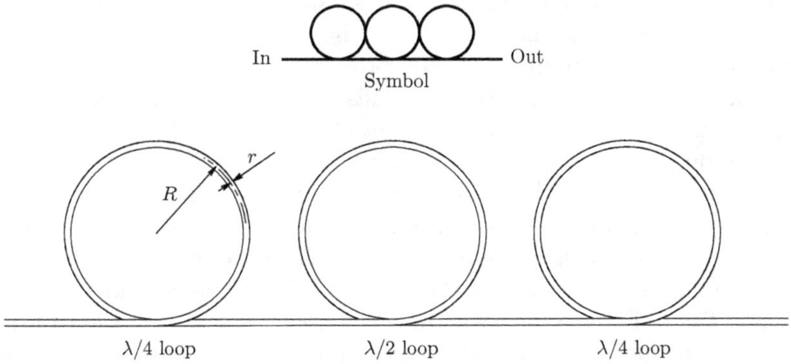


**Fig. 10.29** Polarization controller consisting of three fiber loops which act as wave plates and can be independently rotated about the horizontal axis.

force and which can be rotated independently of the remainder of the fiber. This technique also relies on stress-induced birefringence.

## 10.9 The physics of rare earth ions in glasses

In Chapter 7 we provided a somewhat shallow survey of some of the more popular laser media, such as organic dyes, a helium–neon gas mixture, argon ions and neodymium ions embedded in a crystalline substrate. The last item can be considered to be a prototype of the gain media in virtually all fiber lasers: triply ionized rare earth ions embedded in a glass substrate. In this section we will provide a slightly deeper look into the atomic physics of these systems. Despite the presence of the glass substrate, the states of the active ions embedded in the glass are remarkably similar to the states of isolated ions. We will therefore begin with a brief introduction to the atomic structure of multi-electron atoms and will assume that the reader has some familiarity with a quantum-mechanical treatment of the hydrogen atom.

The Hamiltonian for an atom having $N$ electrons is:

$$H = \sum_{i=1}^{N} \left\{ -\frac{\hbar^2}{2m}\nabla_i^2 - \frac{Ze^2}{4\pi\epsilon_0 r_i} + \sum_{j>i}^{N} \frac{e^2}{4\pi\epsilon_0 r_{ij}} \right\} + \sum_{i=1}^{N} \xi(r_i)\boldsymbol{l}_i \cdot \boldsymbol{s}_i \tag{10.201}$$

where $Z$ is the atomic number of the atom, $r_i$ is the distance of the $i$th electron to the nucleus, $r_{ij}$ is the distance between the $i$th and $j$th electrons and $\nabla_i^2$ is the Laplacian expressed as a function of the coordinates of the $i$th electron. The last term describes the *spin–orbit* interaction, where $\boldsymbol{l}_i$ is the orbital angular momentum of the $i$th electron and $\boldsymbol{s}_i$ is its spin angular momentum. The first two terms are just the sum of the kinetic energies of the electrons and their potential energies in the field of the nucleus. The third term is due to the electrostatic repulsion between pairs of electrons. It is tempting to treat the electrostatic repulsion term as a perturbation to the solution to Schrödinger's equation with a Hamiltonian containing the first two terms. Unfortunately, the third term is usually quite large and cannot be sensibly treated as a small perturbation. Instead, we use an electrostatic Hamiltonian which is rewritten in a manner which allows us to use the *central field approximation.*

In the central field approximation, we write the electrostatic Hamiltonian as the sum of two parts. The first part ($H_0$) contains the kinetic energy of the $N$ electrons and their potential energy ($U(r_i)$) in a *central electrostatic field.* (A central field is one which depends only on the distance, $r_i$, between the $i$th electron and the nucleus.) The potential energy term includes both the potential energy of each electron in the field due to the nucleus and the *central part* of the electrostatic repulsion term. The latter depends upon the charge distribution of all of the electrons (except the $i$th) and therefore requires knowledge of the wavefunction. The second part ($H_1$) of the Hamiltonian is the difference between the exact electrostatic repulsion term, and the potential energy in the central field due to the nucleus and the other electrons. Thus, the *electrostatic* Hamiltonian, $H_{el}$, is

$$\begin{aligned} H_{el} &= H_0 + H_1 \\ \text{where } H_0 &= \sum_{i=1}^{N} \left( -\frac{\hbar^2}{2m} \nabla_i^2 + U(r_i) \right) \\ \text{and } H_1 &= \sum_{i>j}^{N} \frac{e^2}{4\pi\epsilon_0 r_{ij}} - \sum_{i=1}^{N} \left( \frac{Ze^2}{4\pi\epsilon_0 r_i} + U(r_i) \right). \end{aligned} \tag{10.202}$$

Note that this Hamiltonian is exactly the same as the one above (without the spin–orbit interaction). The non-central part ($H_1$) is called the *residual electrostatic term.* It turns out that $H_1$ is generally much smaller than $H_0$, allowing the energy shifts due to the residual electrostatic term to be determined using perturbation theory. Our first task is therefore to obtain a solution to Schrödinger's equation using $H_0$.

Since $H_0$ is the sum of $N$ single-electron terms, its eigenfunctions can be written as a product of $N$ single-electron eigenfunctions. The angular momentum operators, $\boldsymbol{l}_i$, commute with $H_0$ (i.e., the central field does not produce any torque on the individual electrons); therefore, the angular eigenfunctions will be the familiar *spherical harmonics*, $Y_{lm}(\theta, \phi)$. The radial dependence of the single-electron eigenfunctions will be different from that of hydrogen, since $U(r_i)$ is not a Coulomb potential. These single electron eigenfunctions will be identified by a *principal quantum number*, $n$, which identifies the radial eigenfunction, together with the orbital angular momentum, $l$, its $z$-component, $m$, and the $z$-component of the electron spin, $m_s$:

$$\text{Single-electron wavefunctions: } \psi_{nlmm_s}(\boldsymbol{r}_i) = R_{nl}(r_i) Y_{lm}(\theta_i, \phi_i) \chi(m_s), \tag{10.203}$$

where $R_{nl}$ is the radial eigenfunction, $\chi$ is the spin eigenfunction and $\boldsymbol{r}_i$ is the position in *spherical coordinates* $(r_i, \theta_i, \phi_i)$ of the $i$th electron.

The multi-electron Hamiltonian possesses one important symmetry which should be mentioned before continuing. We can examine it by introducing an operator, $P_{ij}$, which exchanges the coordinates of the $i$th and $j$th electron. It is clear that $P_{ij}$ commutes with the Hamiltonian, since the electrons are indistinguishable from one another and swapping the coordinates of any two electrons will not change anything. Since $P_{ij}^2 = I$ (the identity operator), the eigenvalues of $P_{ij}$ are $\pm 1$. An eigenfunction ($\psi$)

of $H$ will also be an eigenfunction of $P_{ij}$ and the possible (non-degenerate) states of the atom are therefore either *symmetric* ($P_{ij}\psi = \psi$) or *antisymmetric* ($P_{ij}\psi = -\psi$). However, electrons are *fermions*, whose wavefunctions must be *antisymmetric* upon the interchange of any pair of coordinates, according to the *Pauli exclusion principle*. A linear combination of product solutions which is antisymmetric upon exchange of particle coordinates is the *Slater determinant*:

$$\psi = (N)^{-1/2} \begin{vmatrix} \psi_1(r_1) & \psi_1(r_2) & \cdots & \psi_1(r_N) \\ \psi_2(r_1) & \psi_2(r_2) & \cdots & \psi_2(r_N) \\ \vdots & \vdots & & \\ & & & \psi_N(r_N) \end{vmatrix}, \tag{10.204}$$

where $\psi_j(r_i)$ is the $j$th single-electron wavefunction for the $i$th electron. It should be clear from the properties of determinants that the wavefunction vanishes if any two electron states are the same and is correctly antisymmetric upon exchange of any pair of coordinates.

The procedure for obtaining the states and energies of the atom is to start with an initial approximation for $U(r_i)$ and numerically solve Schrödinger's equation using the unperturbed ($H_0$) Hamiltonian and a Slater determinant wavefunction. The resulting single-electron states are then used to construct a new spherically averaged $U(r_i)$, which is used to obtain a set of new wavefunctions. The equations are solved iteratively in this manner until two successive potentials ($U(r_i)$) are nearly the same (i.e., the procedure has *converged*). One can then include the effects of the residual electrostatic Hamiltonian ($H_1$) and the spin–orbit term using perturbation theory.

A *configuration* is the specification of the $4N$ quantum numbers for all $N$ electrons in the atom. For each electron, the state specification is:

$$\text{Configuration for each electron: } nl^{\langle\text{number of electrons}\rangle} \tag{10.205}$$

where $l$ is written using *spectroscopic notation:*

$$\begin{aligned} l = 0 &\rightarrow s \text{ state} \\ l = 1 &\rightarrow p \text{ state} \\ l = 2 &\rightarrow d \text{ state} \\ l = 3 &\rightarrow f \text{ state} \end{aligned} \tag{10.206}$$

and the superscript is the number of electrons having the indicated $l$ value. The complete configuration is a list of these single-electron states in order of increasing $n$ and $l$. Thus, the configuration of xenon (which will appear in our discussion of rare earth structures) is:

$$\text{Xenon (Xe) electron configuration: } 1s^2 2s^2 2p^6 3s^2 3p^6 3d^{10} 4s^2 4p^6 4d^{10} 5s^2 5p^6. \tag{10.207}$$

The *total* orbital angular momentum ($\boldsymbol{L}$) and *total* spin ($\boldsymbol{S}$) are obtained by summing the individual angular momenta:

$$\begin{aligned} \boldsymbol{L} &= \sum_{i=1}^{N} \boldsymbol{l}_i \\ \boldsymbol{S} &= \sum_{i=1}^{N} \boldsymbol{s}_i \\ \boldsymbol{J} &= \boldsymbol{L} + \boldsymbol{S}. \end{aligned} \tag{10.208}$$

Since the internal *electrostatic* forces in the atom cannot change the *total* $\boldsymbol{L}$ and *total* $\boldsymbol{S}$, the $\boldsymbol{L}^2$, $L_z$, $\boldsymbol{S}^2$ and $S_z$ operators commute with the electrostatic Hamiltonian, $H_{el}$, and $L$ and $S$ are *good quantum numbers.* In the absence of external fields (including those from the host) and neglecting the nuclear magnetic moment, $J$ and $m_J$ are good quantum numbers. When we can neglect the spin–orbit interaction compared to the residual electrostatic interaction, we have so-called *LS coupling* and the electronic state of the atom can be described by the quantum numbers $L$, $S$ and $J$. These three quantities comprise a *term*, which is written as:

$$\text{Term value:} \quad {}^{2S+1}L_J \tag{10.209}$$

where the superscript is the number of possible $J$ levels (occasionally called the *spin multiplicity*) and spectroscopic notation is used for the $L$ values (with upper-case letters):

$$\text{Term values:} \quad \begin{aligned} L = 0 &\rightarrow \; S \text{ state} \\ L = 1 &\rightarrow \; P \text{ state} \\ L = 2 &\rightarrow \; D \text{ state} \\ L = 3 &\rightarrow \; F \text{ state} \\ L = 4 &\rightarrow \; G \text{ state} \\ L = 5 &\rightarrow \; H \text{ state} \\ L = 6 &\rightarrow \; I \text{ state} \end{aligned} \tag{10.210}$$

The set of electrons with a given $n$ and $l$ is called a *shell* and a shell is *closed* if it contains the maximum number $(2(2l+1))$ of electrons allowed by the Pauli exclusion principle. It can be shown using theorems involving the spherical harmonics that the total orbital angular momentum and spin of a closed shell are zero. For example, the xenon configuration (eqn 10.207) consists entirely of closed shells and its ground-state term is therefore a ${}^1S_0$ (pronounced "singlet S zero"). Due to the presence of the electrostatic repulsion term in the Hamiltonian, the term energies are all *different* from one another. (If the only potential energy were the Coulomb term involving the nucleus, we would have an *accidental degeneracy* – as in hydrogen – and different $l$ values for the same $n$ would have the same energy.)

An additional property of a term is its *parity*, which is important when the selection rules for radiative transitions are considered. As is well known from the theory of the hydrogen atom, single-electron wavefunctions have an angular dependence given by a spherical harmonic, $Y_{lm}(\theta, \phi)$, whose parity under a coordinate inversion $(\boldsymbol{r} \rightarrow -\boldsymbol{r})$ is determined by $l$: the wavefunction is multiplied by $(-1)^l$ under an inversion. Since the various terms corresponding to an unfilled shell are made up of products of the *same* single-electron wavefunctions, the parity of a term is given by $(-1)^{\sum_i l_i}$ where the sum is over the $l$-values of the electrons in the unfilled shell. It is important to realize that all of the $LS$ terms for a given configuration *have the same parity.*

The spin–orbit contribution to the Hamiltonian is the result of the interactions involving a number of pairs of magnetic dipoles. Each electron has an orbital angular momentum and associated magnetic moment as well as an intrinsic spin angular momentum and spin magnetic moment. One can have the following interactions: spin–spin, spin–own-orbit, spin–other-orbit and orbit–orbit (whose meanings should be self-explanatory). It turns out that the spin–spin interaction is only important in light atoms (particularly helium) and the orbit–orbit interaction contributes only to a shift of a term and not to its splitting (into different $J$ levels). For atoms with moderate $Z$, only the spin–own-orbit interaction is significant and the spin–orbit contribution to the Hamiltonian is:

$$\text{Spin–orbit interaction: } H_{so} = \sum_{i=1}^{N} \xi(r_i)\boldsymbol{l}_i \cdot \boldsymbol{s}_i. \tag{10.211}$$

When $LS$ coupling is valid, one can use the *vector model* (or the Wigner-Eckart theorem) to write the spin–orbit term as a function of $\boldsymbol{L}, \boldsymbol{S}$ and $\boldsymbol{J}$:

$$H_{so} = \zeta \boldsymbol{L} \cdot \boldsymbol{S} = \frac{\zeta}{2}(\boldsymbol{J}^2 - \boldsymbol{L}^2 - \boldsymbol{S}^2), \tag{10.212}$$

where $\zeta$ is a constant. If one uses perturbation theory, the $J$-dependence of the term energy ($E(J)$) can be obtained by taking diagonal matrix elements of $H_{so}$ between states of a given $L$, $S$ and $J$:

$$E(J) = \langle LSJ|\zeta \boldsymbol{L} \cdot \boldsymbol{S}|LSJ\rangle = \frac{\hbar^2\zeta}{2}(J(J+1) - L(L+1) - S(S+1)) \tag{10.213}$$

and the *splitting*, $\Delta E(J)$, between adjacent $J$-values for a given $L$ and $S$ is

$$\Delta E(J) = E(J) - E(J-1) = \hbar^2\zeta J. \tag{10.214}$$

The proportionality between $\Delta E(J)$ and the larger of the two $J$ values is called the *Landé interval rule.* The interval rule holds fairly well for atoms with small or moderate $Z$ for which $LS$ coupling is valid. As $Z$ increases, the residual electrostatic interaction increases but the spin–orbit interaction increases even more. The spin–orbit interaction is therefore often *larger* than the residual electrostatic interaction in high-$Z$ atoms, and for these atoms one can ignore the residual electrostatic interaction and treat each electron as if it were moving in its own independent orbit with a spin–orbit interaction similar to that in hydrogen. In this case, we have $j$-$j$ coupling, where terms are specified by the $l$ and $j$-values for each electron not in a closed shell. Between the two extremes of $LS$ and $j$-$j$ coupling, one has *intermediate coupling,* for which the spin–orbit interaction is similar in magnitude to the residual electrostatic interaction.

Concluding our introduction to the physics of multi-electron atoms, we will mention *Hund's rules,* which allow one to determine the ground state term in $LS$ coupling. These rules apply only to the *ground states* of atoms with a *single unfilled shell.* The unfilled shell consists of a number of *equivalent electrons:* electrons with the same values of $n$ and $l$. Due to the Pauli exclusion principle, only a few terms are available to equivalent electrons. For example, for two $p$ electrons in *different* shells, one can form $^1S, ^1P, ^1D, ^3S, ^3P, ^3D$ terms. If they are in the same shell, only the $^1S, ^3P, ^1D$ terms are available. In general, with two equivalent electrons, the terms alternate between singlet ($S = 0$) and triplet ($S = 1$), starting with a $^1S$ state and incrementing $L$ by one for each successive term. Hund's rules state (from Herzberg(1944)):

1. Of the terms given by equivalent electrons, those with the greatest multiplicity lie deepest, and of these the lowest is that with the greatest $L$.
2. Multiplets formed from equivalent electrons are regular when less than half the shell is occupied, but inverted when more than half the shell is occupied

The second rule refers to the sign of the spin–orbit interaction: when a shell is less than half filled, the energy will increase with increasing $J$, but will decrease with $J$ when the shell is more than half filled. The set of $J$ levels for a given $L$ and $S$ form a *fine structure multiplet*, which is *normal* when the energy increases with $J$ and *inverted* when it decreases with $J$. The $L$ and $S$ values in the term specification of the shell which is more than half filled will be the same as those of a shell less than half filled where the number of vacancies in the former is equal to the number of electrons in the latter. Thus, a pair of $f$ ($l = 3$) electrons will form the same set of terms as 12 $f$ electrons (the maximum number of $f$ electrons is 14), but the ordering of the fine structure multiplets will be opposite in the two cases.

We now turn to the atomic structure of the rare earth elements. There are two sets of rare earths: the *lanthanides*, which are the fourteen elements between cerium (Ce, $Z = 58$) and lutetium (Lu, $Z = 71$), and the *actinides*, which are the fourteen elements between thorium (Th, $Z = 90$) and lawrencium (Lw, $Z = 103$). These two sets of elements are characterized by having unfilled inner $f$ shells: the $4f$ shell for the lanthanides and the $5f$ shell for the actinides. All of the fiber lasers we will discuss use elements from the lanthanides, whose configurations are listed in Table 10.2. All of the lanthanide rare earths have a core configuration which is the same as that of xenon (eqn 10.207). Only the shells outside of the xenon core (together with the $5s$ and $5p$ shells) are shown in the table.

Table 10.3 shows the configurations of the triply ionized elements most frequently used in fiber lasers: $Nd^{3+}$, $Er^{3+}$, $Tm^{3+}$ and $Yb^{3+}$. The ions are missing the two outer $6s$ electrons and one of the inner $4f$ electrons. The neodymium ion has three equivalent $f$ electrons, whose possible terms are $^4I, ^4G, ^4F, ^4D, ^4S, ^2L, ^2K, ^2I, ^2H, ^2G, ^2F, ^2D, ^2P$ (taken from Slater (1960); for more than two equivalent electrons, things can get complicated). Using Hund's rules, the ground-state term should have a multiplicity of four (quartet) and an $L$ equal to six with the minimum $J$, which is 9/2 (i.e., a $^4I_{9/2}$ term), and this turns out to be correct. The erbium ion should be similar to neodymium, but with an inverted fine structure, (i.e., $^4I_{15/2}$); again, Hund's rules are borne out. Thulium is a simpler case, with two vacancies in the $4f$ shell. The possible terms are the same as for two $f$ electrons and are $^1S, ^3P, ^1D, ^3F, ^1G, ^3H, ^1I$, using the alternating multiplicity rule mentioned above (for two equivalent electrons). Again, Hund's rules correctly identify the ground state as a $^3H_6$. Finally, the simplest ion is $Yb^{3+}$, which has a single $f$ vacancy and therefore has $^2F$ terms with the largest $J$

**Table 10.2** Lanthanide shell structure (number of electrons in each shell).

| Z | Element | Core | 4f | 5s | 5p | 5d | 5f | 5g | 6s |
|---|---|---|---|---|---|---|---|---|---|
| 58 | Ce | Xe | 1 | 2 | 6 | 1 | | | 2 |
| 59 | Pr | Xe | 3 | 2 | 6 | | | | 2 |
| 60 | Nd | Xe | 4 | 2 | 6 | | | | 2 |
| 61 | Pm | Xe | 5 | 2 | 6 | | | | 2 |
| 62 | Sm | Xe | 6 | 2 | 6 | | | | 2 |
| 63 | Eu | Xe | 7 | 2 | 6 | | | | 2 |
| 64 | Gd | Xe | 7 | 2 | 6 | 1 | | | 2 |
| 65 | Tb | Xe | 9 | 2 | 6 | | | | 2 |
| 66 | Dy | Xe | 10 | 2 | 6 | | | | 2 |
| 67 | Ho | Xe | 11 | 2 | 6 | | | | 2 |
| 68 | Er | Xe | 12 | 2 | 6 | | | | 2 |
| 69 | Tm | Xe | 13 | 2 | 6 | | | | 2 |
| 70 | Yb | Xe | 14 | 2 | 6 | | | | 2 |
| 71 | Lu | Xe | 14 | 2 | 6 | 1 | | | 2 |

**Table 10.3** Shell structure and ground terms for rare earth ions used in lasers.

| Z | Element | Core | 4f | 5s | 5p | 5d | 5f | 5g | 6s | Term |
|---|---|---|---|---|---|---|---|---|---|---|
| 60 | $Nd^{3+}$ | Xe | 3 | 2 | 6 | | | | | ${}^4I_{9/2}$ |
| 68 | $Er^{3+}$ | Xe | 11 | 2 | 6 | | | | | ${}^4I_{15/2}$ |
| 69 | $Tm^{3+}$ | Xe | 12 | 2 | 6 | | | | | ${}^3H_6$ |
| 70 | $Yb^{3+}$ | Xe | 13 | 2 | 6 | | | | | ${}^2F_{7/2}$ |

(7/2) the lowest. Figure 10.30 displays the level structure together with the pumping and laser transitions in $Nd^{3+}$, $Er^{3+}$, $Tu^{3+}$ and $Yb^{3+}$ rare earth ions. Although Hund's rules apply only to ground states, one observes a tendency for the lower energy terms to have larger multiplicities and larger values of $L$.

Although the focus of our previous discussion has been on atoms which obey $LS$ coupling, the spin–orbit term in the lanthanide rare earths is actually larger than the residual electrostatic interaction and the states are better described using an *intermediate coupling* picture. The usual practice is to still use the $LS$ coupling labeling scheme, although the $LS$ terms are actually mixed together due to the spin–orbit interaction. In the intermediate coupling scheme, $L$ and $S$ are no longer good quantum numbers, but $J$ and $m_J$ are. An example of the breakdown of $LS$ coupling is shown in Fig. 10.31, which displays the various term energies of a thulium ion ($Tm^{3+}$) as a function of the spin–orbit interaction. The states are labelled by the term values for a zero spin–orbit interaction. The size of the spin–orbit interaction, ($\chi$), is written in units of a *Slater integral*, $F_2$, which describes the electrostatic interaction between electrons. When $\chi < 1$, the fine structure splitting seems to follow the interval rule, but as $\chi$ gets larger the interval rule is increasingly violated and the levels even cross each other. The actual value of $\chi$ in $Tm^{3+}$ is $\chi = 3$.

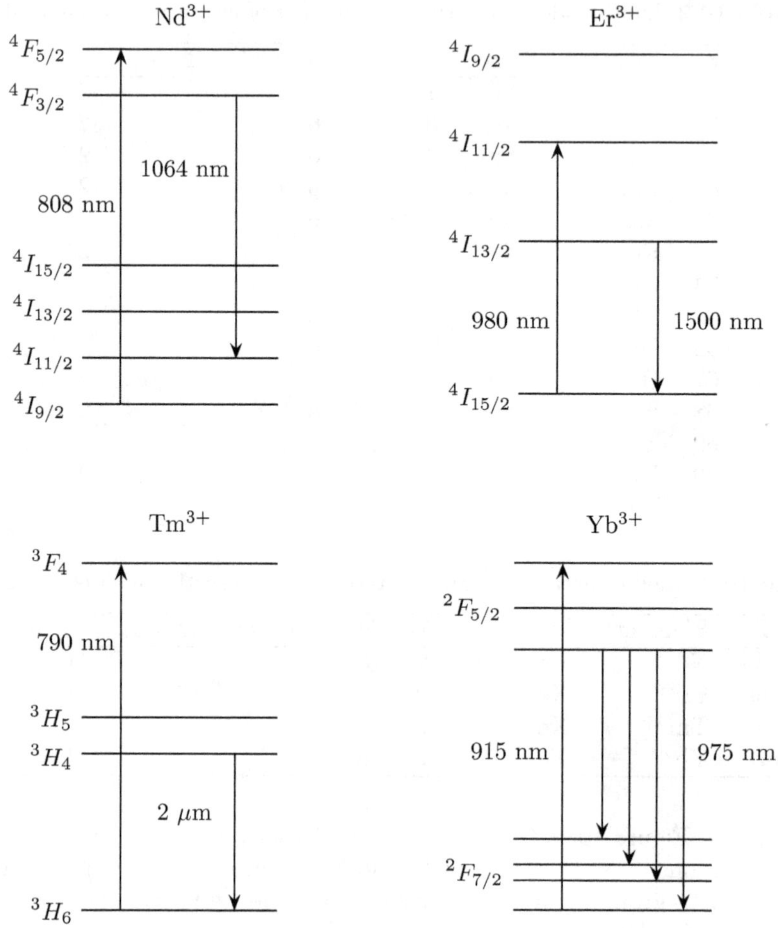


**Fig. 10.30** Energy levels, pumping transitions and laser transitions in thulium, erbium, neodymium and ytterbium rare earth ions. The several lines in each of the two $Yb^{3+}$ terms are due to the *Stark* splitting from the glass host (these lines are not shown in the energy-level diagrams for the other ions).

One of the several advantages of using rare earth ions in a laser is the insensitivity of their energy levels to the presence of the surrounding glass host. In the lanthanide rare earth ions, all of the optical transitions involve the electrons in the unfilled $4f$ shell. These electrons are actually fairly well shielded from their environment by the electrons in the outer filled $5s$ and $5p$ shells. This is illustrated in Fig. 10.32, which displays the radial wavefunction of triply ionized cerium (Ce, $Z = 58$), which has a single $4f$ electron. The peak of the $4f$ wavefunction is somewhat closer to the nucleus than the peaks of the $5s$ and $5p$ wavefunctions. This property becomes more pronounced as $Z$ increases and is called the *lanthanide contraction*. By the time one

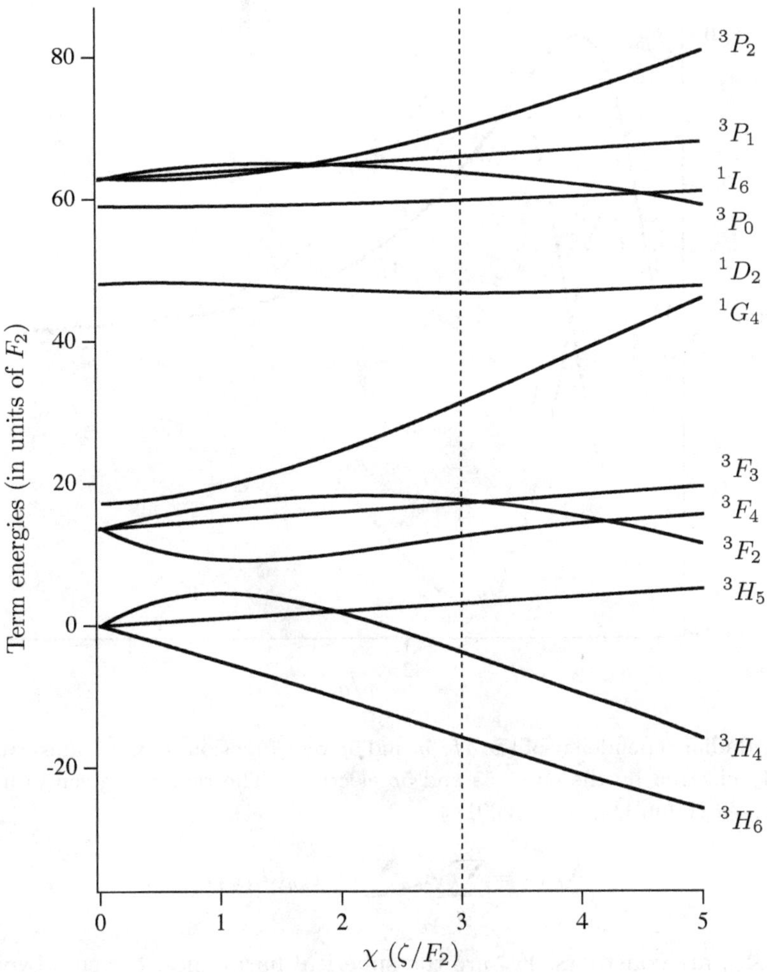


**Fig. 10.31** The various term energies of $Tu^{3+}$ as a function of the spin–orbit interaction, parameterized using $\chi$, which is the ratio of $\zeta$ to the Slater integral, $F_2$. The actual value is $\chi = 3$, shown by the dashed line. The $^1S_0$ term is off the top of the graph with an energy greater than $140 \times F_2$. (From Gruber (1960).)

reaches neodymium (Nd, $Z = 60$), the maximum of the $4f$ wavefunction is well inside the $5s^2 5p^6$ closed shells.

So far, we have made no mention of the influence of the glass substrate on the energy levels of the rare earth ions. The interaction due to the glass host is somewhat smaller than the spin–orbit or residual electrostatic interactions in the free ions and can be treated using perturbation theory. It is customary to represent the fields due to the glass substrate as a sort of multipole expansion using spherical harmonics. The perturbation potential, $V_{host}$, due to the host is:

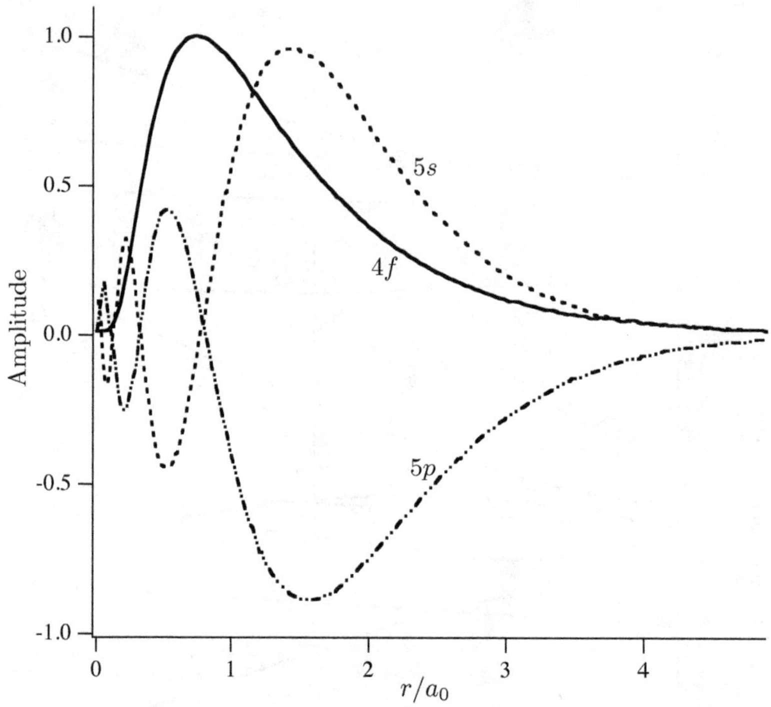


**Fig. 10.32** Radial dependency of the $4f$, $5s$ and $5p$ wavefunctions of $Ce^{3+}$ illustrating shielding of the $4f$ electron by the outer $5s$ and $5p$ electrons. The radius is given in units of the Bohr radius, $a_0$. (From Digonnet (2001).)

$$V_{host} = \sum_{kq} A_{kq} \sum_i r_i^k Y_{kq}(\theta_i, \phi_i), \tag{10.215}$$

where the $A_{kq}$ are constants, $Y_{kq}$ are the spherical harmonics, the sum over $i$ is over each electron in the $4f$ shell and $r^k Y_{kq}(\theta, \phi)$ is a solution to Laplace's equation which is finite at $r = 0$. The perturbation calculations are actually quite complicated and result in the splitting of the free ion terms into a number of *Stark* levels (i.e., levels which are split due to electrostatic interactions with the glass substrate). The important thing to realize is that $V_{host}$ can mix in different configurations which can have different values of $L$ and, most importantly, *different parity* from the unperturbed configuration. The latter is very important and explains the presence of strong *electric dipole* optical transitions in the rare earths. Using first-order perturbation theory of the wavefunction, the mixing due to $V_{host}$ can be expressed as

$$|\psi'\rangle = |\psi> + \sum_i |\psi_i\rangle \frac{\langle\psi_i|V_{host}|\psi\rangle}{E - E_i} \tag{10.216}$$

where $|\psi\rangle$ is the original wavefunction (a linear combination of Slater determinants), $|\psi'\rangle$ is the new wavefunction, $|\psi_i\rangle$ are the possible *configurations* which can be mixed in due to the perturbation, $E$ is the energy of $|\psi\rangle$ and $E_i$ is the energy of $|\psi_i\rangle$. Thus, the matrix element, $\langle\psi_i|V_{host}|\psi\rangle$, can mix configurations with different $L$ and $J$ as well as with different parity, depending upon the parity and angular dependence of $V_{host}$. In order to mix in *opposite parity* configurations, the value of $k$ in eqn 10.215 *must be odd.*

As mentioned above, *optical* transitions in the rare earth ions involve transitions of electrons which reside in (and remain in) the unfilled $f$ level. We summarize in Table 10.4 some of the properties of the three principal types of transition that can take place: electric dipole, magnetic dipole and electric quadrupole.

**Table 10.4** The three most common radiative interactions. The strength is relative to electric dipole transitions and $\alpha \approx 1/137$ is the *fine structure constant. Rigorous LS* coupling is assumed ($\Delta S = 0$).

| Type | Transition operator | Selection rules | Strength |
|---|---|---|---|
| Electric dipole | $\sum_i^N e\boldsymbol{r_i}$ | $\Delta L = 0, \pm 1 (L = 0 \not\to 0)$<br>$\Delta J = 0, \pm 1\ (J = 0 \not\to 0)$<br>$\Delta S = 0$<br>parity change | 1 |
| Magnetic dipole | $\left(\frac{\omega}{c}\right)\left(\frac{e\hbar}{2m}\right)\sum_i^N \boldsymbol{l}_i + 2\boldsymbol{s}_i$ | $\Delta L = 0$<br>$\Delta J = 0, \pm 1\ (J = 0 \not\to 0)$<br>$\Delta S = 0$<br>no parity change | $\alpha^2$ |
| Electric quadrupole | $\frac{\omega^2}{2c}\sum_i^N e x_i y_i$ | $\Delta L = 0, \pm 1, \pm 2\ (L = 0 \not\to 0, 1)$<br>$\Delta J = 0, \pm 1, \pm 2$<br>$(J = 0 \not\to 0, 1; J = \frac{1}{2} \not\to \frac{1}{2})$<br>$\Delta S = 0$<br>no parity change | $\alpha^2$ |

As can be seen in the table, the electric dipole transitions, which require a parity change, are about $10^4$ times stronger than the magnetic dipole or electric quadrupole transitions, which take place between states of the same parity. It has been experimentally determined that most transitions in rare earths are due to electric dipole transitions; in a few cases the transitions are magnetic dipole. However, an initial glance at Fig. 10.30 indicates that the electric dipole selection rules are flagrantly violated, with some $\Delta L = 3$ transitions and transitions between terms with the same parity. The principal mechanism which allows electric dipole transitions to take place is the mixing of opposite parity configurations due to the odd $k$ terms in $V_{host}$. The odd $k$ terms will be zero if the field at the site of the ion due to the host has *inversion symmetry*; most glasses lack this symmetry.

The spectral breadth of rare earth ions embedded in glass substrates is due in roughly equal parts to both homogeneous and inhomogeneous broadening mechanisms. As is seen in Fig. 10.33, the discrete Stark components are resolved in the highly regular geometry of a crystalline (YAG) substrate but not in the amorphous glass host. This is due to the spatial randomness of the perturbations in the glass host – a pure form of *inhomogeneous* broadening. It turns out that the dominant cause of *homogeneous*

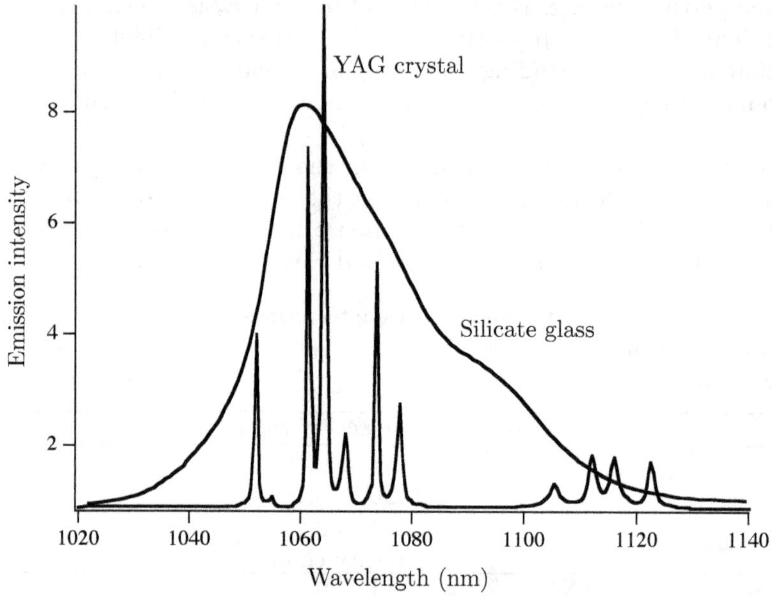


**Fig. 10.33** Comparison of emission spectrum of $Nd^{3+}$ ions in a YAG crystal to the emission from ions embedded in an amorphous glass substrate. The broadening due to the glass is very evident (From Digonnet (2001)).

broadening in both crystalline and glass hosts is due to phonon-induced transitions between different Stark components of the *same* term, where phonons are quantized vibration states in a solid and behave somewhat like particles. This phenomenon is similar to *elastic* scattering in diffuse vapors and results in a randomization of the *phase* of the wavefunction. For the $Nd^{3+}$ ions in a silicate glass (Fig. 10.33), the homogeneous breadth is about 110 $cm^{-1}$ (about 10 nm at 1 $\mu$m) and the inhomogeneous breadth is about 50 $cm^{-1}$ (about 5 nm).

## 10.10 Some specific fiber lasers

We will discuss several specific fiber laser systems in this section, limiting our descriptions to single-frequency, continuous-wave lasers based upon erbium, ytterbium, neodymium and thulium ions. Space limitations prevent us from studying some interesting lasers employing other lanthanide ions or alternative approaches to laser gain, such as fiber brillouin amplification. We will also not discuss fiber ring lasers, which do not seem to be as widely used as standing-wave lasers.

### 10.10.1 Fiber laser resonators

Before describing specific lasers, we will examine a few of the possible resonator geometries, which are shown schematically in Fig. 10.34. Since standing-wave fiber lasers are subject to various forms of inhomogeneous broadening, they all have some frequency selecting element in the laser cavity to ensure single-mode operation. The length of

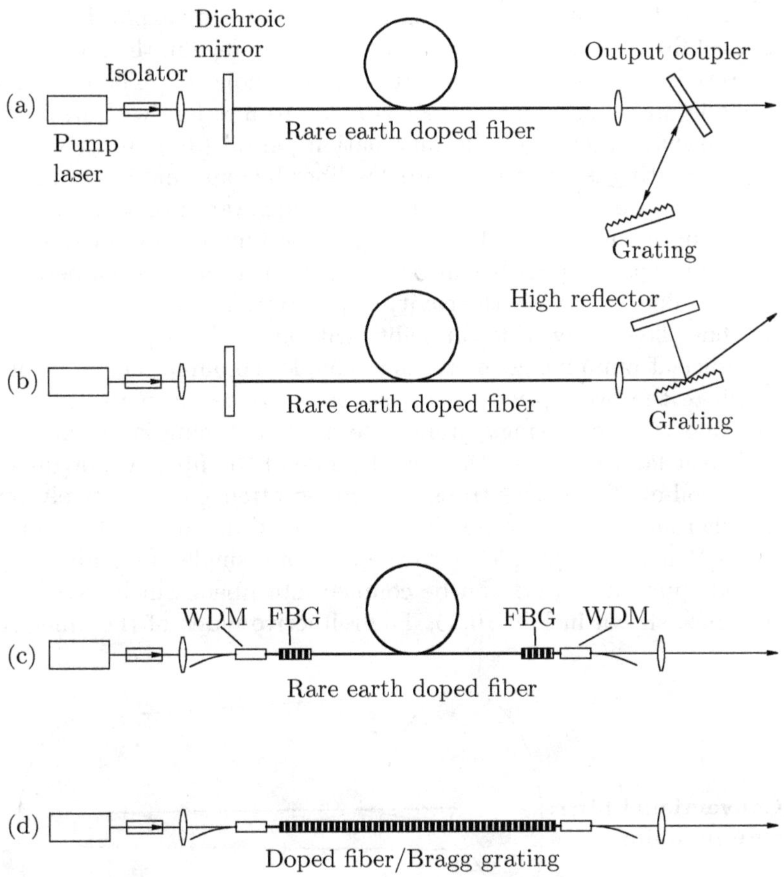


**Fig. 10.34** Several possible single-frequency standing-wave fiber laser resonator geometries. The upper two are presented mainly for historical reasons; the bottom two drawings apply to the most commonly used single-frequency fiber lasers. The fiber grating in the lower drawing has a $\pi/4$ phase shift in the middle to ensure a unique resonant frequency. The WDM element is a wavelength division multiplexor and the FBG is a fiber Bragg grating.

the active fiber is fairly short (10 cm or less) to increase the free spectral range of the laser cavity and make it easier for the filters in the cavity to discriminate against all but a single mode. In diagrams (a) and (b), the frequency selective element is a grating used in the *Littrow* configuration in (a) and in the *Littman* configuration in (b). Currently, most single-frequency fiber lasers use either fiber Bragg grating reflectors (c) or a single distributed Bragg grating (d) to ensure single-mode oscillation. The two wavelength division multiplexors (WDMs) shown in the figures are used to separate the laser radiation from the pump radiation: the WDM on the left passes the pump radiation with little loss, and sends the laser radiation to the coupled fiber arm; the WDM on the right dumps the residual pump radiation into the coupled arm and passes the laser radiation to the output.

The gratings in distributed feedback lasers are created by exploiting the *photosensitivity* of optical fibers: the creation of a permanent change in the index of refraction of certain materials by exposing them to ultraviolet radiation. The ultraviolet radiation from a moderate to high-power laser is made to interfere with itself, generating fringes whose spacing determines the modulation period (and Bragg wavelength) of the grating. The grating is "written" onto the fiber by exposing it to the interference pattern; the amplitude of the resulting permanent index variation is from $10^{-3}$ to $10^{-5}$. The laser wavelength can be tuned by changing the temperature of the structure or by compressing or expanding it. It can be shown that a uniform temperature change or stressing of the fiber will allow the cavity mode to track the center frequency of the gratings and thus allow fairly wide tunability without mode hops.

The most efficient pumping geometry is to couple the pump radiation into the core of the fiber where the overlap with the laser mode is nearly perfect. This is called *core pumping* and requires a single transverse mode pumping laser with good beam quality for efficient launching into the lowest mode of the fiber. Unfortunately, diode lasers with a good-quality single transverse mode often have relatively low power. Furthermore, thermal effects, the possibility of optical damage and nonlinear effects limit the power that can be coupled into the core of a single-mode fiber. High-power multimode diode pumping lasers can be coupled into fibers which have a single core and two claddings, shown in Fig. 10.35. The refractive index of the inner cladding is

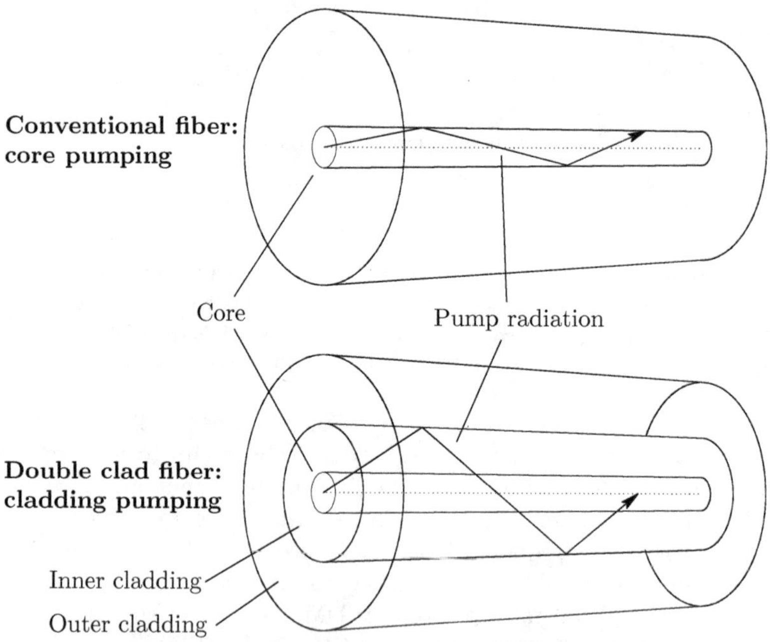


**Fig. 10.35** Illustration of two pumping geometries: *core pumping* which couples the single–mode laser into the fiber core and *cladding pumping* which uses a fiber with two claddings and couples a multimode laser into the inner cladding.

less than that of the core and the index of the outer cladding is lower than that of the inner cladding. Thus, both the core and the inner cladding can support guided-wave propagation. As might be expected, the pumping efficiency is somewhat lower with cladding pumping, since the overlap with the laser mode is much smaller than in core pumping. Most low-power single-frequency fiber lasers use core pumping.

### 10.10.2 Erbium and erbium/ytterbium lasers

Erbium-based fiber lasers emit radiation at about 1550 nm, which is in one of the "communication windows" for fiber-optics systems and largely accounts for the popularity of this gain medium for amplifiers and lasers. Early single-frequency erbium lasers delivered very little power (less than 1 mW), since the pump radiation was only weakly absorbed in the the short resonators required for single-frequency oscillation. Increasing the erbium density in the fiber led to instabilities and a loss of efficiency due to "ion clustering". Moreover, erbium is a three-level system and therefore has a somewhat greater threshold than four-level gain media. The currently accepted solution to these problems is to use a fiber co-doped with $Yb^{3+}$ and $Er^{3+}$, with the ytterbium molar concentration twenty to thirty times as great as the erbium concentration. Since the ytterbium ion absorption at 980 nm is at least an order of magnitude greater than that of erbium and the excitation is efficiently transferred to the erbium ions, the pump absorption in the co-doped fiber is much greater than in single-doped erbium fibers and the resulting slope efficiency (derivative of laser power with respect to pump power) can be greater than 25%. The 980 nm pumping light can be generated by strained-layer quantum-well InGaAs/AlGaAs laser diodes. These diodes are available with single-mode fiber pigtails and can be directly attached to the fiber laser. The level diagrams of erbium and ytterbium ions are shown in Fig. 10.36. The excited-state

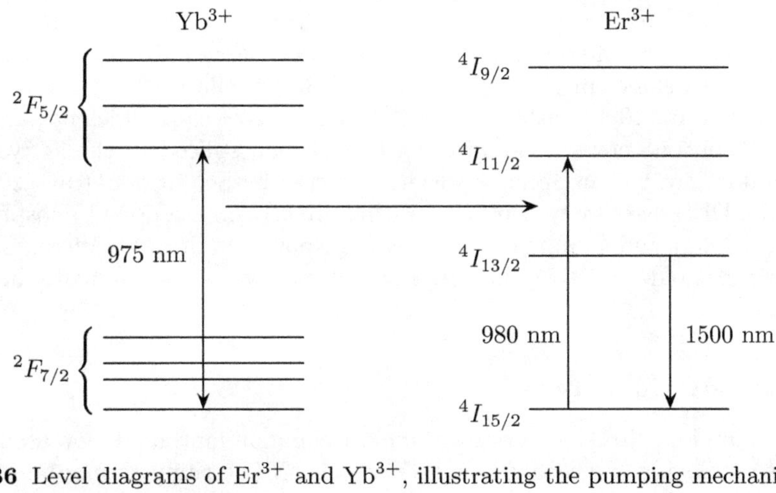


**Fig. 10.36** Level diagrams of $Er^{3+}$ and $Yb^{3+}$, illustrating the pumping mechanism: due to the presence of ytterbium, the pump absorption is increased by an order of magnitude and the excitation is very efficiently transferred to the erbium ions.

lifetime of the upper laser level of erbium is 10 ms.

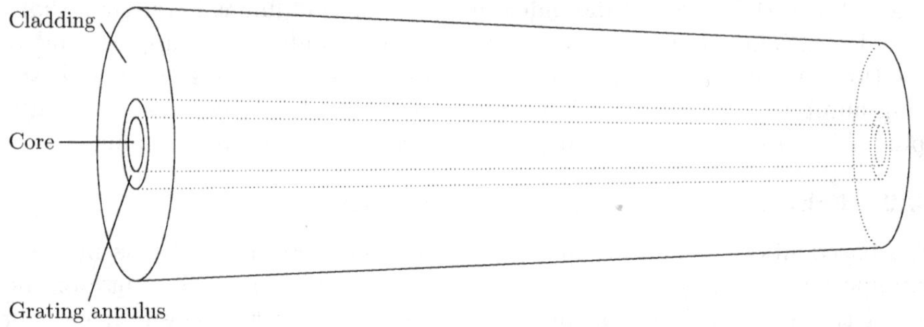


**Fig. 10.37** Section of fiber showing annular photosensitive region in which a distributed Bragg grating can be written.

Rare earth doped fibers have a large proliferation of levels because of their numerous fine-structure multiplets and Stark levels; furthermore, as a result of the various broadening mechanisms, these levels can be spectrally broad. The result of this is the high probability of *excited-state absorption* (ESA): the absorption of pump light from the *excited state*, which can reduce the pumping efficiency. One of the useful features of the 980 nm pumping transition in erbium is that it is completely free from ESA.

There are a number of successful distributed feedback (DFB) fiber lasers and distributed Bragg reflector (DBR) lasers which use the co-doping approach. One of the practical problems encountered in constructing a DFB laser is to find a fiber with both the desired gain properties and sufficient photosensitivity for the creation of a grating. A common solution to this is to use a photosensitive *annular* ring around the active co-doped core for the grating (Fig. 10.37). This procedure was followed by Dong et al. (1997), who constructed both distributed Bragg reflector and distributed feedback lasers with up to 18 mW output and a 25% pumping slope efficiency. Loh et al. (1998) used similar techniques and obtained 4 mW from a 5 cm long DFB laser with an 18 kHz linewidth and 40% slope efficiency. A distributed Bragg reflector laser was constructed by Spiegelberg et al. (2003) with a linewidth of 2 kHz, 24% slope efficiency and greater than 200 mW output power. The gratings were fusion spliced to the active fiber. A laser which does not use an annulus was constructed by Schülzgun et al. (2008), who constructed a DFB laser by writing the grating directly onto a doped phosphate glass fiber (3.5 cm long) and pumped it using a high-power multimode diode. This laser had a high threshold ($\approx 1$ W) and low slope efficiency ($< 3\%$) and produced up to 160 mW.

### 10.10.3 Neodymium lasers

The neodymium ion, $Nd^{3+}$, is perhaps the most common lanthanide ion used in solid-state lasers employing a *crystalline* substrate (usually yttrium–aluminum–garnet, or YAG). Its use in fiber lasers appears to be less common than erbium or some other lanthanides. This is slightly ironic, since the first fiber laser was constructed using $Nd^{3+}$ in 1961 and is described in a paper by Snitzer (1961); this was during the era when lasers were called *optical masers*. The level structure appears in Fig. 10.38. The

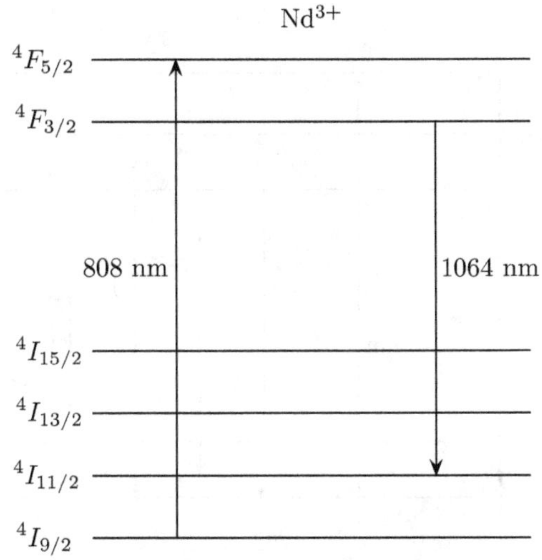


**Fig. 10.38** Level structure of $Nd^{3+}$.

usual pump wavelength is 808 nm and there are three possible oscillation wavelengths: 0.92 $\mu$m, 1.06 $\mu$m and 1.34 $\mu$m. The strongest line is at 1.06 $\mu$m and is due to a four-level laser scheme. The lifetime of the excited state ($^4F_{3/2}$ term) is about 0.5 ms.

The first single transverse-mode $Nd^{3+}$ fiber laser was reported by Mears et al. (1985). This laser used a 2 m long $GeO_2/SiO_2$ fiber lightly doped with $Nd^{3+}$ and employed external, highly reflecting mirrors for laser feedback (the laser probably oscillated in several *longitudinal* modes). Jauncey et al. (1988) described an $Nd^{3+}$ laser using a 5.1 cm long fiber with a dichroic high-reflecting mirror at one end and an etched grating at the other end. The laser produced up to 0.78 mW at 1082 nm with 40 mW of 594 nm pump light (from a dye laser) and had a measured linewidth of 1.3 MHz.

### 10.10.4 Ytterbium lasers

The level diagram for $Yb^{3+}$ is shown in Fig. 10.39. The level structure consists of a fine-structure doublet which is split into several Stark levels. The usual laser transition is at 975 nm and is between the lowest Stark levels of each member of the doublet. The pumping radiation at 915 nm excites the highest Stark level of the $^2F_{5/2}$ state. The excited-state lifetime is 1.8 ms. Due to the simplicity of the level structure, there is no excited-state absorption. Despite the three-level laser system, ytterbium fiber lasers with 25% slope efficiency have been demonstrated. Ytterbium fibers can be highly doped (unlike erbium fibers), leading to good pump absorption in short fibers.

A single-frequency distributed Bragg reflector laser was constructed by Zhu et al. (2012) and generated 100 mW of radiation at 976 nm with a linewidth of less than 3 kHz and a slope efficiency of 25%. The laser consisted of a 2 cm long highly doped phosphate glass fiber with two distributed Bragg reflectors fusion spliced at each end; the FBGs had reflectivities of 99% and 60%. The latter FBG was made with a polarization maintaining fiber and allowed single-polarization operation of the laser.

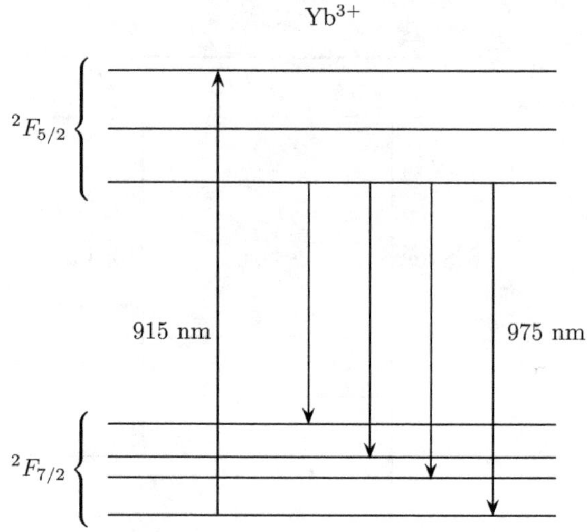


**Fig. 10.39** Level structure of $Yb^{3+}$. The structure is simple: a fine-structure doublet split into a number of Stark levels.

Guan et al. (2007) demonstrated a somewhat similar system which generated 30 mW on the line at 1029 nm with a 490 mW pump at 976 nm.

### 10.10.5 Thulium lasers

The level diagram for $Tm^{3+}$ appears in Fig. 10.40. One should take note of the unusual ordering of the $J$-levels for the $^3H$ terms: the failure to follow the interval rule (also shown in Fig. 10.31) is an indication of the breakdown of LS coupling in $Tm^{3+}$. Thulium lasers can oscillate at wavelengths from about 1.7 $\mu$m to 2.1 $\mu$m; the usual

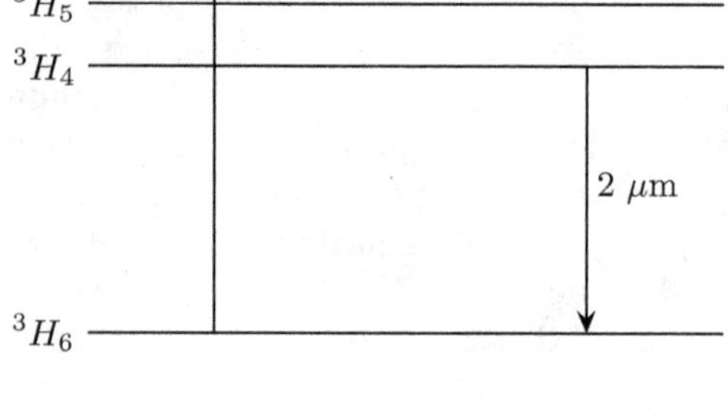


**Fig. 10.40** Level structure of $Tm^{3+}$.

wavelength is about 2 $\mu$m and the pump wavelength is 790 nm, allowing efficient pumping using AlGaAs diode lasers. The non-radiative decay from the upper pumping level ($^3F_4$) to the upper laser level occurs in about 20 $\mu$s. The lifetime of the upper laser level ($^3H_4$) is about 0.2 ms. Fiber lasers using short thulium-doped silica fibers suffer from low pump absorption and produce very little power; less than 1 mW. This situation is dramatically improved by co-doping the fiber core with germanium and aluminum or by using a germanate glass fiber (one containing $GeO_2$ in addition to $SiO_2$), and efficiencies substantially greater than 25% have been reported. The large difference between the pumping wavelength (790 nm) and the laser wavelength (2000 nm) makes it impossible to obtain pumping efficiencies of 40% or greater: if every pump photon caused a stimulated decay (i.e., 100% quantum efficiency), the pumping efficiency would still be slightly less than 40%.

Using a silica fiber with 14.5% Ge concentration, Agger et al. (2004) constructed a 5 cm long distributed feedback laser which produced slightly less than 1 mW at 1735 nm with about 600 mW of pumping radiation (from a Ti–Sapphire laser at 790 nm). The writing of the grating increased the birefringence of the fiber and allowed the laser to oscillate in a single predictable polarization. Both cladding- and core-pumped thulium lasers were constructed by Geng et al. (2009). Their lasers employed distributed Bragg reflectors at each end of the fibers, whose lengths were 2 cm and 6 cm and were composed of a multicomponent silicate glass in order to improve the pump absorption. The longer cladding-pumped laser produced up to about 190 mW output (with a 10 W pump laser) and the shorter core-pumped laser produced about 40 mW with a 500 mW pump. Laser linewidths of less than 3 kHz were observed.

## 10.11 Further reading

There is an enormous literature on the subject of fiber optics and related subjects. We will list a few that were useful in our treatment of guided-wave optics. Our favorite book on electromagnetic theory is Jackson's well-known textbook (1999); it includes several useful sections on planar waveguides and fibers, though its emphasis is on metallic waveguides and cavities. Perhaps the most comprehensive book on optical fibers is by Snyder (1983); it treats the modes of a fiber using the rigorous theory and provides all of the relevant formulas in a series of convenient tables. An excellent treatment of fiber modes using the scalar theory is found in Ghatak and Thyagarajan's book (2000); much of this material is also found in the book on optical electronics (1989) by the same authors. Useful discussions of waveguide and fiber theory are found in the quantum electronics books by Davis (1996) and Yariv (1989). The *Bell Systems Technical Journal* has a number of classic articles by Marcuse on coupled-mode theory (1975), fiber splicing losses (1977) and fiber couplers (1971). Excellent textbooks on atomic theory (for the section on rare earth ions) are by Foot (2005) and Woodgate (2000) as well as the older classics by Condon and Shortley (1964) and Herzberg (1944). A complete treatment of fiber lasers is the book by Digonnet (2001).

## 10.12 Problems

(10.1) Using the *ray approach*, where a wave travels in a zig-zag fashion down a slab waveguide, obtain the fields in the *core* of the waveguide and show that they are the same as are obtained using the wave approach. Assume that the angle of the ray with respect to the waveguide axis is $\theta_t$, which satisfies eqn 10.6. Do this for both TE and TM modes.

(10.2) Using the vector wave equations in an inhomogeneous medium (eqns 10.13), show that the wave equations for the $z$-components of the fields are *decoupled* when
 (a) the waveguide is a *slab* and
 (b) the waveguide is a cylindrical fiber and the fields are azimuthally invariant.

(10.3) It was stated in the text that the cavity mode would track the center frequency of a Bragg mirror in a fiber laser employing Bragg grating reflectors if the *entire structure* were subject to a uniform expansion or contraction due to temperature changes or mechanical stress. Demonstrate the truth of this claim.

(10.4) Calculate the cutoff frequencies for the $LP_{11}$ and $LP_{21}$ modes in a fiber with $n_2 = 1.5$, $\Delta = 0.004$ and $a = 3\ \mu m$.

(10.5) For the fiber described in the preceding problem, design a *wavelength division multiplexer* (WDM) which can be used to couple pump radiation at 980 nm into an erbium fiber laser operating at 1550 nm. Assume that the core separation between the fibers in the coupler is 4 $\mu$m and that the fibers are operating in the lowest ($LP_{01}$) mode. Use the approximate expression for the coupling constant that appears in the text.

(10.6) Assume that we have two fibers whose parameters are the same as those in the two preceding problems. We would like to determine the transmission loss when (a) the fiber separation is one half the core diameter and (b) the fiber axes are displaced from each other by one half the core diameter. The significant loss for a positioning error of 1.5 $\mu$m indicates the need for very precise fiber alignment when a splice is made.

(10.7) Using the Gaussian beam model, derive the equation for the transmission between two fibers whose cores diameters are different (eqn 10.130).

# 11 Mode-locked lasers and frequency metrology

## 11.1 Introduction

All of the lasers discussed so far have emitted continuous-wave (CW) radiation. Despite a promise in the Preface that we would consider CW lasers exclusively, there is one type of laser which has aspects of both a CW laser and a pulsed laser and has recently become so important to the frequency metrology community that we would be remiss in not describing it in some detail. This is the *mode-locked laser*, which is capable of generating such extremely short pulses that in some realizations it is called a *femtosecond* laser, implying that it emits pulses containing only *a few cycles of optical radiation*. We will present the theory of mode-locked lasers, describe a common implementation (the mode-locked Ti–sapphire laser) and discuss the application of these lasers in frequency metrology. The latter application led to the Nobel Prize in Physics for John L. Hall and Theodor W. Hänsch in 2005.

## 11.2 Theory of mode locking

Our treatment of mode locking will not be very rigorous; for more rigor, one can refer to a large number of excellent books and papers on the subject, some of which are mentioned at the end of this chapter. Initially, we will assume there is no *dispersion* in the laser cavity: the propagation speed of a wave packet (group velocity) is the same as the phase velocity. We begin by making the (somewhat unrealistic) simplifying assumption that the spectrum of an inhomogeneously broadened laser consists of a series of lines of constant amplitude spaced by the cavity free spectral range, $\nu_c$. If there are $N$ lines, the total width, $\Delta\nu_{inhom.}$, is

$$\Delta\nu_{inhom.} = (N-1)\nu_c. \tag{11.1}$$

Figure 11.1 displays the rectangular spectrum of this laser. If the center frequency is $\omega_0$, the total electric field is

$$E_{total} = E_0 \sum_{-(N-1)/2}^{+(N-1)/2} \exp[i(\omega_0 + n\omega_c)t + i\phi_n(t)], \tag{11.2}$$

where $\phi_n(t)$ is a random function of time uniformly distributed between 0 and $2\pi$ and $E_0$ is the field for each line. It is easy to see that the time-averaged intensity is

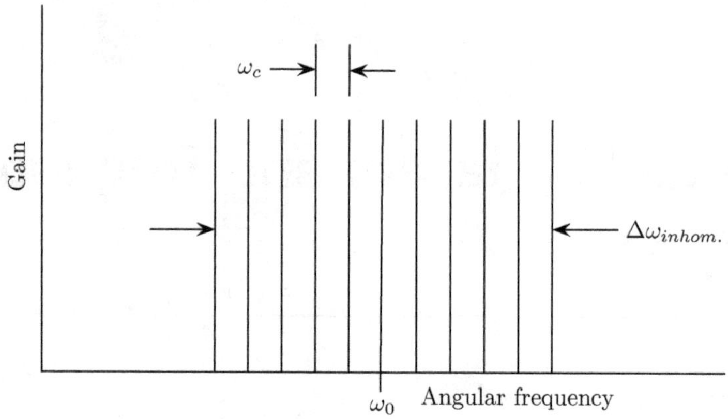


**Fig. 11.1** Simplified rectangular spectrum of inhomogeneously broadened laser.

$$\langle I \rangle = NI_0 \qquad I_0 = \text{intensity of each line,} \tag{11.3}$$

since the cross-terms average to zero when one takes the square modulus of the field. The behavior with time of the intensity would be extremely noisy due to the random phases (left side of Fig. 11.2), though any laboratory measurement of the intensity would average for a sufficient length of time to remove most of the noise. Note that the noise is entirely due to the *cross-terms* when one takes the square modulus of the field: it is a random *interference* among the laser modes. Now let us assume that we have a mechanism for *locking* the phases of all of the lines together, so that we may take $\phi_n(t)$ out of the sum. The total field is then

$$E_{total} = E_0 e^{i\omega_0 t + i\phi(t)} \sum_{-(N-1)/2}^{+(N-1)/2} e^{in\omega_c t} = E_0 e^{i\omega_0 t + i\phi(t)} \frac{\sin N\omega_c t/2}{\sin \omega_c t/2} \tag{11.4}$$

and the intensity is

$$I = I_0 \left[ \frac{\sin N\omega_c t/2}{\sin \omega_c t/2} \right]^2, \tag{11.5}$$

where $I_0$ is again the intensity in each line. Using l'Hôpital's rule, the peak intensity of this function is

$$I_{peak} = N^2 I_0. \tag{11.6}$$

The curve of intensity versus time when the modes are locked is shown on the right side of Fig. 11.2. The temporal width, $\Delta t_p$, of each pulse is roughly

$$\Delta t_p \approx \frac{1}{N\nu_c} \qquad \nu_c = \omega_c/2\pi. \tag{11.7}$$

We therefore conclude that this process increases the peak power for each pulse over the CW laser average power by a factor of $N$ and generates pulses with a duty cycle $\nu_c \Delta t_p = 1/N$. The average power is unchanged over that of the CW laser but the peak

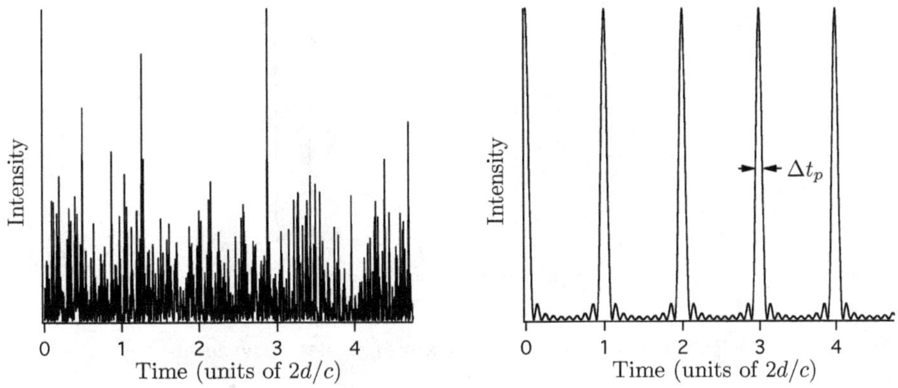


**Fig. 11.2** Intensity versus time for laser with rectangular frequency spectrum and ten modes. On the left the modes have random phase and on the right the phase is constant. The time scale is in units of the round-trip time ($2d/c$) which is about 10 ns. Thus, the noise on the left would be normally averaged out with most laboratory detectors.

power can be extremely high when $\Delta t_p$ is in the femtosecond ($10^{-15}$ $s$) range. Finally, the pulse width is very simply

$$\Delta t_p \approx \frac{1}{\Delta\nu_{inhom.}} \qquad \Delta\nu_{inhom.} = \Delta\omega_{inhom.}/2\pi. \tag{11.8}$$

The reader should recognize eqn 11.5 as essentially the same function as that describing the interference pattern from $N$ slits uniformly illuminated by coherent light (except that the $N$-slit function depends upon a spatial coordinate). We have on several occasions used a temporal analogy to help illustrate a spatial behavior; we now reverse this and use a familiar spatial situation ($N$-slit interference) to help explain a temporal one. The important point is that the $N$ waves interfere with one another to produce noise if they have random phases (equivalent to illuminating the slits with *incoherent light*) and a regular pattern of sharp lines if the phases are locked together (equivalent to using *coherent* light to illuminate the slits).

To illustrate how the phases can be locked together, we use a conceptual diagram of a mode-locked laser (Fig. 11.3). All mode-locked lasers use some variation of this technique. The key element which distinguishes a mode-locked laser from a simple CW laser is an intracavity *shutter* which is arranged to cyclically open and close with a period exactly equal to the round-trip time of the intracavity wave. The only field distribution which can survive in such a laser is a regular series of pulses with a repetition rate equal to the shutter cycling frequency (which is also equal to the free spectral range of the cavity and therefore to its mode separation). One can work backwards (using Fourier theory) and show that the only field frequency dependence which can generate these regular pulses is a series of uniformly spaced modes whose separation is the cavity free spectral range. The amplitudes might be different from the ones we have used, but the phases of all the components are locked together. We should mention that the phases need not be all identical to obtain mode locking; they can also depend linearly on the summation index, $n$. Finally, judicious placement of

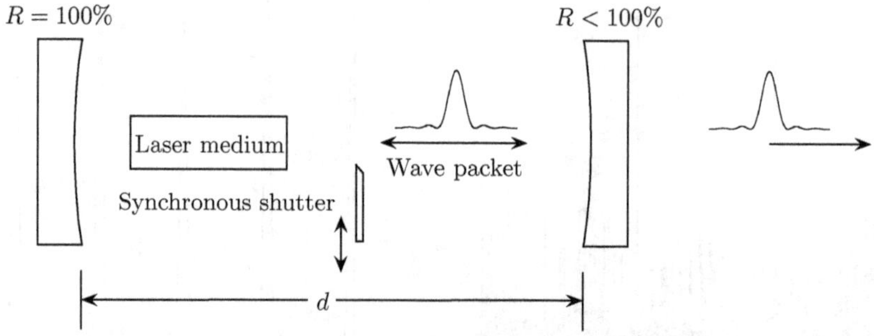


**Fig. 11.3** Conceptual diagram of mode-locked laser.

the shutter in the cavity allows mode locking at *harmonics* of the mode spacing ($c/2d$); this is called *harmonic mode locking.*

A much more realistic mode distribution in an inhomogeneously broadened laser is shown in Fig. 11.4, where the envelope of modes is a Gaussian. The amplitude of the electric field at the $n^{th}$ mode is

$$E_n = E_0 \exp\left[-2\ln 2\left(\frac{n\omega_c}{\Delta\omega}\right)^2\right], \tag{11.9}$$

where $E_0$ is the field at the peak of the Gaussian and $\Delta\omega$ is the inhomogeneous width in angular units. In the time domain, the field is

$$\begin{aligned} E(t) &= \sum_{-(N-1)/2}^{+(N-1)/2} E_n e^{i\omega_0 t + in\omega_c t} \\ &= E_0 e^{i\omega_0 t} \sum_{-(N-1)/2}^{+(N-1)/2} \left\{\exp\left[-2\ln 2\left(\frac{n\omega_c}{\Delta\omega}\right)^2\right]\right\} e^{in\omega_c t}. \end{aligned} \tag{11.10}$$

This series can be summed when $N$ is very large by converting it into an integral using the following substitutions:

$$x = n\omega_c \tag{11.11}$$

$$dx = (n + 1/2)\omega_c - (n - 1/2)\omega_c = \omega_c. \tag{11.12}$$

The field is then

$$E(t) = \frac{E_0 e^{i\omega_0 t}}{\omega_c} \int_{-\infty}^{\infty} \left\{\exp\left[-\frac{2(\ln 2)x^2}{(\Delta\omega)^2}\right]\right\} e^{ixt}\, dx. \tag{11.13}$$

(We introduced the $dx$ factor by multiplying the exponential by $dx/\omega_c = 1$.) The integral should be readily recognized as the Fourier transform of the expression in the curly brackets, which is a Gaussian. The transform is also a Gaussian:

$$E(t) = E_0 e^{i\omega_0 t}\left(\frac{\pi}{2\ln 2}\right)^{1/2}\left(\frac{\Delta\omega}{\omega_c}\right)\exp\left[-\left(\frac{\Delta\omega t}{2\sqrt{\ln 2}}\right)^2\right]. \tag{11.14}$$

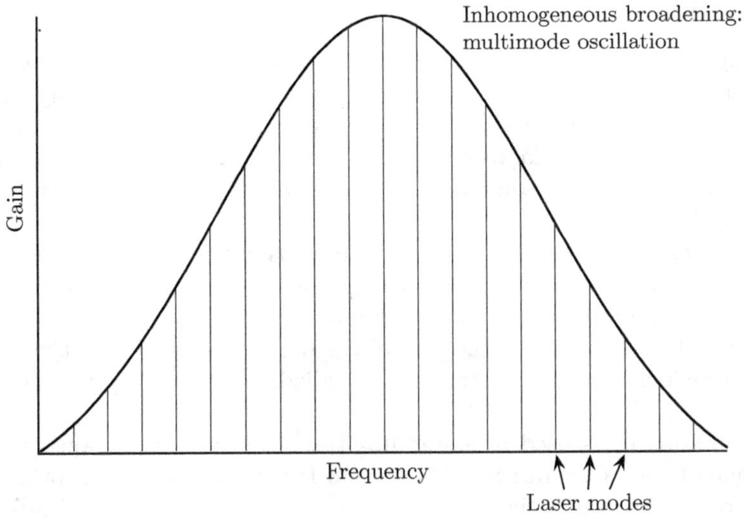


**Fig. 11.4** Inhomogeneously broadened laser modes (Gaussian envelope).

The intensity is

$$I(t) = I_0 \exp\left[-2\left(\frac{\Delta\omega t}{2\sqrt{\ln 2}}\right)^2\right], \tag{11.15}$$

where $I_0$ is the peak intensity. This is a Gaussian-shaped pulse of width

$$\Delta t_p = \frac{2\ln 2}{\pi}\frac{1}{\Delta\nu} = \frac{0.44}{\Delta\nu} \qquad \Delta\nu = \Delta\omega/2\pi. \tag{11.16}$$

This equation only describes a *single* pulse, since the periodic structure was lost when we converted the sum to an integral: the mode spacing approached zero in the frequency domain while the pulse separation became infinite in the time domain.

We began our theoretical analysis with an inhomogeneously broadened laser to avoid a possible paradox: since a homogeneously broadened laser only oscillates on a *single mode*, one might conclude (erroneously) that it cannot be mode locked. Here the frequency domain picture leads us astray: viewed from the time domain, it should be clear that the use of a shutter allows only a repetitive pulse emission regardless of the type of broadening in the amplifying medium. A Fourier analysis of the laser field would give the same result as for an inhomogeneously broadened laser. The resolution of the paradox lies in the behavior of the shutter. The shutter, interrupting the gain at the rate $c/2d$, actually *imposes the needed mode structure* on the homogeneously broadened laser, allowing it to behave like the inhomogeneously broadened laser described above. Thus, the results for a homogeneously broadened laser are essentially the same as for an inhomogeneously broadended laser, and the former laser type can be mode-locked using the same techniques as are used in the latter. Since the modulation in the homogeneously broadened laser serves two purposes (imposing a mode structure and coupling the modes together), a somewhat larger modulation amplitude is needed to achieve mode locking than in an inhomogeneously broadened laser.

## 11.3 Mode-locking techniques

Mode locking is implemented in two general ways. *Active mode locking* uses an externally operated shutter which opens and closes at the mode spacing frequency, as was suggested in the conceptual model introduced above. A more common set of techniques, called *passive mode locking*, uses the fields themselves to operate an internal "shutter" which is based on some nonlinear optical effect (such as saturation) in the intracavity medium.

The shutter in an actively mode-locked laser is usually either an amplitude or phase modulator placed inside the cavity. A third type of shutter which is externally operated is the *pump laser*: a mode-locked pump laser with the same mode spacing (and cavity length) will amplitude-modulate the gain of the pumped laser in just the right manner to achieve mode locking. This is occasionally called *synchronous pumping*, for obvious reasons.

When a modulator is used in a standing-wave cavity, it is advantageous to place it close to one of the laser mirrors. The reason for this is to prevent other pulse configurations from developing; these can occur if the pulse passes the shutter *twice* per round trip with similar delays between each pass. If the shutter is near the mirror, the two shutter passes will occur with very different delays. In a ring laser, the shutter placement is not critical. There is one situation where a shutter is deliberately placed halfway between the mirrors in a standing-wave cavity. Such a configuration can support two counter-propagating pulses and is called the *colliding pulse* scheme. These lasers can actually generate very short pulses.

The theory of amplitude-modulated (AM) mode locking is essentially the same as that presented in the conceptual model. The modulator provides a synchronism between adjacent modes and ensures that the only internal field distribution supported by the laser is a narrow pulse which passes through the modulator when it allows the maximum transmission. Amplitude-modulated mode locking is usually accomplished with an intracavity acousto-optic modulator (AOM). The modulator is a bit different from that used outside of a laser: the acoustical termination on the far side of the crystal from the transducer is removed so that there are acoustical standing waves inside the modulator medium. The loss from the modulator is greatest when the acoustical standing-wave amplitude is greatest, and this happens at twice the modulation frequency. Thus, the modulator is driven at one half of the mode separation frequency. Figure 11.5 is a schematic of an AM mode-locked laser.

Frequency-modulation (FM) mode locking works on a somewhat different principle from AM mode locking. A phase modulator (usually an electro-optic modulator or Pockel cell) is inserted into the cavity very near to one of the cavity mirrors. Phase modulation is equivalent to a modulation of the optical path length in the cavity. A useful model for this behavior is to consider one of the cavity mirrors as moving back and forth at the modulation frequency. A wave packet incident on the moving mirror will be Doppler shifted after reflection by an amount which depends upon the speed of the mirror when the wave reaches it. Thus, such a wave will be subject to a succession of Doppler shifts which will cumulatively change the frequency of the intracavity wave by an amount which is sufficient to prevent laser oscillation. On the other hand, if the wave impinges on the mirror when it is reversing direction, there will be no Doppler

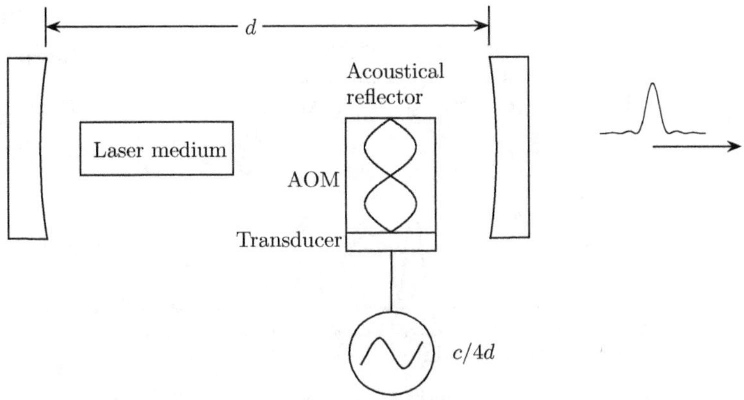


**Fig. 11.5** Mode locking using an acousto-optic modulator driven at one half the cavity mode spacing ($c/4d$). The acoustical standing waves in the AOM are shown.

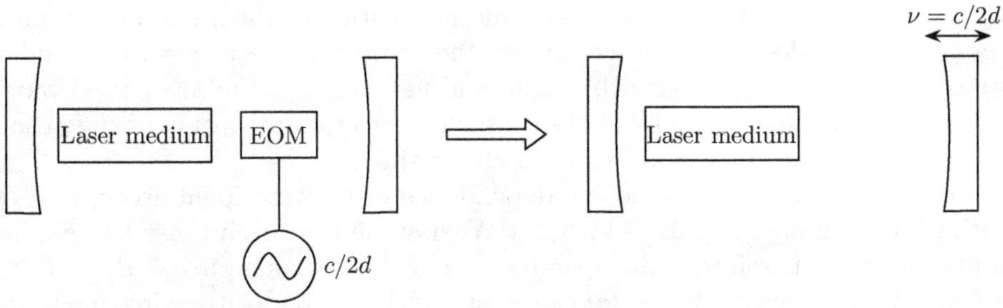


**Fig. 11.6** FM mode locking using an electro-optic modulator. The equivalent oscillating mirror model is shown on the right.

shift and laser oscillation can take place. A schematic of an FM mode-locked laser together with the equivalent *moving mirror* model appears in Fig. 11.6.

FM mode locking is *bistable*: there are two possible times in the mirror motion (at the two turning points) which can lead to laser oscillation, and the laser can jump between these two possible modes leading to an irregular or noisy pulse train. The pulses are also slightly frequency chirped since the effective mirror is never completely at rest. These are two disadvantages of FM mode locking and it is employed somewhat less frequently than AM active mode locking.

The most common mode-locking scheme exploits some nonlinear behavior of the intracavity medium and is called *passive mode locking.* We will discuss two passive approaches: the use of a *saturable absorber* and the *Kerr-lens* technique. These schemes are simpler to implement than active schemes and are routinely generating very short (femtosecond) pulses.

A saturable absorber is one whose absorption, due to saturation, *decreases* with increasing intensity. We will present a simplified theory of mode locking using a saturable absorber (from Garmire (1967)). We will assume without proof that a saturable absorber results in mode-locked behavior. The motivation for this is suggested

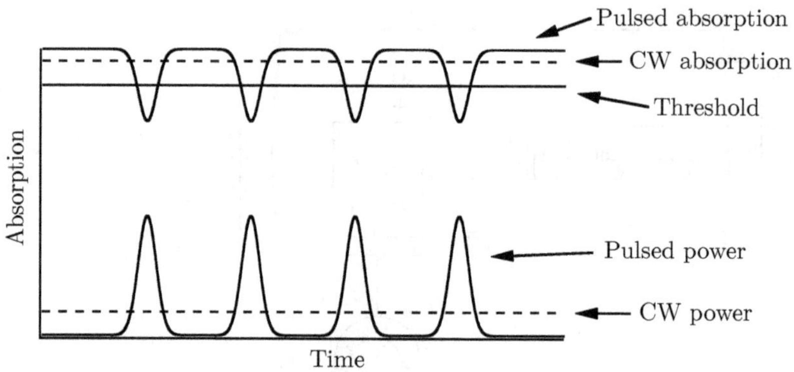


**Fig. 11.7** Behavior of saturating medium in laser cavity. The dotted lines are for CW operation with the same average power as pulsed. Only the pulsed operation dips below threshold absorption.

by Fig. 11.7, which plots the pulse power and absorption of the medium as a function of time and shows that a CW wave having the same average power (area under the curve) as the pulses will be below threshold and not oscillate, but the pulsed wave will be above threshold near the peak of the pulses (due to the saturation of the absorbing medium) and the laser will therefore oscillate at those times.

Our calculation will determine the dependence of the fractional absorption of the saturating medium on the pulse width, $t_p$. We assume that each pulse has $E$ photons per unit area and therefore a flux per pulse of $E/t_p$. For simplicity, the pulse time dependence is assumed to be *rectangular*, of width $t_p$. The saturating medium will be modeled as a two-level system with level population densities $N_2$ and $N_1$. The treatment is very similar to the derivation of laser gain in Chapter 5. The population densities satisfy a rate equation at each point in space:

$$\frac{dN_1}{dt} = R(N_2 - N_1) + N_2/\tau, \tag{11.17}$$

where $R$ is the stimulated rate and $\tau$ is the lifetime of level 2. We will assume that the wave propagates along the $z$-axis. The loss in photon flux per unit length is equal to the number of absorptions per unit time per unit volume:

$$\frac{d}{dz}\left(\frac{E}{t_p}\right) = -R(N_1 - N_2). \tag{11.18}$$

We assume that all of the molecules (the saturating medium is usually a molecular dye) are in the ground state at $t = 0$ and the initial population is $N_{10}$. The above equations can be readily solved with these initial conditions. The solution provides the fractional absorption per unit length, $\alpha$

$$\alpha = \frac{N_{10}\sigma s t_p}{2\sigma E + st_p} + \frac{2N_{10}\sigma^2 E}{(2\sigma E + st_p)^2}[1 - e^{-(2\sigma E + st_p)}], \tag{11.19}$$

where $s = 1/\tau$ and $\sigma = Rt_p/E$. Two conclusions can be drawn from this result. First, $\alpha$ is a *monotonically decreasing* function of $1/t_p$: as the pulses get shorter the loss

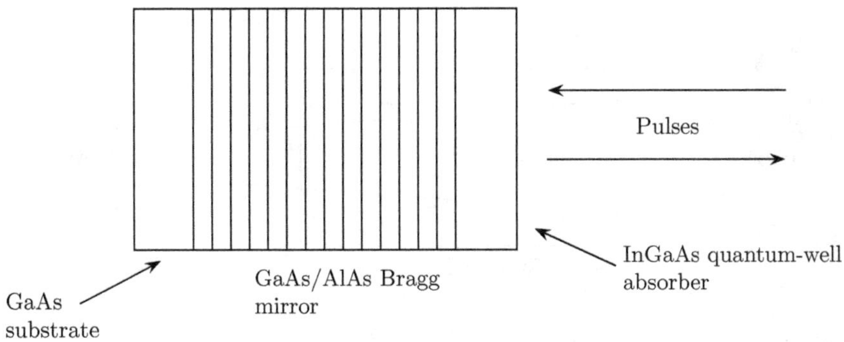


**Fig. 11.8** Schematic of semiconductor saturable absorber mirror (SESAM).

*always* decreases. This is the reason that passive mode locking favors very short pulses and is therefore the scheme invariably used in femtosecond lasers. Second, the loss is nearly independent of $t_p$ when $st_p < 1$: there is little improvement in transmission of the molecules when the pulse width is smaller than the molecular lifetime. Thus, the pulse width will tend to approach the molecular lifetime, which is often some tens of femtoseconds. Of course, this assumes that the bandwidth, $\Delta\nu$, of the gain medium is sufficient to amplify all of the modes which make up this short pulse: $\Delta\nu \gg s$.

Although historically the saturable medium has often been a molecular dye, another effective saturable medium is a *semiconductor*. It is often combined with one of the cavity mirrors to form a *semiconductor saturable absorber mirror* (SESAM). A schematic of a SESAM is shown in Fig. 11.8. The mirror is constructed on a GaAs substrate. The reflection is due to Bragg reflection from a periodic structure consisting of alternating layers of GaAs and AlAs. The saturable region faces the intracavity radiation and consists of an InGaAs quantum-well absorber whose band gap is tuned to absorb radiation at the desired wavelength (usually around 1 $\mu$m). The region between the Bragg mirror and the front surface forms an optical cavity which can adversely affect the functioning of the device; the device is usually designed to be *antiresonant* at the desired wavelength (since the absorber can be very thin – as little as 10 nm thick – the free spectral range of this cavity is very large). The time behavior of the absorber shows a very rapid saturation (in less than 1 ps) followed by a longer recovery (typically several picoseconds to hundreds of picoseconds). The advantages of these mirrors is their simplicity; their disadvantage is their possibly insufficient bandwidth for the shortest pulses.

Kerr-lens mode locking exploits the *optical Kerr effect*, which is the second-order electro-optic effect. Thus, the refractive index is a linear function of the electric field squared (the intensity, $I$):

$$\text{Optical Kerr effect:} \qquad n = n_0 + n_2 I. \tag{11.20}$$

Since this effect is *even* in the electric field, most media display a Kerr effect (unlike the first-order Pockel effect, which requires a non-centrosymmetric medium). The radial intensity distribution of a Gaussian beam is

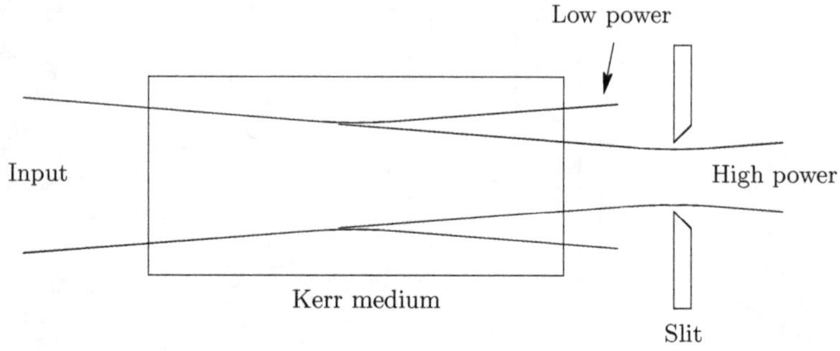


**Fig. 11.9** The Kerr effect combined with a slit can be used to favor high intensities and therefore pulsed behavior.

$$I = I_0 e^{-2r^2/\omega_0^2}. \tag{11.21}$$

For a medium of length $L$, the overall phase shift, $\phi$, is

$$\phi = \frac{2\pi n L}{\lambda}. \tag{11.22}$$

Thus, the change, $d\phi$, in the phase shift due to the intensity distribution is

$$d\phi = \frac{2\pi L}{\lambda} dn = \frac{2\pi L n_2 I_0}{\lambda} e^{-2r^2/\omega_0^2} \approx \frac{2\pi L n_2 I_0}{\lambda}(1 - 2r^2/\omega_0^2) \tag{11.23}$$

near the center of the beam ($r \ll \omega_0$). This *quadratic* dependence of the phase shift with radial distance is characteristic of a *positive lens.* Since the strength of the lens is proportional to the intensity, intense beams will be focused and weak beams will not. By placing a slit downstream from the Kerr medium, one can cause the loss for weak (and therefore non-focused) beams to be much greater than that for strong (focused) beams. This has the same effect as a saturable medium in favoring a pulsed time-dependence instead of a continuous one. Since the Kerr effect is often not sufficiently discriminatory, the optical cavity is frequently designed to be *close to unstability* to increase the effectiveness of the Kerr effect. Figure 11.9 illustrates the use of the Kerr effect for mode locking.

## 11.4 Dispersion and its compensation

In optics, *dispersion* refers to the dependence of the index of refraction, $n(\omega)$, on the frequency, $\omega$. The term is usually only employed when there is a non-trivial dependence: $n \neq$ constant. Thus, in a dispersive system, the wave vector, $k$, will no longer have the simple linear dependence on $\omega$ ($k = n\omega/c$) that a non-dispersive system has. Until now, we have only been tangentially interested in dispersion and have only mentioned it when discussing laser frequency pulling and in semiconductor diode lasers. The presence of dispersion in the laser medium or cavity optics can, however, make it very difficult to obtain very narrow pulses in a mode-locked laser since it can cause pulse broadening. The pulse is composed of a spread of frequencies and each frequency travels with a slightly different phase velocity causing the pulse shape to change as the pulse propagates. We will discuss this phenomenon and several techniques for eliminating it in the following.

When describing *wave packets* (i.e., pulses), it is useful to distinguish between two velocities: the *group velocity*, which is the velocity of the packet (and usually energy transport) and the *phase velocity*, which is the velocity of the constant phase surfaces (*wavefronts*) of the wave. The usual procedure for describing a wave packet is to consider it to be a continuous sum of components whose frequencies differ only slightly from each other and which are arranged to cancel outside of some finite time interval, $\Delta t_p$. One can sum over $\omega$ (temporal frequency) or $k$ (spatial frequency); we will use the former. The sum is the Fourier integral,

$$E(z,t) = \int_{-\infty}^{\infty} A_\omega(\omega - \omega_0) e^{ik(\omega)z - i\omega t} d\omega, \tag{11.24}$$

where we assume the wave depends upon only one spatial coordinate ($z$) and $\omega_0$ is the center frequency of the wave.

We will first expand the wave vector around $\omega_0$:

$$k(\omega) = k_0 + \left(\frac{\partial k}{\partial \omega}\right)_{\omega_0} (\omega - \omega_0) + \cdots . \tag{11.25}$$

Substituting this into the Fourier integral,

$$E(z,t) = e^{i(k_0 z - \omega_0 t)} \int_{-\infty}^{\infty} A_\omega(\Delta\omega) \exp\left\{ -i\Delta\omega \left[ t - \left(\frac{\partial k}{\partial \omega}\right)_{\omega_0} z \right] \right\} d\Delta\omega, \tag{11.26}$$

where $\Delta\omega = \omega - \omega_0$. The integral, which we will call $A()$, depends only upon $t - (\partial k/\partial \omega)z$ and the field is

$$E(z,t) = A\left[ t - \left(\frac{\partial k}{\partial \omega}\right)_{\omega_0} z \right] e^{i(k_0 z - \omega_0 t)}. \tag{11.27}$$

The field is therefore the product of a plane wave propagating to the right with speed $\omega_0/k_0$ and an *envelope function*, $A(t - (\partial k/\partial \omega)z)$, propagating to the right with (possibly different) speed $\partial\omega/\partial k$. When the dispersion function is *linear*, these two speeds are the same. As mentioned previously, the speed of the envelope is called the *group velocity*, $v_g$, defined as

$$\text{Group velocity:} \qquad v_g \equiv \left(\frac{\partial \omega}{\partial k}\right)_{\omega_0} . \tag{11.28}$$

In all that follows, we assume that the wave vector depends *weakly* on the frequency and that the index of refraction is a monotonically increasing function of the frequency (normal dispersion). With these assumptions, the group velocity has a simple interpretation as the velocity of energy transport, and we avoid certain non-physical behaviors, such as the case of a group velocity that is greater than $c$. In a medium with refractive index $n(k)$, the surfaces of constant phase travel at the *phase velocity*, $v_\phi$:

$$\text{Phase velocity:} \qquad v_\phi = \frac{\omega(k)}{k} = \frac{c}{n(k)}. \tag{11.29}$$

Differentiating the second equality with respect to $k$ yields the following relation for $v_g$:

$$v_g = \frac{c}{n(k) + \omega\frac{dn(\omega)}{d\omega}}. \tag{11.30}$$

One often defines the *group index of refraction* as $n_g = c/v_g$; it is related to the ordinary index ($n$) by

$$n_g = n + \omega\frac{dn(\omega)}{d\omega}. \tag{11.31}$$

Except in the vicinity of an absorption, most materials show *normal dispersion*, defined as $dn/d\omega > 0$. In these cases, the group velocity is less than the phase velocity. Near an absorption, one can have *anomalous dispersion* ($dn/d\omega < 0$) which can result in the group velocity being greater than the phase velocity. While a group velocity which differs greatly from the phase velocity is possible (even one greater than $c$), we exclude such curiosities in this book as being somewhat non-physical and not consistent with our assumption of *weak* dispersion.

In a dispersive medium ($n$ not constant), a pulse whose temporal width at $z = 0$ is $t_p$ will spread an additional $\Delta t_p$ at $z = l$ by approximately

$$\Delta t_p \approx \frac{l}{(v_g)_{min}} - \frac{l}{(v_g)_{max}} \approx l\frac{\Delta\left(\frac{1}{v_g}\right)}{\Delta\omega}\Delta\omega = l\frac{\Delta\left(\frac{\partial k}{\partial\omega}\right)}{\Delta\omega}\Delta\omega = l\left(\frac{\partial^2 k}{\partial\omega^2}\right)_{\omega_0}\Delta\omega, \tag{11.32}$$

where $\Delta\omega$ is the spread of frequencies in the wave.

In the absence of dispersion, $k$ depends linearly on $\omega$ and the group and phase velocities are the same. The simplest description of dispersion is to assume an additional quadratic dependence on $\omega$:

$$k(\omega) = k_0 + \left(\frac{\partial k}{\partial\omega}\right)_{\omega_0}(\omega - \omega_0) + \frac{1}{2}\left(\frac{\partial^2 k}{\partial\omega^2}\right)_{\omega_0}(\omega - \omega_0)^2. \tag{11.33}$$

Equation 11.32 shows that a pulse in a medium described by this simple dispersion relation will spread (if the second derivative $> 0$) as it propagates, and the spread will be proportional to the distance traveled. The second derivative of $k$ with respect to $\omega$ is called the *group velocity dispersion* (GVD):

$$\text{Group velocity dispersion} = \left(\frac{\partial^2 k}{\partial\omega^2}\right)_{\omega_0} = \left(\frac{\partial\frac{1}{v_g}}{\partial\omega}\right)_{\omega_0}. \tag{11.34}$$

If we multiply the GVD by the distance traveled, we obtain the *group delay dispersion* (GDD). Since the *phase*, $\phi$, accumulated by a wave in distance $l$ is $kl$, the GDD is

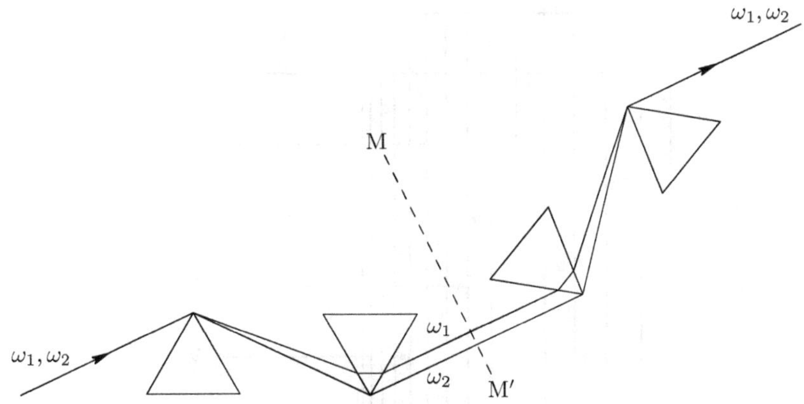


**Fig. 11.10** Dispersion compensation using four mirrors (or two in a standing-wave laser).

occasionally called $\phi''$ (where the two primes signify double differentiation with respect to $\omega$). Most media have a positive $\phi''$ and thus lead to pulse spreading. Dispersion compensation is the addition of an element having a negative $\phi''$ into the laser cavity.

A common approach to dispersion compensation is to use the symmetrical four-prism sequence shown in Fig. 11.10. The figure shows two waves, one at $\omega_1$ and one at $\omega_2$. The prisms are cut and arranged so that the light enters each prism at or near Brewster's angle to reduce the loss and at the minimum deviation angle. Adjacent surfaces in each prism pair are parallel to each other. Analysis by Fork et al. (1984) demonstrated that one can produce a negative group velocity dispersion using this configuration and the size of the dispersion can be changed by changing the separation between the prisms within each pair. In a standing-wave laser, the symmetry plane (MM′) can be replaced with a mirror enabling one to use only two prisms.

A more recent approach to dispersion compensation is to replace at least one cavity mirror with a *chirped mirror*. These are a special kind of *dielectric mirror*, similar to the mirrors used in virtually all laser cavities. Highly reflective coatings using transparent dielectric materials are usually of the *Bragg* variety, which consist of a number (up to 100) of alternating layers of materials with two different refractive indices whose optical thicknesses are one quarter of the central wavelength. If one includes the 180° phase shift when going from a higher index to a lower index, it is easy to see that reflections from alternate layers are all in phase yielding a cumulative high reflectivity. Using a larger number of coatings increases both the bandwidth and the reflectivity.

The layers of dielectric material in a chirped mirror are *chirped in thickness* (i.e., uniformly varying with depth in the mirror). This causes *different* wavelengths to be reflected from *different* quarter-wave stacks. The wavelength-dependent penetration depth results in a wavelength-dependent phase shift which can be tailored to compensate the group delay dispersion of the laser. Figure 11.11 illustrates how the longer wavelengths penetrate the mirror to a greater extent than the shorter wavelengths. The original chirped mirrors based upon these simple ideas unfortunately did not work very well. The group delay had large-amplitude oscillations which prevented the mirrors from being used to generate femtosecond pulses. This problem was resolved

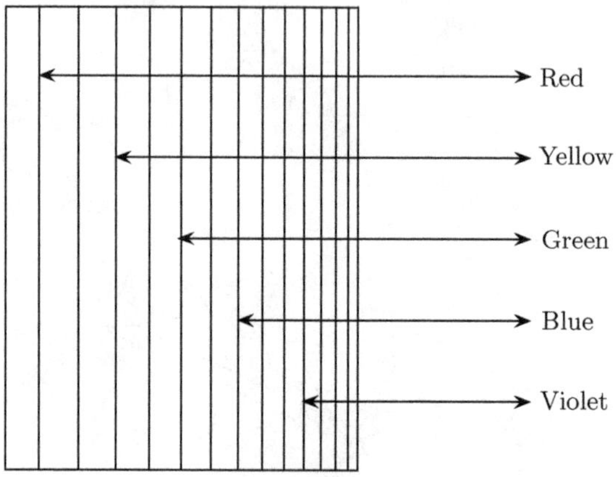


**Fig. 11.11** Operating principle of chirped mirror, providing negative group delay dispersion.

with the *double chirped* configuration. In addition to the *chirp* of the Bragg wavelength, the thickness ratio of the high-index to low-index layers was adjusted to smooth the transition from the highly reflective region to the low-reflectance region. In addition, a broadband antireflective coating was placed over the front surface of the mirror. The double chirped mirrors provide a fairly wavelength-independent negative group-delay correction and have a larger bandwidth than prisms, which have some wavelength dependence of their group delay due to dispersion in the prism material. These mirrors are therefore better suited for correcting dispersion in femtosecond lasers.

## 11.5 The mode-locked Ti–sapphire laser

The Ti–sapphire laser is the pre-eminent source of femtosecond pulses since its gain bandwidth is larger than that of any other laser in the visible or near-infrared. The emission spectrum of a Ti–sapphire mode-locked laser is shown in Fig. 11.12. There are a number of commercial mode-locked Ti–sapphire lasers that deliver respectably narrow pulses at a range of repetition rates from a bit less than 100 MHz to above 1 GHz. The field is rapidly growing and the state-of-the-art is in the non-commercial research laboratories. We will discuss some of the characteristics of a representative sample of the latter group of lasers.

Before discussing a specific system, we need to slightly refine our discussion of mode-locked theory and introduce the *soliton-type* mode-locked laser, which can generate the narrowest pulses. The soliton laser was developed in the 1980s (see, for example, Mollenauer (1984)) and initially used an optical fiber to produce the solitons. A soliton is a short pulse which is a solution to the nonlinear Schrödinger equation and possesses the very useful characteristic of propagating either with unchanging shape (lowest order) or with a shape which is a periodic function of distance (higher order). In a mode-locked laser, solitons can appear when the medium has a *negative* group velocity dispersion combined with a Kerr nonlinearity, in which the index of refraction is a linear function of the intensity. The former condition can be obtained by overcompensating the dispersion using one of the techniques described in the previous

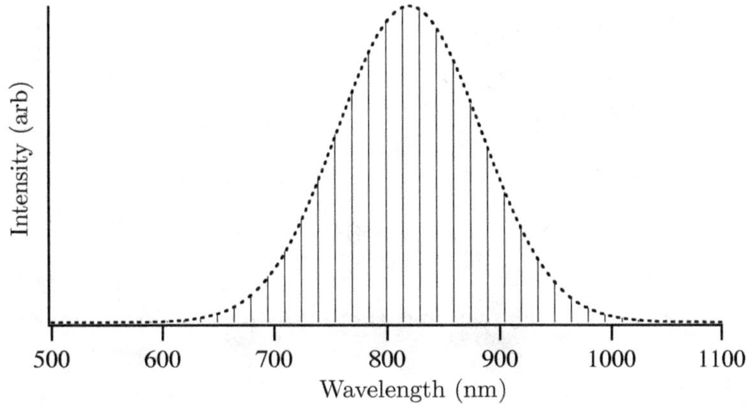


**Fig. 11.12** Emission spectrum of mode-locked Ti–sapphire laser.

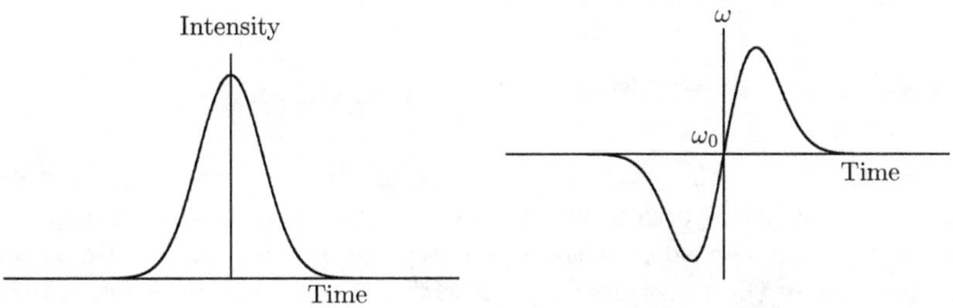


**Fig. 11.13** Time dependence of the pulse frequency (right) for Gaussian pulse (left) due to the Kerr effect.

section. The Kerr nonlinearity occurs in the Ti–sapphire gain crystal and will produce *self-phase modulation* (SPM).

Self-phase modulation is simply a *phase modulation* of the pulse which is due to the Kerr effect and the rapid temporal variation of the pulse intensity. When the Kerr effect is present, the total phase, $\phi(t)$, of a pulse whose carrier frequency is $\omega_0$ is given by

$$\phi(t) = \omega_0 t - k(z)z = \omega_0 t - \frac{n(z)\omega_0}{c}z = \omega_0 t - \frac{(n_0 + n_2 I)\omega_0}{c}z. \tag{11.35}$$

The instantaneous frequency of the pulse is

$$\omega = \frac{d\phi}{dt} = \omega_0 - \frac{\omega_0 n_2 z}{c}\frac{\partial I}{\partial t}. \tag{11.36}$$

The behavior of the pulse frequency as a function of time is shown on the right side of Fig. 11.13. Notice that the pulse acquires a positive-going frequency *chirp* near the peak of the pulse. We will now show that the negative group velocity dispersion due to other elements in the laser cavity will produce a negative-going frequency chirp which will counter the one due to the Kerr effect in the gain medium.

In the presence of a small amount of dispersion, the expression for a Gaussian pulse can be shown to be (Svelto, Appendix G)

$$E(z,t') = \frac{E_0 \tau_p}{(\tau_p^2 + iD_2 z)^{1/2}} \exp\left(-\frac{t'^2}{2(\tau_p^2 + iD_2 z)}\right) e^{i\omega_0 t}, \tag{11.37}$$

where $\tau_p$ is the pulse width, $D_2$ is the group velocity dispersion,

$$D_2 = \frac{\partial^2 k}{\partial \omega^2} \tag{11.38}$$

and $t'$ and $z$ are related by the fact that the pulse envelope propagates at the group velocity

$$t' = t - \frac{z}{v_g}. \tag{11.39}$$

If we calculate the *phase* of this expression (including the $e^{i\omega_0 t}$ factor) and differentiate to obtain the instantaneous frequency, we obtain

$$\omega = \omega_0 + \frac{D_2}{|D_2|} \frac{z/L_D}{1 + (z/L_D)^2} \frac{2t'}{\tau_p^2}, \tag{11.40}$$

where $L_D = \tau_p^2/|D_2|$ is the *dispersion length* of the medium. We thus see that a negative group velocity dispersion will produce a negative-going frequency chirp. If we adjust the laser parameters so that the two chirps cancel, we can obtain a soliton-type pulse behavior, where the pulse propagates with constant shape. From soliton theory, it can be shown that the laser *power*, $P$, and pulse width, $\tau_p$, are now related as

$$\tau_p^2 \propto \frac{|D_2|}{P}. \tag{11.41}$$

A realization of a GHz mode-locked Ti–sapphire ring laser used by workers at the National Institute of Standards and technology (NIST) in Boulder, Colorado (Ye (2005)) is shown in Fig. 11.14. The cavity is the familiar astigmatically compensated ring discussed in some detail in Chapters 2 and 3. The net negative group velocity

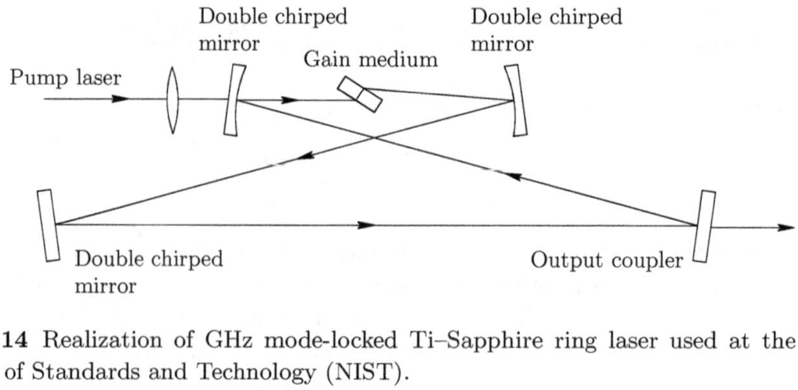


**Fig. 11.14** Realization of GHz mode-locked Ti–Sapphire ring laser used at the National Institute of Standards and Technology (NIST).

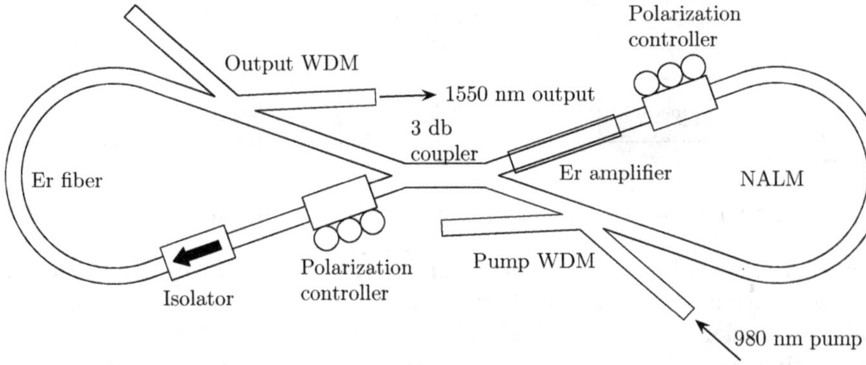


**Fig. 11.15** Schematic of "figure-eight" mode-locked fiber laser using a NALM.

dispersion is provided by the three double chirped cavity mirrors. The Ti–sapphire crystal is fairly short in order to ensure a negative group velocity dispersion since the crystal itself contributes a positive GVD proportional to its length. The repetition rate is 2 GHz for a fairly short cavity. The output coupler has a transmission of 2%. This laser produced pulses as short as 14 fs, while a variant with two additional folded chirped mirrors and a longer Ti–sapphire crystal produced $\approx$ 40 fs pulses.

## 11.6 Mode-locked fiber lasers

Much progress has been recently made in constructing mode-locked fiber lasers with pulse widths as low as 30+ fs. One approach to mode-locking a fiber laser is to use the same components and geometry as other mode-locked lasers with the gain medium replaced by an active fiber. However, in the interest of simplicity, there is a strong motivation to make an *all-fiber* mode-locked laser, using as many fiber components as possible. We will discuss the techniques for *passive mode-locking* of fiber lasers in this section.

As we learned earlier in this chapter, all approaches to passive mode-locking use a *nonlinear* medium to favor the emission of short intense pulses instead of lower-intensity continuous waves. The shortest pulses are produced when the medium utilizes the extremely fast Kerr nonlinearity rather than the somewhat slower atomic or molecular saturation (eqn 11.19). There are two commonly used approaches to an all-fiber passively mode-locked laser based upon the Kerr nonlinearity. The laser can use a *nonlinear optical (or amplifying) loop mirror* (Chapter 10) or it can use *additive-pulse mode-locking* by exploiting *nonlinear polarization evolution* in a slightly birefringent fiber.

A diagram of a "figure-eight" laser employing a nonlinear amplifying loop mirror (NALM) appears in Fig. 11.15. There are two variants of the nonlinear loop mirror (Chapter 10): one with an amplifier and one without. The amplifying version produces a larger fractional modulation of the intracavity power and is therefore preferred, despite its slighter greater complexity. The laser operates in traveling-wave (ring) mode and the loop mirror is adjusted so that the ring laser loop is only *closed* when the

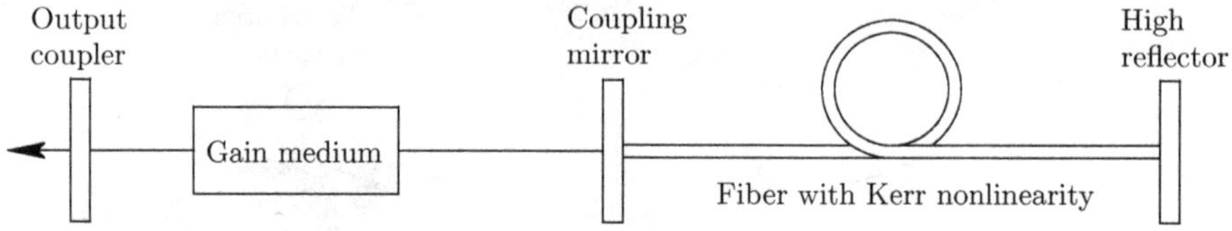


**Fig. 11.16** Conceptual diagram of additive-pulse mode-locked laser.

intracavity intensity is high. The optical isolator enforces single direction operation of the ring (clockwise in the figure) while the polarization controllers correct any adverse polarization behavior in the non-polarization-preserving fibers. The output wavelength division multiplexor (WDM) is set up to couple out a small amount of laser light at 1550 nm while keeping the 980 nm pump radiation in the fiber ring. Similarly, the pump WDM couples the pump radiation into the system while preventing the laser radiation from leaving the NALM. This laser falls slightly short of being an all-fiber laser since the isolator is a discrete component with fiber pigtails. The remaining components are all fiber-based.

One desirable feature of a mode-locked laser is that the oscillation be *self-starting.* Continuous-wave lasers are easily started by amplified spontaneous emission but mode-locked lasers need sufficient intracavity power to activate the Kerr-based amplitude discriminating component which is responsible for mode locking. If a laser is not self-starting, it requires an intracavity modulator to initiate pulsed-laser operation. The NALM switches at very low power and the figure-eight laser is therefore self-starting.

The second approach to fiber-laser mode locking is a variant of *additive-pulse mode-locking* (APM). A schematic of an additive-pulse laser appears in Fig. 11.16. The key to its operation is the presence of a second cavity which contains a nonlinear Kerr medium and which is coupled to the main laser cavity via a slightly transmitting *coupling mirror.* The round-trip time in the second (auxiliary) cavity is made similar to that in the laser cavity. As a result of the Kerr nonlinearity, the index of refraction and therefore the *phase* of the wave in the auxiliary cavity will depend upon the power. Assume that the laser is generating pulses and that the phase of a pulse which is reflected from the auxiliary cavity leads that of the pulse in the main cavity by an angle between 0° and 180°. As a result of the Kerr-induced self-phase modulation, the peak of the pulse will be retarded by a larger angle than the wings of the pulse. The reflected pulse will therefore constructively enhance the peak of the laser pulse to a greater extent than the wings of the pulse. Effectively, the reflectivity from the coupling mirror will be greater for pulses than for continuous-wave radiation and the pulses returned by the auxiliary cavity will tend to shorten the pulses in the laser cavity. This is the criterion for mode locking and for the production of very short pulses, and this technique is called *additive-pulse mode locking.* Note the difference between this approach and one which uses a saturable absorber or Kerr lens. The latter relies on passive *amplitude modulation* of the intracavity environment while the former relies on a combination of nonlinear *phase modulation* and *interference* to favor the production of short pulses over the continuous-wave operation of the laser.

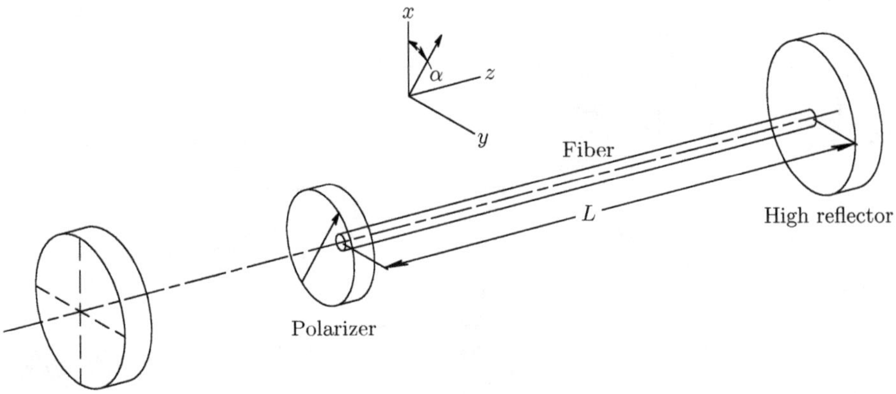


**Fig. 11.17** Conceptual model of standing-wave laser employing nonlinear polarization evolution for mode locking.

It turns out that it is not necessary to use an auxiliary cavity for additive-pulse mode locking; *two orthogonal polarization modes* propagating in the *same cavity* will suffice, provided they have *different* nonlinear phase shifts and can be made to interfere with each other. This is the idea behind *nonlinear polarization evolution* (NPE) mode locking. A conceptual model of a standing-wave laser utilizing NPE appears in Fig. 11.17. The fiber is assumed to be slightly birefringent with its principal axes aligned with the $x$- and $y$-axes. In general, the phase shift, $\delta\phi$, due to Kerr nonlinearity for a wave traveling a distance $z$ is given by

$$\delta\phi = \beta z + \frac{2\pi n_2}{\lambda} I z \tag{11.42}$$

where $I$ is the intensity of the light and $\beta$ is the propagation constant. If we have a wave propagating in a *birefringent* medium, we must include certain geometrical factors, since $n_2$ is derived from a fourth-rank tensor. Furthermore, each field component will have two contributions to its nonlinear phase shift: one from the *same* component and one from the *orthogonal* component. The shift arising from the same polarization is called *self-phase* modulation, and that from the orthogonal polarization is called *cross-phase* modulation. Due to the geometrical considerations mentioned previously, the cross-phase term has a 2/3 coefficient in front of it. If the distance between the polarizer and high reflecting mirror is $L$, the round-trip phase shifts, $\phi_{x,y}$, for both polarizations are:

$$\begin{aligned} \phi_x &= \left\{ \beta_x + k_0 n_2 (|A_x|^2 + \frac{2}{3}|A_y|^2) \right\} 2L \\ \phi_y &= \left\{ \beta_y + k_0 n_2 (|A_y|^2 + \frac{2}{3}|A_x|^2) \right\} 2L \qquad k_0 = \frac{\omega}{c} \end{aligned} \tag{11.43}$$

where $\beta_{x,y}$ are the propagation constants for the two polarizations and the fields are normalized so that the total intensity is $I = |A|^2 = |A_x|^2 + |A_y|^2$. The fields, $A_{x,y}$, for each polarization are

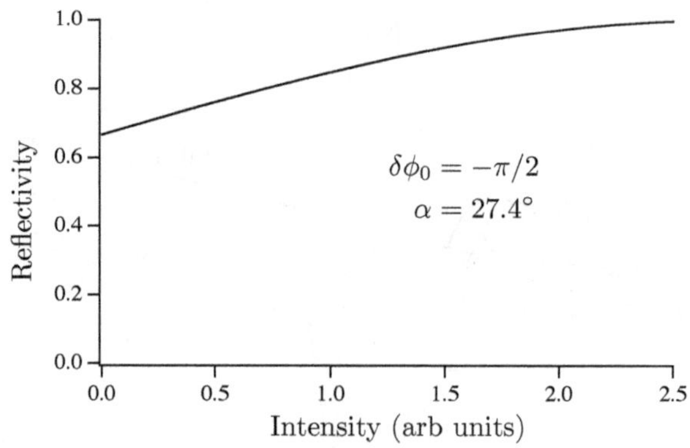


**Fig. 11.18** Graph of reflectivity versus intensity for NPE mode locking.

$$\begin{aligned} A_x(z) &= A(z)\cos\alpha \\ A_y(z) &= A(z)\sin\alpha \end{aligned} \tag{11.44}$$

where $\alpha$ is the angle between the polarizer and the $x$-axis. The *magnitude* of the power reflectivity, $R$, for everything to the right of and including the polarizer is given by:

$$R = \cos^4\alpha + \sin^4\alpha + 2\cos^2\alpha\sin^2\alpha\cos\delta\phi \tag{11.45}$$

where $\delta\phi$ is the phase shift difference for one round trip and consists of a linear part, $\delta\phi_0$, and a nonlinear part, $\delta\phi_{nl}$:

$$\delta\phi = \phi_x - \phi_y = \delta\phi_0 + \delta\phi_{nl}. \tag{11.46}$$

The linear component (occasionally called the "phase bias" and equal to $(\beta_x - \beta_y)2L$) can be adjusted to any value by changing either $L$ or the propagation constant difference. The values of $\alpha$ and $\delta\phi_0$ which maximize $dR/dI$ are $\alpha = 27.4°, \delta\phi_0 = -\pi/2$. Figure 11.18 is a plot of the reflectivity versus the intensity for these settings. In deriving the formula for $R$, we have assumed that each polarization evolves with $z$ independently of the other. It turns out that this is only valid for a somewhat birefringent fiber ($\delta\phi_0 > 3\pi$), where the rotation of the polarization ellipse due to the nonlinear effects can be ignored. When this is not the case, one must solve the coupled *nonlinear Schrödinger equations*, which describe the propagation of waves in a medium which is dispersive and which exhibits a third-order (Kerr) nonlinearity.

To show that the nonlinear reflectivity derived above discriminates in favor of pulsed-laser oscillation, we merely need to include some gain, $\gamma$, in the fiber. If we do so, we find that the expression for $R$ will be multiplied by $e^{2\gamma L}$. When we use this to determine the criterion for laser oscillation (Chapter 6), we find that we have reproduced eqn 6.7, which favors larger values of $R$ (and therefore larger intracavity intensity) for the initiation of laser oscillation.

An early and simple NPE-based laser (Tamura et al. (1992)) is shown in Fig. 11.19. The laser uses a ring configuration with an optical isolator containing polarizers and

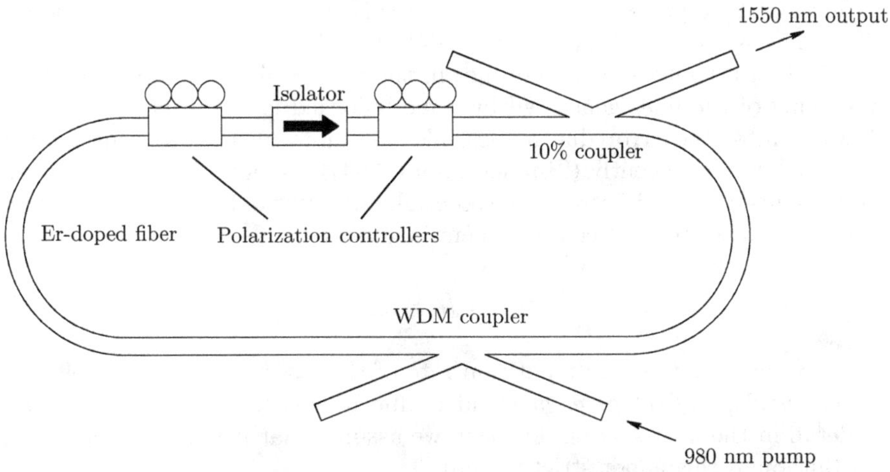


**Fig. 11.19** A realization of an erbium fiber laser which uses NPE for mode locking (Tamura et al. (1992)).

two polarization controllers. The polarization controller downstream from the isolator converts the linear polarization into elliptical polarization. The Kerr nonlinearity rotates the polarization ellipse and the polarization controller on the left provides the needed bias for higher intensities to experience lower loss. Pulse widths slightly less than 500 fs were observed with a 42 Mhz repetition rate. Using a ring geometry allowed the laser to be self-starting. Later NPE lasers have achieved 30+ fs pulse widths using a similar configuration with additional components in the ring to control the fiber dispersion. Despite being much simpler than figure-eight lasers, NPE lasers are more susceptible to temperature-induced drifts and therefore require more frequent adjustment of the polarization controllers.

## 11.7 Frequency metrology using a femtosecond laser

While mode-locked lasers are excellent tools for making time-resolved studies of extremely short-lived phenomena, their principal use in the atomic physics laboratory is in optical-frequency metrology. Formerly, optical frequencies were accurately measured by coherently linking them to a number of microwave standards using an elaborate chain of lasers and microwave oscillators. Some ingenious techniques were suggested for making a coherent link in one step, such as phase locking a relativistic electron in a Penning trap to a laser and detecting the electron's radio-frequency synchrotron radiation. The most practical technique for accomplishing this single-step link, however, is to exploit the extraordinary uniformity of the *comb* of frequencies emitted by a mode-locked laser. The already very broad comb from a femtosecond laser is further broadened using the recently developed *microstructure fiber*, allowing the comb to span more than an octave and enabling direct measurement of optical frequencies. This section will only touch on the high points of this research, since it is being very rapidly developed at present and there is already a large literature on the subject (including some books, e.g. Ye (2005)).

The utility of a femtosecond comb lies in the extraordinary uniformity of the frequency spacing of the lines generated by a mode-locked laser. Originally, by comparison of two combs, the error due to possible non-uniformities was considered to be about $4 \times 10^{-17}$; more recently (Zimmermann (2004)) the error bound was reduced to $6.6 \times 10^{-21}$. Thus, the comb introduces essentially no error in frequency measurements. The frequency, $f_n$, of the $n^{th}$ comb element is given by

$$f_n = nf_r + f_{ceo} \tag{11.47}$$

where $n$ is an integer, $f_r$ is the repetition rate of the laser and $f_{ceo}$ is an offset called the *carrier envelope offset frequency* and is due to dispersion. We will discuss this term in detail in the next section; for now, we assume that it is a stable and constant quantity (although this is not strictly true).

Femtosecond combs were originally used to *compare* two optical frequencies with great accuracy using an apparatus similar to that shown in Fig. 11.20. At the top of the figure is a spectrum of the laser including the two optical frequencies being compared. The quantities $\delta_1$ and $\delta_2$ are the differences between the laser frequencies and the nearest comb line. Since $f_r$ is $\approx$ 100 MHz and $\delta_1, \delta_2 < f_r$, these two frequencies are easily manipulated using standard radio-frequency practices. The difference frequencies are measured by making the three beams collinear and directing them to a grating which separates the composite beam into two beams centered around $f_1$ and $f_2$. The difference frequencies appear in each diode photocurrent since the latter is proportional to the *square* of the incident electric field. The difference frequencies, $\delta_1$ and $\delta_2$, can be mixed together to obtain $\delta_1 + \delta_2$. The difference between $f_1$ and $f_2$ is then

$$f_1 - f_2 = nf_r + \delta_1 + \delta_2, \tag{11.48}$$

where the integer $n$ can be obtained if the frequencies are known to the accuracy of a common digital wavemeter (about 30 MHz). The repetition rate, $f_r$, is locked to some subharmonic of a microwave frequency standard using a piezo-electric driven mirror in the laser cavity to control the mode spacing and a photodiode to sample the laser output and detect the mode spacing. A simple feedback system will make $f_r$ an exact submultiple of the frequency of the microwave standard. The ultimate fractional error can be equal to the error in the microwave standard since the error in the difference frequencies is negligible compared to $n$ times the error in the subharmonic of the microwave standard ($n$ can be $10^6$ or more).

Although many of the earlier Ti–sapphire mode-locked lasers had a repetition rate of about 100 MHz, there has recently been a drive to increase this to 1 GHz or greater. A larger $f_r$ has two advantages. First, since there are fewer teeth in the comb, each will have more power, improving the signal-to-noise of the various beat signals used in a frequency measurement. Second, ambiguities in the determination of the mode number, $n$, are eased when the repetition rate is very large. A 100 MHz $f_r$ pushes current digital frequency meters to their accuracy limit in obtaining the mode number.

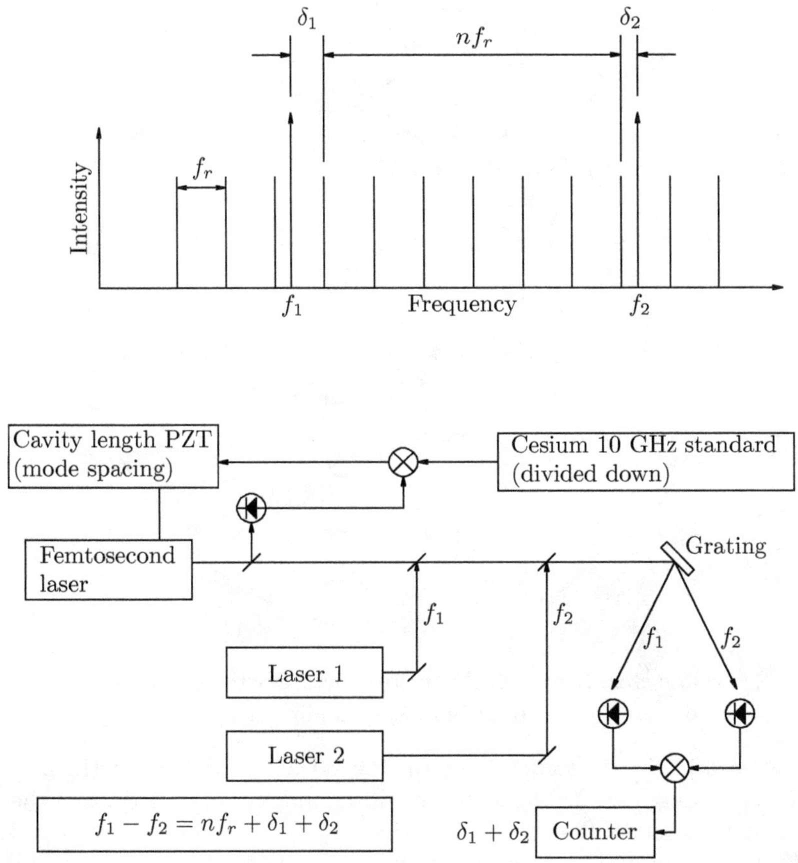


**Fig. 11.20** Femtosecond comb apparatus used to very accurately compare two optical frequencies.

A breakthrough occurred when the *microstructure* fiber was developed and the comb was made to span more than an octave. Assuming that the fiber did not degrade the uniformity of the mode spacing, an absolute frequency measurement could be made by doubling the unknown frequency, $f$, and comparing $f$ to $2f$. In this case, $f_1 - f_2 = f$, providing the unknown frequency directly. This scheme is called a *self-referenced* frequency measurement or a "$\nu$-to-$2\nu$ measurement". Difference frequency measurements and $\nu$-to-$2\nu$ measurements solve the problem of the unknown $f_{ceo}$ by measuring the *difference* between two frequencies, which cancels the inhomogeneous term, $f_{ceo}$. Another approach is to *control* $f_{ceo}$, usually forcing it to some fixed frequency. The latter scheme will be discussed in the next section.

The key to the microstructure fiber is its very small core diameter: as little as 1.7 $\mu$m, compared to the $\approx 5$ $\mu$m diameter in a conventional single-mode fiber. This greatly increases the intensity of the guided wave and thereby enhances the nonlinearities which broaden the comb width by generating numerous intermodulation products of the comb frequencies. From fiber theory, a reduction in the core diameter necessi-

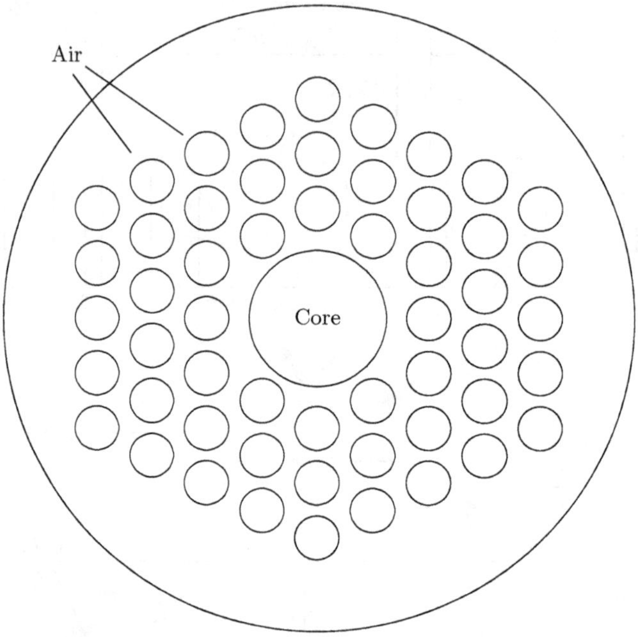


**Fig. 11.21** Schematic cross section of the microstructure with very small core diameter and low group velocity dispersion due to air holes surrounding the core.

tates an increase in the difference between the refractive indices of the core and the cladding (surrounding layer). The lowest cladding index is obtained when the cladding is *air.* The microstructure fiber surrounds the core by a number of hollow tubes, making the effective cladding index close to unity (Fig. 11.21). In addition to a small core diameter for the highest nonlinearity, it is also necessary to have very small group velocity dispersion so that the pulse does not broaden and lose intensity as it travels down the fiber. By careful design, the microstructure fiber shown in the figure can possess both of these useful properties.

Although the ability to obtain an octave spanning comb using a microstructure fiber is a remarkable achievement, it has some drawbacks. Because of the small core diameter and consequent high intensity of the guided light, optical damage is a problem with these fibers and their lifetime can be as short as a few hours. There has been some success in spanning an octave directly with a properly designed Ti–sapphire laser, using the nonlinearity of the laser active medium.

Finally, it should be pointed out that the definition of an octave span can be operationally relaxed quite a bit, since it is possible to obtain useful beats when the intensities of components at the edge of the gain curve are four orders of magnitude below the peak intensity. Thus, a comb whose full width at half maximum is well under an octave can still be considered to be "octave spanning" and used in the *self-referenced* frequency measurements described above.

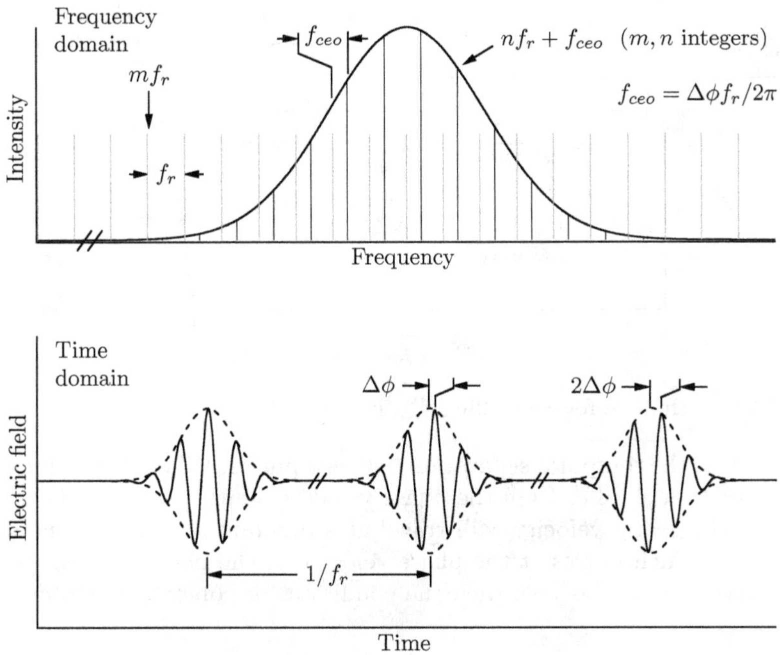


**Fig. 11.22** Illustration of "carrier envelope slippage" which results in non-zero $f_{ceo}$.

## 11.8 The carrier envelope offset

Most of the frequency measurement schemes which utilize femtosecond combs are dominated by the need to cope with the inhomogeneous offset, $f_{ceo}$. In this section, we will discuss the origin of this quantity and a technique for controlling it.

An explanation of the origin of the carrier envelope offset will exploit the analogy with multiple slit interference which was used earlier in this chapter. The usual technique for analyzing multislit interference is to consider the accumulated *phase difference* of the field between adjacent slits as the field propagates to a screen where the interference pattern can be observed. Constructive interference occurs when the phase difference is an integral multiple of $2\pi$. From this requirement, we obtain the possible values of the wave number, $\boldsymbol{k}$ (a *spatial frequency*), which give constructive interference. The treatment of the mode-locked laser is exactly the same, except that it is in the *time* domain rather than the *spatial* domain of multislit interference. Figure 11.22 illustrates the concepts used in our treatment. In order to determine the *frequencies* which result in constructive interference, we determine the amount by which the phase has evolved between successive pulses and require that it be an integral multiple of $2\pi$. In the absence of dispersion, constructive interference occurs when

$$\begin{aligned} \omega\tau &= 2\pi n \qquad \tau = 1/f_r, \quad \text{integral } n \\ \Longrightarrow f_n &= nf_r, \end{aligned} \tag{11.49}$$

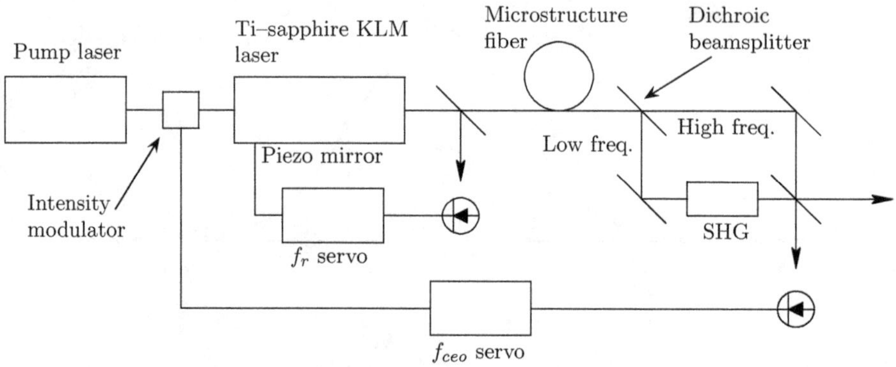


**Fig. 11.23** Technique for controlling $f_{ceo}$ in a Kerr-lens mode-locked (KLM) laser.

where $\tau = 1/f_r$ is the temporal separation between pulses. When dispersion is present, the group velocity will differ from the phase velocity. Thus, the pulse envelope, which propagates at the group velocity, will travel at a different speed from the underlying sinusoidal wave, which moves at the phase velocity. If the *additional* phase difference between successive pulses is $\Delta\phi$, then the condition for constructive interference will be

$$\begin{aligned} \omega\tau &= 2\pi n + \Delta\phi \\ \Longrightarrow f_n &= nf_r + \Delta\phi f_r/2\pi. \end{aligned} \tag{11.50}$$

Thus,

$$f_{ceo} = \Delta\phi f_r/2\pi. \tag{11.51}$$

As shown in the figure, there is a *slippage* between the wave ("carrier") and its envelope.

When dispersion is present, the mode spacing of a mode-locked laser cavity is *different* from the pulse separation, since the mode separation is determined by the phase velocity while the pulse separation is determined by the group velocity, and the two are not exactly the same. The very fact that the pulses are extremely narrow is a *proof* of the uniformity of the frequency modes: any slight non-uniformity would make it impossible to obtain pulses containing only a few optical cycles since the different Fourier components of a non-uniform spectrum would cause the pulses to broaden. The narrow pulses are, of course, also an indication of the success in eliminating the intracavity dispersion. The remaining dispersion is responsible for a non-zero $f_{ceo}$.

A standard technique for controlling $f_{ceo}$ is shown schematically in Fig. 11.23. It is a variant of the self-referenced $\nu$-to-$2\nu$ approach but applied to the comb rather than the unknown frequency. The laser output is broadened to about an octave using a microstructure fiber and separated into low frequencies and high frequencies using a dichroic beamsplitter. The low-frequency part is frequency doubled and beat with the high frequencies, yielding $f_{ceo}$ directly when $2n = m$ ($n$, $m$ are integers):

$$\text{beat frequency} = 2(nf_r + f_{ceo}) - (mf_r + f_{ceo}) = (2n - m)f_r + f_{ceo}. \tag{11.52}$$

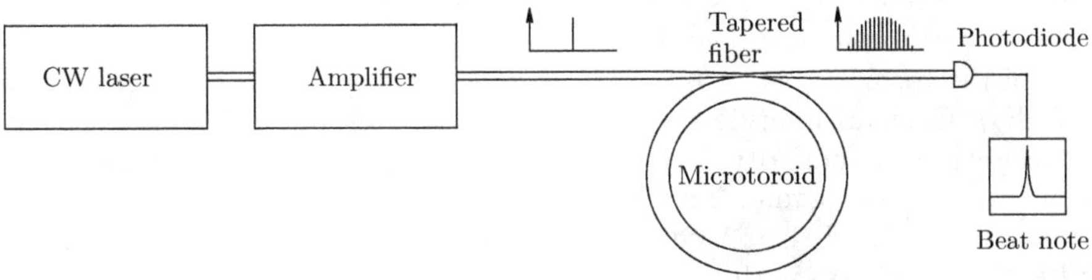


**Fig. 11.24** Optical components used to generate a comb using a microtoroid.

The Mach–Zehnder interferometer shown in the figure ensures that the high-frequency pulses and the frequency-doubled low-frequency pulses overlap. The carrier offset frequency is locked to a fixed value using a phase-locked loop with a loop bandwidth of about 10 kHz since the *noise* in $f_{ceo}$ is in the kHz region (it might be higher if an argon laser is used for a pump). The value of $f_{ceo}$ can be controlled either through the pump power or through the tilt angle of one of the cavity mirrors; the figure shows the former method.

## 11.9 Comb generation in a microresonator

Despite rapid progress being made in fiber-based mode-locked lasers, most researchers currently generate combs using a mode-locked Ti–Sapphire laser pumped with either an argon laser or a diode-pumped solid-state YAG laser with intracavity frequency doubling. Hopefully, this chapter has conveyed to the reader some sense of the complexity of these systems. Laser based combs are also physically large, expensive, power-hungry and often not as reliable as one would like. In an effort to come up with of an approach to comb generation which lacks these drawbacks, a number of researchers have dispensed with the mode-locked laser altogether and are using rapidly-cascading parametric four-wave mixing in an extremely high-$Q$ toroidal microresonator to generate a comb. This technique has the potential to replace an optical table covered with lasers and other gear with a small, inexpensive and very reliable single-chip comb-generator. In this section, we will try to give the reader a taste of some of the current research into these intriguing systems.

The optical components are shown schematically in Fig. 11.24. The modes supported by small spheres and toroids are called “whispering gallery modes”, named after an explanation for the ability of a weak sound wave (a whisper) to be easily heard across the full 32 m diameter of the dome in St. Paul's Cathedral in London. The explanation advanced by Lord Rayleigh in the nineteenth century was that the sound seemed to “stick” to the walls of the dome and travel within a narrow layer along its inner surface with much less attenuation than if the sound went directly across the full diameter of the dome. The optical fields in small smooth structures (usually, but not necessarily, spheres or toroids) behave very similarly to the sound waves and are constrained by total internal reflection to travel in a loop along the inner surface of the confining structure. Mathematically, whispering gallery modes are analyzed by solving the time independent wave equation (Helmholtz's equation) in the resonator. We will not do that here, but will use some of the theory from our treatment of traveling-wave resonators in Chapter 3 to describe the most important characteristics of toroidal resonators.

Most of the toroids used in comb generation are made from silica or calcium fluoride ($CaF_2$). Calcium fluoride resonators have less loss while silica resonators have somewhat greater nonlinearity. The torroids typically have a major radius, $R$, of from about 50 $\mu$m to several mm. The most important parameters describing these resonators are their $Q$ and their *mode volume*. Their free spectral range is equal to $c/nl$ where $l$ is the round trip path length and $n$ is the index of refraction:

$$\text{Free spectral range:} \quad \text{FSR} = \frac{c}{2\pi n R}. \tag{11.53}$$

The mode frequencies are determined in the same way as in any ring cavity: the optical path length for one round trip must be an integral multiple of $\lambda$. Thus, the propagation constant, $\beta_m$, and frequency, $\omega_m$, for the $m$th mode are:

$$\begin{aligned} \beta_m &= \frac{m}{R} \qquad \text{integral } m \\ \omega_m &= \frac{mc}{nR}. \end{aligned} \tag{11.54}$$

The *quality factor* ($Q$) of the toroid is equal to the *finesse* times the number of wavelengths in one round trip (Chapter 3). From our treatment of energy loss in a resonator, we found that the finesse was approximately $\pi$ divided by the the fractional energy loss per round trip. The highest $Q$ is obtained when the only loss mechanism is absorption, described by the absorption constant, $\alpha$. Putting all of this together, we find that:

$$\begin{aligned} \text{Maximum finesse:} \quad \mathcal{F} &= \frac{\pi}{2\pi R\alpha} = \frac{1}{2R\alpha} \\ \text{Maximum } Q\text{:} \quad Q &= \frac{\pi n}{\alpha\lambda}. \end{aligned} \tag{11.55}$$

It is interesting that the maximum $Q$ of the toroid is independent of its geometry, unlike the case of a ring cavity constructed from discrete components. As a result of the very small absorption coefficients of the materials used in the construction of toroids, the maximum $Q$ can be as high as $10^{10}$ or $10^{11}$ and the corresponding finesse can be greater than one million, which is very competitive with the best discrete optical cavities. In addition to the absorption loss, there are radiative losses due to the curvature of the resonator, scattering losses due to surface inhomogeneities, and losses due to surface contaminants. If one defines a $Q_i$ for each of these losses, the total $Q$ is the reciprocal sum of the component $Q$ values:

$$\frac{1}{Q} = \sum_i \frac{1}{Q_i}. \tag{11.56}$$

The mode volume, $V_{eff}$, is defined as the volume which, with the peak energy density, would yield the observed total stored energy:

$$V_{eff} = \frac{\text{Stored energy}}{\text{Maximum energy density}} = \frac{\int_V w(\boldsymbol{r})d^3\boldsymbol{r}}{\max(w(\boldsymbol{r}))}. \tag{11.57}$$

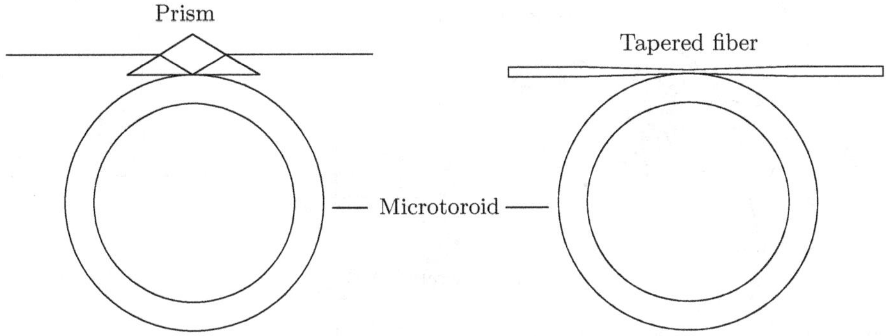


**Fig. 11.25** Coupling into a microtoroid using a prism (left) or a tapered fiber (right).

The mode volume in a *sphere* can be derived analytically; it is roughly equal to (Savchenkov et al. (2004))

$$\text{Sphere:} \quad V_{eff} \approx 2\pi\lambda R^2. \tag{11.58}$$

There is no analytical solution for the mode volume of a microtoroid, but it is somewhat smaller than that of a sphere. A large $Q$ and small mode volume favor the efficient generation of photons due to nonlinear processes.

Two techniques are commonly used to couple light into a microtoroid (Fig. 11.25). The oldest method uses a prism with the surface experiencing total internal reflection placed against the toroid. The nearness of the toroid partially "frustrates" the total internal reflection and allows some of the evanescent wave to "leak" into the toroid. The more recent and generally preferred technique is to use a tapered fiber. The idea behind this scheme is very similar to the coupling mechanism in fiber couplers, described in Chapter 10. By adjusting the degree of taper and length of the portion of the fiber in contact with the toroid, one can adjust the coupling to any desired value. The *optimum coupling* is occasionally called "critical coupling" in the microresonator community (harkening back to the design of radio-frequency circuitry) and is exactly the same as what we called "impedance matching" in Chapter 3. All of the incoming power can be transferred to the toroid when the fraction of power coupled in is equal to the sum of all of the other round-trip fractional losses in the toroid.

The comb is generated by a combination of *degenerate parametric four-wave mixing* and *non-degenerate four-wave mixing.* Both of these processes are due to the third-order Kerr nonlinearity in the material used in the construction of microtoroids; second-order interactions are absent since the material is *centrosymmetric* (has a center of symmetry). Some of these matters are discussed more fully in Chapter 14, where second-order parametric and mixing phenomena are treated quantitatively. That treatment is based upon classical concepts, but the classical *Manley–Rowe* relation (eqn 14.188) allows the various nonlinear interactions to be described using a photon model (which is, of course, supported by a fully quantum-mechanical treatment). We will use a photon model here and avoid a quantitative approach, which can be very elaborate.

The two processes are displayed graphically in Fig. 11.26. Both processes must

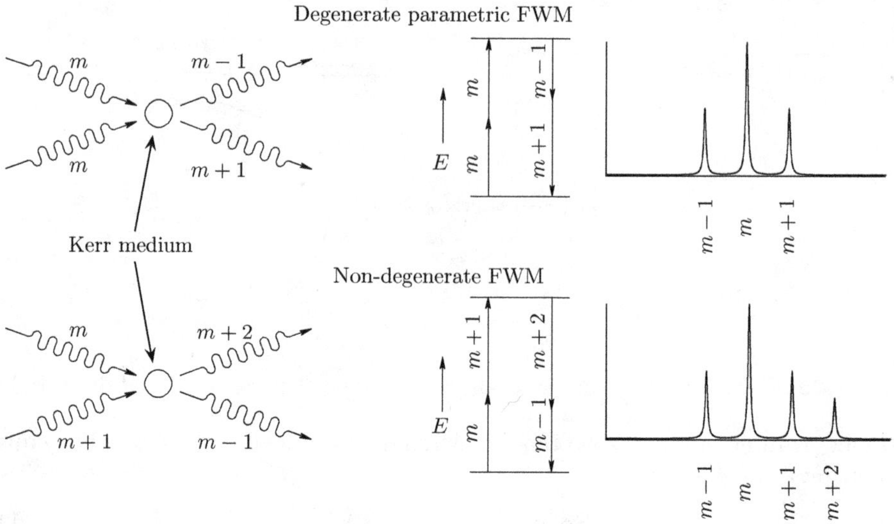


**Fig. 11.26** Depiction of two nonlinear processes responsible for comb generation in a microtoroid. The processes are enhanced when the photon frequencies are the same as the cavity mode frequencies; $m$ is the cavity mode order.

conserve energy and momentum. We begin with two pump photons at the frequency, $\omega_p$, of the pump laser. As a result of degenerate parametric four-wave mixing, two photons with equal frequency can be absorbed, generating two new photons, called the *signal* and *idler*, which are displaced symmetrically about the pump frequency. The process is enormously enhanced when all three frequencies correspond to a cavity resonance. We will assume (as indicated in the figure) that the pump frequency is situated over cavity mode $m$ and the signal is over $m+1$ while the idler is over $m-1$. Momentum and energy conservation require that

$$\begin{aligned} \text{Momentum conservation:} \quad 2\hbar\beta_m &= \hbar\beta_{m-1} + \hbar\beta_{m+1} \\ \text{Energy conservation:} \quad 2\hbar\omega_m &= \hbar\omega_{m-1} + \hbar\omega_{m+1}. \end{aligned} \tag{11.59}$$

Since the propagation constant of the $m$th mode in the toroid is $m/R$, conservation of momentum is automatically satisfied, regardless of *material* dispersion (index of refraction variation with $m$). Conservation of energy can result in a failure of the cavity modes to overlap with the signal and idler frequencies, since various forms of dispersion can make the cavity modes unequally spaced.

Once the signal and idler photons appear, they will interact with the existing pump photons via non-degenerate four-wave mixing, in which two photons with *unequal* frequencies can be absorbed and create two new photons. A likely process is the absorption of a photon at $\omega_m$ and one at $\omega_{m+1}$ to produce photons at $\omega_{m-1}$ and $\omega_{m+2}$. Energy conservation requires that

$$\text{Energy conservation:} \quad \hbar\omega_m + \hbar\omega_{m+1} = \hbar\omega_{m-1} + \hbar\omega_{m+2}. \tag{11.60}$$

The two important outcomes of this process are that it *cascades* – with sufficient power, it occurs many times and photons at many new cavity modes are produced – and that the frequency spacing is rigorously preserved, since, from energy conservation, $\omega_r = |\omega_m - \omega_{m-1}| = |\omega_{m+2} - \omega_{m+1}|$. Numerous experimental tests have confirmed the uniformity of the spacing to several parts in $10^{18}$ (for example, Del'Haye (2007)).

One interesting feature of the four-wave mixing phenomena is that they *conserve photon number* (unlike the three-wave processes discussed in Chapter 14). The total number of photons generated by four-wave parametric processes is equal to the number of pump photons. If the total number of pump (or generated) photons is $N$ and the number in each resonator mode is $n_m$, by conservation of energy and photon number we have

$$\omega_p = \sum_m \omega_m \left(\frac{n_m}{N}\right). \tag{11.61}$$

The bracketed quantity is proportional to the intensity of the $m$th line in the comb. Thus, we have a *conservation of the center of mass* of the comb spectrum: the pump frequency is at the *centroid* of the spectrum.

In order for the comb to have an appreciable width, it is necessary for the toroidal resonator to have very low dispersion, since otherwise the comb frequencies which are distant from the pump frequency might not be situated over a cavity resonance and the cascade would be terminated. There are two contributions to dispersion in a toroid: *geometrical* dispersion and *material* dispersion. We will characterize the effects of dispersion by the change in the free spectral range, $\Delta\omega_{FSR}$, defined by

$$\Delta\omega_{FSR} = (\omega_{m+1} - \omega_m) - (\omega_m - \omega_{m-1}) \approx \frac{\partial^2 \omega_m}{\partial m^2}. \tag{11.62}$$

Geometrical dispersion is due to the $m$-dependence of the round-trip path length and was analyzed in a *microsphere* by Schiller (1993), who obtained the approximate formula:

$$\text{Microsphere:} \quad \omega_m = \frac{c}{nR}\left(m + 1/2 + \eta_1 \left(\frac{m+1/2}{2}\right)^{1/3} + \cdots\right) \tag{11.63}$$

where $\eta_1 \approx 2.34$. From this, we obtain:

$$\text{Microsphere:} \quad \Delta\omega_{FSR} \approx -0.41 \frac{c}{nR} m^{-5/3} < 0. \tag{11.64}$$

Thus, we observe that the geometrical dispersion is always *negative*: the FSR decreases with increasing mode number or frequency. Although this result is for a *microsphere*, the minus sign (at least) will still hold for a microtoroid.

The *material* dispersion is due to the frequency dependence of the index of refraction of the material of which the resonator is constructed. Thus, it is given by:

$$\Delta\omega_{FSR} \approx \frac{\partial^2}{\partial m^2}\left(\frac{mc}{n(m)R}\right) \approx \frac{c^2\lambda^2}{2\pi n^3 R^2} \cdot D \tag{11.65}$$

where $D$ is the *material dispersion parameter* defined by

$$D \equiv \frac{d}{d\lambda}\left(\frac{1}{v_g}\right) \approx -\frac{\lambda}{c}\frac{\partial n^2(\lambda)}{\partial \lambda^2} \tag{11.66}$$

and $v_g$ is the group velocity. The difference between $D$ and the *group velocity dispersion* (GVD) is that the former is the derivative of $1/v_g$ with respect to $\lambda$ and the latter is the derivative of $1/v_g$ with respect to $\omega$. The relation between the two quantities is:

$$D = -\frac{2\pi c}{\lambda^2} \cdot \text{GVD}. \tag{11.67}$$

The units of $D$ are ps/nm-km while the units of the GVD are ps$^2$/km. We observe that the material dispersion can partially cancel the intrinsically negative geometrical dispersion when $D > 0$. In silica, $D$ is negative below 1350 nm and positive above this wavelength. At the pump frequency of an erbium fiber laser (1550 nm), the total dispersion of an 80 $\mu$m diameter sphere is approximately zero. The zero-dispersion wavelength in a toroidal resonator occurs at a somewhat shorter wavelength. For larger resonators, there is a similar shift to shorter wavelengths. Toroids have an additional degree of freedom (the minor radius) which can be adjusted to help in obtaining the desired dispersion.

Due to the small size of the resonator, its high $Q$ and the fairly large power ($\approx 1$ W) needed for combs with a large wavelength range, one must consider the thermal effects due the absorbed pump power. The heating of the toroid by the pump laser will shift the resonant frequency of the cavity; the shift can be more than 100 cavity linewidths. This shift will be in *one direction only* – toward a longer wavelength – regardless of the wavelength of the pump laser. Therefore, when scanning up in wavelength the resonance will be *pushed away* from the laser wavelength, and when scanning down in wavelength the resonance will be pulled toward the laser wavelength. This is typical *hysteretic* behavior and can be used to *thermally lock* the cavity resonances to the laser. This matter is well described in the references (e.g., Carmon (2004)).

In order for a comb to be useful in frequency metrology, both the comb spacing ($\omega_r$) and the *carrier-envelope-offset frequency* ($\omega_{ceo}$) must be stringently controlled and stabilized so that $m$th comb line,

$$\omega_m = \omega_{ceo} + m\omega_r, \tag{11.68}$$

is extremely stable. These parameters are actually much more easily controlled in a microtoroid comb than in a comb generated by a mode-locked laser. As one might expect from the way the microtoroid comb is generated, the $\omega_{ceo}$ is directly controlled by the *frequency* of the pump laser. The mode spacing, $\omega_r$, is controlled by the *power* of the pump laser through changes in the temperature of the toroid due to absorption of the pump power. The comb spacing is a linear combination of the pump power, $P_{in}$, and the pump frequency, $\omega_p$

$$\omega_r = \omega_r^0 + \kappa_\omega \omega_p + \gamma_p P_{in} \tag{11.69}$$

where $\omega_r^0$ is the mode spacing at a specific set-point and $P_{in}$ is the power of the pump laser. The cavity is assumed to be thermally locked to the pump laser. Using a

750 $\mu$m (major diameter) silica toroid, workers (Del'Haye (2008)) found that the two parameters were $\gamma_p \approx -20$ kHz/mW and $\kappa_\omega \approx 0.65$ kHz/MHz. The latter parameter means that the effect of the pump frequency on the mode spacing is suppressed by more than a factor of 1000. Unfortunately, the carrier envelope offset, $\omega_{ceo}$, depends sensitively on $\omega_r$ (and therefore on $P_{in}$). It might be more practical to express the comb frequency as a function of the comb spacing and pump frequency,

$$\omega_l = \omega_p + l\omega_r \qquad \text{integral } l. \tag{11.70}$$

We now need one microwave frequency ($\omega_r$) and one optical frequency ($\omega_p$) to completely characterize the comb. The workers referred to above (De'Haye (2008)) used a second stabilized comb generated by a mode-locked laser to determine the pump frequency. Unfortunately, this is not as convenient as a laser-based comb, which requires two microwave frequencies for comb specification.

Rapid progress is being made in comb generation using a microtoroid, with the recent reports (Del'Haye (2010)) of octave-spanning combs created entirely in a microtoroid without the need for external spectral broadening. The present devices are based upon very small toroids with awkwardly large free spectral ranges (hundreds of GHz) and fairly high pump-power requirements (watts) for adequate spectral breadth. As development of microtoroids based upon improved materials proceeds, one hopes that a truly single-chip comb generator will be soon be possible.

## 11.10 Further reading

The books on lasers and quantum electronics mentioned in previous chapters all have good descriptions of mode-locked lasers. None, however, discuss frequency metrology using a comb. A good reference for femtosecond combs and frequency metrology is the recent book by Ye and Cundiff (2005). The use of a saturable absorber for passive mode locking is described in a paper by Garmire and Yariv (1967). Dispersion compensation using prisms is discussed in the paper by Fork et al. (1984) and double-chirped mirrors for the same purpose are described in a paper by Paschotta et al. (1999). Soliton lasers are described in the paper by Mollenauer and Stolen (1984). Numerous papers discuss the use of femtoseconds lasers in frequency metrology; a recent example is the paper by Zimmermann et al. (2004).

## 11.11 Problems

(11.1) Given two sinusoidal waves (representing the electric field, for example) propagating to the right with equal amplitudes and with frequencies $\omega_1$ and $\omega_2$ and wave vectors $k_1$ and $k_2$, there will be an envelope of "beats" which propagates at the group velocity, $v_g$, which will, in general, be different from the phase velocity, $v_\phi$. For this example, write an expression for both the phase and group velocities. (Hint: the group velocity is usually given by a derivative with respect to $k$. With only two waves, this does not make much sense. You will find that the derivative is replaced by something which applies to two discrete values of $k$).

(11.2) Consider a 623.8 nm He–Ne laser (lasing atom is Ne, which is Doppler broadened at room temperature).
  (a) Estimate the shortest pulse that can be produced by such a laser.
  (b) What is the separation between pulses if the mirror separation is 40 cm?
  (c) Why do liquid-dye and many solid-state lasers produce shorter pulses than typical gas lasers (such as a He–Ne laser)?

(11.3) Show that mode locking can occur if the *phase* of each mode is a *linear* function of the mode number, $n$. What happens if the phase depends *quadratically* on the mode number?

(11.4) What is the minimum FWHM of the (Gaussian) envelope of modes in a mode-locked laser employed in the measurement of an unknown optical frequency using the *self-referential* method if a useable signal can be obtained when the modes which are compared to the unknown frequency are 40 db lower than the peak?

# 12
# Laser frequency stabilization and control systems

## 12.1 Introduction

Lasers have a well-deserved reputation as sources of extremely spectrally narrow optical radiation. For example, a *free-running* (unstabilized) dye or Ti–sapphire ring laser has a short-term linewidth of less than one MHz, which corresponds to a *relative* instability of less than $10^{-8}$. External cavity diode lasers can have linewidths of 50 kHz and the YAG non-planar-ring-oscillator has a *measured* linewidth of about 3 kHz, which corresponds to a relative instability of $10^{-11}$. The dominant source of frequency noise in all of these systems is *technical noise*: noise which is not fundamental in nature and which is due to fluctuations in the properties of the laser medium or support structures. We have seen that monolithic diode lasers are usually dominated by enhanced Shawlow–Townes noise and have linewidths of a bit more than 1 MHz. There are a number of applications which demand even smaller spectral linewidths than are available from a free-running laser and therefore require some form of active *frequency stabilization.*

We will discuss the techniques of frequency stabilization using negative feedback and a stable frequency reference, which is usually a passive optical cavity. These techniques can also be applied, with some modification, to the stabilization of a laser to an atomic or molecular reference. Discriminants from atomic and molecular resonances will be discussed in a later chapter. In this chapter, it will be assumed that the frequency discriminant will be generated using the Pound–Drever–Hall approach discussed in detail in Chapter 4. We will first derive the basic results using fairly self-evident principles. Next, we will review some simple linear system theory using Laplace transforms. We will then use this theory to determine the dynamical behavior of laser stabilization systems. Since the theory can be used for any control system, we will also briefly describe other servo systems often found in the atomic physics laboratory, such as temperature-stabilization systems.

## 12.2 Laser frequency stabilization – a first look

Before resorting to the machinery of linear network theory to analyze a frequency stabilization system, we will use some simple considerations to obtain the important features of such a system. The basic idea is that we use the frequency discriminant to *correct* deviations of the instantaneous laser frequency from the resonance frequency of the cavity. Since we are using *negative* (degenerative) feedback, the correction signal must have a sign which is *opposite* to that of the error so that it *corrects* the error by reducing it. A block diagram of the stabilization system appears in Fig. 12.1 The

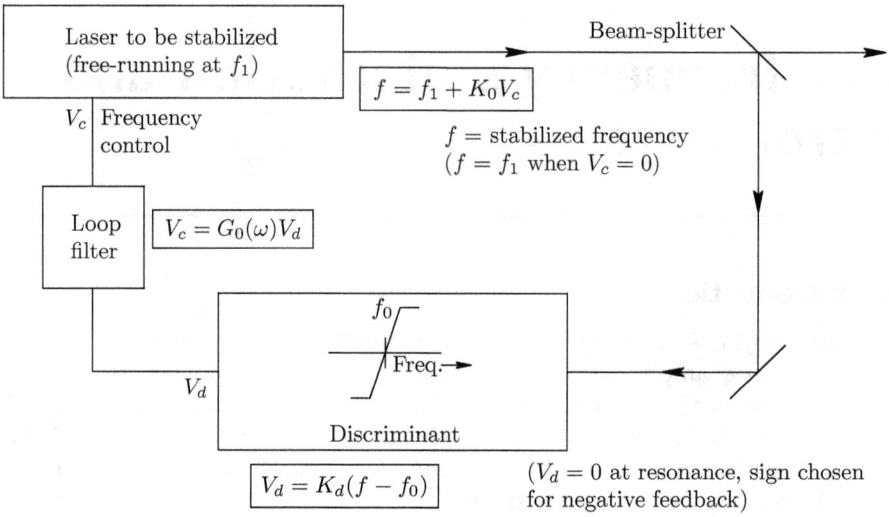


**Fig. 12.1** Block diagram of frequency stabilization system.

laser frequency, $f$, is *linearly* controlled by an external voltage, $V_c$, and the voltage to frequency characteristic is

$$\text{Laser:} \quad f = f_1 + K_0 V_c, \tag{12.1}$$

where the free running frequency ($V_c = 0$) is $f_1$. The frequency discriminator (cavity and Pound–Drever–Hall apparatus) has a response voltage, $V_d$, given by

$$\text{Cavity:} \quad V_d = K_d(f - f_0), \tag{12.2}$$

where $f_0$ is the resonance frequency of the cavity. Between the discriminator output and laser frequency control input, there is a *loop filter*, described by the *transfer function*, $G_0(\omega)$, which depends on the frequency, $\omega$, of the discriminator output. The behavior of the loop filter is described by

$$\text{Loop filter:} \quad V_c = G_0(\omega)V_d. \tag{12.3}$$

Any frequency dependence of the laser frequency control or discriminant can be included in the loop filter response, $G_0(\omega)$.

The *loop equation* can be written down by inspection:

$$f = f_1 + K_0V_c = f_1 + K_0G_0(\omega)V_d = f_1 + K_0K_dG_0(\omega)(f_0 - f), \tag{12.4}$$

where we have chosen a sign for $K_d$ which results in negative feedback. Solving for $f$, one obtains

$$f = \frac{f_1}{1 + G(\omega)} + \frac{G(\omega)f_0}{1 + G(\omega)}, \qquad \text{where} \quad G(\omega) \equiv K_0K_dG_0(\omega). \tag{12.5}$$

The quantity $G(\omega)$ is called the *open-loop gain*: it is the gain for a trip around the loop if one breaks the loop at *any* point. The *closed-loop gain* is the ratio of $f$ to $f_0$ if one ignores $f_1$ (by setting it equal to zero): it is a measure of how well the laser frequency tracks changes in the *reference cavity frequency*:

$$\text{Closed loop gain} = \frac{G(\omega)}{1+G(\omega)}. \tag{12.6}$$

The *closed-loop error* is defined as $f - f_0$ with the loop in effect (closed) and is given by

$$\text{Closed loop error} \equiv f - f_0 = \frac{f_1 - f_0}{1+G(\omega)}. \tag{12.7}$$

Assume that we *independently* change $f_0$ by $\Delta f_0$ and $f_1$ by $\Delta f_1$. Then the change in the laser frequency, $\Delta f$ is

$$\Delta f = \frac{\Delta f_1}{1+G(\omega)} + \frac{G(\omega)\Delta f_0}{1+G(\omega)}. \tag{12.8}$$

From this we see that, when $|G(\omega)| \gg 1$, the *free running jitter* $(\Delta f_1)$ is reduced by $G(\omega)$ while the laser accurately tracks changes in the reference frequency $(\Delta f_0)$ since the closed loop gain is nearly unity. Thus, making the open-loop gain as large as possible *reduces noise due to the laser* and improves the *tracking of the reference frequency*. We further see that, at frequencies outside the bandwidth of $G(\omega)$ (i.e., where $G(\omega) \leq 1$), the laser is dominated by its own frequency noise and poorly tracks the reference cavity. Thus, one can consider the frequency, $\omega$, for which $G(\omega) \approx 1$ as the point below which the laser's frequency characteristics will be dominated by the reference and above which it will be dominated by its intrinsic behavior. These characteristics are common to all control systems which use negative feedback.

## 12.3 The effect of the loop filter

We will assume that the loop filter has a frequency behavior which is characterized by extremely high gain at very low frequencies and a gain which decreases with increasing frequency. We will call it an *integrator*, since this is the behavior of a single integrator (we will discuss this in detail in our section on linear system theory). We will then perform a simple experiment: we will inject a small signal, $V_i$, at one of two places: either before the integrator or after the integrator. The signals will be small sinusoidal voltages at a frequency for which $G(\omega) \gg 1$. This procedure is shown in Fig. 12.2. One can determine the loop behavior due to each of the two injected signals by observing the response at the output of the discriminator (with the loop closed). If one injects the signal *before the integrator*, the laser frequency will change from its former value, $f$, to

$$\text{Before integrator:} \quad f \longrightarrow f + \frac{V_i}{K_d} \quad (G(\omega) \gg 1). \tag{12.9}$$

This is equivalent to a *modulation of the cavity resonance frequency*, and the laser will track it very well. On the other hand, if one adds the same signal *after the integrator*, the change in the laser frequency will be

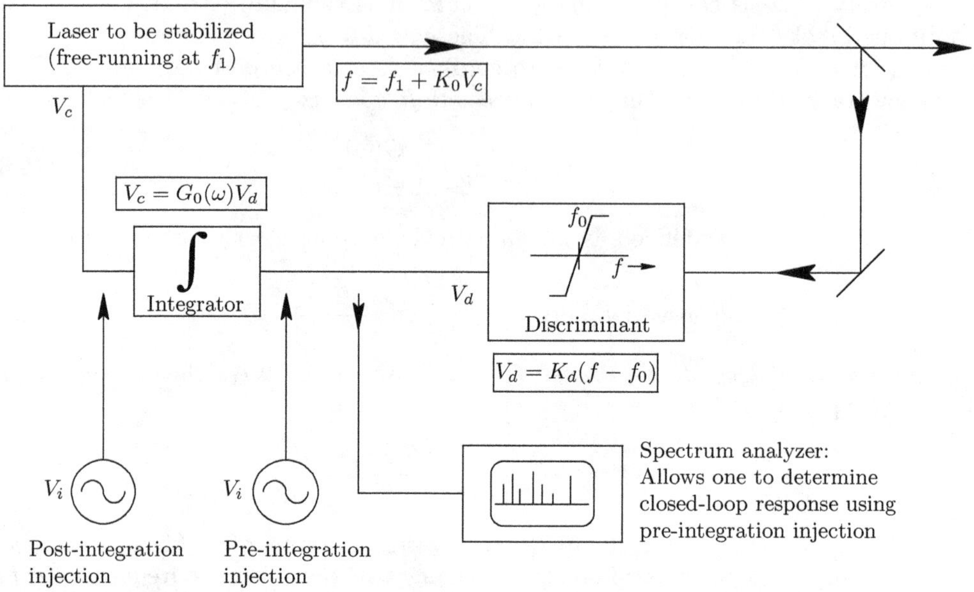


**Fig. 12.2** Injection of small signals before and after the loop filter (integrator).

$$\text{After integrator:} \quad f \longrightarrow f + \frac{K_0V_i}{1+G(\omega)} \longrightarrow f \quad (G(\omega) \gg 1). \tag{12.10}$$

Unlike the previous case, the signal will be *suppressed by the loop*, since it will be treated as an addition to the laser frequency jitter. These two very different behaviors are due to the extremely high gain of an integrator at low frequencies. One can use this procedure to determine the frequency response of the closed loop if one uses a sweeping source after the integrator and observes the output of the discriminator as the frequency is swept (one can inject the signal before the integrator if care is taken to isolate the injected signal from the observed response). The bandwidth of the loop will be the frequency of $V_i$ at which a signal just appears at the discriminator output. Under normal functioning of the closed loop, the signal *after the integrator* can be interpreted as the *correction signal needed to eliminate the free-running laser jitter* and the signal *before the integrator* can be interpreted as a *direct measure of the residual frequency jitter not removed by the loop* (assuming a perfectly stable cavity).

## 12.4 Elementary noise considerations

The ultimate performance of any frequency stabilization system is limited by the shot noise of the signal at the output of the discriminator. A practical system will, of course, have other noise contributions, but these can usually be reduced or eliminated by good design practices. The shot noise is fundamental and cannot be eliminated. We can make some crude estimates of the noise-limited performance with the aid of Fig. 12.3. The noise source will be modeled as a noise voltage, $V_n$, injected into the

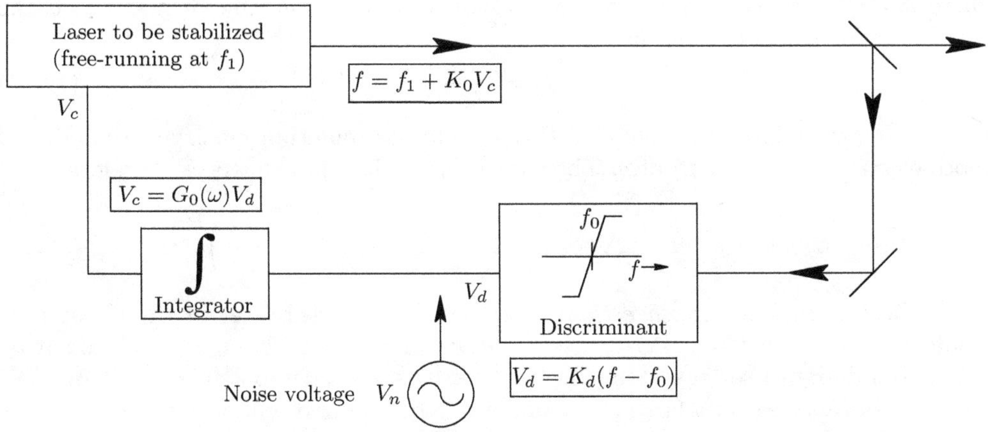


**Fig. 12.3** Block diagram of stabilization system for noise analysis.

loop immediately after the discriminator. The frequency jitter, $\Delta f_{laser}$, within the bandwidth of the loop due to this noise is

$$\Delta f_{laser} = \frac{V_n}{K_d}. \tag{12.11}$$

The *discriminator slope*, $K_d$, is obtained from eqn 4.51:

$$\text{Slope} = 8GP_0J_0(\beta)J_1(\beta)\frac{1}{\delta\omega}, \tag{12.12}$$

where $G$ is the responsivity of the detector in volt/watt and $P_0$ is the incident power. Evaluating the Bessel functions at the optimum $\beta$ (1.08), we obtain

$$\text{Slope} = \frac{2.71GP_0}{\delta\omega}. \tag{12.13}$$

Ignoring the factor $2.71/2\pi$ ($\delta\omega = 2\pi\Delta f_{cavity}$), the value of $K_d$ can be written approximately as

$$K_d \approx \frac{\text{signal}}{\text{cavity width}}, \tag{12.14}$$

since the cavity reflectivity is close to unity and the signal voltage is $\approx GP_0$. This is a general result which holds for all types of discriminants (with slightly different coefficients). Thus,

$$\Delta f_{laser} \approx \frac{\text{noise}}{\text{signal/width}} = \frac{\Delta f_{cavity}}{\langle \text{SNR} \rangle}, \tag{12.15}$$

where SNR is the *signal-to-noise ratio.* From this, we obtain the general rule that the ultimate stabilized frequency jitter is equal to the *cavity width divided by the signal-to-noise ratio.* We occasionally say we can *split* the cavity width by the SNR. The shot-noise-limited SNR is

$$\text{SNR} = \frac{I_s}{I_n} = \frac{I_s}{\sqrt{2eI_s\Delta\nu}}, \tag{12.16}$$

where $\Delta\nu = 1/2\pi\tau$ is the noise bandwidth for an integration time of $\tau$ and $I_s$ is the signal current, which is given by

$$I_s = \frac{eP\eta}{\hbar\omega} \tag{12.17}$$

for a laser power into the cavity of $P$ and detector quantum efficiency (number of photo-electrons per photon) of $\eta$. The shot-noise-limited laser jitter is therefore

$$\text{Laser frequency jitter:} \quad \Delta f_{laser} = \Delta f_{cavity} \left(\frac{\hbar\omega}{\pi P\eta}\right)^{1/2} \left(\frac{1}{\tau}\right)^{1/2}. \tag{12.18}$$

All of these results are *rough estimates* and will be refined later in the chapter. It should be emphasized that $\Delta f_{laser}$ is a *lower bound* on the laser jitter obtained by a very well designed servo system. In practice, it is almost impossible to achieve this level of performance over a large range of noise frequencies. It will be useful to estimate the stabilized laser linewidth if we could actually realize shot-noise-limited behavior. Ultra-high-finesse cavities are available with $\Delta f_{cavity} = 2$ kHz. If the power is 0.1 mW and $\eta = 1$, the shot-noise-limited laser jitter at $\lambda = 1\ \mu$m is $\Delta f_{laser} = 5 \times 10^{-5}$ Hz for an integration time ($\tau$) of 1 s, which is very narrow indeed. The relative instability in 1 second is $2.5 \times 10^{-19}$.

Before leaving this topic, it would be instructive to investigate the actual linewidth of a frequency-locked laser. The jitter calculated above is just the *measured* frequency noise over time $\tau$. If one had an extremely sharp optical spectrum analyzer, what would the laser line look like? It turns out that a laser whose frequency noise is *spectrally uniform* (such as the shot-noise-limited laser being discussed) will have a *Lorentzian* lineshape whose width, $\Delta f$, is given by

$$\text{Linewidth (white noise):} \quad \Delta f = \pi S_{\Delta f}, \tag{12.19}$$

where $S_{\Delta f}$ is the *power* spectral density of the laser frequency noise (in $\mathrm{Hz}^2/\mathrm{Hz}$). This can be justified with a calculation of the autocorrelation function of the frequency noise and an application of the Wiener–Khintchine theorem (the calculation is carried out in Elliot et al. (1982)). Using this result, the laser linewidth due to shot noise is

$$\Delta f = \pi \frac{2eI_s}{(K_d)^2} = \frac{2\pi(\Delta f_{cavity})^2 h\nu}{\eta P}, \tag{12.20}$$

where we used eqn 12.17 for the current and estimated $K_d$ (in A/Hz) as

$$K_d \approx \frac{e\eta P}{h\nu \Delta f_{cavity}} \qquad \text{(in A/Hz)}. \tag{12.21}$$

(The A/Hz units for $K_d$ are appropriate since the shot noise is represented as a *current*.) If one compares $\Delta f$ to the Schawlow–Townes expression for the quantum limit on the laser linewidth (eqn 6.84), one observes that the linewidth of a frequency-locked laser is described by essentially the same formula, replacing the laser cavity width with the external cavity width. Since $\Delta f$ is proportional to the *square* of the cavity linewidth, the improvement can be many orders of magnitude – a factor of $10^{12}$ for diode lasers! Again, this is a theoretical limit which is never realized in practice.

## 12.5 Some linear system theory

From the preceding section we concluded that the best results are obtained when one constructs a laser stabilization system with the largest open loop gain possible over the largest possible range of frequencies. In order to realize this, one needs to examine the dynamical properties of servo systems, since the maximum gain and bandwidth will be limited by the requirement that the loop be *stable*. This section will provide a background in linear system theory as a prelude to a discussion of loop stability and other matters.

A *linear system* is one which satisfies the following *linearity conditions*:

$$f_1(t) \longrightarrow r_1(t) \tag{12.22}$$

$$f_2(t) \longrightarrow r_2(t) \tag{12.23}$$

$$\Rightarrow \alpha f_1(t) + \beta f_2(t) \longrightarrow \alpha r_1(t) + \beta r_2(t), \tag{12.24}$$

where $\alpha, \beta$ are complex constants, $r_1$ and $r_2$ are the *responses* to *driving functions* $f_1, f_2$, and the arrows connect the responses to the driving function. A general response, $r(t)$, is connected to a driving function, $f(t)$, by the differential equation

$$a_n \frac{d^n r}{dt^n} + a_{n-1}\frac{d^{n-1} r}{dt^{n-1}} + \cdots + a_1 \frac{dr}{dt} + a_0 r(t) = b_m \frac{d^m f}{dt^m} + b_{m-1}\frac{d^{m-1}}{dt^{m-1}} + \cdots + b_1 \frac{df}{dt} + b_0 f(t), \tag{12.25}$$

where the coefficients are independent of time (a system described by a differential equation with constant coefficients is called a *time-invariant* system: its properties do not change with time). We will first assume a simple exponential dependence for both $r(t)$ and $f(t)$:

$$f(t) = f_0 e^{st} \tag{12.26}$$

$$r(t) = r_0 e^{st}, \tag{12.27}$$

where $s$ is a *complex number* and is called the *complex frequency.* By including a real part in $s$, we allow for exponentially damped and exponentially increasing signals. Substituting these functions into the differential equation and solving for $r_0/f_0$, we obtain

$$H(s) = \frac{r_0}{f_0} = \frac{b_m s^m + b_{m-1}s^{m-1} + \cdots + b_1 s + b_0}{a_n s^n + a_{n-1}s^{n-1} + \cdots + a_1 s + a_0}. \tag{12.28}$$

The quantity $H(s)$ is called the *transfer function* for the system and is a complex-valued rational fraction. It is simply the ratio of the response to the driving function. It should be fairly obvious that the overall transfer function of *cascaded systems* (systems placed one after the other) is simply the *product* of the transfer functions of the individual systems. We see that an exponential driving function will give rise to an exponential response with the *same* complex frequency.

Most driving and response functions are not exponentials. In order to describe a *general* function, we decompose it into a *continuous sum of damped, eternal sinusoidal functions* using the *Laplace transform.* For a function of time, $f(t)$, the Laplace transform is $F(s)$; by convention, the time function is in lower case and the Laplace transform is in upper case using the same letter. The Laplace transform and its inverse are defined by

$$F(s) = \mathcal{L}[f(t)] = \int_{0^-}^{\infty} f(t)e^{-st}dt \tag{12.29}$$

$$f(t) = \mathcal{L}^{-1}[F(s)] = \frac{1}{2\pi i}\int_{\sigma - i\infty}^{\sigma + i\infty} F(s)e^{st}d\omega, \tag{12.30}$$

where $s = \sigma + i\omega$, and the inverse transform integration takes place in the complex plane along a vertical line whose real intercept is $\sigma$; any line which is to the right of the singularities of $F(s)$ can be used. The lower limit ($t = 0^-$) in the first integral refers to a time which is infinitesimally before $t = 0$ to include in the integration a possible *impulse* at $t = 0$. The Laplace transform is a generalization of the *Fourier transform* to include damped exponentials, which makes this transform easier to apply to certain problems. Due to the presence of a damped factor, the Laplace transform is more likely to exist than the Fourier transform. The Laplace transform of $f(t)$ exists for $s = \sigma + i\omega$ if the integral

$$\int_{-\infty}^{\infty} |f(t)|e^{-\sigma t}dt \tag{12.31}$$

is finite. One can show that the latter condition is satisfied for all $\sigma > \alpha$ if there exists a real positive number $\alpha$ satisfying

$$\lim_{t\to\infty} [e^{-\alpha t}f(t)] = 0. \tag{12.32}$$

The following is a list of some of the properties of the Laplace transform:

**Linearity**: If $f_1(t) \leftrightarrow F_1(s)$ and $f_2(t) \leftrightarrow F_2(s)$, then:

$$a_1 f_1(t) + a_2 f_2(t) \leftrightarrow a_1 F_1(s) + a_2 F_2(s)$$

**Time-differentiation:** If $f(t) \leftrightarrow F(s)$, then

$$\frac{df}{dt} \leftrightarrow sF(s) - f(0^-)$$

**Time-integration:** If $f(t) \leftrightarrow F(s)$, then

$$\int_0^t f(t)dt \leftrightarrow \frac{F(s)}{s}$$

**Time-shifting:** For $t_0 > 0$,

$$f(t - t_0) \leftrightarrow F(s)e^{-st_0}$$

**Time-convolution:** If $f_1(t) \leftrightarrow F_1(s)$ and $f_2(t) \leftrightarrow F_2(s)$, then:

$$\int_{-\infty}^{\infty} f_1(\tau)f_2(t - \tau)d\tau \leftrightarrow F_1(s)F_2(s)$$

**Initial-value theorem:** If $f(t)$ and $df/dt$ have Laplace transforms, then the *initial value* of $f(t)$ is given by:

$$f(0^+) = \lim_{s \to \infty} [sF(s)]$$

**Final-value theorem:** If $f(t)$ and $df/dt$ have Laplace transforms, then the *final value* of $f(t)$ is given by:

$$f(\infty) = \lim_{s \to 0} [sF(s)]$$

The double arrow ($\leftrightarrow$) identifies a *Laplace transform pair* and the symbols $0^+$ and $0^-$ indicate a time infinitesimally after and infinitesimally before $t = 0$. A list of the most important Laplace transform pairs is given in Table 12.1.

**Table 12.1** Laplace transforms.

| $f(t) = \mathcal{L}^{-1}[F(s)]$ | $F(s) = \mathcal{L}[f(t)]$ |
|---|---|
| $\delta(t)$ (delta function) | $1$ |
| $u(t)$ (unit step at $t = 0$) | $\dfrac{1}{s}$ |
| $tu(t)$ (unit ramp) | $\dfrac{1}{s^2}$ |
| $u(t)\sin(\omega_0 t)$ | $\dfrac{\omega_0}{s^2 + \omega_0^2}$ |
| $u(t)\cos(\omega_0 t)$ | $\dfrac{s}{s^2 + \omega_0^2}$ |
| $u(t)e^{at}$ | $\dfrac{1}{s - a}$ |

Functions of time, $f(t)$, which have the property that $f(t) = 0$ when $t < 0$, are called *causal functions.* Since, by causality, the response cannot anticipate the driving function, the response must also be zero for $t < 0$. Most functions we will encounter are causal.

If we assume that the driving function is causal and take the Laplace transform of both sides of eqn 12.25, we obtain the same result for the transfer function as was obtained for an exponential driving function:

$$H(s) = \frac{R(s)}{F(s)} = \frac{b_m s^m + b_{m-1} s^{m-1} + \cdots + b_1 s + b_0}{a_n s^n + a_{n-1} s^{n-1} + \cdots + a_1 s + a_0}, \tag{12.33}$$

where we assume the system is initially de-energized and set the initial values equal to zero. Note that, from the initial value theorem, the transfer function at $t = 0$ will be infinite unless $m < n$; such a rational fraction is called a *proper fraction.*

The response function, $r(t)$, is obtained in three steps: first, we determine the transfer function, then take the Laplace transform of the driving function, $f(t)$, and finally invert the product:

$$r(t) = \mathcal{L}^{-1}[H(s)F(s)]. \tag{12.34}$$

The beauty of this approach is that the driving function transform can usually be obtained from a table and the transfer function can be obtained using the *standard rules for combining impedances*, where some elementary impedances are

$$\text{Capacitor:} \quad Z_C = \frac{1}{sC} \tag{12.35}$$

$$\text{Inductor:} \quad Z_L = sL \tag{12.36}$$

$$\text{Resistor:} \quad Z_R = R. \tag{12.37}$$

In order to use the table of Laplace transforms, it is usually necessary to simplify $H(s)$ by factoring the denominator and expanding $H(s)$ into *partial fractions.* If we divide out $a_n$ and factor the denominator,

$$H(s) = \frac{b_m s^m + b_{m-1}s^{m-1} + \cdots + b_1 s + b_0}{(s-s_1)(s-s_2)\cdots(s-s_l)(s-s_0)^p}, \tag{12.38}$$

where there are $l$ *simple roots* and one root with multiplicity $p$ ($n = l + p$). The partial fraction expansion yields

$$H(s) = \frac{c_1}{s-s_1} + \frac{c_2}{s-s_2} + \cdots + \frac{c_l}{s-s_l} \quad \text{(simple roots)} \tag{12.39}$$

$$+ \frac{k_0}{(s-s_0)^p} + \frac{k_1}{(s-s_0)^{p-1}} + \cdots + \frac{k_p}{s-s_0} \quad \text{(multiple root).} \tag{12.40}$$

There are numerous procedures for obtaining partial fraction expansions; we will briefly describe one for simple roots and one for multiple roots. First assume that the above partial fraction expansion contains only simple roots. Then, the coefficient $c_r$ is

$$\text{Simple roots:} \quad c_r = H(s)(s-s_r)|_{s=s_r}, \tag{12.41}$$

where $s_r$ is one of the roots. The explanation for this should be self-evident. For a multiple root with *multiplicity* (maximum power) equal to $p$, the expansion is

$$H(s) = \frac{k_0}{(s-s_0)^p} + \frac{k_1}{(s-s_0)^{p-1}} + \cdots + \frac{k_j}{(s-s_0)^{p-j}} + \cdots + \frac{k_p}{s-s_0} \tag{12.42}$$

and the $j$th coefficient is

$$k_j = \frac{1}{j!}\frac{d^j}{ds^j}\left[H(s)(s-s_0)^p\right]\bigg|_{s=s_0}. \tag{12.43}$$

An example should illustrate the power and simplicity of the use of Laplace transforms to analyze an electrical network. The circuit appears in Fig. 12.4. We seek an expression for the current as a function of time if the switch is closed at $t = 0$. The

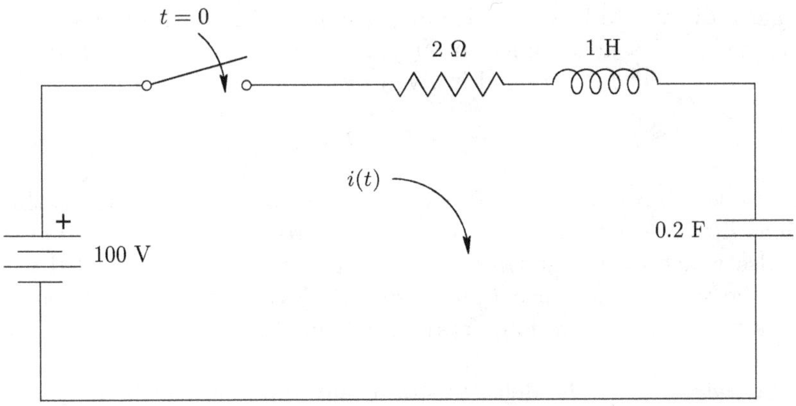


**Fig. 12.4** RLC circuit analyzed using Laplace transforms.

driving function is a step of magnitude equal to 100 V and which occurs at $t = 0$. From the table, the transform of the driving function is

$$F(s) = \mathcal{L}[100u(t)] = \frac{100}{s}. \tag{12.44}$$

The transfer function is the reciprocal of the total impedance

$$H(s) = \frac{1}{Z(s)} = \frac{1}{2 + s + 1/0.2s}, \tag{12.45}$$

where we used the expressions for the individual impedances of the elements and combined them using the well-known rule for series impedances. The current is

$$\begin{aligned} i(t) &= \mathcal{L}^{-1}[H(s)F(s)] \\ &= \mathcal{L}^{-1}\left[\frac{100}{s^2+2s+5}\right] = 25i\mathcal{L}^{-1}\left(\frac{1}{s+1+2i} - \frac{1}{s+1-2i}\right) \\ &= 25i(e^{(-1-2i)t} - e^{(-1+2i)t}) = 50e^{-t}\sin 2t. \end{aligned} \tag{12.46}$$

## 12.6 The stability of a linear system

The most important and severe constraint on a control system is the need to prevent undesired sustained *oscillation*. In order to examine the conditions for oscillation (instability), we need to understand the behavior of the *natural modes* in a system; these are the responses which occur in the *absence of a driving function*. We begin by discussing the *impulse response* of a linear system.

A *unit impulse* which occurs at time $t'$ can be represented by a *delta function*, $\delta(t - t')$, defined by

$$\delta(t - t') = 0, \quad \text{when } t \neq t' \tag{12.47}$$

$$\int_{-\infty}^{\infty} \delta(t)dt = 1. \tag{12.48}$$

This is a *pulse* of infinite height and infinitesimal width which occurs at $t = t'$. The Laplace transform of a delta function, $\delta(t)$, is 1 (from the table). Therefore, the *response* to a *unit impulse* is $h(t)$ and is given by

$$h(t) = \mathcal{L}^{-1}[H(s)], \tag{12.49}$$

for an impulse which occurs at $t' = 0$. Since the system is time-invariant, this can also be written as $h(t - t')$, which is interpreted as the response at observation time $t$ due to an impulse which occurs at time $t'$. The quantity $h(t - t')$ is called the *impulse response* of the system and, for a *time-invariant* system, depends only on $t - t'$. One can consider the transfer function, $H(s)$, to be the *Laplace transform of the impulse response.*

From the definition of the delta function, any function can be expanded in unit impulses:

$$f(t) = \int_{-\infty}^{\infty} f(t')\delta(t - t')dt'. \tag{12.50}$$

Since $h(t - t')$ is the response at $t$ to a single impulse at $t'$, by *superposition* (a consequence of linearity), the response to a general driving function can be written as a *continuous sum* of the impulse response, $h(t - t')$, multiplied by the driving function evaluated at time $t'$:

$$r(t) = \int_{-\infty}^{\infty} h(t - t')f(t')dt'. \tag{12.51}$$

This is a *convolution integral,* and the use of the *time-convolution property* yields the familiar relation, $R(s) = H(s)F(s)$.

An impulse driving function will excite the *natural modes* of a linear system. To see why this is so, we expand the transfer function in partial fractions:

$$h(t) = \mathcal{L}^{-1}[H(s)] = \mathcal{L}^{-1}\left[\frac{c_1}{s - s_1} + \frac{c_2}{s - s_2} + \cdots + \frac{c_n}{s - s_n}\right] \tag{12.52}$$

$$= c_1 e^{s_1 t} + c_2 e^{s_2 t} + \cdots + c_n e^{s_n t}, \tag{12.53}$$

where we assumed that all of the roots of the denominator are simple. Thus, the response to an impulse will be a series of exponentials at frequencies which are the *roots of the denominator of the transfer function.* Since the delta function is active only for an infinitesimal duration at $t = 0$, these responses occur in the *absence* of a sustained drive and the frequencies are called the *natural frequencies of a system.* (The sound of a *bell* which is briefly struck is a familiar example of this.) This result can also be obtained by solving the *homogeneous* equation which is obtained by setting the right-hand side of eqn 12.25 equal to zero (i.e., removing the driving function). The roots of the denominator of the transfer function are called the *poles* of $H(s)$, and their locations in the complex plane determine the stability of a linear system. The roots of the *numerator* of the transfer function are called the *zeros* of $H(s)$.

Figure 12.5 displays the natural responses of a system for a number of pole locations in the complex $s$-plane. We observe that the poles in the left-hand plane will have damped behaviors and those in the right-hand plane will have exponentially increasing

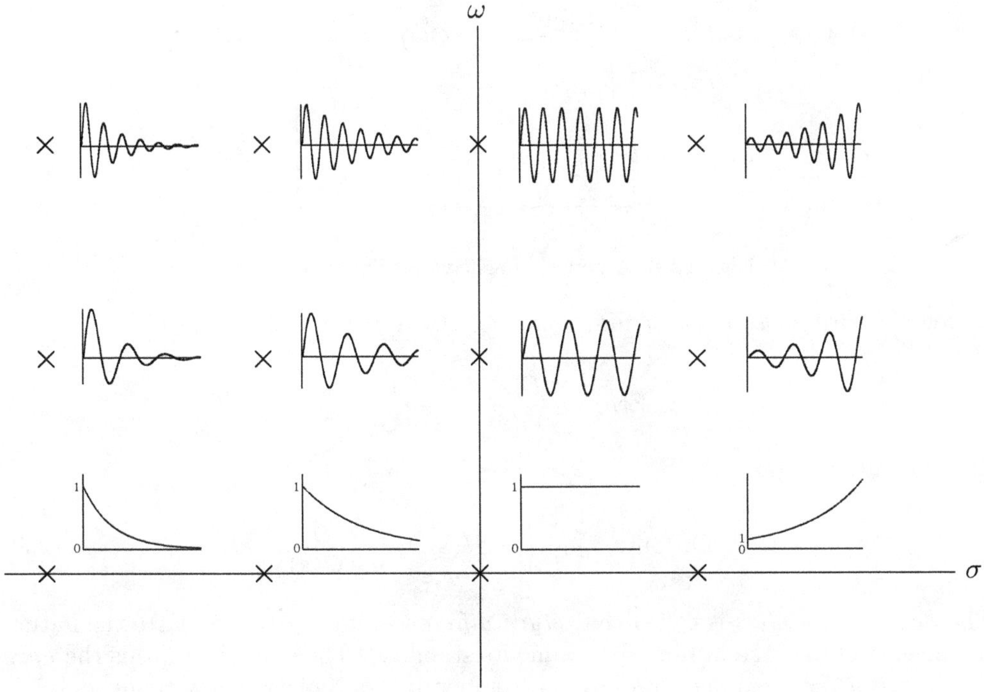


**Fig. 12.5** Natural responses of a system for various pole locations (the "×"s).

behaviors. Clearly, the latter are undesirable. The *simple* poles on the imaginary axis will correspond to a stable sinusoid; while this is not usually desirable, it is not considered an unstable behavior. Finally, the existence of poles with multiplicity greater than one on the imaginary axis results in at least a linearly increasing sinusoid and therefore is unstable behavior. To see why this is so, we evaluate the inverse transform of $1/(s - i\omega)^2$:

$$\mathcal{L}^{-1}[1/(s - i\omega)^2] = te^{i\omega t}u(t). \tag{12.54}$$

Poles with higher multiplicity will be associated with a response that increases even more rapidly with time. We thus arrive at the following rule. *A system is stable if and only if the poles of the transfer function are either all in the left-hand plane or are on the imaginary axis and are simple.*

There are a number of techniques for determining the stability of a system using the behavior of the poles. We will not discuss these but will instead use a more easily understood approach which is uniquely applicable to feedback systems. It is called the *Bode plot* and will be presented after a general discussion of feedback.

## 12.7 Negative feedback

A block diagram of a generic system employing negative feedback appears in Fig. 12.6. In this configuration, the output is made to *track* the input by comparing the output to the input, taking the difference, and applying the difference (the error signal, $E(s)$)

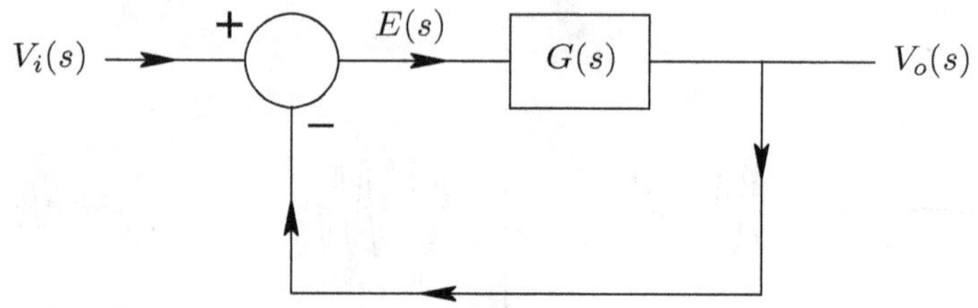


**Fig. 12.6** A generic negative feedback system.

to the element whose transfer function is $G(s)$. After some simple algebra, the error signal is

$$\text{Error:} \quad \frac{E(s)}{V_i(s)} = \frac{1}{1 + G(s)} \tag{12.55}$$

and the output signal is

$$\text{Output:} \quad \frac{V_o(s)}{V_i(s)} = H(s) = \frac{G(s)}{1 + G(s)}. \tag{12.56}$$

The quantity $1 + G(s)$ is called the *return difference*: its *zeros* (roots) are the natural frequencies of the system (and determine its stability). The quantity $G(s)$ is the *open-loop transfer function*: the gain around the loop if it is broken at any point.

Some treatments of feedback use other network topologies. These are easily transformed into the geometry depicted in Fig. 12.6 using *block diagram equivalences*, some of which appear in Fig. 12.7. For example, the top-left diagram is used in some other treatments of feedback. With the aid of the indicated equivalence, one can immediately write down the transfer function for this geometry:

$$\text{(Alternative geometry)} \quad H(s) = \left(\frac{1}{F(s)}\right) \frac{G(s)F(s)}{1 + G(s)F(s)} = \frac{G(s)}{1 + G(s)F(s)}. \tag{12.57}$$

This geometry is more general than the original one and can describe a system where the output does *not* track the input. We will use the original geometry in all that follows, since laser stabilization systems track the reference.

An important characteristic of a tracking feedback system is its *dynamic behavior*: how well the tracking performs with a *changing* input. We usually consider inputs which vary as powers of $t$: a step input (constant), a ramp input ($\propto t$) and an accelerated input ($\propto t^2$). We would like the *error to approach zero after long times.* To analyze the behavior as $t \to \infty$ we use the final value theorem. Thus, we take the limit

$$e_s(\infty) = \lim_{s \to 0} [sV_i(s)H_e(s)], \tag{12.58}$$

where $V_i(s)$ is the transform of the input and $H_e(s)$ is the transfer function for the *error signal*, $e_s(t)$:

$$H_e(s) = \frac{1}{1 + G(s)}. \tag{12.59}$$

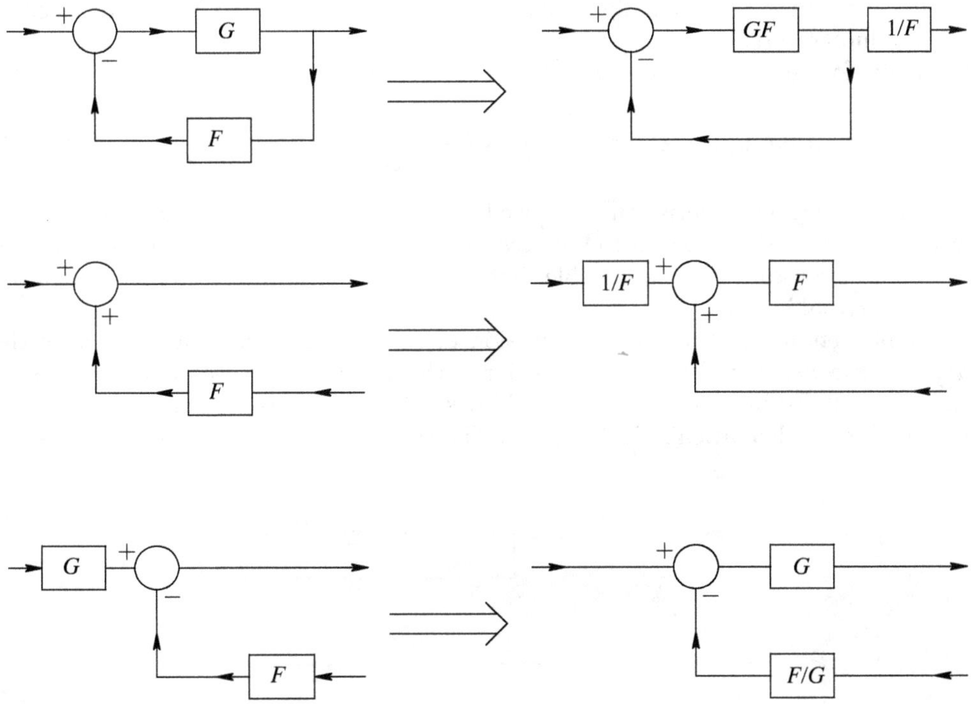


**Fig. 12.7** A few block diagram equivalences.

We first model the function $G(s)$ (the open loop transfer function) as one or more cascaded integrators. For a single integrator, $G(s)$ is

$$\text{Single integrator:}\quad g(t) = \int_0^t [\ ]dt' \Longrightarrow G(s) = \frac{1}{s} \tag{12.60}$$

from the integration rule for Laplace transforms (the integrand is not shown since the integral is treated as an *operator*). Clearly, for $n$ cascaded integrators, the transform is

$$n \text{ integrators:}\quad G(s) = \frac{1}{s^n}. \tag{12.61}$$

For a *step input*, the transform of the input is $V_i(s) = 1/s$. The error is

$$\text{Step input:}\quad e_s(\infty) = \lim_{s\to 0}\left(s\frac{1}{s}\frac{1}{1+G(s)}\right) = \frac{1}{1+G(0)}. \tag{12.62}$$

In order to force $e_s(\infty) \to 0$, it is necessary for $G(0) \to \infty$. This requires at least one integrator.

The transform of a *ramp input* is $V_i(s) = 1/s^2$. The error is

$$\text{Ramp input:}\quad e_s(\infty) = \lim_{s\to 0}\left(s\frac{1}{s^2}\frac{1}{1+G(s)}\right) = \lim_{s\to 0}\frac{1}{sG(s)}. \tag{12.63}$$

To reduce the error (occasionally called the *velocity* error) to zero, we clearly need at least two integrators.

Finally, for an *acceleration input*, $V_i(s) = 1/s^3$ and

$$\text{Acceleration input:}\ \ e_s(\infty) = \lim_{s\to 0}\left(s\frac{1}{s^3}\frac{1}{1+G(s)}\right) = \lim_{s\to 0}\frac{1}{s^2G(s)}. \tag{12.64}$$

It should come as no surprise that we need at least three integrators to eliminate the acceleration error at $t = \infty$. In the interest of stability, it is desirable to minimize the number of integrators, so it is possible that some of the tracking capabilities might be lost in an actual system.

Complementary to the above discussion of steady-state errors at $t = \infty$ is the *transient response*: the short-term behavior of the output of a closed loop with a step input at $t = 0$. To investigate this, we will describe a *canonical* second-order system, which is shown schematically in Fig. 12.8. The transfer function for the output is

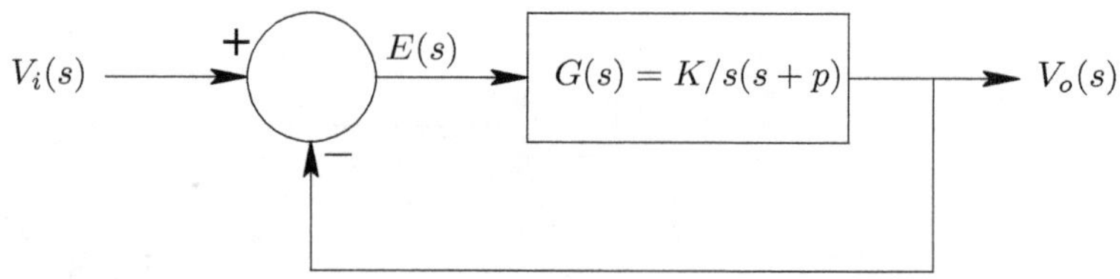


**Fig. 12.8** A canonical second-order loop.

$$H(s) = \frac{G(s)}{1+G(s)} = \frac{K}{s^2+ps+K} = \frac{\omega_n^2}{s^2+2\zeta\omega_n s+\omega_n^2}. \tag{12.65}$$

The constants $p$ and $K$ are positive and real. We have introduced two new parameters: the *natural frequency*, $\omega_n$, and the *damping constant*, $\zeta$. These are given by

$$\omega_n = \sqrt{K} = \ \text{natural frequency} \tag{12.66}$$

$$\zeta = \frac{p}{2\sqrt{K}} = \text{damping constant}\ \begin{cases} \zeta > 1 & : \ \text{overdamping} \\ \zeta = 1 & : \ \text{critical damping} \\ \zeta < 1 & : \ \text{underdamping}. \end{cases} \tag{12.67}$$

The poles of $H(s)$ are

$$\text{Poles:}\ \ s_1, s_2 = -\zeta\omega_n \pm \omega_n\sqrt{\zeta^2-1}. \tag{12.68}$$

Note that the poles are real when $\zeta > 1$ (overdamping) and there is therefore no oscillatory component to the step response in this case. Furthermore, the system is *unconditionally stable*, since the poles are in the left-hand plane for all finite $\zeta$ and $\omega_n$.

The transient response is obtained by taking the inverse transform of $F(s)H(s)$. The result, for $\zeta \leq 1$, is

$$r(t) = \mathcal{L}^{-1}\left[\frac{\omega_n^2}{s(s^2 + 2\zeta\omega_n s + \omega_n^2)}\right] = 1 - \frac{1}{\beta}e^{-\zeta\omega_n t}\sin(\omega_n \beta t + \theta) \quad (12.69)$$

$$\text{where } \beta = \sqrt{1 - \zeta^2} \quad (12.70)$$

$$\theta = \tan^{-1}\beta/\zeta. \quad (12.71)$$

For $\zeta > 1$, the result is

$$r(t) = 1 - \frac{1}{\beta}e^{-\zeta\omega_n t}\sinh(\omega_n \beta t + \theta) \quad (12.72)$$

$$\text{where } \beta = \sqrt{\zeta^2 - 1} \quad (12.73)$$

$$\theta = \tanh^{-1}\beta/\zeta. \quad (12.74)$$

For $\zeta \ll 1$, the response of $H(s)$ to a *sinusoidal input* is strongly peaked at resonant frequency $\omega_n$. The response can be described by a $Q$ value:

$$Q = \frac{1}{2\zeta}. \quad (12.75)$$

The response to a step input for a number of damping factors ($\zeta$) appears on the left of Fig. 12.9. The pole locations for the same values of $\zeta$ are shown in the graph on

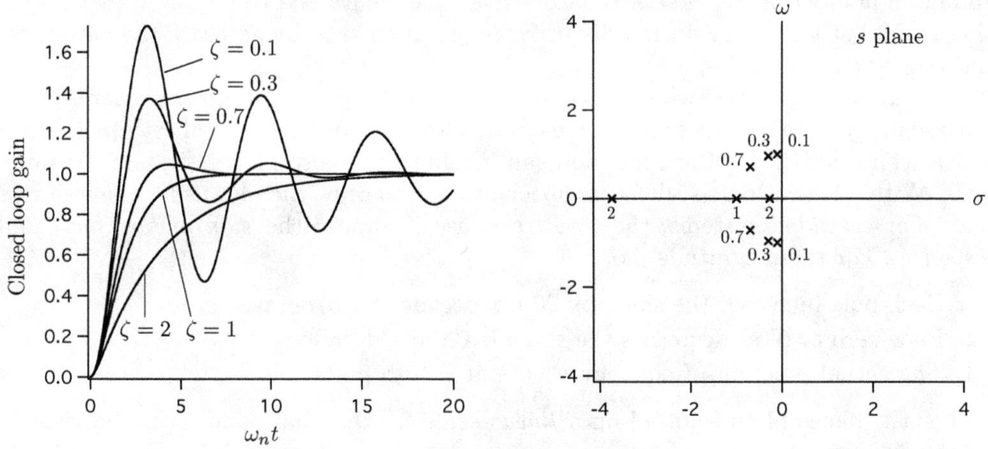


**Fig. 12.9** Transient response of a second-order loop. The curve on the left is the response curve for a number of damping factors, and the pole locations for the same values of $\zeta$ are shown in the right-hand graph.

the right. This sequence of poles plotted as a function of $\zeta$ is an example of the *root locus* technique of stability analysis, where the *locus* of the poles of the closed-loop transfer function is plotted as a function of some parameter and one can determine whether or not the poles are confined to the left hand of the complex $s$-plane. The second-order loop is simple enough that the pole locations can be easily determined from the algebraic expression (eqn 12.68) for the poles.

The *Bode plot* method of stability analysis is perhaps the most familiar and easiest to apply. It begins with a fairly obvious observation about the transfer function, $H(s)$,

$$H(s) = \frac{G(s)}{1 + G(s)}. \tag{12.76}$$

For the loop to be stable, the return difference, $1 + G(s)$, must be *non-zero* for *all frequencies.* This means that the *open-loop gain,* $G(s)$, must never have a magnitude of unity and a phase of $-180°$ *at the same time.* We usually design a system with stability in mind by starting at the *unity gain frequency* of $G(s)$ and making sure that the phase shift is less than $\pi$ radians at this frequency. The utility of this method is due to the fact that the unity gain frequency is often limited by inescapable time delays and, for best servo performance, one designs the loop filter to allow the maximum unity gain frequency. Furthermore, $|G(s)|$ is almost always a monotonically decreasing function of frequency and therefore, if one establishes stability when $|G(s)| = 1$, the possibility of instability will *never* be revisited at lower frequencies. We can then include a sufficient number of filter stages to provide adequate gain at *lower* frequencies in order to satisfy the error and tracking requirements without adding destabilizing phase shifts at the unity gain frequency. The unity gain frequency of the open-loop gain has another important characteristic: it is the frequency where the *closed-loop gain* begins to fall off (it is only the $-3$ dB point of the closed-loop gain). One is cautioned to use the minimum number of stages to avoid excessive time delays and to be careful with large phase shifts at gain $> 1$, since a disturbance can momentarily saturate the amplifiers and trigger oscillation.

We can see at a glance whether a system is stable by plotting both the *magnitude* of the open-loop gain and its *phase* versus frequency. The frequency is always plotted on a logarithmic scale, and the open-loop gain magnitude is also plotted on a logarithmic scale. With a logarithmic scale, it is especially easy to draw the plot using a few simple rules. For cascaded systems, the total response is simply the *sum of the individual responses.* For the magnitude plot:

1. Each pole increases the slope by 20 db/decade (6 db/octave).
2. Each zero of $G(s)$ decreases the slope by 20 db/decade.
3. The actual pole/zero frequency occurs at $\pm 3$ db point.

For the phase plots (plotted on a *linear* scale for the phase and logarithmic scale for the frequency),

1. Each pole provides a $-90°$ phase shift (*lag*), asymptotically.
2. Each zero provides a $+90°$ phase shift (*lead*), asymptotically.
3. The lag (lead) at the exact pole (zero) frequency is $45°$.

It is interesting to note that each 20 dB/decade increase or decrease in *slope* is accompanied by a $90°$ increase or decrease in phase shift. This characteristic is a consequence of a general *dispersion relation* which applies to all physically realizable linear systems and has its origin in the requirement of *causality*: every response must *follow* its cause.

Typical Bode magnitude and phase plots are shown in Fig. 12.10. One can easily

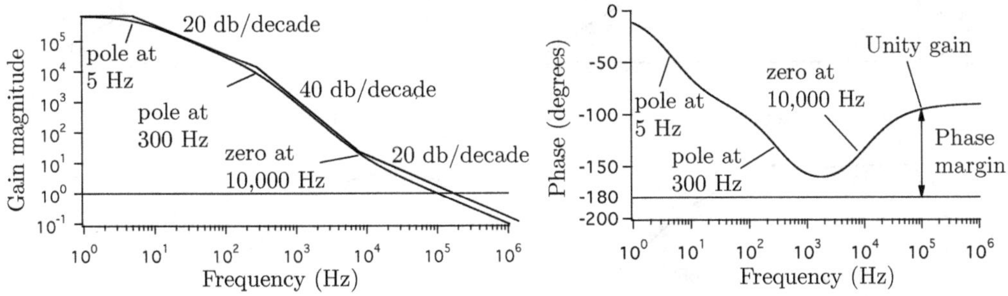


**Fig. 12.10** Bode magnitude and phase plots.

draw the plots by approximating them using straight lines with the slopes and values given by the above rules. Although Fig. 12.10 was for an arbitrary system, it does illustrate a technique for obtaining increased low-frequency gain with the judicious use of one or more *zeros* (provided by a *circuit element which differentiates*). While a system with two poles will not oscillate, it can have a *transient response* characterized by significant *ringing*, which can be thought of as a condition of *near instability*. By introducing a zero slightly below the unity gain frequency, one can greatly improve the transient response and still have adequate gain at lower frequencies.

The nearness to instability can be gauged by the *phase margin*: the difference between the actual phase shift and $-180°$ at the unity gain frequency The smaller the phase margin, the closer the loop is to being unstable. In order to better illustrate the significance of phase margin, we will display Bode plots for a second-order loop with two different values of the damping constant, $\zeta$. The Bode plots on the left of Fig. 12.11 are for a damping constant of 0.707, and the plots on the right are for a damping constant of 0.1. The larger damping constant provides a comfortable 65° phase margin while the smaller damping constant provides a fairly small 11° phase margin. If one refers back to Fig. 12.9, it should be evident that the small phase margin is accompanied by a very poor transient behavior: the response to a step input takes several cycles before it settles down. The larger phase margin, on the other hand, quickly settles after a slight *overshoot*. For these reasons, the usual practice is to require a phase margin of 30° ($\pi/6$), which provides a good protection against instability while ensuring reasonable transient behavior.

Before continuing, we should caution the reader about a source of possible mistakes for the beginner. We have been discussing two different transfer functions: the *open-loop gain* and the *closed-loop transfer function*. The Bode plot analysis of stability *only applies to the open-loop gain*. The closed-loop gain is useful for determining stability from the *pole positions* (must be in the left half-plane) and one can determine the transient response to a unit step by inverting the quantity $H(s)/s$, but the Bode plot approach to stability analysis can be used only with the open-loop gain. Finally, the poles of the open-loop transfer function are used in making the Bode plot according to

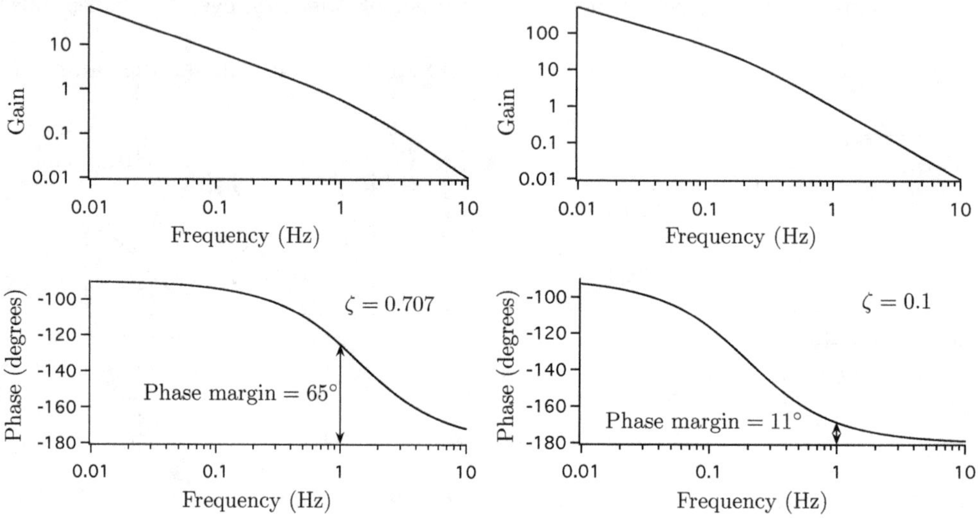


**Fig. 12.11** Bode magnitude and phase plots for the second-order loop for $\zeta = 0.707$ (left) and $\zeta = 0.1$ (right).

the rules discussed earlier, and the poles of the closed-loop transfer function are used for an analysis of stability and the transient response.

As a final illustration of Bode plots, we will provide an explanation for the well-known unity-gain instability of uncompensated operational amplifiers, familiar to anyone who has worked with them. An operational amplifier with feedback in the non-inverting configuration is shown schematically in Fig. 12.12. The *feedback factor*, $T(s)$,

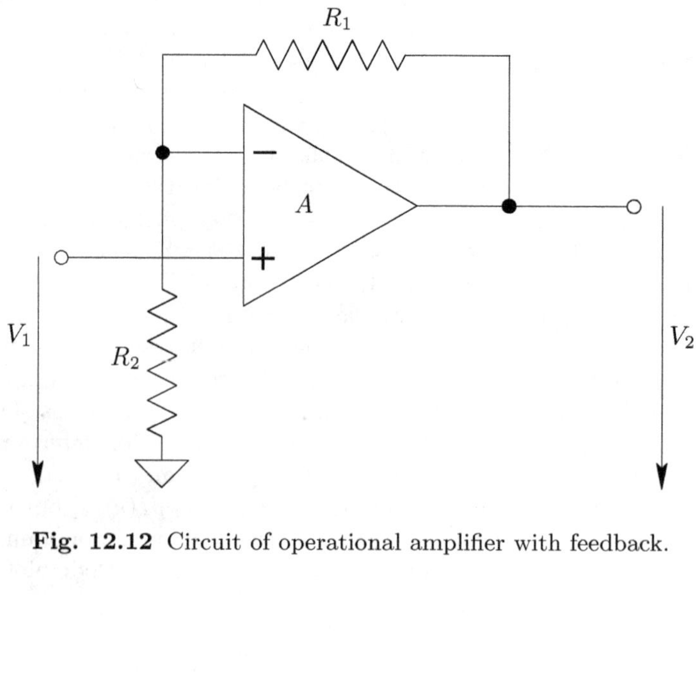


**Fig. 12.12** Circuit of operational amplifier with feedback.

is equal to the voltage divider ratio:

$$T(s) = \frac{R_2}{R_1 + R_2}. \tag{12.77}$$

It is easy to show that the transfer function for the circuit is

$$H(s) = \frac{A(s)}{1 + A(s)T(s)}, \tag{12.78}$$

where $A(s)$ is the gain of the amplifier without feedback (the open-loop gain is $A(s)T(s)$). The closed-loop gain at frequencies for which $A(s)T(s) \gg 1$ is simply $1/T(s)$: the reciprocal of the feedback factor. This is one of the chief advantages of negative feedback: one can force desired properties of the circuit (the gain in this case) to be dependent upon *passive* elements which are much more stable and less prone to nonlinearities (which cause distortion) than active ones.

Bode plots for the bare gain of an operational amplifier without feedback (left) and the open-loop gain with two different levels of feedback (right) appear in Fig. 12.13. The bare operational amplifier is modeled as having a DC gain of $10^6$ and having

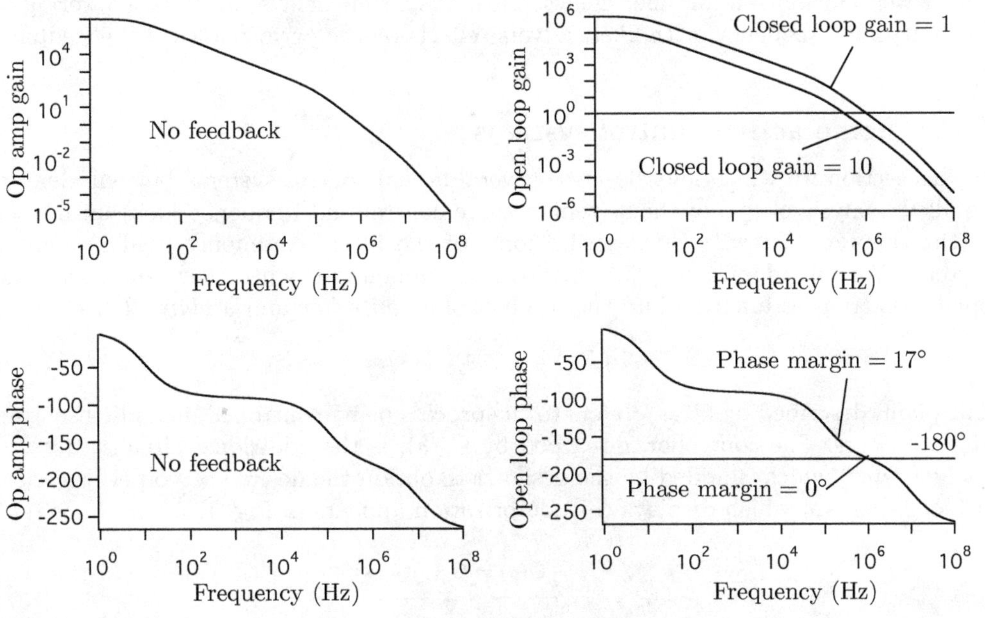


**Fig. 12.13** Bode plots for bare operational amplifier (left) and operational amplifier with a gain of 1 and gain of 10 (right). Note that the phase plot is the same for the two gains.

poles at 10 Hz, 100 kHz and 10 MHz. The open-loop gain with a feedback factor of 1 ($R_2 \gg R_1$) is the same as the gain ($A(s)$) of the bare amplifier, and with a feedback factor of 0.1 (closed-loop gain=10) the open-loop gain is $0.1A(s)$. When one considers the phase margin at both unity gain frequencies, the reason for the instability should be clear: the phase margin at a closed-loop gain of 1 is 0° and at a closed-loop gain of 10 is 17°. One can use similar arguments to explain the instability of an amplifier with a capacitive load (it adds another pole due to the output impedance of the amplifier and the load capacitance).

We conclude this section with a discussion of the adverse effect on stability of *time delays*. From the time-shifting property of Laplace transforms, a time shift of $\tau$ changes the transform by a factor of $e^{-s\tau}$:

$$f(t-\tau) \longrightarrow F(s)e^{-s\tau}. \tag{12.79}$$

Thus, there is a phase shift at angular frequency $\omega$ of $-\omega\tau$ radians. From this, we arrive at the effect of time delays on stability: *a time delay of $\tau$ seconds reduces the phase margin at the unity gain frequency, $\omega_0$, by $\omega_0\tau$ radians.* As an example, we will consider a loop filter consisting of four operational amplifiers plus 10 feet of coaxial cable. Assume the unity gain frequency of the loop is 1 MHz. A fast operational amplifier will have a time delay of about 10 ns. The delay in typical coaxial cables is about 1.5 ns per foot. The total delay is 55 ns which corresponds to a phase change of 20° at the unity gain frequency. This is quite significant and can cause an otherwise stable system to oscillate. The only cure is to use shorter cables and circuit topologies with fewer and shorter amplifier delays. Ultimately, time delays can force a lowering of the unity gain frequency, with likely adverse effects on the performance of the feedback system.

## 12.8 Some actual control systems

In this section we will briefly describe several actual control systems but will defer a detailed analysis of two of them (temperature control and frequency lock) until later in the chapter. We will also describe some of the filters commonly used in control systems. We will adopt a convention which is common in the literature and factor the open-loop transfer function into the product of a *controller* and a *plant*. Thus,

$$G(s) = G_c(s)G_p(s). \tag{12.80}$$

The plant, described by $G_p(s)$, refers to the process being controlled, including sensors and actuators. The controller, described by $G_c(s)$, is the additional circuitry (essentially a *filter*) which is added by the designer to obtain the desired servo performance. A block diagram which displays this factorization appears in Fig. 12.14.

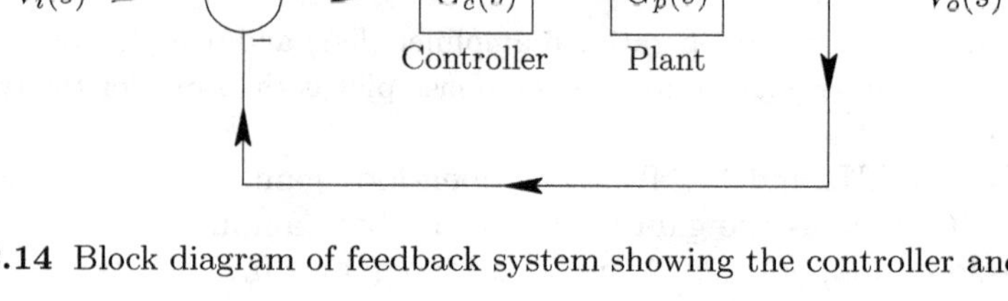


**Fig. 12.14** Block diagram of feedback system showing the controller and plant.

**Phase-locked loop:** The block diagram for a phase-locked loop appears in Fig. 12.15. The variable being controlled in a phase-locked loop is the *phase* of the output of a

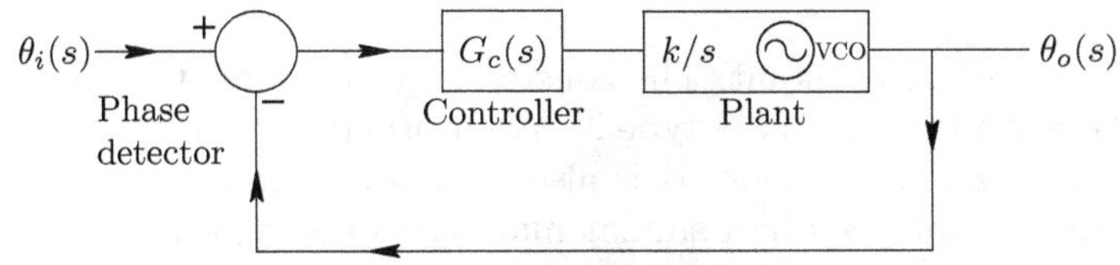


**Fig. 12.15** Block diagram of phase-locked loop.

voltage-controlled oscillator (VCO). Since the *frequency* of the VCO is controlled by its input, the plant contains an *implicit* integrator (phase is the time integral of frequency):

$$G_p(s) = \frac{k}{s}. \tag{12.81}$$

The phase of the VCO output is compared to that of an input *reference* signal in a *phase detector*, whose output is proportional to the phase difference between the two inputs.

Phase-locked loops are described by *type* and by *order*. The *type* is the number of poles at the *origin* of $G(s)$ (the open-loop gain) and essentially characterizes the loop's tracking behavior. The minimum type is 1, since there is at least one integrator (in the plant). Using the results of our discussion of the steady-state errors, we list the errors by loop type in Table 12.2.

**Table 12.2** Steady-state errors for various types of phase-locked loops.

| | Type 1 | Type 2 | Type 3 |
|---|---|---|---|
| Step position | zero | zero | zero |
| Step velocity | constant | zero | zero |
| Step acceleration | unbounded | constant | zero |

The *order* of a loop is the degree of the numerator of the return difference $(1 + G(s))$. It is equal to the number of poles in the closed-loop transfer function. A common configuration has a *controller* whose gain is

$$G_c(s) = \frac{K}{s+p}. \tag{12.82}$$

When combined with the plant gain, this is just the open-loop gain of the second-order loop which we recently studied. Since it is only type 1, it might not be suitable for changing signals. Another common design uses a *lead–lag* controller, whose gain is

$$G_c(s) = K\frac{s+a}{s}, \tag{12.83}$$

where $a$ is a positive constant. The constant $a$ is all-important: it provides a zero in the left-hand plane. This is a type 2 second-order loop which has better tracking behavior than the type 1 loop. It is also unconditionally stable and its transient behavior can be analyzed in a similar manner to the type 1 second-order loop.

**Temperature control:** The block diagram for a temperature controller appears in Fig. 12.16. The plant consists of a sensor and heat generator. The sensor is usually

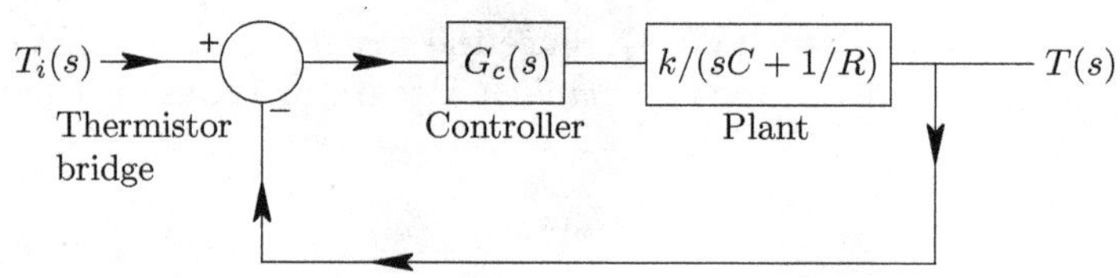


**Fig. 12.16** Block diagram of temperature controller.

a *thermistor bridge*, which will be described in detail later. It basically compares a temperature-sensitive resistor (thermistor) to a temperature-insensitive resistor and generates an output voltage proportional to the difference in resistance. The heat generator (often a simple resistive heater) accepts a power, $P$, and changes the temperature, $T$, according to the differential equation

$$P = C\frac{dT}{dt} + \frac{T - T_0}{R}, \tag{12.84}$$

where $C$ is the heat capacity of the system being controlled, $R$ is its thermal resistance to the environment and $T_0$ is the ambient temperature. If $T_0$ is constant, we can consider the controlled variable to be $\Delta T = T - T_0$. Taking the Laplace transform of this equation yields

$$\Delta T(s) = \frac{P(s)}{sC + 1/R}. \tag{12.85}$$

If the heating element is a *thermoelectric cooler* (which can also heat), the power, $P(s)$, would be replaced with a current, since the output of this type of heater is proportional to the current, not the power. This behavior has a number of advantages, which will be discussed later. Thus, the plant is described by

$$G_p = \frac{k}{sC + 1/R}, \tag{12.86}$$

which is strictly valid only for a heater whose heat output is proportional to the drive current or voltage (e.g., a thermoelectric heater).

**Laser frequency stabilization:** A block diagram of a laser stabilization system appears in Fig. 12.17. The plant consists of the laser electronic frequency control

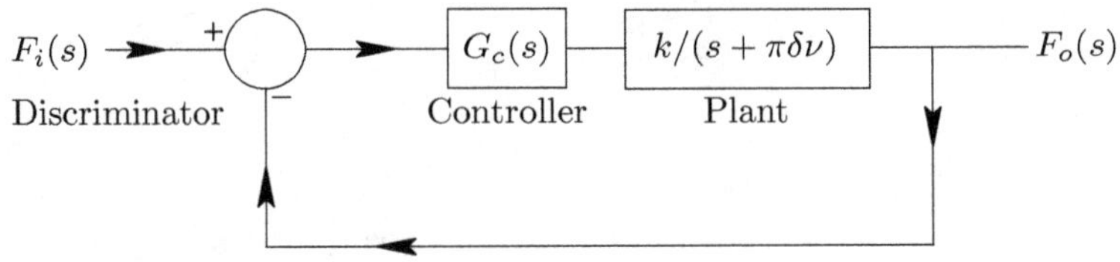


**Fig. 12.17** Block diagram of laser stabilization system.

and the discriminator, which was discussed in great detail in Chapter 4. It was shown in that chapter that the transfer function for an optical cavity of width $\delta\nu$ employing the Pound–Drever–Hall method for generating a discriminant is

$$G(\omega) = \frac{1}{1+i\omega\tau}, \qquad \text{where } \tau = \frac{2}{\delta\omega} = \frac{1}{\pi\delta\nu}. \tag{12.87}$$

Extending this to the whole $s$-plane,

$$G(s) \propto \frac{1}{s+\pi\delta\nu}. \tag{12.88}$$

There are a number of possible response functions for the electronic control of the laser frequency. For example, directly controlling the laser diode current has essentially unlimited bandwidth (from our point of view), but controlling the frequency in an external cavity diode laser using a PZT has a very limited bandwidth.

We will conclude this section with a description of some of the actual circuits used in loop filters, including the increasingly popular *proportional–integral–differential* (PID) controller.

**Integrator:** Perhaps the most important loop-filter function is integration. An operational amplifier integrator together with its Bode plots is shown in Fig. 12.18. The transfer function, $G(s)$, of this circuit is

$$G(s) = k\frac{1}{1+s\tau} \qquad \text{where } k = \frac{R_2}{R_1},\ \tau = R_2C. \tag{12.89}$$

An ideal integrator would have $R_2 = \infty$; we have included a finite value to account for capacitor leakage and a finite (but very large) impedance across the operational amplifier. For an ideal integrator:

$$G(s) = \frac{k}{s} \qquad \text{where } k = \frac{1}{R_1C}. \tag{12.90}$$

**Differentiator:** The inverse function to integration is differentiation and its schematic and Bode plots appear in Fig. 12.19. The transfer function is

$$G(s) = k(1+s\tau) \qquad \text{where } k = \frac{R_2}{R_1},\ \tau = R_1C. \tag{12.91}$$

An integrator circuit is the basic method for inserting *one pole* at $s = -1/\tau$ into the filter and a differentiator inserts *one zero* at $s = -1/\tau$.

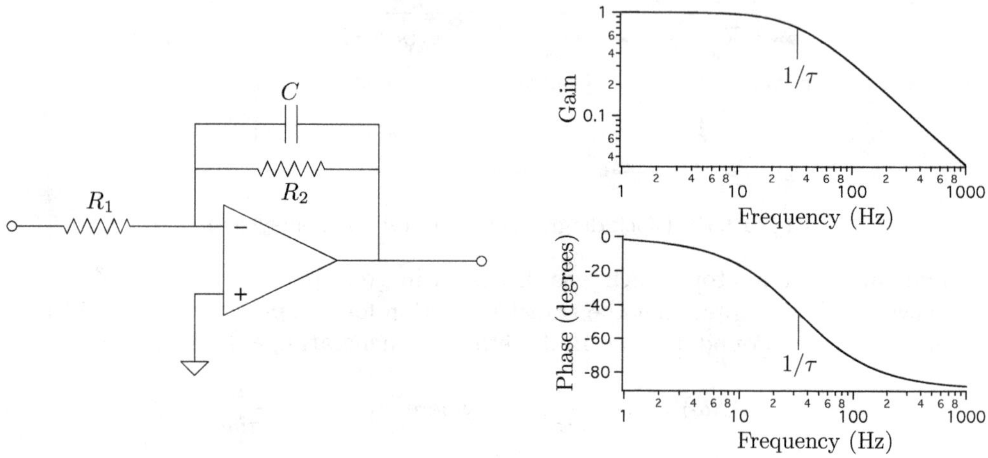


**Fig. 12.18** Integrator and its Bode plots.

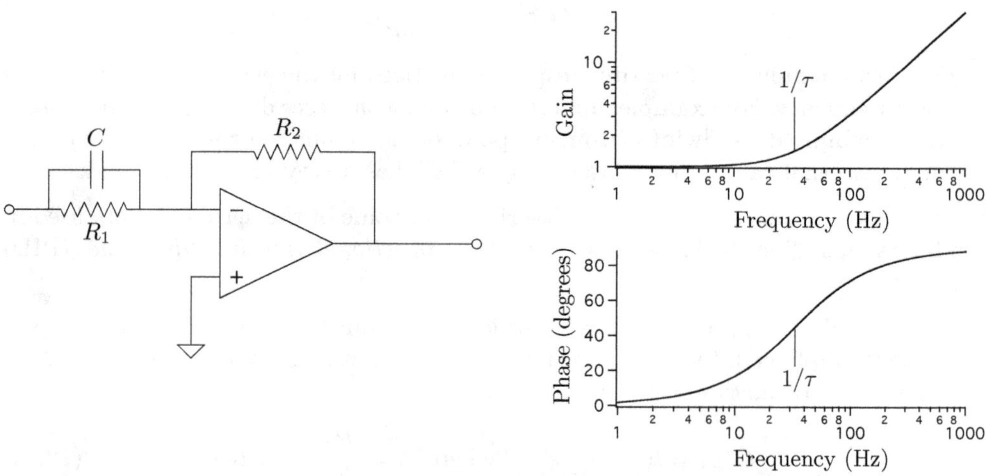


**Fig. 12.19** Differentiator and its Bode plots.

**Lead–lag:** A *lead–lag* circuit combines an integrator and a differentiator by inserting both a pole and a zero into the filter. Its transfer function is

$$G(s) = k\frac{1 + s\tau_1}{1 + s\tau_2}. \tag{12.92}$$

There are three circuit realizations of a lead–lag circuit, shown in Fig. 12.20 together with their Bode plots. As with the integrator, an *ideal* lead–lag filter would not have the 1 in the denominator; its transfer function would be

$$G(s) \propto \frac{1 + s\tau_1}{s}. \tag{12.93}$$

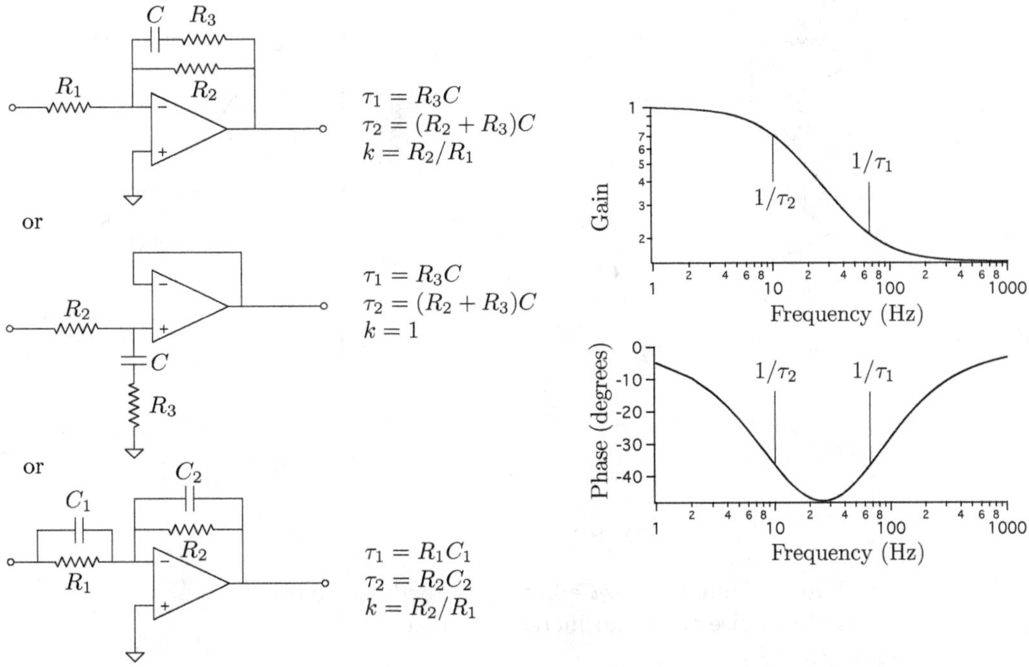


**Fig. 12.20** Lead–lag circuits and their Bode plots.

**Proportional–integral–differential:** The *proportional–integral–differential* controller has a single pole at the origin and two zeros. Its circuit and Bode plots are shown in Fig. 12.21. The transfer function is

$$G(s) = K_p + \frac{K_i}{s} + K_d s = \frac{K_d s^2 + K_p s + K_i}{s}. \tag{12.94}$$

The constants are

$$K_p = \frac{R_{p2} R_f}{R_{p1} R_p} \tag{12.95}$$

$$K_i = \frac{R_f}{R_{i1} R_i C_i} \tag{12.96}$$

$$K_d = \frac{R_{d2} C_d R_f}{R_d}. \tag{12.97}$$

The pole at $1/(R_{d1} C_d)$ in the differentiator is assumed to be well above the highest zero and is ignored. This controller circuit is occasionally used without the differentiator stage, in which case it is called a *proportional–integral* (PI) controller and is merely another realization of a lead–lag circuit. This circuit allows the three response functions to be controlled *independently*, in some cases with three "knobs" on the front panel of the controller box. The behaviors of the three components are:

- **Proportional ($K_p$):**

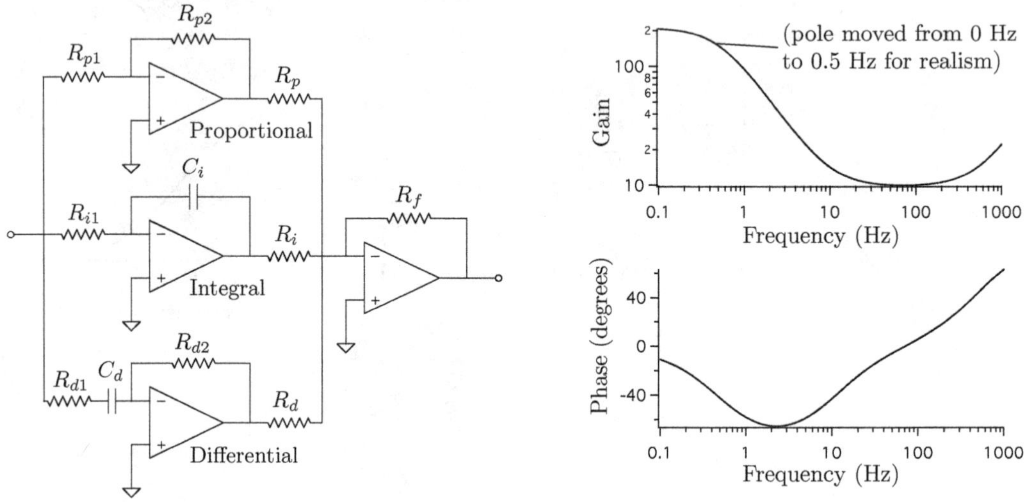


**Fig. 12.21** PID circuit and its Bode plots.

  * Reduces (but *does not* eliminate) steady-state error
  * Reduces rise time and increases overshoot
- **Integral ($K_i$):**
  * Eliminates steady-state error and reduces rise time
  * Worsens transient response (overshoot, settling time)
- **Differential ($K_d$):**
  * Improves transient response (overshoot, settling time)
  * Little effect on steady-state error or rise time

In summary, one would use the proportional and integral components to reduce the steady-state error as much as possible and correct the resulting poor transient response with the differential component.

## 12.9 Temperature stabilization

We reproduce the block diagram for a temperature control system below in Fig. 12.22. Substituting the product, $G_c(s)G_p(s)$, into the expression for the closed-loop gain, we

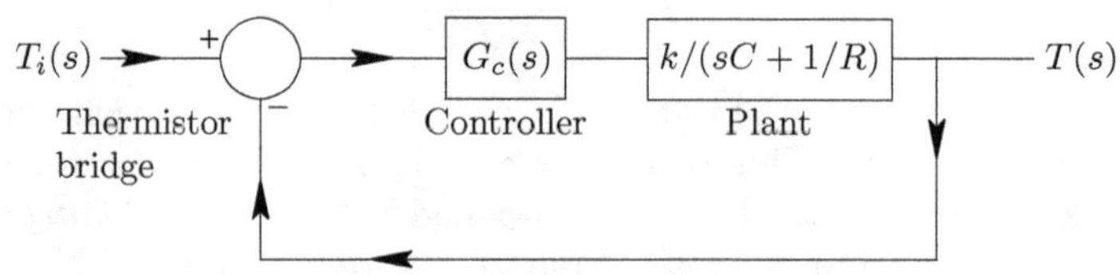


**Fig. 12.22** Block diagram of temperature controller.

obtain an expression for the final system temperature as a function of the reference, $T_i$:

$$T = \frac{G_c(s)\frac{k}{sC+1/R}}{1+G_c(s)\frac{k}{sC+1/R}} T_i = \frac{kG_c(s)}{sC+1/R+kG_c(s)} T_i, \tag{12.98}$$

where the expression for the controller function, $G_c(s)$, is yet to be determined.

An examination of the steady-state behavior of the system to a *step increase in ambient temperature* will help assess the minimum requirements for $G_c(s)$. We determine the response with the aid of Fig. 12.23. The figure requires some explanation.

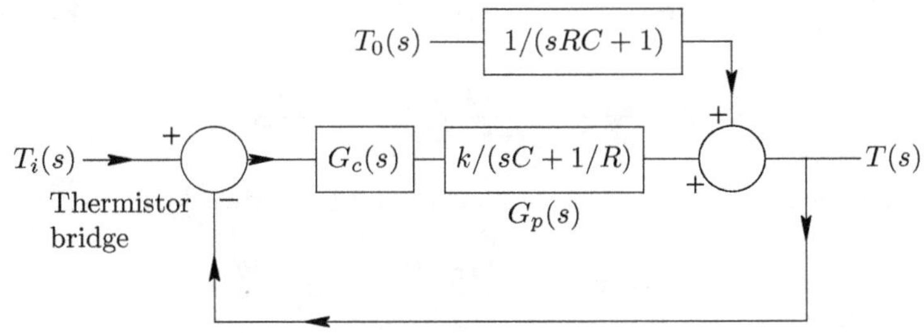


**Fig. 12.23** Block diagram of temperature controller with step change in ambient temperature.

Formerly, we assumed a *constant* $T_0$ and derived the plant response accordingly. If $T_0$ can vary with time, the differential equation and Laplace transform are

$$P = C\frac{d}{dt}T + \frac{T - T_0}{R} \tag{12.99}$$

$$T(s) = \frac{P(s)}{sC+1/R} + \frac{T_0(s)/R}{sC+1/R}. \tag{12.100}$$

Since the input to the plant is a *voltage* (we assume a thermoelectric cooler and driver), the *heating power* is related to the voltage via a constant of proportionality, $k$, and the plant transfer function is

$$G_p(s) = \frac{k}{sC+1/R}. \tag{12.101}$$

We include the environmental influence by placing a summing junction *after* the plant, as shown in the figure, first multiplying $T_0(s)$ by the low-pass function, $1/(sRC+1)$. Then, the equation for the controlled temperature, $T$, is

$$T(s) = \frac{G_c(s)G_p(s)}{1+G_c(s)G_p(s)} T_i(s) + \frac{G_p(s)/(Rk)}{1+G_c(s)G_p(s)} T_0(s). \tag{12.102}$$

The steady-state response, $\Delta T$, to a step in ambient temperature, $\Delta T_0$, is (with the aid of the final value theorem)

$$\Delta T = \frac{1}{1+RkG_c(0)} \Delta T_0, \tag{12.103}$$

where we assume that $T_i(s)$ is constant. Thus, we see that we must have at least one integrator to eliminate errors due to changes in the ambient temperature.

The usual practice is to use a PID controller or lead–lag controller (sometimes called a "PI controller") for $G_c(s)$. The analysis of the loop behavior using a lead–lag controller begins with the open-loop transfer function

$$G(s) = G_c(s)G_p(S) = \left(\frac{s\tau_c + 1}{s}\right)\left(\frac{Rk}{s\tau_p + 1}\right), \tag{12.104}$$

where $\tau_p = RC$ is the thermal time constant of the system and $\tau_c$ is the time constant of the lead–lag filter. From this, we can write down the closed-loop transfer function, $H(s)$

$$H(s) = \frac{(Rk/\tau_p)(s\tau_c + 1)}{s^2 + [(Rk\tau_c + 1)/\tau_p]s + Rk/\tau_p}. \tag{12.105}$$

This can be written in a form similar to that of the second-order loop already studied,

$$H(s) = \frac{(s\tau_c + 1)\omega_n^2}{s^2 + 2\zeta\omega_n s + \omega_n^2}, \tag{12.106}$$

where

$$\omega_n = \sqrt{\frac{Rk}{\tau_p}} \tag{12.107}$$

$$\zeta = \frac{1 + Rk\tau_c}{2\sqrt{Rk\tau_p}}. \tag{12.108}$$

The design procedure is to use sufficient gain ($k$) to obtain acceptable error and response time ($\propto 1/\omega_n$) and adjust the controller parameter ($\tau_c$) for a damping factor $\approx 1$ (since the *transient response* is determined mainly by the denominator of $H(s)$, the transient behavior of this loop would be very similar to that of the second-order loop discussed earlier). One can measure $\tau_p$ by manually turning on the heater until the system temperature is several degrees above ambient, after which the heater is turned off and the temperature is plotted versus time. The time constant in the exponential decay of the temperature is $\tau_p$.

If a PID controller is used, there are a number of *algorithms* for setting the three knobs. The *Ziegler–Nichols* procedure is as follows:

- First set the I and D gains to zero.
- Increase the P gain until the system just begins to oscillate. This *critical gain* is called $K_c$ and the oscillation period is $T$.
- Set the P gain to $K_p = 0.6K_c$.
- Set the I gain to $K_i = 2K_p/T$.
- Set the D gain to $K_d = K_pT/8$.

At the beginning of this section, we alluded to a problem which occurs when one uses a resistive heating element. The problem is the *nonlinearity* of the system which is due to the fact that the temperature sensor produces a voltage proportional to the temperature error and the heater power is proportional to the voltage input squared.

There are three solutions. One is to linearize the heater power relation by expanding it around the operating point. The heater power is

$$P = \alpha^2 V_{in}^2 / R_h, \tag{12.109}$$

where $R_h$ is the heater resistance, $\alpha$ is the voltage gain of the driver and $V_{in}$ is the input voltage. If the desired temperature is $T_i$, the steady-state power ($dT/dt = 0$) is

$$P = (T_i - T_0)/R, \tag{12.110}$$

where $R$ is the *thermal resistance* to the environment and should not be confused with the heater resistance, $R_h$. For *small changes*, $\Delta T$, around the *output temperature*, $T$,

$$\Delta P = \Delta T / R. \tag{12.111}$$

From the heater power relation, this $\Delta P$ is

$$\Delta P = 2P\Delta V_{in} / V_{in}. \tag{12.112}$$

We thus obtain a linear relation

$$\Delta T = \left( \frac{2PR}{V_{in}} \right) \Delta V_{in}. \tag{12.113}$$

The problem with this is that the gain depends upon the operating point. It is much more desirable to use a *linear heater.* One can either use a thermoelectric cooler or a *pulse-width modulation* approach to the heater. In the latter system, a series of pulses whose width is proportional to the input voltage is delivered to the resistive heater; the heater averages the pulses and therefore the heating is proportional to the input voltage.

Finally, we will say a few words about the temperature sensor. It is standard practice to use a thermistor bridge, shown schematically in Fig. 12.24. The bridge is

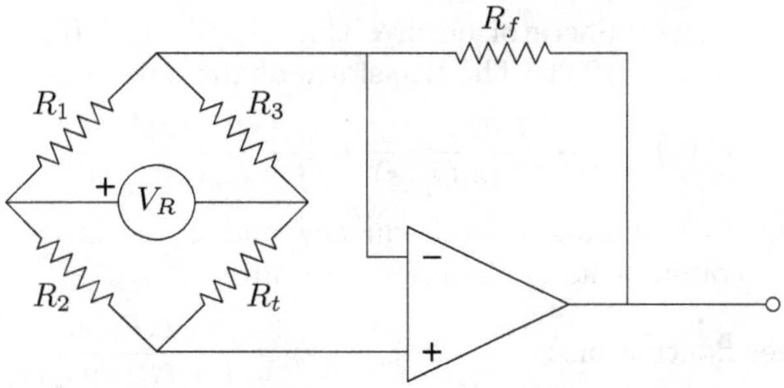


**Fig. 12.24** Thermistor bridge used as a temperature sensor.

*balanced* and the output is zero when opposite pairs of resistors have the same ratio. Three of the resistors ($R_1, R_2$ and $R_3$) have very small temperature coefficients, and

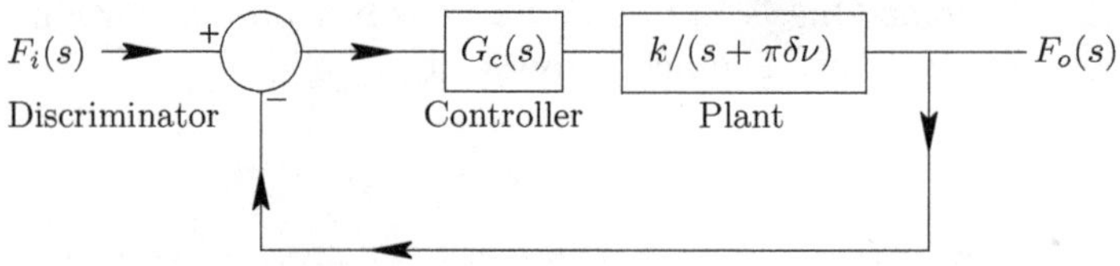


**Fig. 12.25** Block diagram of laser stabilization system.

$R_t$ (a *thermistor*) has a large temperature coefficient. The voltage, $V_R$, is derived from a very stable source. The output voltage is

$$V_{out} = \frac{V_R R_f}{R_2 + R_t}\left(\frac{R_t}{R_3} - \frac{R_2}{R_1}\right). \tag{12.114}$$

If we assume that $R_t/R_3 \approx R_2/R_1$ and consider only *small changes* in $R_t$, the output is approximately

$$\Delta V_{out} \approx \frac{V_R R_f}{R_t + R_2}\left(\frac{R_2}{R_1}\right)\frac{\Delta R_t}{R_t}. \tag{12.115}$$

This can be converted into a temperature-to-voltage function if the temperature-to-resistance characteristic of the thermistor is known. The errors due to the small temperature coefficients of $R_{1-3}$ can be eliminated by mounting these resistors in good thermal contact with the thermistor or by placing them on a heat sink and controlling its temperature using a simple commercial monolithic temperature controller.

## 12.10 Laser frequency stabilization

We finally return to the subject which was touched on at the beginning of the chapter: the frequency stabilization of a laser. We begin by summarizing the earlier results. The block diagram is reproduced (for the third time) in Fig. 12.25. Recalling the definitions of $K_0$ (laser frequency-to-voltage ratio) and $K_d$ (discriminator voltage-to-frequency ratio), the plant function is

$$G_p(s) = K_0 K_d \pi\delta\nu_c \frac{1}{s + \pi\delta\nu_c}. \tag{12.116}$$

The factor of $\pi\delta\nu_c$ in the numerator ensures that $G_p(0) = K_0 K_d$. The Bode plots of the plant are shown in Fig. 12.26. The transform of the output frequency is

$$F_o(s) = \frac{F_1(s)}{1 + G_c(s)G_p(s)} + \frac{G_c(s)G_p(s)F_0(s)}{1 + G_c(s)G_p(s)}, \tag{12.117}$$

where $F_1(s)$ is the free-running laser frequency and $F_0(s)$ is the cavity resonance frequency. The two components of the laser jitter are

$$\text{Laser fluctuations:} \qquad \Delta F_{laser}(s) = \frac{\Delta F_1(s)}{1 + G_c(s)G_p(s)} \tag{12.118}$$

$$\text{Cavity fluctuations:} \qquad \Delta F_{cavity}(s) = \frac{G_c(s)G_p(s)\Delta F_0(s)}{1 + G_c(s)G_p(s)}, \tag{12.119}$$

where $\Delta F_{laser}$ is a change in the laser frequency *due to* the laser and $\Delta F_{cavity}$ is a change in the laser frequency *due to* the cavity.

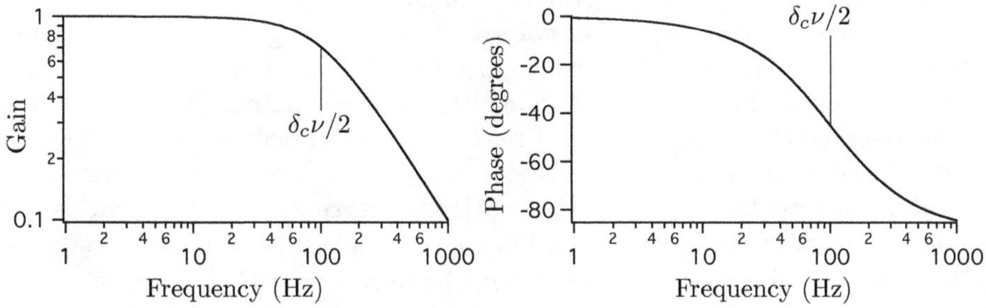


**Fig. 12.26** Bode plots for optical cavity used as a discriminator. The cavity linewidth is $\delta\nu_c$.

The choice of $G_c(s)$ (controller) is strongly dependent on the type of *frequency actuator*, which of course depends upon the type of laser being stabilized. The following is a list of possible actuators:

- **Laser diode current:** The laser diode current has an essentially unlimited bandwidth from the point of view of laser stabilization; the bandwidths are in the GHz range, which is far greater than any likely unity gain frequency. Controlling the current has the disadvantage of introducing amplitude modulation, which can increase the residual jitter of the stabilized laser.
- **Laser diode temperature:** The diode temperature allows a very large change of laser frequency at the expense of extremely low bandwidth. Even a fast temperature controller will still be limited to settling times of a second or so. It is often useful to use a separate loop to control the temperature; this keeps the laser frequency *close to the cavity resonance* and reduces the dynamic range requirements of the fast loop.
- **Piezo-electric positioner for ECDL mirror or grating:** The frequency of an extended-cavity diode laser can be controlled either via the diode current (fast, but introduces AM) or the piezo-electrically actuated grating or mirror. The latter provides a very large sweep range but is only moderately fast. It is limited by acoustical resonances in the piezo crystal, which can be as low as a few kHz and as high as 100 kHz. The resonances introduce significant phase shifts at frequencies within a factor of two of the resonance, and these can destabilize the loop. The resonances can be reduced by using a *notch filter*, allowing unity gain frequencies well above the resonance frequencies. It requires considerable expertise in the art of laser stabilization to achieve success in this endeavor.
- **Piezo-electric element for monolithic lasers:** Monolithic, optically pumped lasers are, almost by definition, difficult to electronically tune. Non-planar ring oscillators (NPROs) are normally tuned with a small piezo-electric element bonded to the YAG crystal – this stresses the crystal when a voltage is applied, changing the geometry and laser frequency. Since the piezo-electric actuator is very small, the resonant frequencies are quite high, in the 100 kHz range. The NPRO laser is intrinsically quiet and the piezo-electric device will often be sufficient to obtain good stabilized performance without consideration of its resonances.

- **Galvonometrically driven plate:** Dye and Ti–sapphire lasers often use a small galvonometrically driven plate of glass at Brewster's angle in the internal beam to change the effective cavity length and therefore the laser frequency. This method of frequency control has a modest bandwidth – several hundred Hz. It can also cause small laser *pointing fluctuations.* Brewster plates are often used in conjunction with a faster frequency control (often a piezo-driven mirror) – the Brewster plate stabilizes against large frequency deviations occurring at low frequencies and the piezo-driven mirror eliminates small-deviation higher-frequency fluctuations.
- **Electro-optic modulators (EOMs):** The technique with the largest bandwidth for controlling the frequency of a dye or Ti–sapphire (or any laser with a large open beam path) is an electro-optic modulator. Although these devices occasionally have piezo-like resonances, they allow small deviation frequency control at frequencies of several MHz. They can require substantial voltages, but this is usually not a problem in most lasers since the high-frequency noise usually has fairly small deviation and requires only a modest voltage applied to the EOM for its correction.

A number of the devices mentioned above require that modest voltages be applied to capacitive loads at a very high *slew rate.* Since $I = CdV/dt$, a significant current is needed to meet the slew-rate requirement. Conventional high-voltage operational amplifiers usually do not have large current drive capacities together with fast response. There are a number of commercial parts, called *power boosters*, which do combine these two useful properties, and these are recommended for driving piezos or EOMs.

We seek a controller which provides shot-noise-limited performance over as wide a frequency range as possible while maintaining stability (a phase margin of 30°). We will begin our analysis by summarizing the results of Mor and Arie (1997) for a lead–lag (PI) filter whose transfer function is

$$G_c(s) = \frac{k_p s + k_i}{s}. \tag{12.120}$$

The laser noise *spectral density*, $S_{laser}$, (in Hz$^2$/Hz) is the sum of three components: white noise, *flicker* (1/f) noise and *random walk* noise, which is proportional to $1/f^2$, where $f$ is the frequency of the noise.

$$S_{laser} = \frac{\Delta\nu}{\pi} + \frac{k_f}{f} + \frac{k_r}{f^2}. \tag{12.121}$$

The laser lineshape due to white noise only is a Lorentzian with width $\Delta\nu$.

Three types of lasers will be considered:

- **Single-mode diode:** This laser is characterized by white noise only with a width of 50 MHz.
- **Distributed-feedback diode:** DFB lasers are also characterized by a white-noise spectrum; we will assume the width is 1 MHz.
- **NPRO:** Non-planar ring oscillators are dominated by flicker and random walk noise; the white-noise can be ignored. We will assume that $k_f = 1.4 \times 10^6$ Hz$^2$ and $k_r = 5.7 \times 10^8$ Hz$^3$.

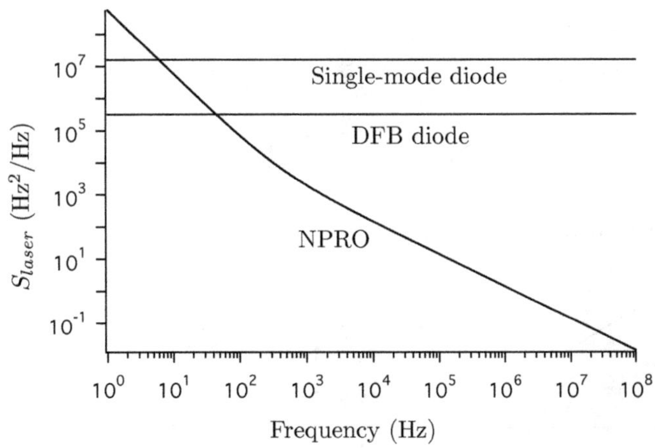


**Fig. 12.27** The frequency noise spectra for a single-mode diode, DFB diode and NPRO.

The noise spectra for the three lasers are plotted in Fig. 12.27.

The total closed-loop laser-noise spectral density ($S$) will be the quadratic sum of the residual intrinsic laser jitter ($S_{laser}$) and the shot noise from the discriminator ($S_{disc}$) *referred to the laser output*, since the two noise sources are *uncorrelated*. From eqn 12.118, $S$ is

$$S = \frac{S_{laser} + |K_0 G_c|^2 S_{disc}}{|1+G|^2}, \tag{12.122}$$

where $G = G_c G_p$, the discriminator noise is related to $\Delta F_0$ by the factor $K_0 G_c$ (proceeding from the discriminator output to the laser output) and we have dropped the $s$-dependence of all of the terms to reduce clutter. In order for the locked laser to be shot noise limited, we require that

$$\text{Shot noise limited:} \quad |K_0 G_c(s)|^2 > \frac{S_{laser}}{S_{disc}}. \tag{12.123}$$

The open-loop gain is

$$G(s) = \left(\frac{k_p s + k_i}{s}\right)\left(\frac{K_0 K_d \pi \delta\nu_c}{s + \pi\delta\nu_c}\right) = \frac{K(s+s_0)}{s(s+s_1)}, \tag{12.124}$$

where we have absorbed the multiplicative constants into $K = K_0 K_d k_p \pi \delta\nu_c$ and have a zero $s_0 = -k_i/k_p$ and a pole $s_1 = -\pi\delta\nu_c$. We note that this is formally very similar to the open-loop gain of a temperature-stabilization loop with a PI (or lead–lag) filter. The only two parameters available for optimization are the overall gain ($K$) and the zero ($-s_0$) in the lead–lag filter.

In order to obtain the largest gain at lower frequencies, it is desirable to have the largest possible unity gain frequency for the open-loop gain. For the single-mode diode and DFB diode, which have noise components into the MHz range, we can demonstrate that time delays set a limit on the unity gain frequency. The NPRO has less noise at

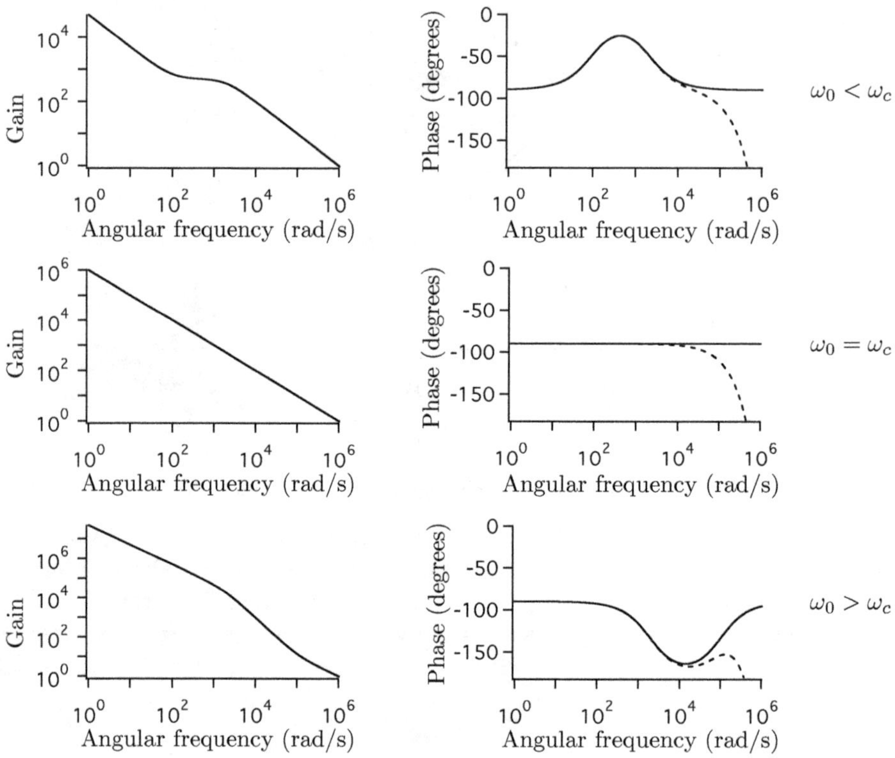


**Fig. 12.28** Bode plots for three locations of the zero ($\omega_0$) relative to the pole due to the cavity ($\omega_c$). The dashed curve includes 200 $\mu$s of time delay, which is greatly exaggerated to illustrate its effect on the phase. The pole frequency is $2 \times 10^3$ rad/s and the zero frequencies are 100, 2000 and 100,000 rad/s.

higher frequencies and its stabilization system unity gain frequency is likely to be limited by transducer resonances rather than time delays.

We first substitute $s = i\omega$, $s_0 = \omega_0$ and $s_1 = \omega_c$ ($\omega_0$ is the zero location and $\omega_c$ is the pole due to the cavity) into eqn 12.124:

$$G(\omega) = \frac{K(\omega - i\omega_0)}{i\omega(\omega - i\omega_c)}. \tag{12.125}$$

Figure 12.28 contains Bode plots for three possible locations of $\omega_0$ with respect to $\omega_c$: $\omega_0 < \omega_c$, $\omega_0 = \omega_c$ and $\omega_0 > \omega_c$. As can be seen from the figure, increasing the frequency of the zero increases the gain at lower frequencies. The risk is the increased possibility of instability, particularly if time delays are present. We will estimate the frequency range for which one can expect shot-noise-limited performance assuming that there is time delay, $\tau$, and that the zero frequency is somewhat above the cavity pole and about a factor of 5 below the unity gain frequency.

The magnitude and phase of the open-loop gain are

$$|G(\omega)| = \frac{K}{\omega}\sqrt{\frac{\omega^2+\omega_0^2}{\omega^2+\omega_c^2}} \tag{12.126}$$

$$\angle G(\omega) = \tan^{-1}\left(\frac{\omega_c\omega_0+\omega^2}{\omega(\omega_0-\omega_c)}\right). \tag{12.127}$$

If we include the effects of time delay ($\tau$) and insist on a phase margin of 30°, the following equation must be solved for $\omega$:

$$\tan^{-1}\left(\frac{\omega_c\omega_0+\omega^2}{\omega(\omega_0-\omega_c)}\right) - \omega\tau = -\tfrac{5}{6}\pi. \tag{12.128}$$

The procedure used by Mor and Arie (1997) was to first neglect time delays and determine the frequency which produced a phase margin of 45°; this is equivalent to setting the argument of the $\tan^{-1}$ equal to 1. The resulting quadratic equation has two roots whose location can be seen in the lower right phase plot in Fig. 12.28: one is a little above the second pole and the other is a little above the zero (they are at the intersection of the line $\phi = -135°$ with the curve, and are on either side of the bowl-shaped region of rapid phase change). The authors of this work found the value of $\omega_0$ that provided the desired phase margin (the larger root), and placed the unity gain frequency approximately at $\omega_0$ for the highest possible gain at lower frequencies. While this procedure would probably work, it suffers from the presence of a large and rapidly varying phase very near the unity gain frequency, which can cause instability.

We will use a simpler and more conservative approach: we will place $\omega_0$ at a factor of 5 below the unity gain frequency, $\omega_{UG}$. Then the phase is varying very slowly at the unity gain frequency and is very nearly $-90°$ (from the asymptotic behavior of two poles and one zero). The factor of 5 is a bit arbitrary; a smaller factor could be used, though it should be larger than 2 to avoid rapid phase changes near the unity gain frequency. The unity gain frequency will be determined from time-delay considerations and $\omega_0$ will be set using $\omega_0 = \omega_{UG}/5$. Finally, the parameter $K$ will be determined from the requirement that $|G(\omega_{UG})| = 1$.

At $\omega_{UG}$, we have $-90°$ of phase shift; we are allowed an additional $-60°$ (due to time delays) to satisfy the phase margin requirement. The unity gain frequency is therefore

$$\omega_{UG} = \frac{\pi}{3\tau}. \tag{12.129}$$

From our assumption that $\omega_0$ is five times smaller than the unity gain frequency,

$$\omega_0 = \frac{\pi}{15\tau}. \tag{12.130}$$

Finally, $K$ is chosen from $|G(\omega_{UG})| = 1$:

$$K = \omega_{UG}\sqrt{\frac{\omega_{UG}^2+\omega_c^2}{\omega_{UG}^2+\omega_0^2}}, \tag{12.131}$$

where $\omega_0$ and $\omega_c$ are determined from the previous equations.

We will estimate the behavior of a loop used to stabilize a DFB diode laser with the presence of 55 ns of delay. The unity gain frequency is 3 MHz ($19 \times 10^6$ rad/s). The zero ($\omega_0$) is placed at 600 kHz ($3.8 \times 10^6$ rad/s). We will assume a cavity linewidth of 2 kHz (typical of the highest-finesse cavities); thus $\omega_c = 6.2 \times 10^3$ rad/s, which is much lower than $\omega_0$ as required. Since $\omega_c \ll \omega_{UG}$ and $\omega_0 = \omega_{UG}/5$, one can estimate $K$ as

$$K \approx \omega_{UG} \qquad (\omega_c \ll \omega_{UG},\ \omega_0 = \omega_{UG}/5). \tag{12.132}$$

From eqn 12.123, shot-noise-limited performance will occur at and below the frequency which satisfies

$$\left[\frac{\omega_{UG}}{K_d\omega_c}\right]^2 \frac{\omega^2 + \omega_0^2}{\omega^2} \geq \frac{S_{laser}}{S_{disc}} \Longrightarrow \frac{\omega^2 + \omega_0^2}{\omega^2} \geq \left[\frac{\omega_c K_d}{\omega_{UG}}\right]^2 \frac{S_{laser}}{S_{disc}}, \tag{12.133}$$

where we divided $G(\omega)$ by $|G_p(\omega)/K_0|$ to obtain $|K_0 G_c(\omega)|$ on the left side (note that $\omega_c = \pi\delta\nu_c$). Anticipating that the right-hand side will be much greater than 1, we can simplify this to obtain

$$\omega_{snlp} = \omega_0 \left[\frac{\omega_{UG}}{\omega_c K_d}\right] \sqrt{\frac{S_{disc}}{S_{laser}}} \tag{12.134}$$

where $\omega_{snlp}$ is a frequency below which the loop is shot-noise-limited. For the DFB laser, $S_{laser} = 3 \times 10^5$ Hz$^2$/Hz. The shot-noise current per $\sqrt{\text{Hz}}$ at the output of the discriminant is

$$i_{noise} = \sqrt{2ei}. \tag{12.135}$$

At resonance, the carrier will be totally absorbed and the reflected power from the two sidebands will be $2J_1^2(\beta)P_i$ where $P_i$ is the incident power. The photocurrent will have an additional factor of $\sqrt{2}$, since the mixer will respond to signals and noise both above and below the reference frequency. The result is

$$\sqrt{S_{disc}} = i_{noise} = \sqrt{8eJ_1^2(\beta)\mathcal{R}P_i} \qquad (\text{in A}/\sqrt{\text{Hz}}), \tag{12.136}$$

where $\mathcal{R}$ is the *responsivity* (in A/watt) of the photodiode. In units of A/Hz, the expression for $K_d$ is

$$K_d = \frac{2.71\mathcal{R}P_i}{\delta\omega} = \frac{1.35\mathcal{R}P_i}{\omega_c}, \tag{12.137}$$

where $\delta\omega = 2\omega_c$ is the cavity width. If we assume that $P_i = 0.1$ mW and $\mathcal{R} \approx 1$, then $i_{noise} = 5.3 \times 10^{-12}$ A/$\sqrt{\text{Hz}}$ and $K_d = 2.2 \times 10^{-8}$ A/Hz. Putting all of these numbers into eqn 12.134, we have $\omega_{snlp} = 5.1 \times 10^3$ rad/sec (809 Hz). The fact that $\omega_{snlp} \ll \omega_0$ justifies our recent simplification. Note that the shot-noise-limited frequency range is proportional (up to a point) to $\omega_0$ and one would therefore like to make $\omega_0$ as large as possible without destabilizing the loop. The bare diode has 50× as much noise as the DFB laser; its stabilization procedure would be similar to the above and the shot-noise-limited performance would be somewhat worse than that of the DFB laser.

We mentioned above that the stabilization system for an NPRO would be subject to different design constraints from the DFB laser discussed above. The main difference is the actuator: the piezo-electric actuator has mechanical resonances which would

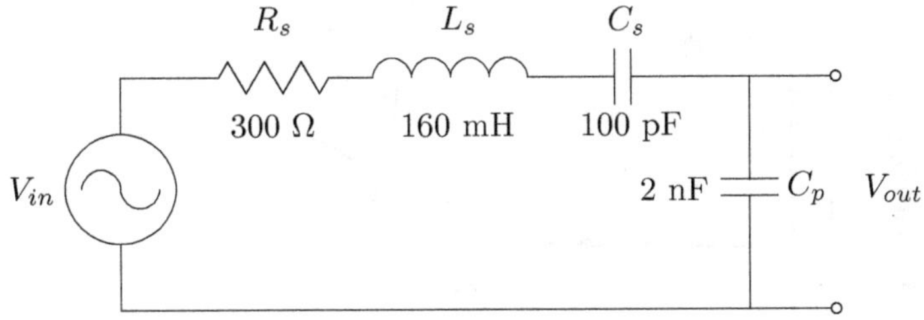


**Fig. 12.29** Equivalent circuit of a piezo-electric transducer. Typical values are indicated for a resonant frequency of 40 kHz.

produce destabilizing phase shifts and would be a much greater constraint on the unity gain frequency than time delays. This would be the case with any laser system which depends upon a piezo-electric actuator for fast frequency control.

A piezo-electric transducer is a *resonant, two-pole* electromechanical system which can be modeled with the equivalent circuit shown in Fig. 12.29. The mechanical mass is represented by the inductor, the stiffness by the series capacitor and the damping by a resistor (a quartz crystal for frequency control has the same equivalent circuit). The mechanical response to a drive voltage $V_{in}$ is proportional to the voltage, $V_{out}$, across $C_p$. The response can be written in the familiar form used in the analysis of a second-order loop:

$$\frac{V_{out}}{V_{in}} = \left(\frac{C_s}{C_p}\right) \frac{\omega_n^2}{s^2 + 2\zeta\omega_n s + \omega_n^2}, \tag{12.138}$$

where

$$\omega_n = \frac{1}{\sqrt{L_s C_s}} \tag{12.139}$$

$$\zeta = \frac{R}{2\omega_n L} \tag{12.140}$$

$$Q = \frac{1}{2\zeta}. \tag{12.141}$$

The bode plots for a variety of damping factors appear in Fig. 12.30. Although the $Q$ of the transducer shown in the diagram is $> 100$, many piezo-electric devices have lower $Q$ values; tubular actuators can have a $Q$ of about 10 with a response corresponding approximately to the $\zeta = 0.1$ ($Q = 5$) curve. We note from the plot that the phase change is not that large until one is fairly close to the resonance. However, instability can still unexpectedly occur due to inadequate *gain margin.* The gain margin is the amount by which the gain is below unity (usually expressed in decibels) when the phase shift is 180°. The highly peaked response of the high-$Q$ transducers can destabilize the loop near the resonance by increasing the gain until it is equal to or greater than unity.

The usual conservative approach with piezo-electric actuators is to roll off the servo gain sufficiently rapidly so that the unity gain frequency is well below the mechanical resonance of the transducer, which can be as high as 100 kHz but is often much lower.

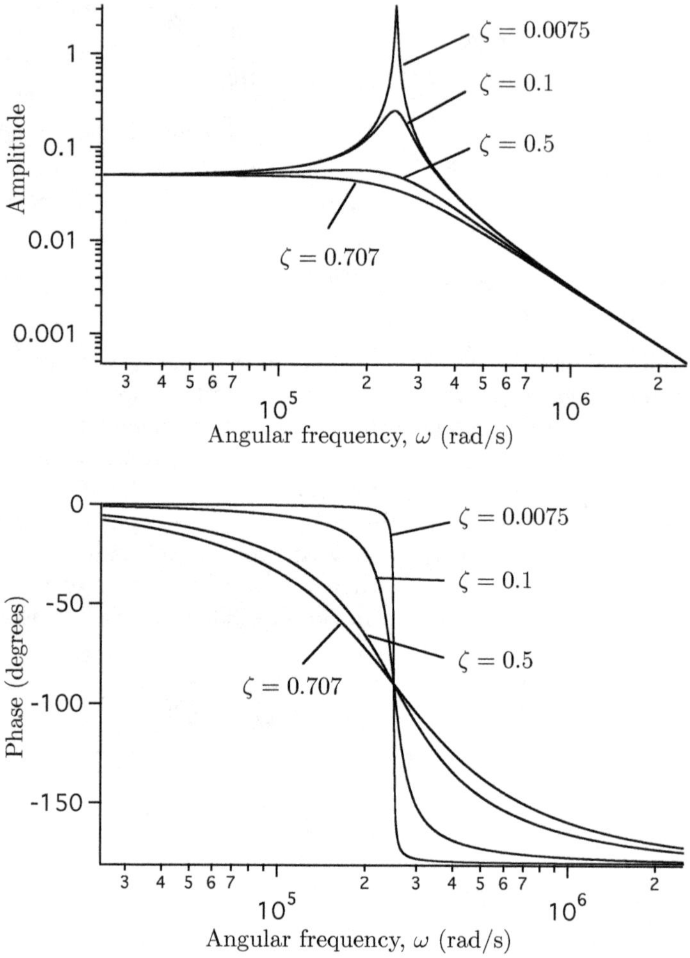


**Fig. 12.30** Bode plots for second-order system. The highly peaked piezo response is shown for $\zeta = 0.0075$.

Unfortunately, this sets a severe limit on the performance of the stabilization system. Some very expert workers in the field have sought to find a strategy for transcending this limitation and have reported on their successful efforts in the *Handbook of Optics, Volume IV* (Hall et al. (2000)). We will summarize their findings.

The first attempt to surmount the limit due to the piezo-electric resonance was to use a PID circuit: the derivative adds a *phase lead* which can reduce the adverse effect of the phase shift due to the actuator. While this allowed a unity gain frequency somewhat above the piezo-electric resonance (by a factor of 2), the lead circuit reduces the low-frequency gain and little is accomplished by using it.

The next step was to use a PI controller and a *notch filter* to suppress the resonance. The notch approximates the inverse of the resonance peak. This eliminated the

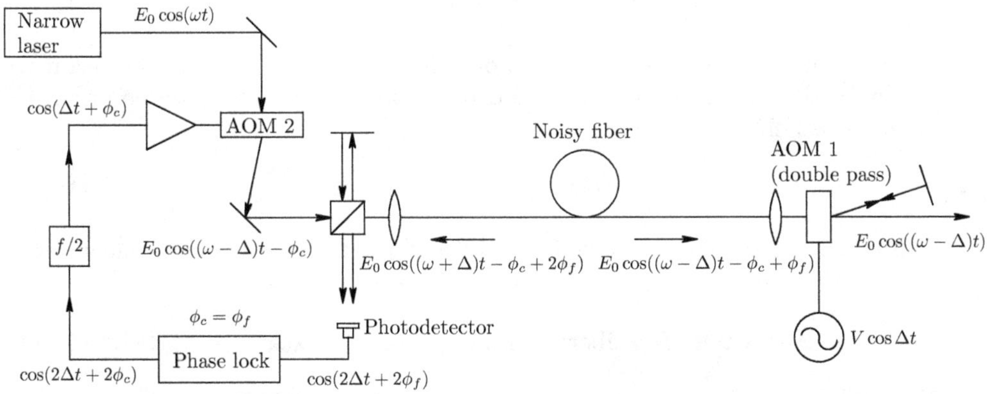


**Fig. 12.31** Setup for canceling phase noise introduced by a fiber.

undesired effects of the resonance and provided increased low-frequency gain compared to the PID circuit. The unity gain was slightly lower with the notch than with the PID (and no notch) but still above the resonant frequency. The best results were obtained using a *second* PI stage together with the notch. This provided good transient response and a gain improvement of more than 20 dB below 1 kHz (the resonance was at 25 kHz). This would be the best strategy for stabilizing an NPRO or other intrinsically stable monolithic lasers whose frequency is controlled by a piezo-electric actuator.

## 12.11 Optical-fiber phase noise and its cancellation

Single-mode optical fibers provide a very convenient means of transporting laser beams from one place to another and can easily be made to preserve the beam properties and polarization. However, ambient acoustical noise can impose several radians of phase noise onto a beam transported by a fiber, broadening the laser by several kHz. For many applications, this is not a serious problem, but it is certainly undesirable for a laser whose linewidth has been reduced to 1 Hz or less. A number of methods for canceling this phase noise have been devised: they are all based upon the useful fact that the phase noise in a fiber is *independent* of the propagation direction. Thus, if one can retransmit the beam in a direction opposite to the original direction, the total phase noise will be exactly doubled. The cancellation will only work if the *coherence time* of the fiber noise ($\geq 0.1$ ms) is greater than the round-trip travel time in the fiber: this limits the application of this technique to fiber lengths of about 10 km or less.

We will describe in detail one technique for fiber phase noise cancellation employed by workers at the Joint Institute for Laboratory Astrophysics (Ma (1994)). The scheme is shown schematically in Fig. 12.31. Although it is a conceptually simple idea, it appears rather complicated in the figure. In the following, we will trace the path of the beam through each step.

1. The beam emitted by the laser is

$$\text{Initial laser beam:} \quad E_0 \cos \omega t \tag{12.142}$$

2. After passing though the first acousto-optic modulator, AOM 2, the beam will acquire a *correction phase*, $-\phi_c$ and a frequency shift by $-\Delta$, whose origin will be described shortly.

$$\text{After AOM 2:} \quad E_0 \cos((\omega - \Delta)t - \phi_c) \tag{12.143}$$

3. The beam crosses the beam-splitter and enters the fiber, where it acquires an additional phase, $\phi_f$, due to the phase noise from the fiber.

$$\text{Upstream, after fiber:} \quad E_0 \cos((\omega - \Delta)t - \phi_c + \phi_f) \tag{12.144}$$

4. The effect of the second acousto-optic modulator, AOM 1, is to send a small portion of the beam back through the fiber shifted in frequency by $+2\Delta$, since the modulator is operated in the *double-pass* configuration and is driven at frequency $\Delta$. The quantities $\phi_c$ and $\phi_f$ are not affected by this AOM.

$$\text{Downstream, before fiber:} \quad E_0 \cos((\omega + \Delta)t - \phi_c + \phi_f) \tag{12.145}$$

5. After passing through the fiber for a second time, the fiber contribution to the phase noise will be doubled.

$$\text{Downstream, after fiber:} \quad E_0 \cos((\omega + \Delta)t - \phi_c + 2\phi_f) \tag{12.146}$$

6. The photodetector will receive two beams: one is from the fiber (downstream) and the other is a sample of the output of AOM 2, obtained from the beam-splitter and mirror on the left side of the diagram. These beams are:

$$\text{Beam 1:} \quad E_0 \cos((\omega + \Delta)t - \phi_c + 2\phi_f) \tag{12.147}$$

$$\text{Beam 2:} \quad E_0 \cos((\omega - \Delta)t - \phi_c) \tag{12.148}$$

The signal (photocurrent) at the output of the detector will be proportional to the square of the total electric field incident on the detector. The operation of squaring will generate sum and difference frequencies. The *difference frequency* signal is

$$\text{Output of photodetector:} \quad \cos(2\Delta t + 2\phi_f) \tag{12.149}$$

7. The phase-locked loop acts essentially like a filter, generating the low-frequency correction phase, $\phi_c$, which is equal to $\phi_f$. This operation eliminates external noise from various electronic sources and preserves the high frequency $\Delta$.

$$\text{Output of phase locked filter:} \quad \cos(2\Delta t + 2\phi_c) \tag{12.150}$$

8. The *divide by two* circuit digitally divides the frequency (and phase) by two.

$$\text{Output of f/2:} \quad \cos(\Delta t + \phi_c) \tag{12.151}$$

9. Finally, the signal is amplified and inverted and applied to AOM 2. We observe the source of the upstream $\Delta$ modulation and the fact that the beam downstream of the fiber is shifted down by $\Delta$ and lacks the fiber-induced phase noise (since $\phi_c = \phi_f$).

$$\text{Fiber output (downstream):} \quad E_0 \cos((\omega - \Delta)t) \tag{12.152}$$

The scheme has been successful in reducing the fiber-induced phase changes to less than 0.3 radians, essentially eliminating the fiber as a source of phase noise.

## 12.12 Characterization of laser frequency stability

The techniques described in this chapter allow one to considerably reduce the frequency fluctuations of a laser by frequency locking it to a stable reference. In assessing the success of one's frequency stabilization efforts, it is worthwhile to establish some "yardstick" with which to measure frequency fluctuations. In this section, we will provide a summary of the standard measures of laser frequency stability. Before doing so, it is important to distinguish *accuracy* from *stability.* Accuracy is a measure of how close the measurement is to a conceptual (but fictitious) *exact value of a quantity.* It is usually expressed as a maximum fractional frequency error with no associated averaging time. The errors which contribute to an oscillator's inaccuracy are called *systematic errors.* Stability is a measure of how well the frequency has been determined, without reference to its "true value". An oscillator's stability is usually limited by *random fluctuations*, which can be due to the intrinsic measurement process or due to environmental effects such as temperature changes or vibration. Stability includes *drift*, though this can often be removed by comparing the oscillator to a more stable one. Stability is usually measured by comparing the number of oscillator cycles in adjacent measurement periods and therefore depends strongly on the measurement time. A stable oscillator is not necessarily accurate: there are oscillators with excellent stability but, due to systematic errors, are insufficiently accurate to be candidates for a primary frequency standard. On the other hand, common sense dictates that an accurate source should have an equivalent stability, since it would be very difficult to assess the accuracy of an unstable oscillator. We will be interested only in oscillator stability in the following.

As with all time-varying quantities, there is both a *time domain* and *frequency domain* approach to the measurement of frequency stability. Stability can be measured in the time domain by making repeated measurements of the number of cycles of the oscillator over fixed time intervals and obtaining some kind of variance of the deviations from a mean. The frequency domain approach might use a discriminator to convert frequency fluctuations to amplitude fluctuations and obtain the power spectrum of the latter. Since we are interested in the frequency noise in a *laser*, it is of course currently impossible to directly count the cycles of its radiation and we therefore usually make a frequency domain measurement of the laser's frequency noise. It is fortunate that it is fairly easy to convert the frequency domain measurement to one in the time domain.

We begin with the usual description of the output of an oscillator with phase noise

$$E(t) = E_0 \sin(2\pi\nu_0 t + \phi(t)), \tag{12.153}$$

where $\phi(t)$ is a random variable which depends upon time and $E(t)$ is the electric field. Since we are not interested in *amplitude noise*, we assume that $E_0$ is constant. The *instantaneous frequency*, $\nu(t)$, is defined as

$$\nu(t) \equiv \frac{1}{2\pi}\frac{d}{dt}(2\pi\nu_0 t + \phi(t)) = \nu_0 + \frac{1}{2\pi}\frac{d\phi}{dt}. \tag{12.154}$$

We are interested in the *fluctuations* of the frequency about the *fixed* $\nu_0$; we call this quantity $\Delta\nu(t)$:

$$\Delta\nu(t) = \nu(t) - \nu_0 = \frac{1}{2\pi}\frac{d\phi}{dt}. \tag{12.155}$$

We usually work with the *fractional frequency and phase fluctuations*, so we divide by $\nu_0$ to obtain $y(t)$ and $x(t)$:

$$\text{Frequency:} \quad y(t) \equiv \frac{\Delta\nu(t)}{\nu_0} \tag{12.156}$$

$$\text{Phase:} \quad x(t) \equiv \int_0^t y(t')dt' = \frac{\phi(t)}{2\pi\nu_0}. \tag{12.157}$$

Finally, the actual *measured* quantity is the *phase* (proportional to the number of oscillator cycles or the number of ticks of a clock) averaged over time $\tau$. The $i^{th}$ measurement is called $\bar{y}_i$ and is defined as

$$\bar{y}_i \equiv \frac{x(t_i+\tau) - x(t_i)}{\tau}. \tag{12.158}$$

In the time domain, the usual variance of the quantities $\{\bar{y}_i\}$ is given by

$$\text{Common variance:} \quad \sigma^2(N,\tau) = \frac{1}{N-1}\sum_{i=0}^{N}(\bar{y}_i - \langle\bar{y}\rangle)^2, \tag{12.159}$$

where $N$ measurements are made and $\langle\bar{y}\rangle$ is the *mean* of these measurements. This variance is fairly well behaved when the $\phi(t)$ is spectrally uniform "white noise". It is, however, divergent for many of the noise sources encountered by lasers and is therefore an inadequate measure of laser stability. Instead, it is customary to use the *Allan variance*, defined by

$$\text{Allan variance:} \quad \sigma_y^2(\tau) = \left\langle \tfrac{1}{2}(\bar{y}(t+\tau) - \bar{y}(t))^2 \right\rangle, \tag{12.160}$$

where the angle brackets denote an infinite time average. One obtains the error by taking the difference between *adjacent samples* rather than the difference of a sample from a mean; this emphasizes random fluctuations and is insensitive, to first order, to slowly varying frequency changes. Of course, an infinite time average is not possible with experimental data; the Allan variance is very well approximated by taking $N$ samples and using the following formula:

$$\text{Allan variance:} \quad \sigma_y^2(\tau) \approx \frac{1}{2(N-1)}\sum_{i=0}^{N-1}(\bar{y}_{i+1} - \bar{y}_i)^2. \tag{12.161}$$

This variance is actually quite well behaved with most noise sources encountered in the laboratory. In all of the above definitions, we assume there is no *dead time*: the measurements are of duration $\tau$ and are separated by an interval equal to $\tau$.

We are often interested in the variation of $\sigma_y^2(\tau)$ with $\tau$. It turns out that this is easily obtained without retaking the data. In order to obtain the variance for $2\tau$, one simply adds adjacent measurements in pairs and applies the formula. This "trick" only works for the Allan variance of $y$ (with no dead time) and allows one to obtain $\sigma_y^2(n\tau)$ for integral $n$ from a single dataset. The Allan variance obtained in this way is only reliable when $n\tau < T/m$, where $T$ is the total duration of the data and $m$ is an integer between 2 and 8.

In addition to providing a universal measure of oscillator performance, the Allan variance allows one to obtain very useful information about the character of the noise. Before discussing this, we will introduce the frequency domain measure of oscillator noise: the *spectral density of fractional frequency fluctuations*: $S_y(f)$. We begin with the spectral density of the phase noise, $S_\phi(f)$. This is simply the power spectrum of $\phi(t)$ and can be obtained using the Wiener–Khintchine theorem, which states that the power spectrum of some function of time is the Fourier transform of its autocorrelation function. If the autocorrelation function of the phase is $R_\phi(\tau)$,

$$R_\phi(\tau) = \lim_{T\to\infty} \frac{1}{T} \int_{-T/2}^{T/2} \phi(t+\tau)\phi(t)dt \tag{12.162}$$

$$S_\phi(\omega) = \frac{1}{2\pi} \int_{-\infty}^{\infty} R_\phi(\tau)e^{i\omega\tau} d\tau. \tag{12.163}$$

If the quantity being analyzed were a voltage or current (or electric field), $S_\phi(\omega)$ would be a true spectral density of the *power* whose units would be watt/Hz. It is still customary to call $S_\phi(\omega)$ a spectral density even though its units are radian$^2$/Hz. The spectral density of *frequency fluctuations*, $S_{\dot\phi}(\omega)$, is then

$$S_{\dot\phi}(\omega) = \omega^2 S_\phi(\omega). \tag{12.164}$$

This is simply a consequence of the fact that the Fourier transform of a time derivative is $i\omega$ times the original transform (applied twice, since $S_\phi$ is a *spectral density*, which is quadratic in the transform of $\phi$). From the autocorrelation relations at $\tau = 0$ and the inverse Fourier transform, we obtain

$$\langle \dot\phi^2 \rangle = R_{\dot\phi}(0) = \int_{-\infty}^{\infty} S_{\dot\phi}(\omega)d\omega, \tag{12.165}$$

which justifies our characterization of $S_{\dot\phi}(\omega)$ as a spectral density. Finally, transforming to frequencies in cycles per second, the spectral density of $y$ is easily shown to be

$$S_y(f) = \left(\frac{f}{\nu_0}\right)^2 S_\phi(f). \tag{12.166}$$

The spectral density of oscillator noise is conventionally modeled as a *power series*, where each power has a different physical significance:

$$S_y(f) = \sum_{\alpha=-2}^{+2} h_\alpha f^\alpha. \tag{12.167}$$

The five types of noise are

$$\begin{aligned} \text{White phase:} \quad & \alpha = 2 \\ \text{Flicker phase:} \quad & \alpha = 1 \\ \text{White frequency:} \quad & \alpha = 0 \\ \text{Flicker frequency:} \quad & \alpha = -1 \\ \text{Random-walk frequency:} \quad & \alpha = -2. \end{aligned} \tag{12.168}$$

The familiar noises are white frequency and flicker frequency (also called $1/f$ noise). A remarkable property of the Allan variance is that, for the last four noise types, the Allan variance satisfies the following simple power law:

$$\sigma_y^2(\tau) \propto \tau^{-\alpha-1} \qquad -2 \leq \alpha \leq 1. \tag{12.169}$$

The *Allan deviation* ($\sigma_y(\tau)$) is the square root of the Allan variance and provides the fractional frequency stability of the oscillator. From the above considerations, we obtain the well-known result that an oscillator which is limited by shot noise (white frequency) has a fractional frequency instability which is proportional to $\tau^{-1/2}$.

The effect of drift on the fractional frequency offset, $y(t)$, is to add a term,

$$\text{Drift:} \quad x(t) = \tfrac{1}{2}Dt^2 \Longleftrightarrow y(t) = Dt, \tag{12.170}$$

where $D$ is a constant (the drift rate). For an oscillator subject to only drift, $\bar{y}_{i+1} - \bar{y}_i = D\tau$ for all $i$ and the Allan deviation is obtained immediately from eqn 12.161:

$$\text{Drift:} \quad \sigma_y(\tau) = \frac{D\tau}{\sqrt{2}}. \tag{12.171}$$

If the power-series expansion of the noise is valid, the curve of $\log(\sigma_y(\tau))$ versus $\log(\tau)$ should have a concave shape with each successively *smaller* value of $\alpha$ becoming dominant as $\tau$ increases. Such a curve for a fictitious oscillator is shown in Fig. 12.32. One can use a curve of this kind together with a little detective work to determine the nature of the sources of oscillator frequency noise and drift.

The frequency-domain measures can be used to directly determine the oscillator *linewidth* and can also be converted into an Allan variance. Experimentally, the spectral density of frequency fluctuations, $S_{\Delta f}(f)$, can be measured with a discriminator, which converts the frequency noise to amplitude noise after which the power density can be obtained using a radio-frequency spectrum analyzer. One can use the output of the frequency-locking discriminator for this purpose. Alternatively, one can use an external interferometer. The assumption is that the reference cavity or interferometer has very little noise over the frequencies of interest; if this assumption fails, the measurement becomes one which is relative to the cavity. The quantity $S_{\Delta f}(f)$ is defined via an autocorrelation function in the same way as are other similar quantities which are discussed above. The *mean square frequency deviation*, $\langle(\Delta f)^2\rangle$, is

$$\langle(\Delta f)^2\rangle = \int_0^\infty S_{\Delta f}(f)df, \tag{12.172}$$

where we use a *one-sided spectral density* which is defined only for positive frequencies. Taking the square root of this quantity provides one with the root-mean-square

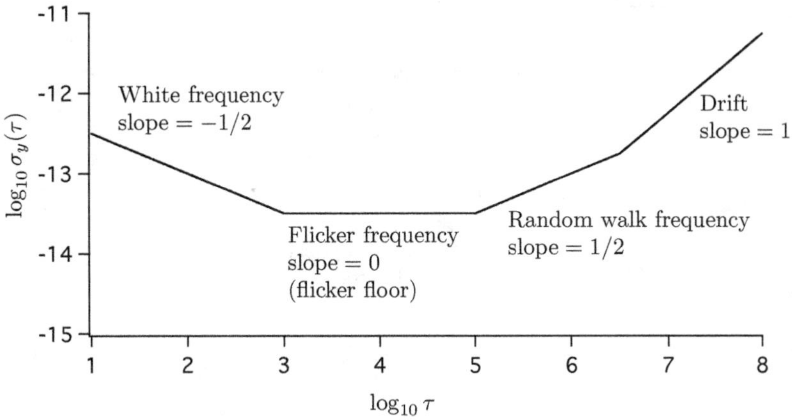


**Fig. 12.32** Typical Allan deviation curve showing how three different noise types and drift become dominant as $\tau$ increases.

linewidth of the oscillator. For white frequency noise, this integral diverges and one uses instead eqn 12.19:

$$\text{Linewidth with white frequency noise:} \quad \Delta f = \pi S_{\Delta f}. \tag{12.173}$$

It is more common to characterize the laser stability using the Allan variance, which can be obtained from $S_y(f)$ by evaluating an integral which is similar to eqn 12.172 but with a different weighting function:

$$\sigma_y^2(\tau) = 2\int_0^\infty S_y(f)\frac{\sin^4(\pi f\tau)}{(\pi f\tau)^2}df. \tag{12.174}$$

Again, the Allan deviation is just the square root of this quantity and, together with its dependence on $\tau$, provides a fairly complete description of the stability of an oscillator. Note that the linewidth ($\Delta f$) is independent of the averaging time since it is a property of the laser beam and not of the measurement process. The linewidth can be measured, in principle, by analyzing the beam with an extremely narrow optical spectrum analyzer (Fabry–Pérot). The Allan variance, on the other hand, depends upon the averaging time since it is an estimate of the error in determining the laser frequency after a measurement time equal to $\tau$.

We suggested that the linewidth of a laser can be measured by obtaining the power spectrum of the *error signal* from the discriminator used for frequency locking and calculating the RMS width from eqn 12.172. There is a frequency-domain alternative which yields the linewidth more directly and is probably the most reliable approach when the laser is extremely narrow (particularly when it is locked to an atomic resonance). This is to build a second *identical* oscillator and direct both sources to a fast photodetector, observing the spectrum of the beat note at the output of the detector. The beat note will reflect the *phase noise* spectrum of the laser if $|\phi(t)| \ll 1$, which is usually true for narrow sources. To see why this is so, we expand the electric field when $\phi(t)$ is small:

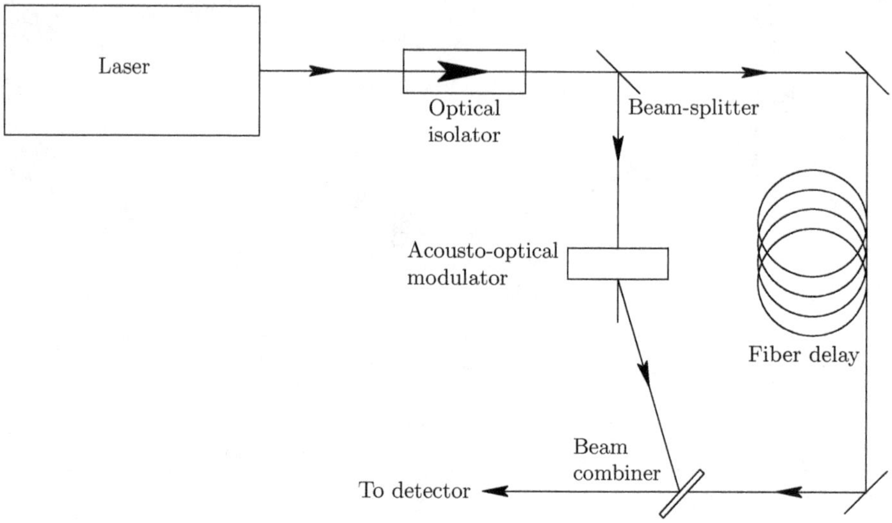


**Fig. 12.33** Block diagram of delayed heterodyne apparatus used to measure laser linewidths.

$$\begin{aligned} E(t) &= E_0 \sin(2\pi\nu_0 t + \phi(t)) \\ &= E_0(\sin(2\pi\nu_0 t)\cos\phi(t) + \cos(2\pi\nu_0 t)\sin\phi(t)) \\ &\approx E_0(\sin(2\pi\nu_0 t) + \phi(t)\cos(2\pi\nu_0 t)). \end{aligned} \tag{12.175}$$

The beat note for two uncorrelated sources, each of which is described by this expression, will have a low-frequency cross term that contains $\phi(t), \phi'(t)$ and is proportional to

$$(\phi(t) - \phi'(t))\sin(2\pi(\nu_0 - \nu_0')t), \tag{12.176}$$

where the primed quantities refer to the second source and we ignore the $\phi(t)\phi'(t)$ term. The magnitude of the total phase fluctuations will be the square root of the quadratic sum of the two phase terms, since the two sources are uncorrelated. If the sources are "identical", the width of the spectrum will be $\sqrt{2}$ greater than the width of *either* oscillator. This approach is often impractical, since the stabilized laser might be extremely complicated; a similar approach which is nearly as good is to use the *delayed homodyne* or *delayed heterodyne* technique, where we mix the laser beam with a *time-delayed* replica of itself. This is shown schematically in Fig. 12.33. If the delay time is greater than the coherence time of the laser, the beat note will have the same characteristics as one generated from two identical uncorrelated oscillators and the linewidth can be obtained as described above. It is usually more practical to shift the frequency of one of the beams: then the technique is called a time-delayed self-heterodyne measurement and the beat note will be at the shift frequency. A drawback of this approach is apparent when it is applied to the narrowest (sub-Hz) lasers: the length of the optical fiber can be impractically large. For example, some workers (Richter (1986)) have concluded that an accurate measurement requires a delay time which is six times the laser's coherence time. An external-cavity diode laser with an expected 10 kHz linewidth would require 36 km of fiber to make an accurate linewidth

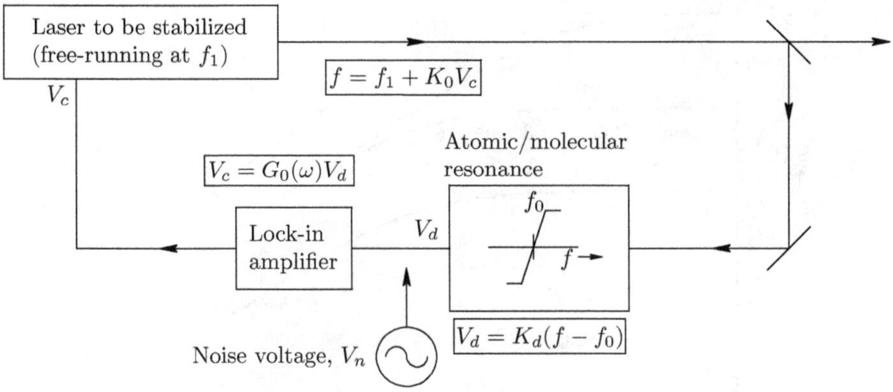


**Fig. 12.34** Block diagram of apparatus used to frequency-lock to an atomic or molecular resonance using a lock-in amplifier.

measurement! The linewidth measurement of a sub-Hz laser would, of course, be impossible using this technique. These workers have shown how the measurement can be made using a variety of fibers whose delay times are less than the coherence time. Essentially, one is directly obtaining the autocorrelation function at several values of delay using this modified technique. Another approach is the use a *multipass* fiber, where the effective delay is increased by making the fiber into a *ring resonator* (see, for example, Yue et al. (1988)). For the reasons mentioned above, the delayed homodyne and heterodyne approaches are most usefully employed on lasers with linewidths of $\approx$1 MHz or greater.

## 12.13 Frequency locking to a noisy resonance

It is a common practice in the laboratory to frequency-lock a laser to an atomic or molecular resonance by modulating some aspect of the resonance and obtaining the discriminant using a lock-in amplifier. The latter includes a simple low-pass filter with one or more poles after the phase-sensitive detector and has front panel controls which allow the operator to conveniently adjust the gain, time constant, number of poles, phase and modulation frequency. Thus, the lock-in very conveniently contains nearly all of the electronics needed to lock the laser to the resonance. We seek a strategy which optimizes the quality of the frequency lock. A block diagram of the apparatus appears in Fig. 12.34. The discriminant signal (possibly generated in the lock-in) is represented by voltage $V_d$ and the lock-in filter (including the gain circuitry) is represented by $G_0(\omega)$. These assignments allow us to use the previous theory to determine the noise characteristics of the loop.

One approach to adjusting frequency-locking systems is to acquire a lock with a relatively low loop gain and then turn up the gain until something adverse occurs. There are two possible scenarios: either the loop will begin to oscillate or the locked laser will become very noisy due to the application of amplified source or electronic noise to the laser frequency control input. The former is the most common scenario

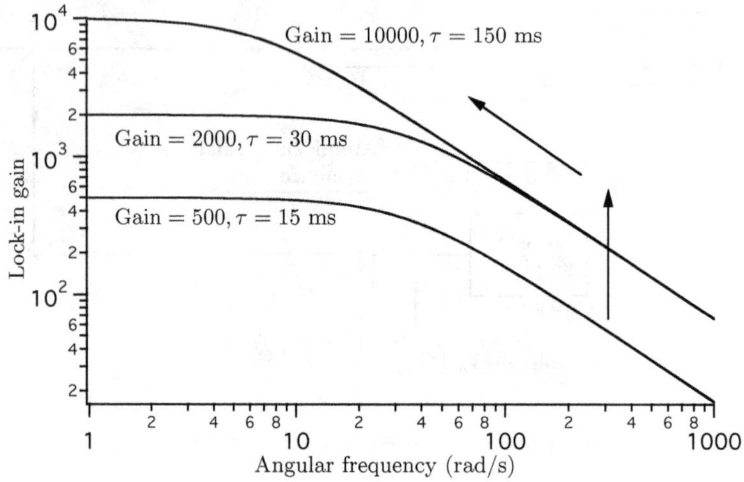


**Fig. 12.35** Gain versus frequency of lock-in as the gain is first increased without changing the time constant (vertical arrow) and then increased in proportion to the time constant (slanted arrow) preserving the high-frequency gain characteristic.

when locking to a passive cavity and the latter when locking to an atomic or molecular resonance.

If $V_n$ is the noise carried by the resonance, one can work backwards in the loop and observe that this will introduce a laser jitter equal to $V_n/K_d$ at frequencies where the loop gain is somewhat greater than unity. At low frequencies, the free-running intrinsic laser noise, $\Delta f_1$, will be reduced to

$$\text{Laser frequency noise reduction:} \quad \Delta f = \frac{\Delta f_1}{1 + K_0 K_d G_0(\omega)} \longrightarrow \frac{\Delta f_1}{K_0 K_d G_0(\omega)}. \tag{12.177}$$

The criterion for the source noise being the dominant contribution to the laser noise is then

$$\text{Source noise dominance:} \quad \frac{V_n}{K_d} \geq \frac{\Delta f_1}{K_0 K_d G(\omega)}. \tag{12.178}$$

When $V_n$ is quite large, as in an atomic or molecular resonance, this inequality will be satisfied only with a fairly small value of $G(\omega)$ and the source noise will probably dominate well before the loop oscillates.

The gain is then set to a value which is just below the point where the source noise increases the laser jitter. This can be determined by either observing the lock-in output and looking for a sudden increase in the amplitude of the "correction" signal or by observing the laser's spectrum with an optical spectrum analyzer and noticing a sudden increase in the jitter of the resonance. The final step is to improve the low-frequency lock by increasing the lock-in time constant, $\tau$, and increasing the gain, $G_0$, by the same factor until the maximum time constant is reached. Plots of the lock-in gain versus frequency for three settings are shown in Fig. 12.35. As shown in the figure, first the gain is increased until the source noise is observed on the laser output (vertical arrow) and then both the gain and time constant are increased in the same ratio (slanted arrow). The latter operation preserves the medium- to high-frequency gain of the lock-in and should have little or no effect on the laser spectrum, since most of the source noise is in this frequency range. The very large DC gain will reduce the low-frequency loop error.

## 12.14 Further reading

There are a large number of books which cover servo and control systems with varying degrees of formalism. A classic text favored by the author is *Modern Control Systems* by Dorf (1967). An excellent short summary of control systems is in the book on phase-locked loops by Gardner (2004). There is likewise an enormous literature on linear system analysis and synthesis. The book used by the author is an undergraduate engineering text by Lathi (1965). Many of the references for frequency stabilization of lasers were given in the chapter on frequency discriminants. Three important papers which deserve being mentioned again are Drever et al. (1983), Helmcke et al. (1982) and Day et al. (1992). A more recent paper in the *Handbook of Optics, Volume IV* (Hall (2000)) contains much useful information on more recent laser stabilization techniques. There is a large and extremely useful literature on oscillator frequency stability freely available online (currently) from NIST and indexed by TN 1337. Included are many classical papers on the subject, including some by D. W. Allan.

## 12.15 Problems

(12.1) Obtain the transfer function (ratio of $v_{out}$ to $v_{in}$) for the circuit shown.

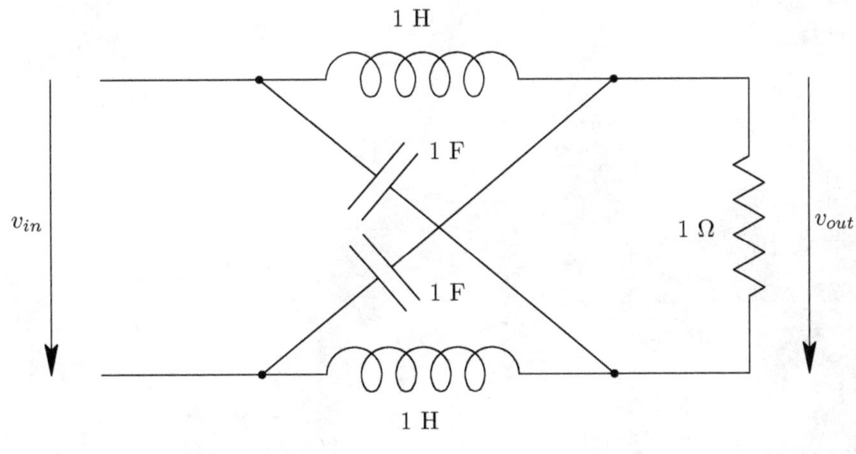


(12.2) Determine the output, $v_{out}(t)$, of the circuit in Problem 11.1 when the input is $e^t u(-t)$ (a rising exponential from negative $t$ and which is shut off at $t \geq 0$). To accommodate negative times, use the *bilateral Laplace transform* (sometimes called the *complex Fourier transform*) whose lower limit is $t = -\infty$ and which, for our purposes, has the same properties as the (unilateral) Laplace transform.

(12.3) Using the equation for $T(s)$ (eqn 12.102) for a temperature-stabilization loop, find an expression for the temperature as a function of *time* when there is a step increase of $\Delta T_0$ in the *ambient* temperature. Assume that the controller is a lead–lag circuit.

(12.4) Using the technique described in the section on laser stabilization, determine the maximum frequency for shot-noise-limited behavior for the *single-mode diode*, whose noise spectrum is characterized in the text. Use a lead–lag filter and assume that there are 25 ns of time delay and that the zero of the filter is a factor of 2 below the unity gain frequency.

# 13 Atomic and molecular discriminants

## 13.1 Introduction

Chapter 4 described in some detail the use of a *passive cavity* to generate a discriminant for laser stabilization. Passive cavities are fairly simple to use and can be very stable: a cavity constructed from Zerodur or ULE glass can have a drift rate as low as 0.1 Hz/second. However, it is often desirable to use an atomic or molecular resonance as a frequency reference. The long-term stability of such a reference can often be better than that of a passive cavity and, in some favorable cases, a rich molecular spectrum can also be used as a *poor man's frequency meter* to find and set the laser frequency much more accurately than is possible with a digital wavemeter. An atomic or molecular resonance will usually have a much smaller signal-to-noise ratio than a cavity resonance and will therefore usually be used only for long-term stabilization as the final stage in a hierarchy of increasingly stable references. The most extreme example of this is in the area of optical atomic frequency and time standards, where a very narrow resonance in a single ion or in a group of atoms trapped in an optical lattice is used to lock the laser frequency in the long term.

In this chapter we will discuss some techniques used to obtain an atomic or molecular discriminant. The focus will be on the molecular spectra of $Te_2$ and $I_2$ obtained using sub-Doppler saturation spectroscopy. These molecules are very useful since they have rich vibrational–rotational spectra in the ranges 385–523 nm and 500–900 nm respectively, and it is relatively easy to find a molecular resonance which is fairly close to any desired frequency in their ranges. For atomic resonances, the allied techniques of *polarization spectroscopy* and *dichroic atomic vapor laser lock* (DAVLL) will also be discussed. Finally, an example of the side-of-line approach with an atomic ion will be given.

## 13.2 Sub-Doppler saturation spectroscopy

In saturation spectroscopy, a laser beam is split into two (unequally intense) beams which are delivered to the atoms or molecules of interest using mirrors arranged so that the beams are counter-propagating and overlapping inside the sample. The stronger beam (*saturating beam*) burns a *hole* in the Doppler profile and the weaker beam (*probe beam*) is sent into a detector which measures the absorption as a function of frequency. Since the beams have opposite $k$-vectors, they interact with *Doppler subsets of the absorbing species* which are displaced symmetrically about the center of the velocity profile of the absorbers (the Doppler shift is $\boldsymbol{k} \cdot \boldsymbol{v}$). They only interact with the *same Doppler subset* when the detuning from resonance is less than the homogeneous width.

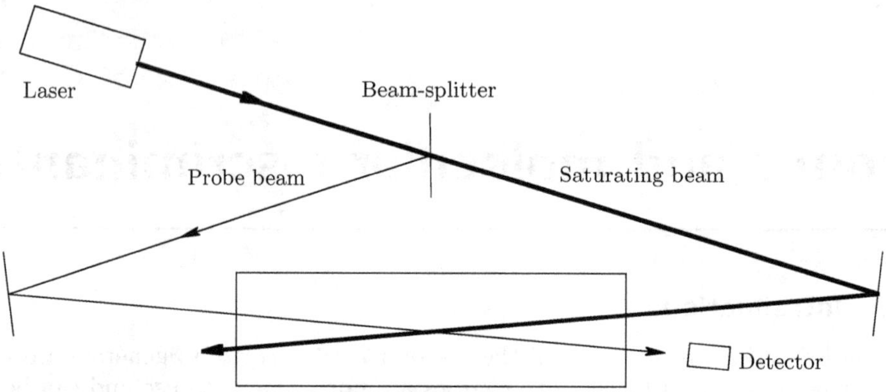


**Fig. 13.1** Schematic of apparatus for saturation spectroscopy.

The absorption of the probe beam versus laser frequency will show a broad Doppler profile with a pronounced *dip* when the two beams are interacting with the same atoms or molecules (those with approximately zero velocity along $\boldsymbol{k}$). The dip is Doppler-free and has a width equal to the homogeneous width of the sample species. The apparatus is shown in Fig. 13.1 and the resonant and off-resonant absorption profiles appear in Fig. 13.2. Not shown in the apparatus figure are lenses used to focus the saturating and probe beams into the sample; they facilitate obtaining the onset of saturation with modest laser power. A confocal parameter (eqn 1.20) of roughly the sample length is close to ideal. Finally, in addition to the resonances that occur at the various transition frequencies, there will be resonances halfway between pairs of transitions which share the same ground state and whose separations are less than the Doppler shift. These are called *cross-over* resonances, and an explanation of their origin is left as an exercise for the reader.

One can obtain an analytic expression for the expected signal from eqn 5.181 in Chapter 5. That equation describes the gain experienced by a weak, tunable probe beam at frequency $\omega$ in the presence of a fixed-frequency, strong saturating beam at $\omega'$. The result (called *hole burning*) will be repeated here for reference:

$$\text{Hole burning:}\quad \gamma(\omega) \propto \int_{-\infty}^{\infty} N(v) g(\omega - \omega_0 - kv) \frac{1}{1 + \frac{I}{I_s}\bar{g}(\omega' - \omega_0 - k'v)} dv,$$

where $I$ is the intensity of the saturating laser and $I_s$ is the value of $I$ which produces saturation. This equation can be used to analyze saturation spectroscopy by setting $\omega = \omega'$ (probe and saturating frequencies are the same) and setting $k' = -k$ (beams are counter-propagating). The result is

$$\text{Saturation spectroscopy:}\quad \gamma(\omega) \propto \int_{-\infty}^{\infty} N(v) g(\omega - \omega_0 - kv) \frac{1}{1 + \frac{I}{I_s}\bar{g}(\omega - \omega_0 + kv)} dv, \qquad (13.1)$$

where $\bar{g}(\omega)$ is a Lorentzian normalized to $g(0) = 1$ and $N(v)dv$ is the number of atoms or molecules whose speed is between $v$ and $v + dv$. The quantity $\gamma(\omega)$ will describe

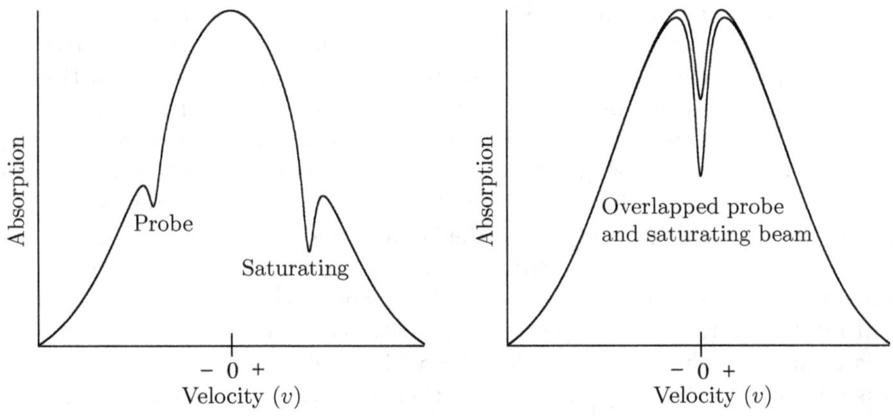


**Fig. 13.2** Doppler profiles in the presence of saturating and probe beams as a function of atomic or molecular velocity. The velocity, $v$, is converted to a frequency, $\omega$, using $\omega = kv + \omega_0$, where $\omega_0$ is the laser frequency. On the left, the lasers are away from rest-frame resonance; on the right they are on resonance.

*absorption* when the inversion ($\Delta N$) is positive (larger population in the ground state), which will always be the case in the apparatus under discussion. For simplicity, we assume that the homogeneous linewidth is much smaller than the inhomogeneous (Doppler) linewidth. In the limit of $I \ll I_s$, we can expand the fraction:

$$\gamma(\omega) \propto \int_{-\infty}^{\infty} N(v) g(\omega - \omega_0 - kv) \left\{ 1 - \frac{I}{I_s} \bar{g}(\omega - \omega_0 + kv) \right\} dv. \tag{13.2}$$

When we multiply the three factors and perform the integration, the first term is just a Doppler lineshape function, where we treat the $g(\omega - \omega_0 - kv)$ as a delta function. In evaluating the second term, we take the $N(v)$ factor out of the integration, since it is slowly varying, and the integral then becomes the convolution of two Lorentzians, each of which has width $\Delta\omega$. This convolution can be evaluated with the aid of the following formula, which can be verified with a contour integration:

$$\int_{-\infty}^{\infty} \frac{1}{a^2 + (y + x)^2} \frac{1}{b^2 + (y - x)^2} dx = \left( \frac{a + b}{ab} \right) \frac{\pi}{(2y)^2 + (a + b)^2}. \tag{13.3}$$

The result is

$$\gamma(\omega) \propto N((\omega - \omega_0)/k) \left\{ 1 - \frac{I}{2I_s} \bar{g}(\omega - \omega_0) \right\}. \tag{13.4}$$

The Lorentzian feature in the second term is thus a good replica of the original homogeneously broadened line multiplied by the slowly varying Gaussian function. Saturation spectroscopy is usually performed with $I \approx I_s$ for the best signals. A somewhat more involved analysis will yield the same qualitative result at higher powers: the observed line will have a Lorentzian lineshape with the homogeneous (power-broadened) breadth.

Normally, one is not interested in the Doppler-broadened baseline and would remove it using one of a number of possible techniques. One way is to use a *thick* beam-splitter which generates two parallel probe beams. If it is arranged that one of the probe beams does not overlap with the saturating beam, this probe beam can be used as a *reference* whose absorption signal would be subtracted from that of the other beam (which does overlap with the saturating beam). Since the reference signal contains *only* the Doppler-broadened baseline, the latter will be removed from the signal. Another method for eliminating the Doppler baseline is to *chop* the saturating beam (usually at some low audio frequency) and extract the Doppler-free signal using phase-sensitive detection. The validity of this approach can be seen from eqn 13.4: modulating the saturating beam will modulate the second term in the equation (which contains $I$) and the phase sensitive-detector will therefore yield the $\bar{g}(\omega - \omega_0)$ factor, which is multiplied by the modulated $I$. Either a mechanical chopper or an acousto-optic modulator can be used.

So far, we have explored the use of saturation spectroscopy to obtain Doppler-free atomic or molecular resonances. If we use the techniques described above, these resonances would be simple Lorentzian signals (eqn 13.4) and would not be very useful in that form for laser frequency stabilization. We will now discuss a technique which will generate a dispersion-shaped signal, which provides a good *discriminant* for laser stabilization. This scheme has some additional refinements which will be discussed below. One can, of course, generate a derivative signal by modulating the frequency of the laser at a low frequency (smaller than the resonance linewidth) and using phase-sensitive detection; the technique to be described has a number of advantages over simple frequency modulation despite its greater complexity.

The apparatus is shown in Fig. 13.3; the several modulation frequencies are specified merely for concreteness and would probably be different in an actual apparatus. The saturating beam is chopped at 10 kHz using an acousto-optic modulator (AOM) driven by a 100 MHz source shown in the upper right of the drawing. The upper mixer is used as a modulator to chop the AOM drive at 10 kHz. The probe beam is frequency-modulated at 4 MHz by an electro-optic modulator and the detected signal is demodulated using a mixer. This is just a low-frequency application of the Pound–Drever–Hall technique, described in detail in Chapter 4. Since the modulation frequency (4 MHz) is on the order of the natural linewidth of tellurium or iodine molecular transitions, the output will be a simple dispersion-shaped signal instead of the slightly more complicated signal which is seen when the modulation frequency is much larger than the resonance linewidth. The Pound–Drever–Hall signal is finally demodulated a second time (since it has a 10 kHz amplitude modulation from the chopped saturating beam) using the mixer near the bottom of the drawing; the demodulated signal is the desired discriminant.

In addition to providing a discriminant, this approach has the following additional features:

- The use of an AOM prevents interference fringes from stray beams. The probe and saturating beams are 100 MHz apart and any stray light from both beams can produce interference fringes which are *amplitude-modulated* at 100 MHz and will likely be outside the passband of the detector and amplifier.

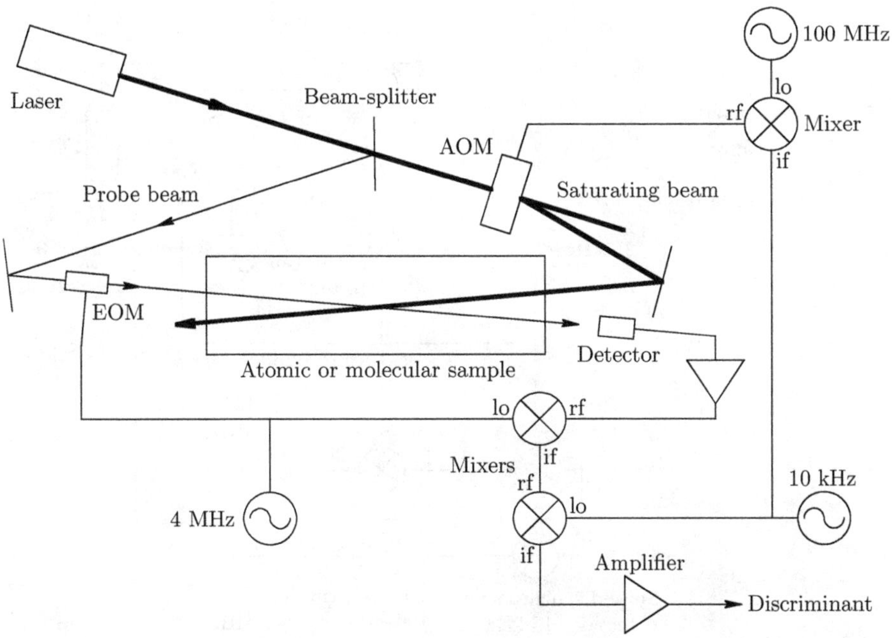


**Fig. 13.3** Schematic of improved saturation set-up.

- The first demodulation is at 4 MHz, which is usually above any amplitude noise in the laser beams and amplifiers. In practice, the author has observed amplitude noise from an external cavity diode laser modulated at 4 MHz; raising the modulation frequency to 20 MHz eliminated most of this noise.
- The amplitude modulation generated by the acousto-optic modulator can be at any frequency below the maximum allowed by the modulator (several hundred kHz for typical AOMs). Normally, one would not use frequencies below 1 kHz, since these would require filters with inconvenient time constants for observation of resonances while the laser is being swept.
- A little thought should convince the reader that the Doppler subset which will be simultaneously excited by both beams will be the one with a Doppler shift equal to one half of the AOM frequency shift (100 MHz in the figure). This allows the frequency of the resonance (and locked laser) to be precisely *tuned* by changing the AOM drive frequency. We will shortly discuss a technique which cancels the angular shift from the AOM so that a relatively wide tuning range is possible.

Since the ultimate signal is the product of a number of factors, including the saturating and probe intensities, one can also obtain the same results by placing the Pound–Drever–Hall phase modulator in the *saturating beam.* This approach is called *modulation transfer spectroscopy.*

We conclude this discussion with the description of a slightly more refined version of the apparatus just discussed. A schematic of the apparatus appears in Fig. 13.4, which shows an actual system used to lock an external-cavity diode laser to resonances

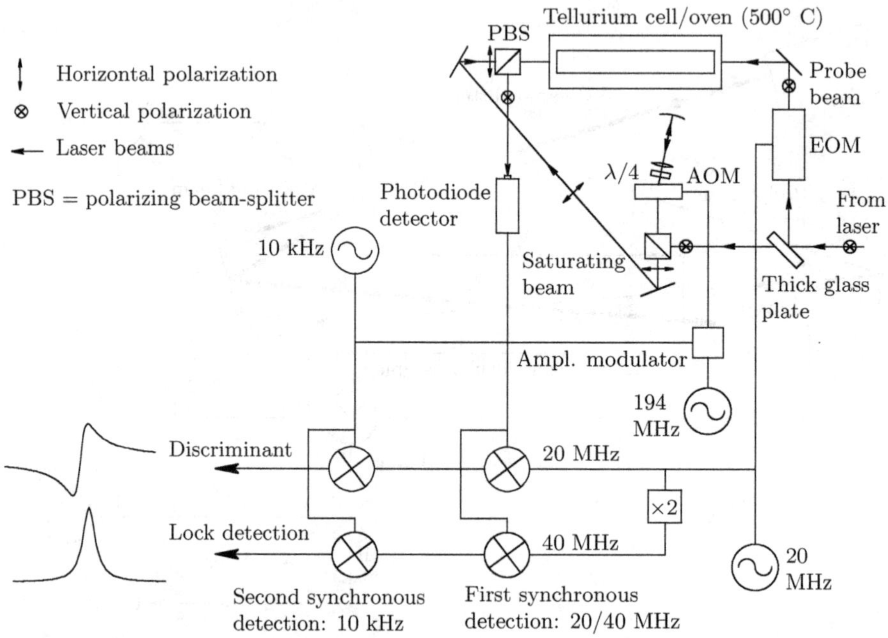


**Fig. 13.4** Schematic of improved saturation setup using a double-pass AOM and generating an additional lock-detection signal.

in tellurium molecules (the laser was frequency doubled). The setup is identical in function to that which was previously discussed: it uses a chopped saturating beam, Pound–Drever–Hall modulation of the probe beam and double demodulation. The following additional features are present:

- By exploiting the polarization manipulations shown in the figure, the saturating and probe beams can be made co-linear with each other with no loss of intensity since dissipative beam-splitters are not used. The only conventional beam-splitter is the thick glass plate which reflects $\approx 9\%$ of the light into the probe beam with the loss of 9% at the second surface.
- The acousto-optic modulator uses the *double-pass with cat's-eye* approach (discussed in Chapter 15) to cancel angular shifts and double the frequency shift. This scheme rotates the polarization by 90°, facilitating the use of the polarizing beam-splitters as discussed in the previous point. Since there are no angular shifts with frequency, the resonance can be *tuned* in frequency by an amount roughly equal to the AOM tuning range (the factor of one half in the frequency shift cancels the factor of 2 due to the double-pass AOM).
- By detecting the second harmonic of the Pound–Drever–Hall signal, an additional *symmetrical* resonance is obtained; this can be used as a *lock detector* to determine when the lock is broken. Additional electronics can be used to automatically sweep the laser over the resonance until the lock is reacquired. The reason for the symmetrical signal is a generalization of eqn 4.2: the signal observed with small

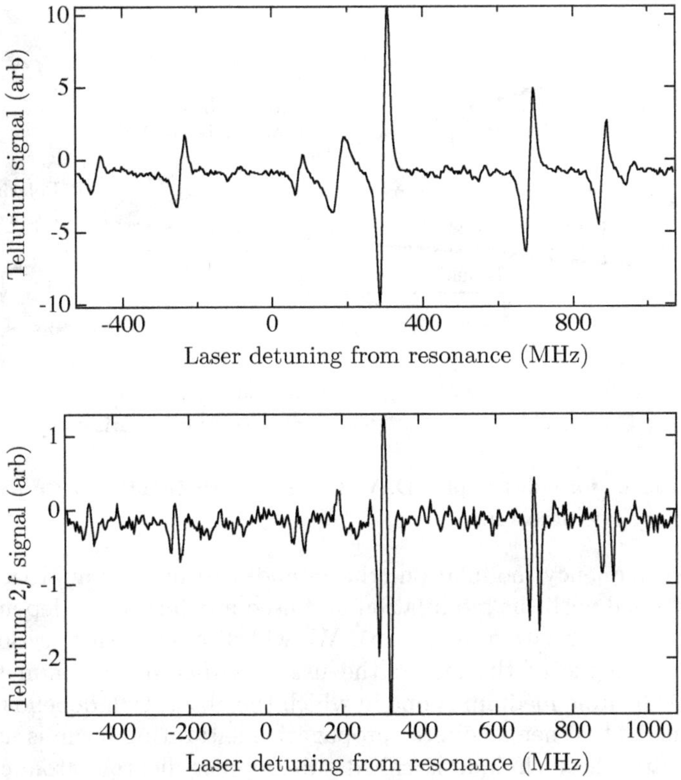


**Fig. 13.5** Saturated resonances in tellurium at 462 nm. The upper plot is the discriminant signal and the lower plot is the lock-detection signal.

deviation FM near a resonance is the $n$th derivative with respect to frequency of the resonance when the signal is detected at the $n$th harmonic of the modulation frequency. Thus, we are observing the *symmetric* second derivative.

Plots of actual discriminant and lock-detection resonances in tellurium are shown in Fig 13.5. A very short time constant was used, allowing a good signal-to-noise to be obtained in an oscilloscope display while sweeping the laser at a rate of about 10 Hz. The characteristic spectrum is readily memorized and serves as a *road map* which allows the operator to use the tellurium spectrometer as a frequency meter to reproducibly set the laser frequency with a precision of about 1 MHz.

## 13.3 Sub-Doppler dichroic atomic vapor laser locking and polarization spectroscopy

In the previous section we discussed the use of the Pound–Drever–Hall technique to generate a discriminant from atomic or molecular resonances. When we discussed techniques for obtaining a discriminant from a passive cavity, we studied two approaches:

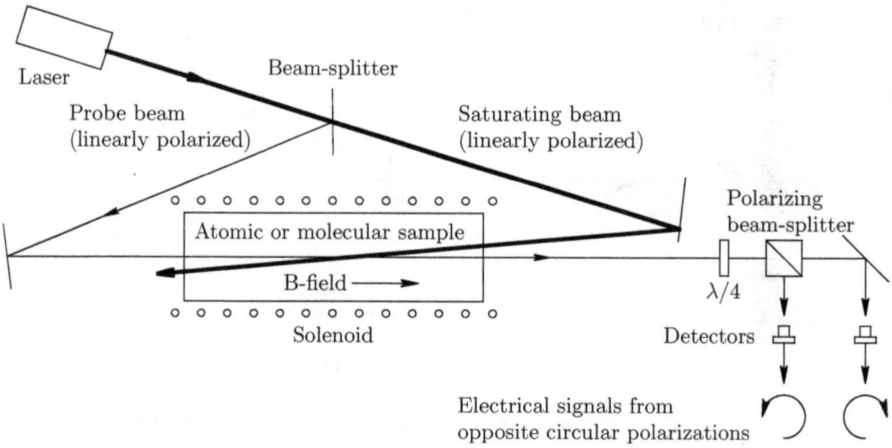


**Fig. 13.6** Apparatus for sub-Doppler DAVLL. The discriminant is the *difference* between the signals from the two detectors.

one which used frequency modulation (the Pound–Drever–Hall approach) and a second which dispensed with the modulation and used a polarization-dependent medium in the cavity (the polarization approach). We will discuss in this section the atomic and molecular analogue of the latter: the use of a *dichroic* medium to generate a discriminant. A dichroic medium is one in which the absorption depends upon the polarization of the light. When a counter-propagating saturating beam is used to remove the Doppler shifts, the technique is called sub-Doppler dichroic atomic vapour laser locking (sub-Doppler DAVLL). The closely allied technique of *polarization spectroscopy* will also be described.

The apparatus is shown in Fig. 13.6. The set-up is very similar to that used for ordinary saturation spectroscopy with the addition of a solenoid to generate a magnetic field parallel to the wave vector of the probe beam and a circular polarization analyzer for the probe beam. The circular analyzer detects both left and right circular polarization using a technique which was described in Chapter 4. The discriminant signal is the *difference* between the two polarization signals.

We will begin our explanation by assuming that the transition is between a $J = 0$ ground state and a $J = 1$ excited state in an atomic vapor. The quantization axis is along the magnetic field. One sense of circular polarization (which we will call $\sigma^+$) will cause a $\Delta m_J = +1$ transition and the other sense ($\sigma^-$) will cause a $\Delta m_J = -1$ transition. We will first consider plain DAVLL, which will not use a saturating beam and will be subject to Doppler broadening. The magnetic field will break the degeneracy between the $m_J = \pm 1$ levels and is large enough to produce Zeeman shifts which are on the order of the Doppler width (this requires a field of $> 100$ G). The probe beam polarization can be decomposed into equal parts of right and left circular polarization. The absorption lines for the two polarizations will be identical in shape but shifted in frequency by the Zeeman shift between the $m_J = \pm 1$ sublevels (there will be no $\Delta m_J = 0$ transition for this polarization). At the zero-field resonant frequency, the absorptions will be equal and their difference will be zero. On either side of resonance, the absorptions will be different and one will therefore obtain a signal which is an odd function of the laser frequency displacement from the zero-field resonance (i.e., a discriminant).

The analysis of saturated, sub-Doppler DAVLL is somewhat more complicated than that of simple saturation spectroscopy and will not be given here (a good reference for this is Maguire (2006)). The principle is, however, the same: the probe and saturating beams only interact with the same atoms when the laser frequency detuning from resonance is less than the homogeneous linewidth of the atoms. The dichroism of the saturated absorption causes a dispersion-shaped signal to be generated when the absorptions from the two senses of circular polarization are subtracted.

It is possible to discuss the optimum parameters without an analytic approach to the sub-Doppler DAVLL signal. The magnetic field need be only large enough to produce a Zeeman shift which is comparable to the homogeneous linewidth; for alkali atoms this is about 10 G. Unfortunately, there will be a Doppler-broadened background, which can be eliminated by chopping the saturating beam, but here we wish to avoid all modulations for the sake of simplicity. This background will make the location of the zero-crossing depend slightly on the magnetic field, which must be considered when locking a laser. A modest stabilization of the field should reduce this effect to a tolerable level. One should have a means of independently adjusting the powers in the saturating and probe beams. A simple method for accomplishing this is to use a half-wave plate and polarizing beam-splitter to split the laser beam into the probe and saturating components. If the wave plate (which rotates the polarization) is between the laser and the beam-splitter, rotating the wave plate will adjust the relative amounts of power in each beam (the orthogonal linear polarizations of the two beams will not be a problem). The saturating power should be near the onset of saturation (about 0.5 mW in a 1 mm diameter beam for alkali atoms): too much will broaden the transition, reducing the slope of the discriminant and too little will reduce the amplitude of the saturated resonance. The probe beam should be about ten times weaker than the saturating beam to avoid power-broadening.

The *polarization spectroscopy* technique uses a circularly polarized pump beam to polarize the atoms and detects the polarization using a linearly polarized probe beam. Historically, the technique used crossed polarizers in the probe beam on either side of the sample: near resonance, the polarization of the probe would be rotated (by the influence of the pump beam on the atoms), producing a signal. Unfortunately, the signal does not have the required shape for a discriminant (it is symmetrical about the resonance frequency). A proper discriminant can be obtained by rotating the polarization of the probe (with a half-wave plate) by 45° before it encounters the sample, and analyzing the polarization of the light after the sample in both the horizontal and vertical planes. The difference signal will clearly be zero away from resonance (the polarization is not rotated) and will be an odd function of the frequency displacement from resonance, as required for a discriminant. The apparatus is shown in Fig. 13.7.

Perhaps the most important distinction between polarization spectroscopy and DAVLL is that the latter detects the *dichroism* produced by a magnetic field and the

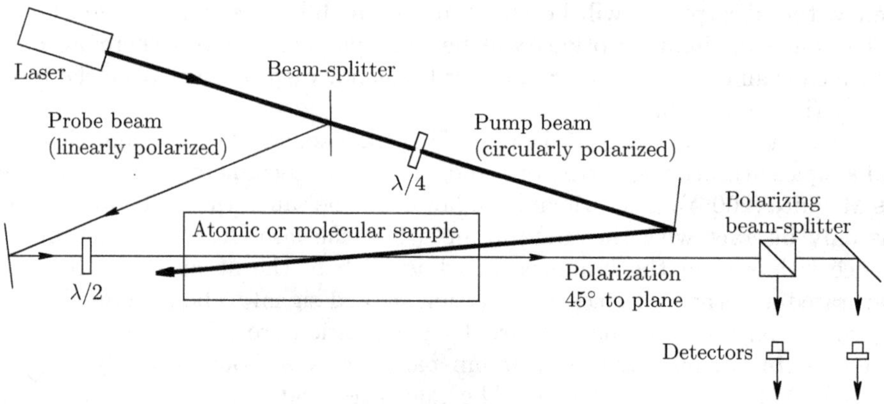


**Fig. 13.7** Apparatus for generating a discriminant using polarization spectroscopy. Again, the discriminant is the *difference* between the signals from the two detectors.

former detects the *birefringence* (polarization-dependent refractive index) produced by a circularly polarized pump beam (it is easy to show that a small birefringence will rotate the polarization of linearly polarized light). One can consider the two techniques to be methods for analyzing the imaginary and real parts of the dielectric susceptibility.

We will derive an analytical expression for the polarization signal in two steps. First, we will determine the signal due to a birefringent medium. Then, we will specify in more detail the influence of the pump beam on the medium, including the effects due the motion of the atoms. Using the Jones calculus (Chapter 4), the polarization of a wave incident on the sample and propagating in the $z$-direction is

$$\boldsymbol{E} = \begin{pmatrix} E_x \\ E_y \end{pmatrix} = E_0 \begin{pmatrix} \cos\phi \\ \sin\phi \end{pmatrix} = \frac{E_0}{2}\left\{ e^{-i\phi}\begin{pmatrix} 1 \\ i \end{pmatrix} + e^{i\phi}\begin{pmatrix} 1 \\ -i \end{pmatrix}\right\}, \tag{13.5}$$

where we have decomposed a beam whose linear polarization makes an angle $\phi$ with the $x$-axis into two opposite circular polarizations. After passing through a sample of length $L$, the field is

$$\boldsymbol{E} = \frac{E_0}{2}\left\{ e^{-i\phi}\begin{pmatrix} 1 \\ i \end{pmatrix} e^{(-ik_+ - \alpha_+/2)L} + e^{i\phi}\begin{pmatrix} 1 \\ -i \end{pmatrix} e^{(-ik_- - \alpha_-/2)L}\right\}, \tag{13.6}$$

where $k_\pm = \frac{\omega}{c}n_\pm$ and $\alpha_\pm$ are the real parts of the $k$-vectors and absorption coefficients for $\sigma^+$ and $\sigma^-$ circular polarization. One can factor out the average refractive index and absorption coefficient

$$\boldsymbol{E} = \frac{E_0}{2} e^{(-i\frac{\omega}{c}n - \frac{\alpha}{2})L} \left\{ e^{-i\phi} \begin{pmatrix} 1 \\ i \end{pmatrix} e^{-i\Omega} + e^{i\phi} \begin{pmatrix} 1 \\ -i \end{pmatrix} e^{+i\Omega} \right\}, \tag{13.7}$$

where

$$\Omega = \frac{\omega}{2c} \Delta n L - i \frac{L}{4} \Delta\alpha \tag{13.8}$$

$$n = \tfrac{1}{2}(n_+ + n_-) \tag{13.9}$$

$$\alpha = \tfrac{1}{2}(\alpha_+ + \alpha_-) \tag{13.10}$$

$$\Delta n = n_+ - n_- \tag{13.11}$$

$$\Delta\alpha = \alpha_+ - \alpha_-. \tag{13.12}$$

The signal intensity is proportional to $|E_x|^2 - |E_y|^2$ which is

$$I_{signal} = I_0 e^{-\alpha L} \cos\left(2\phi + L\Delta n \frac{\omega}{c}\right), \tag{13.13}$$

where $I_0$ is the signal without the sample. The first term in the cosine is twice the angle between the polarizers; we take $\phi$ to be $\pi/4$. The second term is due to the atoms, and we assume that it is much less than one. Expanding the cosine about $\pi/2$, we obtain

$$I_{signal} = I_0 e^{-\alpha L} L \Delta n \frac{\omega}{c}. \tag{13.14}$$

The absorption of the probe has a Lorentzian frequency dependence

$$\Delta\alpha = \Delta\alpha_s \bar{g}(\omega - \omega_0), \tag{13.15}$$

where $\bar{g}(\omega - \omega_0)$ is a Lorentzian of width $\gamma$ normalized to unity at resonance and $\Delta\alpha_s$ is the absorption difference due to the pump beam. One can obtain a *dispersion relation* from eqns 5.114 and 5.115 together with the relations between the real and imaginary components of the susceptibility and the index of refraction and absorption (eqns 5.105 and 5.106). The result is

$$\Delta n = \frac{c}{\omega_0} \frac{\omega_0 - \omega}{\gamma} \Delta\alpha. \tag{13.16}$$

Using this, the signal is:

$$I_{signal} = I_0 e^{-\alpha L} L \frac{\Delta\alpha_s}{\gamma} (\omega_0 - \omega) \bar{g}(\omega - \omega_0), \tag{13.17}$$

which is the expected dispersion-shaped signal obtained when $\phi = \pi/4$.

To complete the calculation, we need to know the effect of the pump beam on the absorption difference, $\Delta\alpha_s$. We will assume that the dependence of $\Delta\alpha_s$ on the *pump beam frequency* is also Lorentzian:

$$\Delta\alpha_s = \Delta\alpha_0 \bar{g}(\omega - \omega_0 - kv), \tag{13.18}$$

where $\Delta\alpha_0$ is the absorption difference when the pump beam is on resonance; it depends upon details of the atomic structure together with the intensity and polarization of the pump beam. We have included the Doppler shift in the above expression. To simplify the calculation, assume that $I \ll I_s$; the width is therefore the same as that due to the probe beam. Finally, we multiply the whole signal expression by the velocity distribution $N(v)$, include Doppler shifts for the probe beam, and integrate over all velocities:

$$I_{signal} \propto \int_{-\infty}^{\infty} N(v)(\omega - \omega_0 + kv)\bar{g}(\omega - \omega_0 + kv)\bar{g}(\omega - \omega_0 - kv)dv, \tag{13.19}$$

where we have left out all of the multiplicative constants such as the intensities, $\Delta\alpha_0$, etc. One should note the different sign of the $kv$ term in the arguments of the probe and pump lineshape expressions. As in the case of saturation spectroscopy, we have a convolution, this time between Lorentzian and dispersion-shaped functions (we factor the Gaussian out of the integral since the inhomogeneous linewidth is much greater than the homogeneous linewidth). A straightforward (but slightly messy) contour integration yields the formula

$$\int_{-\infty}^{\infty} \frac{y+x}{a^2+(y+x)^2}\frac{1}{b^2+(y-x)^2}dx = \frac{2\pi y}{b[(2y)^2+(a+b)^2]}, \tag{13.20}$$

from which we can obtain the result

$$I_{signal} \propto (\omega - \omega_0)\bar{g}(\omega - \omega_0), \tag{13.21}$$

which is the desired dispersion-shaped discriminant.

## 13.4 An example of a side-of-line atomic discriminant

We will conclude this chapter with a description of a side-of-line discriminant obtained using opto-galvanic detection.

The side-of-line locking technique was discussed in Chapter 4. We will briefly mention a variant of this scheme applied to locking a laser to a Doppler broadened *opto-galvanic* line. It was used in the author's laboratory to lock a frequency-doubled external-cavity diode laser to a resonance in a barium discharge lamp; the laser was used to cool a single barium ion and was successfully locked, without interruption, for up to several weeks. The opto-galvanic technique exploits the change in the discharge current–voltage characteristic when a laser beam is resonant with one of the resonance lines of the medium in the discharge (in this case singly ionized barium). A schematic of the apparatus appears in Fig. 13.8.

The acousto-optic modulator (AOM) chops the laser beam (at 10 kHz in the figure). The discharge current is capacitively coupled and *summed* with a sample of the chopped laser intensity using the *undeflected* beam from the AOM. This latter beam is detected with a photodiode to generate a *reference*, which is subtracted from the opto-galvanic signal; this is accomplished by summing the two signals in the operational amplifier since the two outputs of the AOM are modulated 180° out of phase. The output of the operational amplifier is the signal input to the lock-in, and a sample of the modulation voltage is the reference. Subtracting a sample of the laser intensity

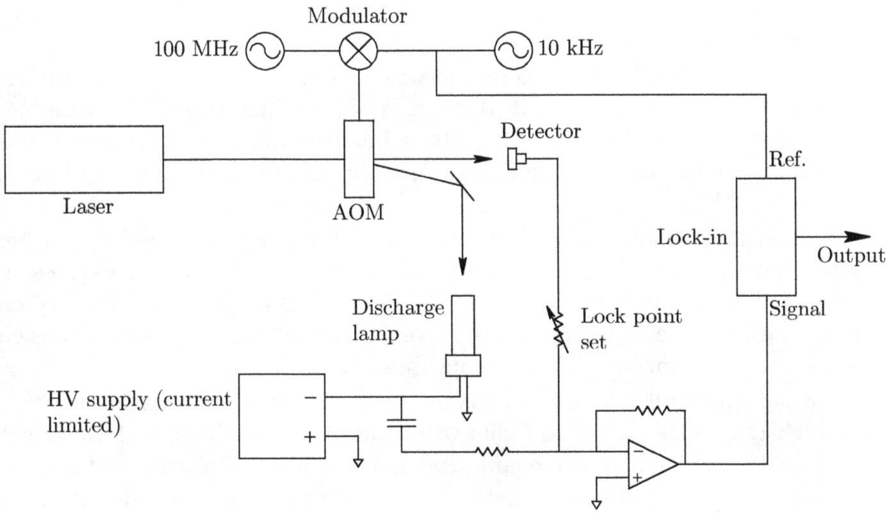


**Fig. 13.8** Apparatus for side-of-line locking to Doppler broadened opto-galvanic line in discharge tube.

makes the lock point fairly robust against changes in the laser power. One would normally lock the laser several hundred MHz below the opto-galvanic resonance (since it is *side-of-line* locking); the AOM frequency shift can be chosen to return the laser frequency to the exact resonance or any desired frequency (limited by the AOM tuning range). The lock point (and laser frequency) can be changed by changing the gain of the reference beam; this moves the lock point to a different place on the resonance line.

## 13.5 Further reading

Saturation spectroscopy is very well described in the book *Atomic Physics* by Foot (2005). The use of the Pound–Drever–Hall technique in saturation spectroscopy appears in a paper by Hall (1981). Sub-Doppler DAVLL is treated qualitatively in Harris, et al. (2008) and an analytical treatment of polarization spectroscopy and its application to laser locking appears in Pearman et al. (2002).

## 13.6 Problems

(13.1) As mentioned in the text, so-called *cross-over resonances* occur in saturated spectroscopy when two allowed transitions share the same ground state and have transition frequencies whose difference is less than the Doppler width of either transition. They occur halfway between the two resonances. Explain the mechanism which allows these *spurious* resonances to occur.

(13.2) The Pound–Drever–Hall saturation technique produces a *symmetric* signal when the first mixer uses the *second harmonic* of the modulation frequency for a reference. Explain why the $n$th derivative of the resonance is obtained when the detection occurs at the $n$th harmonic. (This is strictly true only when the FM deviation and modulation frequency are somewhat smaller than the linewidth of the resonance.)

(13.3) Compare the discriminant slope obtained with the saturated Pound–Drever–Hall technique to that obtained using a side-of-line approach assuming that the linewidths are the same (the side-of-line would use a saturation approach). Assume that the laser in the side-of-line approach is tuned exactly halfway up the side of the line.

(13.4) The polarization scheme will generate a pure dispersion signal only when the polarizer axes are at 45° to each other. A polarization error, expressed as a deviation from 45°, will cause the *lock point* to change. Determine the change in the lock point as a fraction of the atomic linewidth for a small angular change in one of the polarizers from the optimum orientation.

(13.5) Show that a small amount of *circular birefringence* (different indices for right and left circularly polarized light) will rotate the polarization of linearly polarized light. Assume that the beam propagates in the $z$-direction and that, for $\sigma^+$ polarization, the index is $n_+$, while for $\sigma^-$ polarization, the index is $n_-$, where $|n_+ - n_-| \ll n_+$. For simplicity, assume that the laser is originally polarized along the $x$-axis.

# 14
# Nonlinear optics

## 14.1 Introduction

In all of our discussions of electromagnetic wave propagation, we have assumed that the polarization ($\boldsymbol{P}$) is a proportional to the electric field ($\boldsymbol{E}$). This is an excellent approximation for most of the phenomena observed in nature. With the advent of lasers, however, it is possible to generate extremely strong electric fields by tightly focusing the laser beam, and these strong fields can produce a polarization which depends on higher powers of the electric field. These phenomena are the subject matter of *nonlinear optics*. Aside from their intrinsic value as interesting physics, nonlinear interactions have a very practical value in the atomic physics laboratory. They enable one to generate useful amounts of coherent radiation at wavelengths not normally served by commonly available lasers. Thus, via second-harmonic generation and sum-frequency mixing, one can generate radiation in the deep blue and ultraviolet regions of the spectrum. In addition, parametric processes can provide a useful supply of *entangled photons*, enabling experiments in measurement and quantum information theory.

This chapter will discuss the most common nonlinear interactions, beginning with second-harmonic generation and proceeding to sum-frequency mixing and parametric phenomena. The principal techniques of *phase matching* will be discussed in detail. To facilitate this, we will begin with a discussion of *anisotropic crystals*, since crystal birefringence is exploited in the most common technique of phase matching. This introduction will also be useful in understanding the modulation techniques discussed in Chapter 15.

## 14.2 Anisotropic crystals

An anisotropic crystal is one in which the behavior of an electromagnetic wave is dependent upon the direction of the electric field of the wave (i.e., the *electric field polarization*). We will begin our analysis by writing down the time-dependent Maxwell equations and material relations for isotropic media and then modifying the material relations to describe anisotropic media. We assume that there are no free charges or currents and that there is no magnetic material. Ampere's law and Faraday's law are

$$\nabla \times \boldsymbol{H} = \frac{\partial \boldsymbol{D}}{\partial t} \tag{14.1}$$

$$\nabla \times \boldsymbol{E} = -\frac{\partial \boldsymbol{B}}{\partial t}. \tag{14.2}$$

The material relationships are

$$\boldsymbol{D} = \epsilon_0 \boldsymbol{E} + \boldsymbol{P} = \epsilon \boldsymbol{E} \tag{14.3}$$

$$\boldsymbol{B} = \mu_0(\boldsymbol{H} + \boldsymbol{M}) = \mu \boldsymbol{H}. \tag{14.4}$$

Energy relations are described with the aid of the *Poynting vector*, $\boldsymbol{S}$, and the *energy density*, $u$:

$$\boldsymbol{S} = \boldsymbol{E} \times \boldsymbol{H} \tag{14.5}$$

$$u = \tfrac{1}{2}(\boldsymbol{E} \cdot \boldsymbol{D} + \boldsymbol{B} \cdot \boldsymbol{H}). \tag{14.6}$$

These quantities satisfy an *energy continuity equation*,

$$\nabla \cdot \boldsymbol{S} + \frac{\partial u}{\partial t} = 0, \tag{14.7}$$

in the absence of sources.

In an *anisotropic* medium, the polarization, $\boldsymbol{P}$, can point in a different direction from the field and $\boldsymbol{E}$ and $\boldsymbol{D}$ are therefore not necessarily collinear. The dielectric constant, $\epsilon$, is now a *linear operator* represented by the *dielectric tensor*, $\epsilon_{ij}$,

$$D_i = \sum_j \epsilon_{ij} E_j \Longleftrightarrow \begin{pmatrix} D_1 \\ D_2 \\ D_3 \end{pmatrix} = \begin{pmatrix} \epsilon_{11} & \epsilon_{12} & \epsilon_{13} \\ \epsilon_{21} & \epsilon_{22} & \epsilon_{23} \\ \epsilon_{31} & \epsilon_{32} & \epsilon_{33} \end{pmatrix} \begin{pmatrix} E_1 \\ E_2 \\ E_3 \end{pmatrix}. \tag{14.8}$$

The energy density of the *electric field*, $u_e$, and its time derivative, $\dot{u}_e$, are

$$u_e = \frac{1}{2}\boldsymbol{E} \cdot \boldsymbol{D} = \frac{1}{2}\sum_{ij} E_i \epsilon_{ij} E_j \Longrightarrow \dot{u}_e = \frac{1}{2}\sum_{ij} \epsilon_{ij}(\dot{E}_i E_j + E_i \dot{E}_j). \tag{14.9}$$

From the *Poynting theorem*,

$$\nabla \cdot \boldsymbol{S} + \boldsymbol{E} \cdot \dot{\boldsymbol{D}} + \boldsymbol{H} \cdot \dot{\boldsymbol{B}} = 0. \tag{14.10}$$

This implies that

$$\nabla \cdot \boldsymbol{S} + \sum_{ij} E_i \epsilon_{ij} \dot{E}_j + \boldsymbol{H} \cdot \dot{\boldsymbol{B}} = 0 \Longrightarrow \dot{u}_e = \sum_{ij} E_i \epsilon_{ij} \dot{E}_j. \tag{14.11}$$

Equating the two expressions for $\dot{u}_e$, we obtain the important result:

$$\epsilon_{ij} = \epsilon_{ji}. \tag{14.12}$$

The dielectric tensor is therefore represented by a *symmetric matrix*, which can be *diagonalized* yielding three *orthogonal* eigenvectors (principal axes) and *real* eigenvalues (proportional to the refractive indices squared). Henceforth, we will always work in the coordinate system of the principal axes and will use the symbol $\epsilon_i$ for the diagonal elements of $\epsilon_{ij}$:

$$\epsilon_{ii} \longrightarrow \epsilon_i. \tag{14.13}$$

The angular relations among the vectors $\boldsymbol{D}, \boldsymbol{E}, \boldsymbol{H}, \boldsymbol{k}$ and $\boldsymbol{S}$ can be easily obtained for a *plane wave* propagating as $e^{i\boldsymbol{k}\cdot\boldsymbol{r}-i\omega t}$, where the following operator equivalences will be used:

$$\nabla\times \Longrightarrow i\boldsymbol{k}\times \tag{14.14}$$

$$\frac{\partial}{\partial t} \Longrightarrow -i\omega. \tag{14.15}$$

Maxwell's time-dependent equations are

$$\begin{aligned} \boldsymbol{k}\times\boldsymbol{H} &= -\omega\boldsymbol{D} \\ \boldsymbol{k}\times\boldsymbol{E} &= \mu\omega\boldsymbol{H}. \end{aligned} \tag{14.16}$$

From these two equations, one can draw the following conclusions:

- $\boldsymbol{D}$ and $\boldsymbol{H}$ are both perpendicular to $\boldsymbol{k}$.
- $\boldsymbol{D}$ is perpendicular to $\boldsymbol{H}$.
- $\boldsymbol{E}$ and $\boldsymbol{H}$ are perpendicular to each other and to $\boldsymbol{S} = \boldsymbol{E}\times\boldsymbol{H}$.

Thus, there are two sets of three mutually perpendicular vectors:

- $\boldsymbol{E}, \boldsymbol{H}$ and $\boldsymbol{S}$: *energy propagation direction* $= \hat{\boldsymbol{S}}$.
- $\boldsymbol{D}, \boldsymbol{H}$ and $\boldsymbol{k}$: *phase propagation direction* $= \hat{\boldsymbol{k}}$.

Note that $\boldsymbol{D}, \boldsymbol{E}, \boldsymbol{S}$ and $\boldsymbol{k}$ *are coplanar* (they are all perpendicular to $\boldsymbol{H}$). These relationships are illustrated in Fig. 14.1.

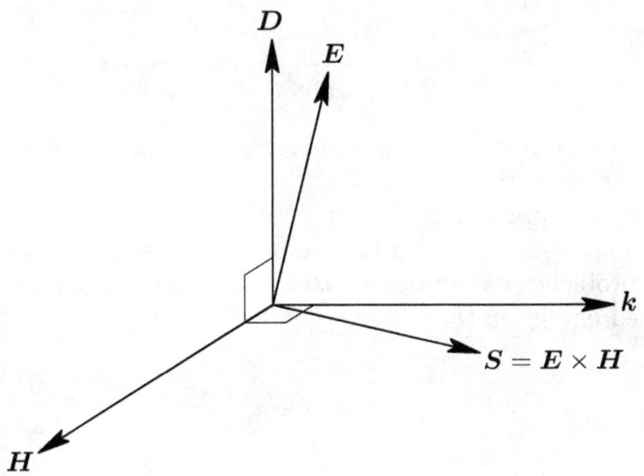


**Fig. 14.1** Angular relations among $\boldsymbol{D}, E, S, H$ and $\boldsymbol{k}$ in an anisotropic crystal.

The allowed values of the refractive index, $n$, can be obtained from the Maxwell equations (eqns 14.16). First, we will write $\boldsymbol{k}$ as

$$\boldsymbol{k} = k\hat{\boldsymbol{k}} = \frac{n\omega}{c}\hat{\boldsymbol{k}}. \tag{14.17}$$

Then, $\boldsymbol{D}$ and $\boldsymbol{H}$ are

$$\boldsymbol{D} = -\frac{n}{c}\hat{\boldsymbol{k}} \times \boldsymbol{H} \tag{14.18}$$

$$\boldsymbol{H} = \frac{n}{\mu c}\hat{\boldsymbol{k}} \times \boldsymbol{E}. \tag{14.19}$$

Substituting the second equation into the first, one obtains

$$\boldsymbol{D} = -\frac{n^2}{c^2\mu}\hat{\boldsymbol{k}} \times \hat{\boldsymbol{k}} \times \boldsymbol{E} = \frac{n^2}{c^2\mu}[\boldsymbol{E} - \hat{\boldsymbol{k}}(\hat{\boldsymbol{k}} \cdot \boldsymbol{E})] = \frac{n^2}{c^2\mu}\boldsymbol{E}_{transverse}, \tag{14.20}$$

where $\boldsymbol{E}_{transverse}$ is the component of $\boldsymbol{E}$ which is perpendicular to $\boldsymbol{k}$. Since we are in the coordinate system of the principal axes of $\epsilon_{ij}$, the dielectric matrix is diagonal and one can solve the above equation for each component of $\boldsymbol{E}$:

$$E_i = \frac{n^2\hat{k}_i(\hat{\boldsymbol{k}} \cdot \boldsymbol{E})}{n^2 - \epsilon_i/\epsilon_0}, \qquad i = x, y, z, \tag{14.21}$$

where we have used $c = 1/\sqrt{\mu_0\epsilon_0}$, $D_i = \epsilon_i E_i$ and $\mu = \mu_0$. Multiplying this equation by $\hat{k}_i$ and summing over $i$ yields

$$\hat{\boldsymbol{k}} \cdot \boldsymbol{E} = \hat{\boldsymbol{k}} \cdot \boldsymbol{E} \sum_{i=x,y,z} \frac{n^2\hat{k}_i^2}{n^2 - \epsilon_i/\epsilon_0}. \tag{14.22}$$

Cancelling the scalar product and dividing by $n^2$, one obtains the *Fresnel equation*:

$$\frac{\hat{k}_x^2}{n^2 - \epsilon_x/\epsilon_0} + \frac{\hat{k}_y^2}{n^2 - \epsilon_y/\epsilon_0} + \frac{\hat{k}_z^2}{n^2 - \epsilon_z/\epsilon_0} = \frac{1}{n^2}. \tag{14.23}$$

This is a quadratic equation in $n^2$ and therefore has *two positive solutions* for the allowed values of the refractive index.

One can pursue this approach further to determine the possible indices of refraction as a function of the propagation direction, but there is a much more useful *graphical* approach to the problem: the *index ellipsoid.* In a coordinate system where $\epsilon_{ij}$ is diagonal, the stored energy in the *electric field* is

$$u_e = \frac{1}{2}(\epsilon_x E_x^2 + \epsilon_y E_y^2 + \epsilon_z E_z^2) = \frac{1}{2}\left(\frac{D_x^2}{\epsilon_x} + \frac{D_y^2}{\epsilon_y} + \frac{D_z^2}{\epsilon_z}\right). \tag{14.24}$$

The index of refraction is given by $n = c/v_\phi$, where $v_\phi$ is the *phase velocity* in the medium. In a non-magnetic medium one can replace the three components of the dielectric tensor by three refractive indices which satisfy $\epsilon_i = n_i^2 \epsilon_0$. The energy density is

$$u_e = \frac{1}{2\epsilon_0}\left(\frac{D_x^2}{n_x^2} + \frac{D_y^2}{n_y^2} + \frac{D_z^2}{n_z^2}\right). \tag{14.25}$$

If we use a Cartesian coordinate system whose axes are the components of $\boldsymbol{r}$ and let

$$\boldsymbol{r} = \frac{\boldsymbol{D}}{\sqrt{2u_e\epsilon_0}}, \tag{14.26}$$

the surface of *constant energy density* is an ellipsoid, called the *index ellipsoid*:

$$\text{Index ellipsoid:} \quad \frac{x^2}{n_x^2} + \frac{y^2}{n_y^2} + \frac{z^2}{n_z^2} = 1, \qquad \left(\frac{\boldsymbol{D}}{\sqrt{2u_e\epsilon_0}} \to \boldsymbol{r}\right). \tag{14.27}$$

An index ellipsoid is portrayed in Fig. 14.2. The three *semi-major axes* are equal to the three indices of refraction. The following three steps describe the use of the index

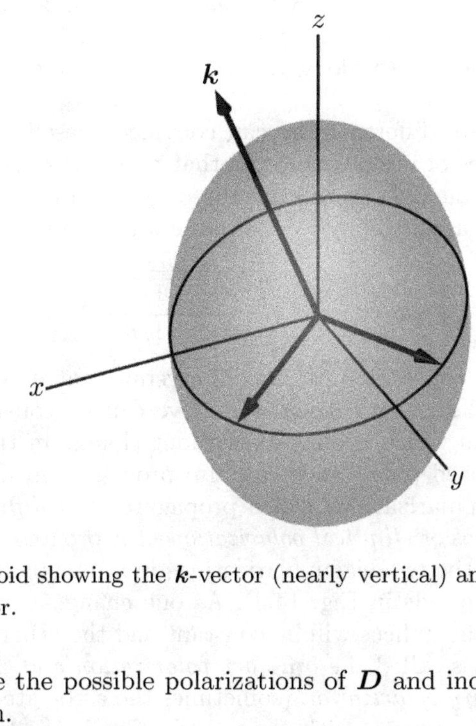


**Fig. 14.2** Index ellipsoid showing the $\boldsymbol{k}$-vector (nearly vertical) and the two possible polarizations of the $\boldsymbol{D}$ vector.

ellipsoid to determine the possible polarizations of $\boldsymbol{D}$ and indices of refraction when the $\boldsymbol{k}$-vector is known.

- First, plot an ellipsoid whose three semi-major axes are equal to three indices of refraction and whose center is at the origin of the coordinate system.
- Then draw a *plane* through the origin perpendicular to $\boldsymbol{k}$ (its equation is $\boldsymbol{r} \cdot \boldsymbol{k} = xk_x + yk_y + zk_z = 0$).
- The *ellipse* at the intersection of the plane and the ellipsoid has *two semi-axes. The allowed directions of* $\boldsymbol{D}$ *are along these semi-axes and the allowed values of* $n$ *are equal to the lengths of the semi-axes.*

The validity of the above can be verified using the Fresnel equation. Implicit in the utility of this graphical approach is the fact that the index ellipsoid is a fixed object for a given crystal and for a given wavelength of light.

Some important conclusions about the behavior of wave propagation in anisotropic media can be immediately obtained from the index ellipsoid. First, it should be obvious that, for any ellipsoid, there is at least one propagation direction for which the ellipse of intersection is a *circle*. For these propagation directions, the index of refraction is *independent* of the polarization. These directions of $\hat{\boldsymbol{k}}$ are called the *optical axes* of the medium. In general, an optical axis is defined as a *propagation direction in which the index of refraction is independent of the polarization.* The number and nature of the optical axes depend upon whether or not any of the refractive indices are *repeated.* If the medium is anisotropic, at least two indices are different. If only two are different, the crystal is called *uniaxial*. Uniaxial crystals for which $n_z > n_x = n_y$ are called *positive uniaxial*, and uniaxial crystals for which $n_z < n_x = n_y$ are called *negative uniaxial.* The ellipsoids for the three types of crystal are shown in Fig. 14.3. In drawings of unaxial index ellipsoids, it is customary for the *different* index to be along the $z$-axis (the two *equal* indices are along the $x$ and $y$ axes). It should be clear that the optical axis for uniaxial crystals is also along the $z$-axis and that there is only one optical axis.

If all three indices are different, one can convince oneself after a little reflection that there are *two* circles of intersection and that these crystals are therefore *biaxial.* The two optical axes of biaxial crystals are tilted symmetrically away from the $z$-axis at angles which depend upon the indices of refraction. The tilt angle, $\theta$, can easily be shown to be

$$\theta = \cos^{-1}\sqrt{\frac{1/n_y^2 - 1/n_z^2}{1/n_x^2 - 1/n_z^2}}, \tag{14.28}$$

where we have used the convention for biaxial crystals that $n_x < n_y < n_z$.

The two allowed polarizations discussed above can be considered to be *eigenpolarizations*: polarizations which propagate without change in the crystal. Of course, a general linearly polarized plane wave can also propagate in the crystal, but it will separate into two eigenpolarizations which propagate with *different phase velocities*, producing various degrees of *elliptical polarization* as it progresses. There is an important difference between the two eigenpolarizations in a uniaxial crystal (which can be seen in the top two ellipsoids in Fig. 14.3). As one changes the angle that $\boldsymbol{k}$ makes with the $z$-axis, one of the indices will be constant and the other will change. The one which remains constant is called the *ordinary polarization* and the one which changes is called the *extraordinary polarization* (sometimes the associated waves are called ordinary and extraordinary waves, with the polarization being the salient difference). The reason for these terms will become clearer when we discuss *ray propagation.*

From the above geometrical considerations, one can calculate the indices of refraction for an arbitrary propagation direction in a uniaxial crystal (also in a biaxial crystal, but the algebra becomes fairly complicated). The calculation makes use of a construction called *normal surfaces*, which are shown in Fig. 14.4 for a positive and negative uniaxial crystal. Normal surfaces are ellipsoids of revolution, which can be reduced to simple ellipses for a uniaxial crystal, and are obtained by plotting, *in a*

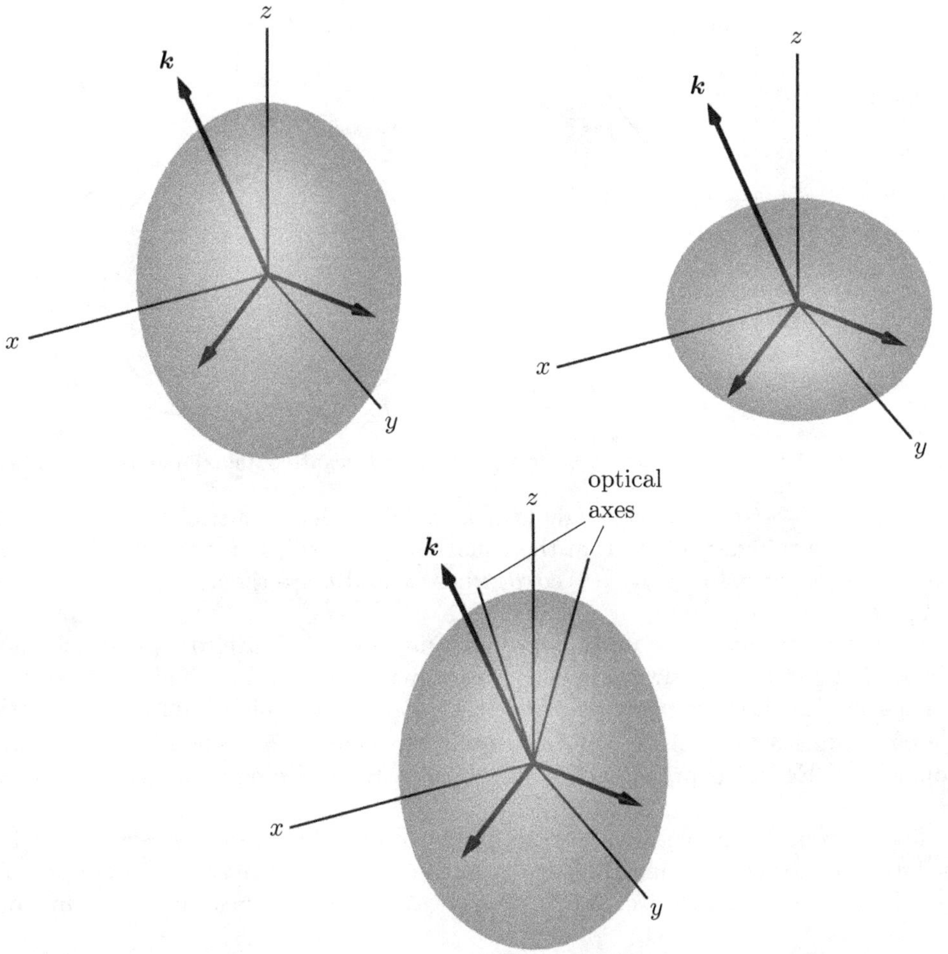


**Fig. 14.3** Index ellipsoid for positive uniaxial crystal (top left), negative uniaxial crystal (top right) and biaxial crystal (bottom). The two optical axes in the bottom figure are the lines without arrowheads.

*polar coordinate* $(\rho, \theta)$ *system*, the two indices $(\rho)$ as a function of the propagation direction $(\theta)$. Normal surfaces should be carefully distinguished from the index ellipsoid, which is a completely different object. The constructions are very simple, since the semi-major and semi-minor axes are just the indices of refraction. Notice that the curve for the ordinary index is a circle and the ellipse is inscribed in the circle for a negative crystal and the circle is inscribed in the ellipse for a positive crystal. Using these constructions, one can immediately write down equations for the indices:

$$n_o(\theta) = n_0 \quad \text{(constant)} \tag{14.29}$$

$$\frac{1}{n_e^2(\theta)} = \frac{\cos^2\theta}{n_o^2} + \frac{\sin^2\theta}{n_e^2}, \tag{14.30}$$

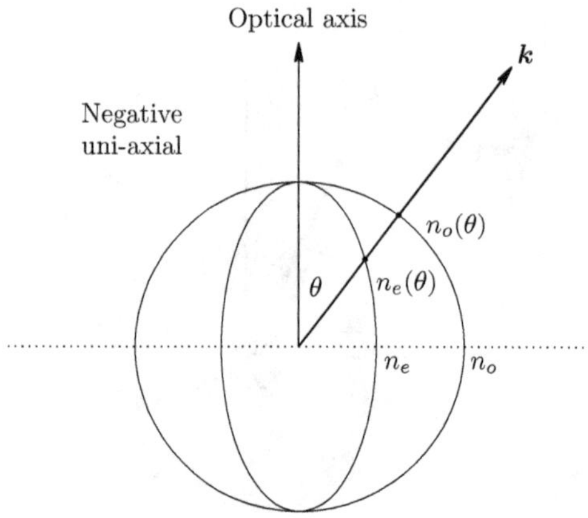


**Fig. 14.4** Normal surfaces for a positive and negative uniaxial crystal.

where $n_e = n_e(0)$ and $\theta$ is the angle that $\boldsymbol{k}$ makes with the optical axis. The second equation follows immediately from the equation for an ellipse together with the relations $x = n_e(\theta)\sin\theta$ and $y = n_e(\theta)\cos\theta$, where $x$ and $y$ are the abscissa and ordinate of the two-dimensional plot.

So far, we have discussed *wave propagation*, in which the wavefront progresses along $\hat{\boldsymbol{k}}$ in a direction that is always normal to the wavefront surface. The speed of wave propagation is called the *phase velocity* ($v_\phi$). In an anisotropic medium, the direction of *energy transport* (along $\hat{\boldsymbol{S}}$) will in general be different from the direction of wave propagation. Energy is propagated along *rays*, with a speed equal to the *ray velocity*, $v_r$.

The difference between wave and ray propagation is illustrated by Fig. 14.5, which displays a construction, using Huygens' principle, of a wave traveling in an anisotropic crystal. For the ordinary wave, the Huygens' wavelets are spheres, since $v_\phi$ is indepen-

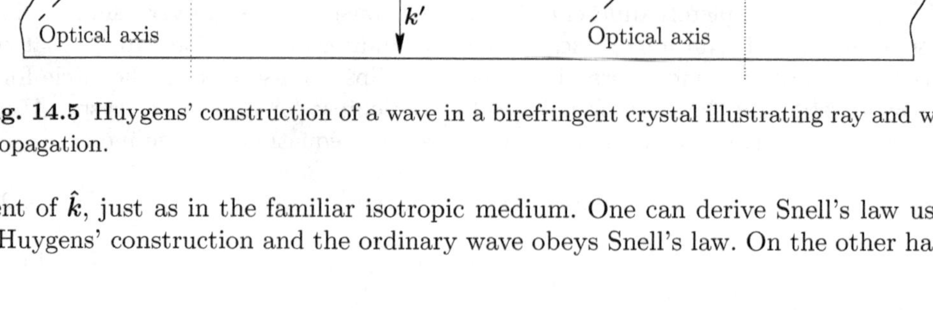


**Fig. 14.5** Huygens' construction of a wave in a birefringent crystal illustrating ray and wave propagation.

dent of $\hat{\boldsymbol{k}}$, just as in the familiar isotropic medium. One can derive Snell's law using a Huygens' construction and the ordinary wave obeys Snell's law. On the other hand, the extraordinary wavelets are ellipsoids, since different portions of the extraordinary wavefront propagate with different speeds due to the dependence of the index of refraction on $\hat{\boldsymbol{k}}$. The extraordinary wave *does not obey Snell's law*, which accounts for the name "extraordinary". If the optical axis is not perpendicular or parallel to the crystal surface, an unpolarized wave normally incident on the crystal will split into two beams: one consisting of ordinary rays and one consisting of extraordinary rays. These two beams will separate spatially as they progress through the crystal. This phenomenon is strikingly observed when a piece of birefringent calcite is placed over printed material and two separate images of the print are observed. The separation into two beams is called *double refraction*. It can be proven that the *ray direction for any wave is normal to the index surface for that wave.* From this, it is easy to show that

$$\tan\phi = \frac{n_o^2}{n_e^2}\tan\theta, \tag{14.31}$$

where $\theta$ is the angle of $\boldsymbol{k}$ with respect to the optical axis and $\phi$ is the angle of the Poynting vector (ray direction) relative to the optical axis.

## 14.3 Second-harmonic generation

We have used the linear electrical susceptibility, $\chi$, several times in this book and will remind the reader that $\epsilon_0\chi$ is the ratio of the polarization to the electric field:

$$\text{Linear susceptibility}: \qquad \boldsymbol{P} = \epsilon_0\chi\boldsymbol{E} \qquad \chi = \text{complex constant}. \tag{14.32}$$

Thus, a time-dependent polarization is a faithful replica (up to a multiplicative complex constant) of the time-varying electric field. If we relax the requirement that $\chi$ be constant and now let it *depend upon* the field, $\boldsymbol{E}$, the polarization is no longer a replica of the field but will contain some *distortion*, as shown in Fig. 14.6.

$$\text{Nonlinear susceptibility:} \qquad \boldsymbol{P} = \epsilon_0\chi(\boldsymbol{E})\boldsymbol{E}. \tag{14.33}$$

If $\chi(\boldsymbol{E})$ is an *odd function* of $\boldsymbol{E}$, the distorted polarization will contain some second harmonic. This will only occur if the crystal is *non-centrosymmetric* (i.e., it lacks a center of symmetry). A centrosymmetric crystal will only generate odd harmonics of the incoming field, as shown in Fig. 14.7.

To treat the problem analytically, we start with the time-dependent Maxwell equations (without sources) with the *polarization* explicitly displayed:

$$\nabla\times\boldsymbol{H} = \frac{\partial\boldsymbol{D}}{\partial t} = \frac{\partial}{\partial t}(\epsilon_0\boldsymbol{E}+\boldsymbol{P}) \tag{14.34}$$

$$\nabla\times\boldsymbol{E} = -\frac{\partial\boldsymbol{B}}{\partial t} = -\frac{\partial}{\partial t}(\mu_0\boldsymbol{H}). \tag{14.35}$$

The polarization has a *linear* and a *nonlinear* contribution:

$$\boldsymbol{P} = \epsilon_0\chi_L\boldsymbol{E} + \boldsymbol{P}_{\text{NL}}, \tag{14.36}$$

where $\boldsymbol{P}_{NL}$ depends *quadratically* on the field:

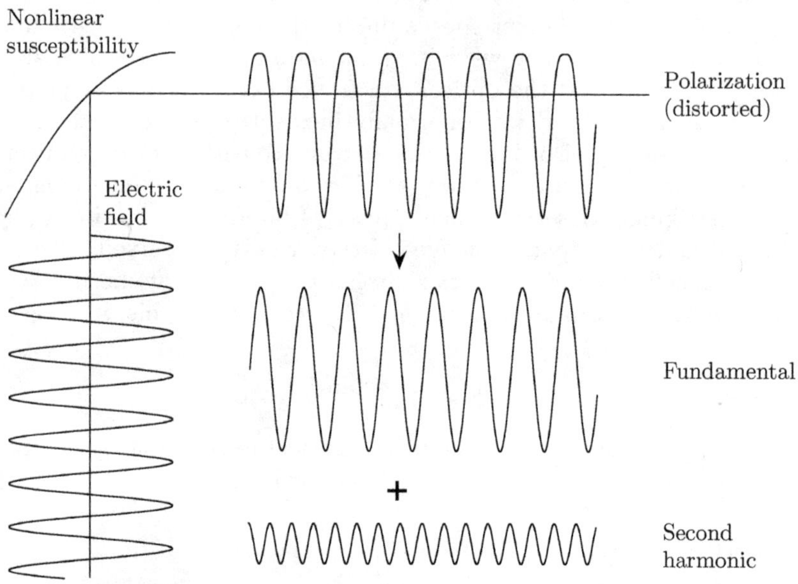


**Fig. 14.6** Illustration of distortion caused by nonlinear susceptibility. The polarization will acquire some second harmonic.

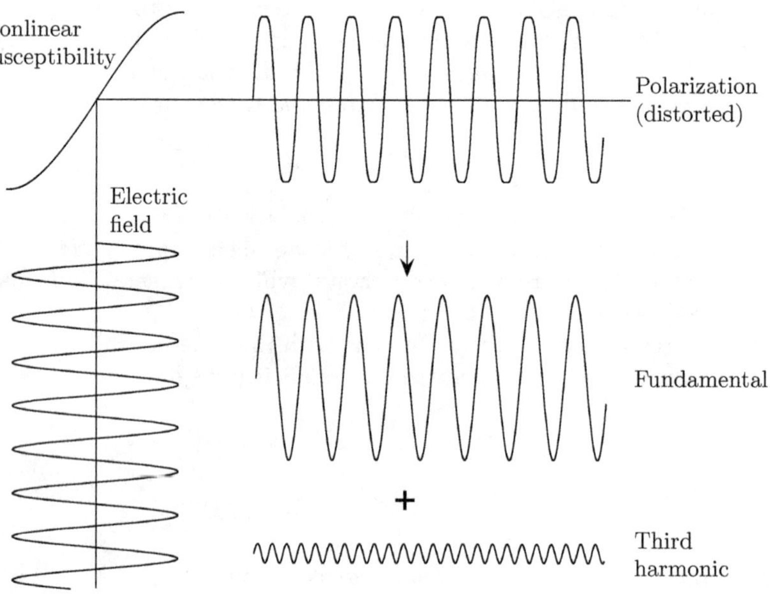


**Fig. 14.7** Illustration of distortion from a centrosymmetric crystal. The polarization will only generate odd harmonics.

$$P_{NL,i} = \sum_{j,k} d_{ijk} E_j E_k \qquad i, j, k = x, y, z \tag{14.37}$$

and $d_{ijk}$ is the nonlinear optical susceptibility tensor. Including the nonlinear polarization in Ampere's law, one obtains

$$\nabla \times \boldsymbol{H} = \frac{\partial}{\partial t} \epsilon \boldsymbol{E} + \frac{\partial \boldsymbol{P}_{NL}}{\partial t}, \tag{14.38}$$

where

$$\epsilon = \epsilon_0 (1 + \chi_L). \tag{14.39}$$

Taking the curl of Faraday's law and substituting the above yields

$$\nabla^2 \boldsymbol{E} - \mu_0 \epsilon \frac{\partial^2 \boldsymbol{E}}{\partial t^2} = \mu_0 \frac{\partial^2}{\partial t^2} \boldsymbol{P}_{NL}. \tag{14.40}$$

This is a wave equation whose source is the nonlinear polarization.

This equation will be solved by assuming that $\boldsymbol{E}$ is a plane wave with frequency $2\omega$ and the source for the nonlinear polarization is a plane wave whose frequency is $\omega$:

$$\text{Output wave:} \quad E_j^{(2\omega)} = \frac{1}{2} E_{2j}(z) e^{i(2\omega t - k_2 z)} + \text{cc} \tag{14.41}$$

$$\text{Input wave:} \quad E_i^{(\omega)} = \frac{1}{2} E_{1i}(z) e^{i(\omega t - k_1 z)} + \text{cc} \quad (i, j = x, y), \tag{14.42}$$

where $k_1$ and $k_2$ are the input and output $k$-vectors respectively and the amplitudes $E_{1i}$ and $E_{2j}$ are *slowly varying* functions of $z$. If we assume that the $x$ and $y$ second derivatives vanish (plane wave) and $\frac{dE_{2j}}{dz} k_2 \gg \frac{d^2 E_{2j}}{dz^2}$, then we can discard the spatial second derivative and the left-hand side of the wave equation becomes

$$-\frac{1}{2} \left[ (k_2)^2 E_{2j}(z) + 2ik_2 \frac{dE_{2j}(z)}{dz} \right] e^{i(2\omega t - k_2 z)} + 4\omega^2 \mu_0 \epsilon \left[ \frac{E_{2j}}{2} e^{i(2\omega t - k_2 z)} \right] + \text{cc}. \tag{14.43}$$

Using $(2\omega)^2 \mu_0 \epsilon = (k_2)^2$ and including the right-hand side (the nonlinear polarization),

$$ik_2 \frac{dE_{2j}}{dz} e^{-ik_2 z} = \mu_0 \omega^2 \sum_{ik} d_{jik} E_{1i} E_{1k} e^{-2ik_1 z}. \tag{14.44}$$

Simplifying,

$$\frac{dE_{2j}}{dz} = -\frac{i\omega}{2} \sqrt{\frac{\mu_0}{\epsilon_2}} \sum_{ik} d_{jik} E_{1i} E_{1k} e^{i\Delta k z}, \qquad \Delta k = k_2 - 2k_1, \tag{14.45}$$

where $\epsilon_2$ is the permittivity of the medium at $2\omega$.

This equation is readily solved if we use the boundary condition that $\boldsymbol{E}_2(z = 0) = 0$ and assume that there is no depletion of $\boldsymbol{E}_1$ along the crystal. Then, a simple integration yields

$$E_{2j}(L) = -\frac{\omega}{2}\sqrt{\frac{\mu_0}{\epsilon_2}}\sum_{ik} d_{jik}E_{1i}E_{1k}\frac{e^{i\Delta kL}-1}{\Delta k}, \tag{14.46}$$

where the length of the crystal is $L$. The second-harmonic *intensity* (power/area) is

$$I^{(2\omega)} = \frac{\text{power}^{(2\omega)}}{\text{area}} = \frac{1}{2}\sqrt{\frac{\epsilon_2}{\mu_0}}\boldsymbol{E}_2 \cdot \boldsymbol{E}_2^*, \tag{14.47}$$

which yields

$$I^{(2\omega)} = \frac{1}{8}\sqrt{\frac{\mu_0}{\epsilon_2}}\omega^2(d_{ijk})^2E_{1i}^2E_{1k}^2L^2\frac{\sin^2(\Delta kL/2)}{(\Delta kL/2)^2}, \tag{14.48}$$

where we assume that only one nonlinear coefficient ($d_{ijk}$) is significant and include a factor of 2 for the sum over $i$ and $k$ in the field expression. Finally, the efficiency is

$$\frac{I^{(2\omega)}}{I^{(\omega)}} = \frac{1}{2}\left(\frac{\mu_0}{\epsilon_0}\right)^{3/2}\frac{\omega^2(d_{ijk})^2L^2}{n^3}I^{(\omega)}\frac{\sin^2(\Delta kL/2)}{(\Delta kL/2)^2}, \tag{14.49}$$

where we assume that $\epsilon \approx \epsilon_2$ and $\epsilon = n^2\epsilon_0$. This displays the characteristic dependence of the efficiency on $L^2$ and $I^{(\omega)}$ (i.e., the second harmonic intensity is proportional to the *square* of the fundamental intensity). We will show later on that the $L^2$ dependence only holds for second-harmonic generation in a guiding structure such as an optical waveguide or fiber; in free space the dependence is at best linear in $L$ when the input beam is focused optimally.

For the purpose of describing second-harmonic generation, the nonlinear optical susceptibility tensor, $d_{ijk}$, is usually written as a rectangular, two-dimensional array whose elements are defined as:

$$\begin{pmatrix} P_{NL,x} \\ P_{NL,y} \\ P_{NL,z} \end{pmatrix} = \begin{pmatrix} d_{11}\ d_{12}\ d_{13}\ d_{14}\ d_{15}\ d_{16} \\ d_{21}\ d_{22}\ d_{23}\ d_{24}\ d_{25}\ d_{26} \\ d_{31}\ d_{32}\ d_{33}\ d_{34}\ d_{35}\ d_{36} \end{pmatrix} \begin{pmatrix} E_x^2 \\ E_y^2 \\ E_z^2 \\ 2E_zE_y \\ 2E_zE_x \\ 2E_xE_y \end{pmatrix}. \tag{14.50}$$

Despite the large number of possible elements, only a few are usually unique, depending upon the crystal geometry. While it would take us into the area of crystal symmetries to show that many of the coefficients are either zero or simply related to each other, there is a simple symmetry that is based upon a principle called *Kleinman's conjecture*. This states that, in the absence of absorption at any of the frequencies, the tensor elements, $d_{ijk}$, are invariant upon any permutation of the indices. Thus, the following index equalities for $d_{ij}$ hold:

$$\begin{aligned} d_{14} &= d_{25} = d_{36} \\ d_{15} &= d_{31} \\ d_{23} &= d_{34} \\ d_{12} &= d_{26} \\ d_{16} &= d_{21} \\ d_{13} &= d_{35} \\ d_{24} &= d_{32}. \end{aligned} \tag{14.51}$$

A list of the non-zero $d$-values for some of the more important crystals is presented here.

- **Tetragonal, class $\bar{4}2m$** – example, ADP, AD*P, KDP, KD*P

$$\begin{pmatrix} 0 & 0 & 0 & d_{14} & 0 & 0 \\ 0 & 0 & 0 & 0 & d_{14} & 0 \\ 0 & 0 & 0 & 0 & 0 & d_{36} \end{pmatrix} \tag{14.52}$$

- **Trigonal, class $3m$** – example, $\beta$-barium borate (BBO), lithium niobate, lithium tantalate

$$\begin{pmatrix} 0 & 0 & 0 & 0 & d_{15} & -d_{22} \\ -d_{22} & d_{22} & 0 & d_{15} & 0 & 0 \\ d_{31} & d_{31} & d_{33} & 0 & 0 & 0 \end{pmatrix} \tag{14.53}$$

- **Hexagonal, class 6** – example, lithium iodate

$$\begin{pmatrix} 0 & 0 & 0 & d_{14} & d_{15} & 0 \\ 0 & 0 & 0 & d_{15} & -d_{14} & 0 \\ d_{31} & d_{31} & d_{33} & 0 & 0 & 0 \end{pmatrix} \tag{14.54}$$

- **Orthorhombic, class** $mm2$ – example, potassium niobate, LBO, KTP

$$\begin{pmatrix} 0 & 0 & 0 & 0 & d_{15} & 0 \\ 0 & 0 & 0 & d_{24} & 0 & 0 \\ d_{31} & d_{32} & d_{33} & 0 & 0 & 0 \end{pmatrix} \tag{14.55}$$

A tabulation of the larger nonlinear coefficients and Sellmeier coefficients (see below) for a number of popular uniaxial crystals appears in Table 14.1. A tabulation for biaxial crystals is in Table 14.2.

**Table 14.1** Nonlinear coefficients and Sellmeier coefficients for some popular uniaxial SHG crystals (Sellmeier coefficients from *Tunable Laser Handbook* (1995), nonlinear coefficients from Roberts (1992), Davis (1996) and Yariv (1989)).

| Crystal | $d_{ij}$ (pm/V) | transparency range (nm) | Sellmeier coefficients ($\lambda$ in $\mu$m) | | | | |
|---|---|---|---|---|---|---|---|
| | | | $A_o$ | $B_o$ | $C_o$ | $D_o$ | $E_o$ |
| | | | $A_e$ | $B_e$ | $C_e$ | $D_e$ | $E_e$ |
| ADP | $d_{36} = 0.41$ | 180-1500 | 1.37892 | 0.91996 | 0.01249 | .15771 | 5.76 |
| | $d_{14} = 0.63$ | | 1.35302 | 0.80752 | 0.01227 | 0.02612 | 3.3156 |
| KDP | $d_{36} = 0.39$ | 180-1550 | 1.41344 | 0.84308 | 0.01229 | 0.26923 | 10.248 |
| | $d_{14} = 0.42$ | | 1.40442 | 0.72733 | 0.01201 | 0.07974 | 12.484 |
| $LiNbO_3$ | $d_{31} = -4.3$ | 370-4500 | 2.33907 | 2.58395 | 0.04588 | 13.8169 | 519.658 |
| | $d_{33} = -27 \pm 1.0$ | | 2.35084 | 2.22518 | 0.04371 | 15.9733 | 741.146 |
| | $d_{22} = 2.1$ | | | | | | |
| BBO | $d_{22} = 2.3$ | 198-2600 | 1.71283 | 1.02790 | 0.01790 | 2.23130 | 138.65 |
| | $d_{31} = 0.1$ | | 1.50569 | 0.86544 | 0.01512 | 0.56478 | 248.36 |

The Sellmeier coefficients are used to calculate the indices of refraction using the Sellmeier equation:

$$n^2 = A + \frac{B\lambda^2}{\lambda^2 - C} + \frac{D\lambda^2}{\lambda^2 - E}. \tag{14.56}$$

There is a set of parameters for each index, where the subscript identifies the index and is $e$ and $o$ for a uniaxial crystal or $x$, $y$ and $z$ for a biaxial crystal. There are many variants of the Sellmeier equation; the one given above provides tolerable accuracy for wavelengths from the ultraviolet to the near-infrared.

## 14.4 Birefringent phase matching

From eqn 14.49, we see that the second-harmonic power is proportional to

**Table 14.2** Nonlinear coefficients and Sellmeier coefficients for some popular biaxial SHG crystals (Sellmeier coefficients from *Tunable Laser Handbook* (1995), nonlinear coefficients from Roberts (1992)).

| Crystal | $d_{ij}$ (pm/V) | transparency range (nm) | Sellmeier coefficients ($\lambda$ in $\mu$m) | | | | |
|---|---|---|---|---|---|---|---|
| | | | $A_x$ | $B_x$ | $C_x$ | $D_x$ | $E_x$ |
| | | | $A_y$ | $B_y$ | $C_y$ | $D_y$ | $E_y$ |
| | | | $A_z$ | $B_z$ | $C_z$ | $D_z$ | $E_z$ |
| KTP | $d_{31} = 3.6$ | 350-4500 | 2.22237 | 0.78681 | 0.04746 | 0.67167 | 54.9 |
| | $d_{32} = 2.0$ | | 2.3059 | 0.72572 | 0.05387 | 1.00870 | 77.50 |
| | $d_{33} = 8.3$ | | 2.35249 | 0.96655 | 0.05812 | 1.24674 | 77.50 |
| LBO | $d_{15} = .85$ | 160-2300 | 2.07557 | 0.38193 | 0.02597 | 2.60858 | 191.04 |
| | $d_{24} = -0.67$ | | 1.61856 | 0.92347 | 0.01355 | 4.48336 | 204.16 |
| | $d_{33} = 0.04$ | | 2.00372 | 0.58147 | 0.02176 | 2.55777 | 155.84 |

$$\left(\frac{\sin(\Delta k L/2)}{(\Delta k L/2)}\right)^2, \qquad \Delta \boldsymbol{k} = \boldsymbol{k}_2 - 2\boldsymbol{k}_1. \tag{14.57}$$

This expression is unity when $\Delta k = 0$ and is zero when $\Delta k L = 2\pi$. In the interests of obtaining a useful amount of second harmonic, we require that the *phase-matching condition* be satisfied:

$$\text{Phase matching:} \qquad \Delta \boldsymbol{k} = 0 \Longrightarrow \boldsymbol{k}_2 = 2\boldsymbol{k}_1, \tag{14.58}$$

which implies that the phase velocities and indices of refraction of the second harmonic and fundamental must be the same:

$$v_{2\omega} = v_\omega \tag{14.59}$$

$$n_{2\omega} = n_\omega. \tag{14.60}$$

If the phase-matching condition is violated, the second-harmonic power will have a $\sin^2(\Delta k z/2)$ dependence on $z$ as one proceeds along the doubling crystal. When $\Delta k \neq 0$, the fundamental and second harmonic will propagate at *different* speeds; after a distance of $\pi/\Delta k$ the two beams will begin to be out of phase and the newly generated second harmonic will *subtract* from the existing second harmonic. It is as though the crystal length were reduced from $L$ to $\pi/\Delta k$, which is a considerable reduction. One can consider the latter length to be a *coherence length*, $l_c$:

$$l_c = \frac{\pi}{\Delta k} = \frac{\lambda}{4(n_{2\omega} - n_\omega)}, \tag{14.61}$$

where $\lambda$ is the *fundamental* wavelength. If no phase-matching techniques are employed, $l_c$ can be 100 $\mu$m, which is two orders of magnitude shorter than a typical crystal.

The phase-matching condition has a finite bandwidth, which can be obtained from the $\text{sinc}^2$ factor in eqn 14.49 together with the dispersion properties of the index. A plot of the dependence of the second-harmonic power on $\Delta k$ appears in Fig. 14.8. The

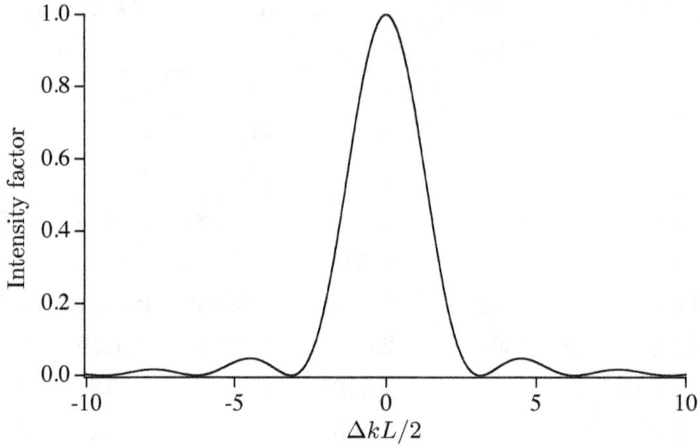


**Fig. 14.8** Plot of frequency dependence of second harmonic due to phase matching.

full width at half maximum of this curve is $\delta(\Delta kL/2) = 0.886\pi$. The wavelength bandwidth (at the fundamental wavelength) is obtained by equating this to the derivative of $L\Delta k/2$ times $\Delta\lambda$:

$$\frac{L}{2}\frac{\partial \Delta k}{\partial \lambda}\Delta\lambda = 0.886\pi. \tag{14.62}$$

The result is

$$\Delta\lambda = \frac{0.443\lambda}{L}\left|\frac{1}{2}\left(\frac{\partial n}{\partial \lambda}\right)_{2\omega} - \left(\frac{\partial n}{\partial \lambda}\right)_{\omega}\right|^{-1}, \tag{14.63}$$

where the derivatives are evaluated at the fundamental and second-harmonic wavelengths, and $\lambda$ is the fundamental wavelength. The derivatives can be obtained from the Sellmeier equation.

Phase matching can also be interpreted as *conservation of momentum.* Assuming that there are no phonon changes in the second-harmonic generation process, phase matching is equivalent to the requirement that the momentum of the *single outgoing photon* at $2\omega$ be equal to the sum of the momenta of the *two incoming photons* at $\omega$.

There are two commonly used methods for phase matching: exploitation of crystal *birefringence* and *quasi-phase matching.* We will discuss the former in this section. The idea behind birefringent phase matching is to use the fact that, under *normal dispersion,* the index of refraction increases with frequency. If the fundamental polarization is aligned with the axis corresponding to the larger of the two birefringent indices (in a uniaxial crystal) and the second harmonic to the smaller, there is a wavelength where the indices of the fundamental and second harmonic are the same. We will shortly show that the optimum condition is where the fundamental beam propagates in a direction perpendicular to the optical axis of a crystal. If the crystal is *negative uniaxial,* the fundamental would be polarized perpendicular to the optical axis (the ordinary wave) and the second harmonic would be polarized along the optical axis (extraordinary wave). The scheme is illustrated in Fig. 14.9 for ADP, which is a negative uniaxial crystal. Phase matching occurs only at a fundamental wavelength of about 0.526 $\mu$m.

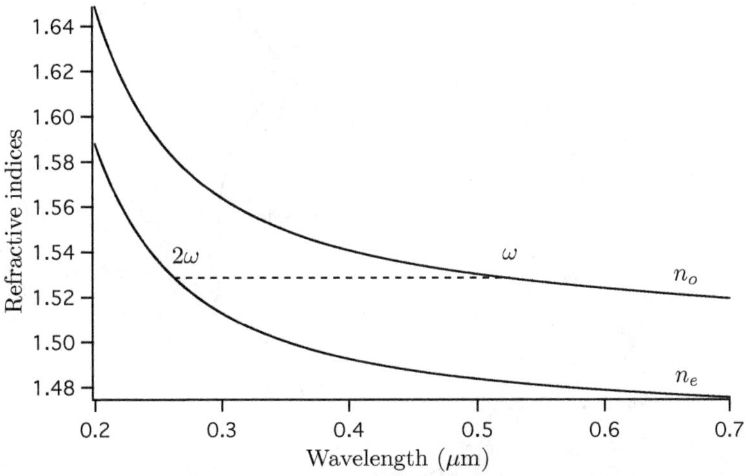


**Fig. 14.9** Index curves for ADP, illustrating phase matching for a fundamental of 0.526 $\mu$m. The indices are calculated using the Sellmeier equation and the fundamental propagates perpendicular to the optical axis.

The scheme described above has a serious drawback: it only works at a single fundamental wavelength. However, we learned that by changing the propagation direction, one can change the *extraordinary index of refraction* (but not the ordinary). Thus, one can *tune* the phase matching by changing the angle that the fundamental $k$-vector makes with the optical axis. This allows phase matching at a range of wavelengths. For a negative uniaxial crystal, the fundamental would be an ordinary wave and the second harmonic would be an extraordinary wave. The functional dependence of the extraordinary index on the angle, $\theta$, between $\boldsymbol{k}$ and the optical axis is derived from the normal surfaces:

$$\frac{1}{n_e^2(\theta)} = \frac{\cos^2\theta}{n_o^2} + \frac{\sin^2\theta}{n_e^2}. \tag{14.64}$$

Setting $n_{o,\omega} = n_{e,2\omega}$,

$$\frac{1}{n_{o,\omega}^2} = \frac{\cos^2\theta}{n_{o,2\omega}^2} + \frac{\sin^2\theta}{n_{e,2\omega}^2}. \tag{14.65}$$

This can be solved for the angle

$$\sin^2\theta = \frac{(n_{o,\omega})^{-2} - (n_{o,2\omega})^{-2}}{(n_{e,2\omega})^{-2} - (n_{o,2\omega})^{-2}}. \tag{14.66}$$

The process is illustrated in Fig. 14.10 using normal surfaces for $\omega$ and $2\omega$ plotted on the same graph. Phase matching takes place where the two solid curves intersect.

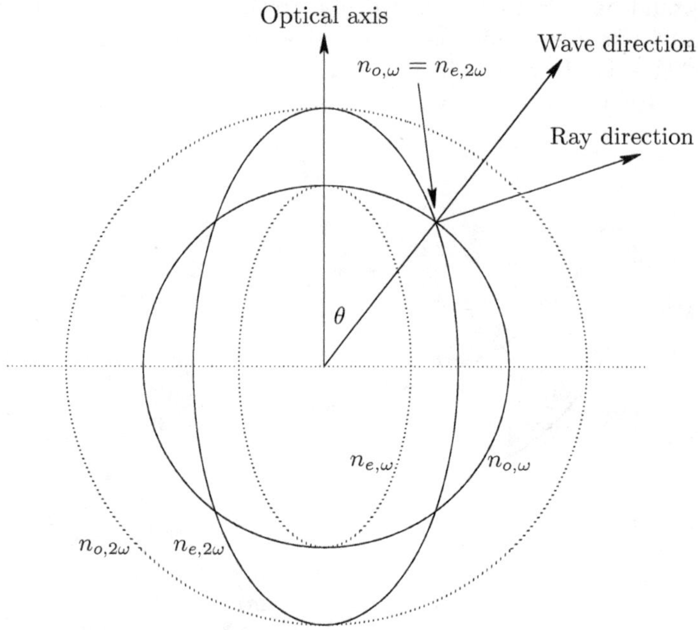


**Fig. 14.10** Illustration of birefringent phase matching in negative uniaxial crystal by plotting the normal surfaces for the fundamental and second harmonic on the same graph. Also shown is *walk-off*: the different directions between the input wave and the output ray (Poynting vector), which is perpendicular to the $2\omega$ extraordinary normal surface.

When one has a *positive uniaxial* crystal, the fundamental would be an extraordinary wave and the second harmonic would be an ordinary wave. Phase matching is obtained by varying the angle and thereby varying the index of the fundamental until it is equal to that of the second harmonic. Proceeding as before, with $n_{o,2\omega} = n_{e,\omega}$, the angle is given by

$$\sin^2\theta = \frac{(n_{o,2\omega})^{-2} - (n_{o,\omega})^{-2}}{(n_{e,\omega})^{-2} - (n_{o,\omega})^{-2}}. \tag{14.67}$$

In our treatment so far, we have assumed that the two input fields in second-harmonic generation have the same polarization. However, in eqn 14.45, the two source field components are independent of each other and second harmonic generation can occur from *orthogonal* source fields if the nonlinear coefficient has non-zero elements which couple two different polarizations. It is customary to refer to SHG from the same polarizations as *type I* and from orthogonal polarizations as *type II*. Phase matching in type II second-harmonic generation is less tunable than in type I, since only one field's refractive index can be used to compensate for the index difference between $\omega$ and $2\omega$ due to normal dispersion. With a negative uniaxial crystal, the output field will be an extraordinary wave and the input field will consist of one ordinary wave and one extraordinary wave. Using a fairly obvious notation, the phase matching requirement is

$$\text{Type II:} \quad \boldsymbol{k}^e_{2\omega} = \boldsymbol{k}^o_\omega + \boldsymbol{k}^e_\omega \Longrightarrow n_{e,2\omega} = \tfrac{1}{2}(n_{o,\omega} + n_{e,\omega}). \tag{14.68}$$

The phase-matching angle can be determined in the same way as for type I, though the equations are fairly cumbersome and might need to be solved using a computer.

In order to satisfy the phase-matching requirements for a given frequency, $\omega$, both the $k$-vector and the polarization are constrained to be in very specific directions. From eqn 14.45, we see that one needs to sum over all of the appropriate field components and nonlinear coefficients to obtain the second-harmonic field. This sum will, of course, depend upon the orientation of the $k$-vector and polarization. It is convenient to define an *effective nonlinear coefficient*, $d_{eff}$, to include these sums, and write the polarization simply as

$$P_{2\omega} = d_{eff} E_\omega E'_\omega, \tag{14.69}$$

where $E_\omega$ and $E'_\omega$ are the two input field magnitudes. One can determine the angular dependence of $d_{eff}$ using geometry. We will illustrate this with a simple example: we will determine the dependence of $d_{eff}$ on the propagation direction for a crystal of class $\bar{4}2m$ (such as ADP). It is useful to first write down *unit vectors* along the ordinary and extraordinary polarization directions when the propagation direction makes an angle $\theta$ with the $z$-axis and an angle $\phi$ with the $x$-axis. We will assume that the phase matching is type I, so $\theta$ and $\phi$ are the same for the two waves (and $E_\omega = E'_\omega$). Using column vectors,

$$\text{Ordinary polarization unit vector:} \quad \boldsymbol{a} = \begin{pmatrix} \sin\phi \\ -\cos\phi \\ 0 \end{pmatrix}, \tag{14.70}$$

$$\text{Extraordinary polarization unit vector:} \quad \boldsymbol{b} = \begin{pmatrix} -\cos\phi\cos\theta \\ -\sin\phi\cos\theta \\ \sin\theta \end{pmatrix}. \tag{14.71}$$

From the nonlinear tensor tabulated above for this class, the polarization is

$$\begin{aligned} P_1 &= 2d_{14}E_3E_2 \\ P_2 &= 2d_{14}E_3E_1 \\ P_3 &= 2d_{36}E_1E_2. \end{aligned} \tag{14.72}$$

We assume that the two input fields are *ordinary waves* and the output field is an extraordinary wave. The input components are $E\boldsymbol{a}$:

$$\begin{aligned} E_1 &= E\sin\phi \\ E_2 &= -E\cos\phi \\ E_3 &= 0. \end{aligned} \tag{14.73}$$

Substituting these into the polarization expressions, we obtain

$$
\begin{aligned}
P_1 &= 0 \\
P_2 &= 0 \\
P_3 &= -2d_{36}E^2 \sin\phi\cos\phi.
\end{aligned}
\tag{14.74}
$$

The output extraordinary polarization, $P_e$, is the projection of the polarization along the extraordinary direction: $\boldsymbol{b}\cdot\boldsymbol{P}$

$$P_e = (-2d_{36}\sin\phi\cos\phi\sin\theta)E^2, \tag{14.75}$$

and the effective nonlinear coefficient is (simplifying $\sin\phi\cos\phi$)

$$d_{eff} = -d_{36}\sin 2\phi\sin\theta = -d_{14}\sin 2\phi\sin\theta, \tag{14.76}$$

where the last equality is due to an application of Kleinman's conjecture. A table of effective nonlinear coefficients for various crystal classes appears in Table 14.3. In

**Table 14.3** $d_{eff}$ for several crystal classes (assumes Kleinman's conjecture).

| Crystal class | $d_{eff}$ for *ooe* | $d_{eff}$ for *eeo* |
|---|---|---|
| Tetragonal, class $\bar{4}2m$ | $-d_{14}\sin\theta\sin 2\phi$ | $d_{14}\sin 2\theta\cos 2\phi$ |
| Trigonal, class $3m$ | $d_{15}\sin\theta - d_{22}\cos\theta\sin 3\phi$ | $d_{22}\cos^2\theta\cos 3\phi$ |
| Hexagonal, class 6 | $d_{15}\sin\theta$ | 0 |

the table, the expression *ooe* refers to two input ordinary waves and a single-output extraordinary wave, and *eeo* refers to two input extraordinary waves and a single-output ordinary wave.

One adverse aspect of birefringent phase matching which is illustrated in Fig. 14.10 is the fact that, in general, the directions of *energy propagation* (ray directions) are different for the fundamental and second harmonic. This will have three consequences. First, the two waves will separate in a distance called the *aperture length* and the conversion will therefore be harmed; this will be discussed in more detail in a later section and is called *walk-off*. The second effect is the greater sensitivity of the phase-matching condition to the input beam propagation direction. Finally, the beam *quality* will suffer due to the walk-off. None of these effects will occur if the input beam propagates perpendicular to the optical axis: there will be no walk-off and the phase-matching conditions will depend *quadratically* on the propagation direction and will therefore be insensitive to it. Phase matching of this kind is called 90° or *non-critical* phase matching and is the desirable state of affairs. Often, one *temperature tunes* the crystal in order to achieve non-critical phase matching.

## 14.5 Quasi-phase matching

In the previous section, we investigated one method of phase matching: the use of crystal birefringence together with normal dispersion to ensure that $n_\omega = n_{2\omega}$. This technique has two glaring flaws. First, there are frequency gaps in which phase matching is impossible. Second, even if phase matching can be accomplished, it often requires

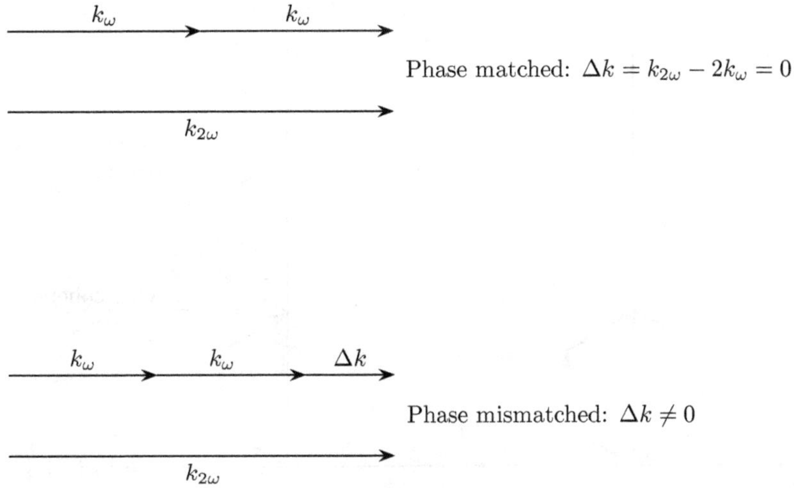


**Fig. 14.11** Vector diagrams depicting phase matching (top) and phase mismatching (bottom)

propagation angles that are far from the optimum *non-critical phase-matching angle* (90° to the optical axis), with adverse effects on beam quality and conversion efficiency. The technique of *quasi-phase matching* avoids both of these drawbacks at the expense of additional complexities in crystal fabrication.

A simple explanation of the technique can be obtained via the notion of *spatial frequency modulation.* We have already introduced this concept when we discussed the feedback mechanism in distributed feedback (DFB) semiconductor lasers. We start with a wave described by the equation

$$E(z,t) = E_0 \sin(\omega t - kz). \tag{14.77}$$

The wave vector, $k$, can considered to be the *spatial frequency* of a wave, just as $\omega$ is its *temporal frequency.* In Chapter 4 we discussed both phase and amplitude modulation by a sinusoid whose temporal frequency is $\omega_m$. For both types of modulation, the *frequency spectrum* will have sidebands on either side of the original *carrier*, displaced from the carrier by $\omega_m$. Just as temporal phase modulation can be achieved by varying the phase of the original wave by a function of time whose angular frequency is $\omega_m$, *spatial phase modulation* can be achieved by directing the wave into a periodic structure whose index of refraction varies with a period of $\Lambda$ and whose *spatial angular frequency* is therefore $2\pi/\Lambda$ (its frequency in *spatial cycles per cm* is $1/\Lambda$).

With this preliminary, we consider the case of a phase-mismatched system, shown in the bottom half of Fig. 14.11. The figure uses the momentum conservation approach to phase matching and depicts both the phase-matched condition (top) and the phase-mismatched condition (bottom). As one can see from the bottom figure, there is an excess of spatial frequency equal to $\Delta k$ at the second harmonic. The *quasi-phase matching* scheme supplies this excess to the nonlinear polarization by generating *spatial frequency sidebands* via a phase modulation which is achieved by alternately reversing

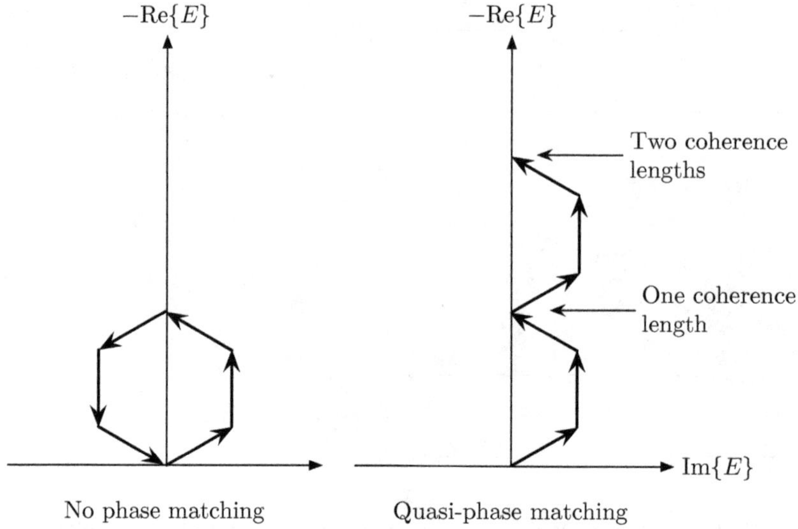


**Fig. 14.12** Phasor diagrams for electric field over two coherence lengths. On the left, complete cancellation of the field is shown when there is no phase matching. On the right, phase reversal after one coherence length in periodic structure allows the field to grow.

the sign of the nonlinear coefficient of the material with a period of $\Lambda$. Phase matching will occur when

$$\frac{2\pi}{\Lambda} + \Delta k = 0. \tag{14.78}$$

If we divide this equation by $\lambda$ and use the fact that $\Delta\lambda/\lambda = -\Delta k/k$, we obtain

$$\Lambda = \lambda \left(\frac{1}{r}\right), \qquad \text{where } r \equiv \frac{\Delta\lambda}{\lambda}, \tag{14.79}$$

and $\Delta\lambda$ is the difference between the desired wavelength and the wavelength at which phase matching is achieved. Thus the spatial period of the structure needed for quasi-phase matching is the wavelength times the reciprocal of the fractional phase mismatch. For example, if there is a 10% phase mismatch ($\Delta\lambda/\lambda$), the *domain period*, $\Lambda$, should be ten times the wavelength.

In Fourier analysis there are two complementary representations for time dependent functions: the time domain and the frequency domain. Spatially dependent functions can be described in an analogous fashion. The *spatial domain* view of quasi-phase matching which is complementary to the above spatial frequency domain view can be made with the aid of Figs. 14.12 and 14.13. In Fig. 14.12 we display on the left a phasor diagram for the electric field over two coherence lengths ($2\pi/\Delta k$). We see that the field adds to zero due to the phase mismatch: the field adds over the first coherence length and subtracts over the second. On the right we show what happens if we abruptly reverse the sign of $d_{eff}$ after *one coherence length*. The phasor diagrams

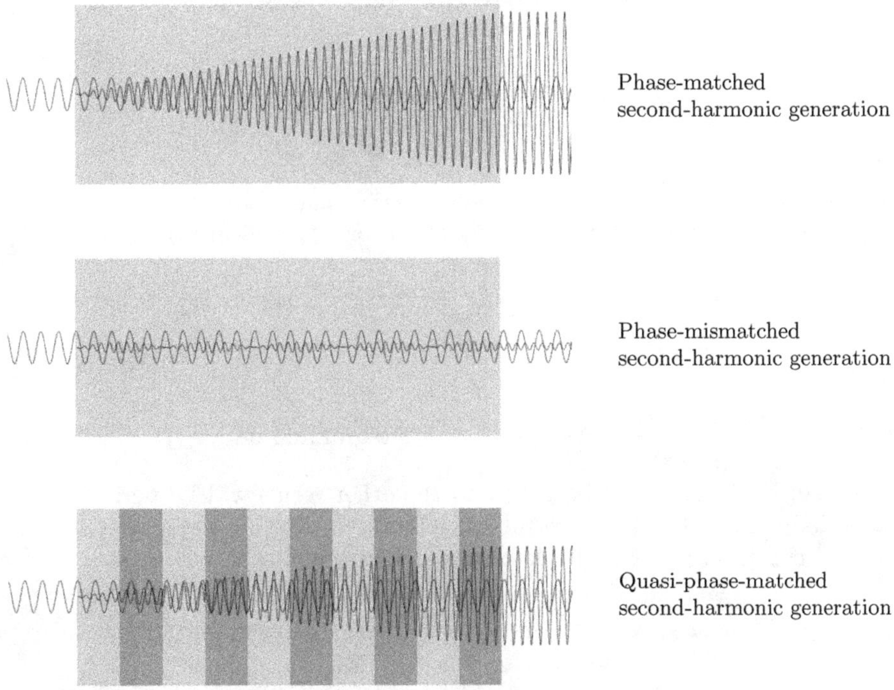


**Fig. 14.13** Snapshots of the fundamental and second-harmonic *electric fields* under the conditions of phase matching (top), phase mismatching (middle) and restoration of the phase match using quasi-phase matching (bottom).

can be more easily interpreted by considering the differential equation for the field (eqn 14.45), which, neglecting unimportant constants, is:

$$\frac{\partial E_{2j}}{\partial z} \propto i d_{eff} e^{i\Delta kz}. \tag{14.80}$$

The field begins along the imaginary axis, and successive phasors follow a circle until one coherence length is reached. At this point, $d_{eff}$ reverses sign and, for a small increment in $z$, the changes in both the real and imaginary parts of $E_{2j}$ also reverse sign; the phasors will now follow the upper semicircle in the diagram. Thus, the field will continue to grow. From the spatial-frequency-modulation model, a complete reversal of $d_{eff}$ at the domain boundaries is not necessary: any abrupt change in the index of refraction will cause a spatial phase modulation and will work. A structure in which $d_{eff}$ alternately reverses sign is, however, the most efficient configuration. The actual waves are depicted in Fig. 14.13 for phase matching, phase mismatching and the restoration of field growth using quasi-phase matching. A careful inspection of the bottom graph should reveal the phase jumps in the second-harmonic field at each domain boundary. One should note by comparing the middle to the bottom plot that the period, $\Lambda$, needed to restore field growth is equal to twice the coherence length.

A more rigorous analysis can be made by expanding the spatially dependent non-linear coefficient, $d(z)$, in a Fourier series:

$$d(z) = d_{eff} \sum_{\substack{n=-\infty \\ n\ odd}}^{\infty} a_n e^{2\pi i n z/\Lambda}, \qquad a_n = \frac{2}{\pi n}, \tag{14.81}$$

which is just the series expansion for a symmetric square wave whose extreme values are $\pm d_{eff}$. Then, from the differential equation for the field growth (eqn 14.45), one has

$$\frac{dE_{2j}}{dz} \propto d(z)e^{i\Delta kz} = \frac{2d_{eff}}{\pi} \sum_{\substack{n=-\infty \\ n\ odd}}^{\infty} \frac{1}{n} e^{i(2\pi n/\Lambda + \Delta k)z}. \tag{14.82}$$

When

$$\frac{2\pi m}{\Lambda} + \Delta k = 0, \qquad m \text{ integral and odd}, \tag{14.83}$$

the $m$th term in the sum will be $1/m$ and the other terms will be periodic functions with a period of $\Lambda$. When these terms are integrated over the crystal length, $L$, they will be zero if $L$ is an exact multiple of $\Lambda$ and very small otherwise. Thus, we have

$$E_{2j} \propto \frac{2d_{eff}L}{m\pi}, \qquad \text{when } \frac{2\pi m}{\Lambda} + \Delta k = 0. \tag{14.84}$$

If the alternating domains are of equal width, $m$ must be odd; otherwise, both odd and even values will work. The second-harmonic intensity, which is proportional to $|E_{2j}|^2$, will be reduced by the factor of $(2/m\pi)^2$ compared to conventional birefringent phase matching. (It is interesting to note that the field reduction factor for $m = 1$ is just equal to $2/\pi$, the ratio of the diameter to the circumference of a semicircle (Fig. 14.12)). On the other hand, the freedom one has in the direction of $\boldsymbol{k}$ will often make up for this modest factor; furthermore, the adverse problems due to double refraction are absent.

The phase-matching bandwidth can be obtained in the same way as for the birefringent case. We assume that only the $m$th term contributes, and obtain, by integrating and squaring,

$$\frac{P^{(2\omega)}}{P^{\omega}} \propto \frac{\sin^2(\Delta kL/2)}{(\Delta kL/2)^2}, \tag{14.85}$$

where $\Delta k$ is the $k$-vector mismatch which includes the effect of the periodic structure:

$$\Delta k \equiv k_{2\omega} - 2k_{\omega} - K_m \qquad \text{where } K_m \equiv \frac{2\pi m}{\Lambda}. \tag{14.86}$$

From this, the bandwidth at the fundamental wavelength is

$$\Delta\lambda = \frac{0.443\lambda}{L} \left| \frac{n_{2\omega} - n_{\omega}}{\lambda} + \left(\frac{\partial n}{\partial \lambda}\right)_{\omega} - \frac{1}{2}\left(\frac{\partial n}{\partial \lambda}\right)_{2\omega} \right|^{-1} \qquad \lambda \text{ at fundamental}. \tag{14.87}$$

This reduces to eqn 14.63 for birefringent phase matching (since $n_{\omega} = n_{2\omega}$ in that case).

The phase-matching condition also depends upon the temperature of the crystal through the temperature dependence of the refractive indices and the size of the period, $\Lambda$. The temperature acceptance, $\Delta T$, can be found in an analogous way as the wavelength bandwidth by solving the following equation for $\Delta T$:

$$\Delta T \frac{\partial}{\partial T}\left(\frac{\Delta k L}{2}\right) = 0.866\pi. \tag{14.88}$$

The result is

$$\Delta T = \frac{0.443\lambda}{L}\left|\frac{\partial \Delta n}{\partial T} + \alpha \Delta n\right|^{-1} \qquad \text{where } \Delta n \equiv n_{2\omega} - n_{\omega}, \tag{14.89}$$

and $\alpha$ is the thermal expansion coefficient of the material: it is defined by

$$\frac{dL}{dT} = \alpha L. \tag{14.90}$$

One can investigate other aspects of quasi-phase matching such as the acceptance angle of the light and the effect of imperfect gratings on the doubling efficiency; such topics are well covered in the paper by Fejer et al. (1992).

The obvious disadvantage of quasi-phase matching is its lack of tunability: each wavelength usually requires a separate grating, which is not trivial to manufacture. For a single grating at a fixed temperature, the phase-matching bandwidth is typically 1–3 nm, depending upon the crystal and its orientation relative to the polarization of the light. Figure 14.14 illustrates two geometrical approaches for addressing this limitation. The use of multiple gratings (middle figure) is a common technique employed by commercial manufacturers for extending the useful wavelength range for quasi-phase matching. Between each grating wavelength, the crystal can be *temperature tuned*. To determine the temperature sensitivity, one solves the equation $\partial(\Delta k L/2)/\partial T = 0$ for $\partial\lambda/\partial T$:

$$\frac{\partial \lambda}{\partial T} = \alpha\lambda + \frac{\lambda}{\Delta n}\frac{\partial \Delta n}{\partial T}, \tag{14.91}$$

where we used the fact that $\partial K_m/\partial T = -\alpha K_m$, which is easily derived from the definitions of $\alpha$ and $K_m$. It is interesting that neither of the two bandwidths derived above nor the temperature sensitivity are dependent upon the grating period, $\Lambda$.

## 14.6 Second-harmonic generation using a focused beam

Our treatment so far has assumed that the beams are all plane waves. A much more realistic model is to use Gaussian beams which are focused into the crystal to increase the local intensity and therefore the conversion efficiency.

We will summarize here the heuristic treatment given by Boyd and Kleinman (1968) in their classic paper. Our approach will only touch on the important points as the theory can be extremely complicated. The salient point made in this section is that the presence of *double refraction* can significantly reduce the conversion efficiency; a secondary point is that there is an optimum waist size (and consequently Rayleigh length) for the maximum conversion efficiency. The latter point is illustrated

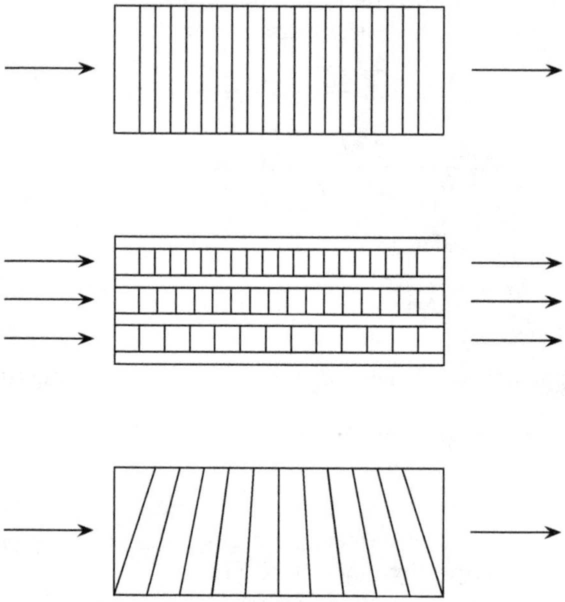


**Fig. 14.14** Two approaches to increasing the tunability of quasi-phase matching. The top figure displays a fixed grating, useable only at a single wavelength. The middle figure shows a crystal with multiple gratings, allowing phase matching at a number of wavelengths. The bottom figures depicts a tapered grating, allowing continuous tuning.

in Fig. 14.15, which depicts three different focusing "strengths". Clearly, one would like to focus as tightly as possible, since the conversion efficiency increases with the intensity of the fundamental. However, as shown in the bottom figure, if one focuses too tightly one will not effectively use the entire length of the crystal, since the beam rapidly expands on either side of the waist and the larger beam area will not contribute much to the second-harmonic generation. Focusing too weakly will use the entire crystal length (since the beam approximates a uniform cylinder) but the conversion will suffer from inadequate intensity. One can show that the optimum conversion efficiency will be obtained when the crystal length is 5.68 times the Rayleigh length for non-critical phase matching.

We will assume that we have a negative uniaxial crystal and the incident radiation is polarized in the ordinary direction and makes an angle $\theta_m$ with the optical axis. In all that follows, we will neglect any absorption of the fundamental or second harmonic. The second harmonic is an extraordinary wave which is phase matched with the fundamental. The coordinate system, whose origin is at the *center* of the crystal, is shown in Fig. 14.16. In order to be consistent with the Boyd and Kleinman symbolism, we will use an expression for a Gaussian electric field which is different from eqn 1.30 but easily shown to be equivalent to it. It is given by

$$E_1 = E_0 \frac{1}{1+i\tau} e^{ik_1 z - \frac{r^2}{\omega_0^2(1+i\tau)}}, \tag{14.92}$$

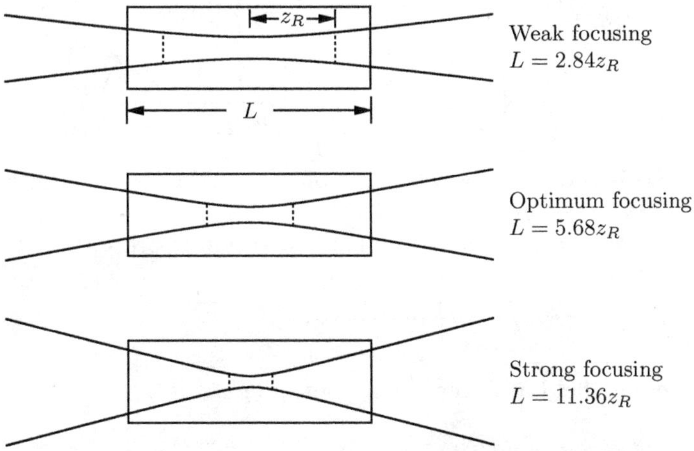


**Fig. 14.15** Three different degrees of focusing into the second-harmonic generation crystal, identified by the ratio of crystal length to Rayleigh length. The vertical dashed lines show the Rayleigh distances to the waists.

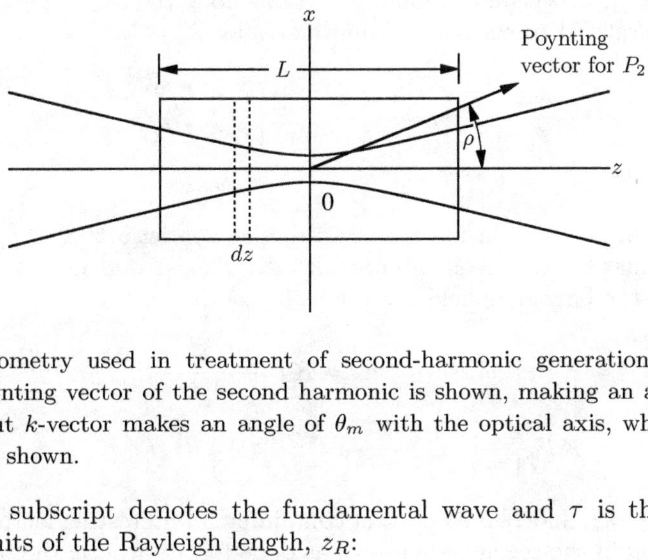


**Fig. 14.16** Geometry used in treatment of second-harmonic generation with Gaussian beams. The Poynting vector of the second harmonic is shown, making an angle $\rho$ with the $z$-axis. The input $k$-vector makes an angle of $\theta_m$ with the optical axis, which is in the $xz$ plane and is not shown.

where the "1" subscript denotes the fundamental wave and $\tau$ is the $z$-coordinate expressed in units of the Rayleigh length, $z_R$:

$$\tau = \frac{z}{z_R}. \tag{14.93}$$

Figure 14.16 shows that the direction of energy propagation makes an angle $\rho$ with the $z$-axis. The phase-matching angle is at the intersection of the normal surface for the fundamental (a sphere since it is an ordinary ray) and the second harmonic (an ellipsoid). From eqn 14.31, it can be shown that the *double refraction* (or *walk-off*) angle, $\rho$, is given by

$$
\begin{aligned}
\tan\rho =& \tfrac{1}{2} n_{o,\omega}^2 \left\{ \frac{1}{(n_{e,2\omega})^2} - \frac{1}{(n_{o,2\omega})^2} \right\} \sin 2\theta_m \\
=& n_{o,\omega}^2 \left\{ \left( \frac{1}{n_{o,\omega}^2} - \frac{1}{n_{e,2\omega}^2} \right) \left( \frac{1}{n_{o,2\omega}^2} - \frac{1}{n_{o,\omega}^2} \right) \right\}^{1/2}
\end{aligned}
\tag{14.94}
$$

where $\rho = \phi - \theta_m$ is the angle between the phase-matched second-harmonic *ray* and the fundamental *wave vector.*

Inside the crystal, the second-harmonic polarization is proportional to $E_1^2$:

$$
P(r, z) = P_0 \frac{1}{(1 + i\tau)^2} e^{2ik_1 z - \frac{2r^2}{w_0^2(1+i\tau)}}. \tag{14.95}
$$

The second-harmonic field at observation point $x', y', z'$ is

$$
E_2(x', y', z') = A_2(x', y', z') e^{ik_2 z'}, \tag{14.96}
$$

where $A_2(x,' y', z')$ is a slowly varying function of the spatial coordinates. We will neglect refraction at the faces of the crystal (or assume that it is embedded in a medium with the same refractive index). We are interested only in the direction of *energy flow* (the polarization and energy flow are along the same *ray*), so the observer coordinates are related to the source coordinates by

$$
\begin{aligned}
&x = x' - \rho(z' - z) \quad (z' < L/2) \\
&x = x' - \rho(L/2 - z) \ (z' > L/2) \\
&y = y'.
\end{aligned}
\tag{14.97}
$$

At a source point, $x, y, z$, inside the crystal, the heuristic theory considers a thin slab, perpendicular to the $z$-axis and of width $dz$. This slab generates an infinitesimal contribution to the harmonic field given by

$$
\begin{aligned}
dA_2(x', y,' z') &= KP(x, y, z) e^{-ik_2 z} dz = KP_0 \frac{1}{(1 + i\tau)^2} e^{i\Delta k z - \frac{2r^2}{w_0^2(1+i\tau)}} dz \\
&= KP_0 \frac{e^{i\Delta k z}}{1 + i\tau} \left\{ \frac{1}{1 + i\tau} e^{-2\frac{x^2 + y^2}{w_0^2(1+i\tau)}} \right\} dz,
\end{aligned}
\tag{14.98}
$$

where $\Delta k = 2k_1 - k_2$, and $K$ is a constant containing uninteresting factors. The slightly odd factorization in the second equation was done for a purpose: the bracketed factor is the *amplitude* of a Gaussian beam with a waist $\sqrt{2}$ smaller and same position as the fundamental beam. This is the quantity which propagates, forming the resultant harmonic wave. We now force the latter to propagate along a ray at angle $\rho$ to the $z$-axis. Prior to summing (integrating) over all the contributions, we will substitute the ray equations, eqn 14.97, into the expression for the contribution of the incremental field to the field at the observation point, $x', y', z'$:

$$
dA_2(x', y', z') = KP_0 \frac{1}{1 + i\tau} \frac{e^{i\Delta k z}}{1 + i\tau'} e^{-2\frac{[x' - \rho(L/2 - z)]^2 + y'^2}{w_0^2(1+i\tau')}} dz \qquad z' > L/2, \tag{14.99}
$$

where

$$\tau' = \frac{z'}{z_R}. \qquad (14.100)$$

We finally integrate over $z$, from $-L/2$ to $L/2$:

$$E_2(x', y', z') = KP_0 \frac{e^{2ik_1z'}}{1+i\tau'} \int_{-L/2}^{L/2} \frac{e^{i\Delta kz}}{1+i\tau} e^{-2\frac{[x'-\rho(L/2-z)]^2+y'^2}{\omega_0^2(1+i\tau')}} dz \qquad z' > L/2. \quad (14.101)$$

The remainder of the calculation consists of changing to a more convenient set of variables, squaring the field to obtain the second harmonic intensity and integrating over $x'$ and $y'$ to obtain the total power. The parameters and their significance are listed as follows:

| | | |
|---|---|---|
| $\xi = L/2z_R$ | crystal length in units of *confocal parameter* | |
| $\sigma = z_R\Delta k$ | phase mismatch | (14.102) |
| $\delta_0 = \omega_0/z_R = (2/z_Rk_1)^{1/2}$ | diffraction half-angle | |
| $\beta = \rho/\delta_0$ | normalized walk-off angle. | |

After a fair bit of algebra, the expression for the second-harmonic power, $P_2$, is written in terms of the function, $h(\sigma, \beta, \kappa, \xi, \mu)$, defined by

$$h(\sigma, \beta, 0, \xi, 0) = \frac{1}{4\xi} \int_{-\xi}^{\xi} \int_{-\xi}^{\xi} \frac{e^{i\sigma(\tau-\tau')-\beta^2(\tau-\tau')^2}}{(1+i\tau)(1-i\tau')} d\tau d\tau'. \qquad (14.103)$$

The third and fifth arguments are the absorption and displacement of the waist from the crystal center respectively and are set equal to zero in our treatment. The second-harmonic power is

$$P_2 = KP_1^2Lk_1 \cdot h(\sigma, \beta, 0, \xi, 0), \qquad (14.104)$$

where $K$ is a constant which depends upon the fundamental frequency, indices of refraction and nonlinear coefficient. A much more useful parameter than $\beta$ (walk-off angle) is the double-refraction parameter, $B$, defined by

$$B = \beta\xi^{1/2} = \rho(Lk_1)^{1/2}/2. \qquad (14.105)$$

This significance of this quantity depends upon the concept of *aperture length*, $L_a$, defined by

$$L_a = \frac{\sqrt{\pi}\omega_0}{\rho}. \qquad (14.106)$$

The aperture length (Fig. 14.17) is the distance over which the fundamental and second-harmonic beams separate. From this, we see that the parameter $B$ is approximately the *ratio of the crystal length to the double-refraction aperture length* (assuming that $\xi \approx 1$, which we will shortly see is close to the optimum value). Putting it another way, $B$ is the inverse of the fraction of the crystal length over which the two beams overlap. Thus, a $B$ of 10 implies that the beams overlap over 10% of the crystal length. A more useful $h$-function than the one defined above is

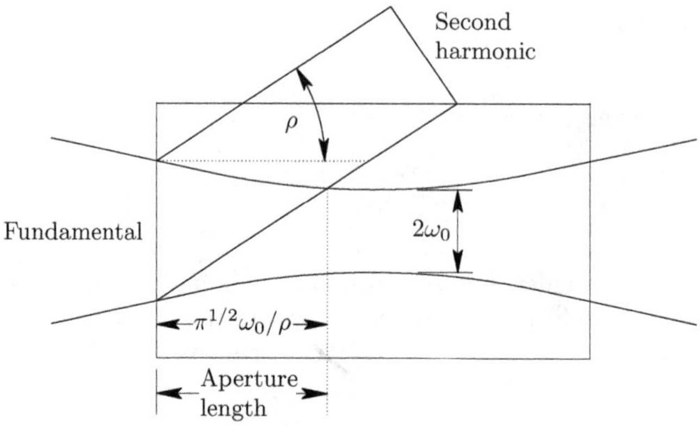


**Fig. 14.17** Illustration of *aperture length* in second-harmonic generation: the axial distance over which the fundamental and second-harmonic beams separate.

$$h(\sigma, B, \xi) = h(\sigma, B\xi^{-1/2}, 0, \xi, 0). \tag{14.107}$$

This function depends upon three important parameters which can be optimized: the phase mismatch, the walk-off and the crystal length (normally, the only control one has over the walk-off is through the crystal length, but it is often useful to consider the former to be an *independent* parameter). We can define $h_m(B, \xi)$ to be the function $h(...)$ which is optimized with respect to $\sigma$ (phase matching) at each value of $B$ and $\xi$:

$$h_m(B, \xi) = h[\sigma_m(B, \xi), B\xi^{-1/2}, 0, \xi, 0). \tag{14.108}$$

Finally, one can optimize $h_m(B, \xi)$ with respect to $\xi$ (crystal length in units of Rayleigh length) to obtain the function $h_{mm}(B)$ defined as

$$h_{mm}(B) = h_m(B, \xi(B)). \tag{14.109}$$

Curves of $h_m(B, \xi)$ have been plotted (Boyd (1968)) versus $\xi$ and display a very broad maximum. The optimum value of $\xi$ for $B = 0$ (no walk-off) is $\xi = 2.84$; this value is often mentioned in books on quantum electronics. For larger values of $B$, the optimum $\xi$ decreases, becoming about 1.4 for $B = 16$. The relative unimportance of the focusing parameter $\xi$ is suggested by the fact that $h_m(B, \xi)$ is within 10% of its maximum over the range $1.52 < \xi < 5.3$ when $B = 0$. For larger values of $B$, the function $h_m(B, \xi)$ is an even broader function of $\xi$. A plot of the optimum focusing parameter, $\xi_m$, as a function of $B$ appears in Fig. 14.18.

While it is somewhat useful to be aware of the optimum value of $\xi$, it is far more important to realize that walk-off can greatly reduce the second-harmonic conversion efficiency. Unfortunately, one has little control over the walk-off, since the parameter $\rho$ is rigidly determined by the phase-matching angle which in turn depends upon the wavelength. The parameter $B$ is, however, determined also by the crystal length, and an appreciation of the diminishing returns possibly obtained by increasing the crystal length can have some benefit. A plot of $h_{mm}(B)$ versus $B$ appears in Fig. 14.19. The

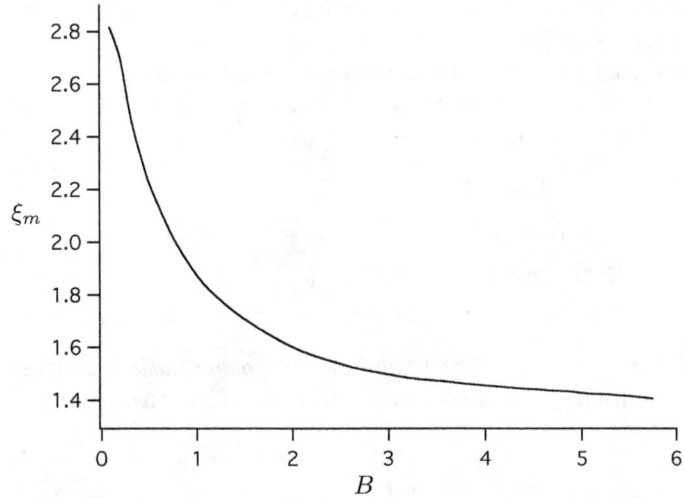


**Fig. 14.18** Plot of optimum focusing parameter, $\xi$, versus double-refraction parameter, $B$. (From Boyd (1968).)

effect of double refraction can be seen in the following asymptotic form for $h_m(B, \xi)$ in the limit of a relatively short crystal ($\xi < 0.4$):

$$h_m(B, \xi) \approx \frac{\sqrt{\pi}}{2}\frac{\xi^{1/2}}{B} \qquad (\xi < 0.4,\ B > \sqrt{6/\xi}). \tag{14.110}$$

The inverse dependence on $B$ is suggested by Fig. 14.19 and is consistent with the previous interpretation of $1/B$ as the *fraction of the crystal length over which the*

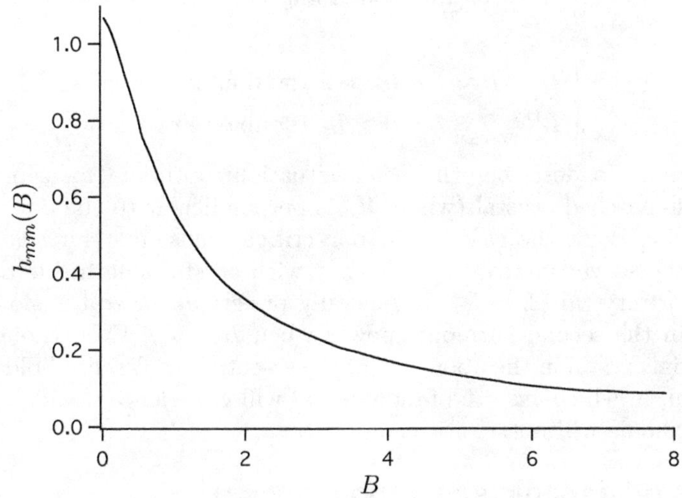


**Fig. 14.19** Plot of optimum $h_{mm}(B)$ versus double-refraction parameter, $B$. (From Boyd (1968).)

*fundamental and second harmonic overlap* (second harmonic generation stops when the beams no longer overlap).

We have presented a number of characteristic lengths and discussed the dependence of the doubling efficiency on the comparison between them and the crystal length. We will now summarize these data. The lengths are:

$$\begin{array}{ll} \text{Crystal length} & L \\ \text{Aperture length} & L_a = \pi^{1/2}\omega_0/\rho \\ \text{Beam radius at waist} & \omega_0 \\ \text{Effective length of focus} & L_f = \pi z_R = n\pi^2\omega_0^2/\lambda. \end{array} \tag{14.111}$$

Boyd and Kleinman (1968) provided the following *asymptotic* behavior for the dependence of the second-harmonic power on these various lengths.

$$P_2 = \frac{KP_1^2}{\omega_0^2}\begin{cases} L^2 & L_a, L_f \gg L \\ LL_a & L_f \gg L \gg L_a \\ L_f L_a & L \gg L_f \gg L_a \\ 4L_f^2 & L \gg L_a \gg L_f \\ 4.75L_f^2 & L_a \gg L \gg L_f \end{cases} \tag{14.112}$$

where the constant $K$ is different from the one used earlier. The first expression is for a very short crystal where the beam is a uniform cylinder and there is no double refraction, and shows the same $L^2$ dependence as a plane wave. The second expression is the one discussed immediately above, where double refraction is the only limitation. Note that the remaining three expressions have no crystal length dependence, which supports the argument made above about using a short crystal when double refraction is significant. When one uses *optimum focusing* ($L \approx L_f$), the first two asymptotic relations yield

$$P_2 \propto \begin{cases} L & L \approx L_f, L_a > L \text{ (no double refraction)} \\ L^{1/2} & L \approx L_f, L > L_a \text{ (double refraction).} \end{cases} \tag{14.113}$$

The first relationship describes the usual situation: optimum focusing into a non-critically phase-matched crystal (where $\boldsymbol{k}_\omega$ is perpendicular to the optical axis). The second expression is for the case where non-critical phase-matching is not available and confirms the slower increase in efficiency with crystal length due to walk-off. In this case, a short crystal ($L \approx L_a$) is generally preferred due to the slow square root improvement in the second-harmonic power when $L > L_a$. Even though we used a negative uniaxial crystal in the above, all of the essential results also hold for a positive uniaxial crystal, in which case the fundamental will experience double refraction and the second harmonic will not.

## 14.7 Second-harmonic generation in a cavity

Second-harmonic generation is an intrinsically inefficient process, since the laser field strength needs to be comparable to the internal atomic field for there to be a significant output. Crystals exist whose conversion efficiency (for $L = 1$ cm) is $\approx 0.01/\mathrm{W}$, but most are in the $10^{-3}/\mathrm{W}$ to $10^{-4}/\mathrm{W}$ range. Fortunately, the efficiency can be greatly increased using a *resonant cavity*, which can increase the power incident on the crystal by a factor of 100 or more. Usually a ring cavity is used, since it has a much smaller (destabilizing) back reflection into the laser and the traveling wave produces a *single* SHG beam instead of the two counter-propagating beams generated in a standing-wave cavity. An enhancement of the fundamental by a factor of 100 will result in an enhancement of the second harmonic by a factor of 10,000, which is usually sufficient to provide a useful amount of the second harmonic. In favorable cases, a large fraction of the fundamental can be converted to the second harmonic. Large conversion efficiencies are also available using laser *intracavity* doubling, but there are many reasons for separating the laser-oscillation process from the frequency-doubling process, particularly when narrow linewidths are sought after.

A ring doubling cavity is shown schematically in Fig. 14.20. All of the components

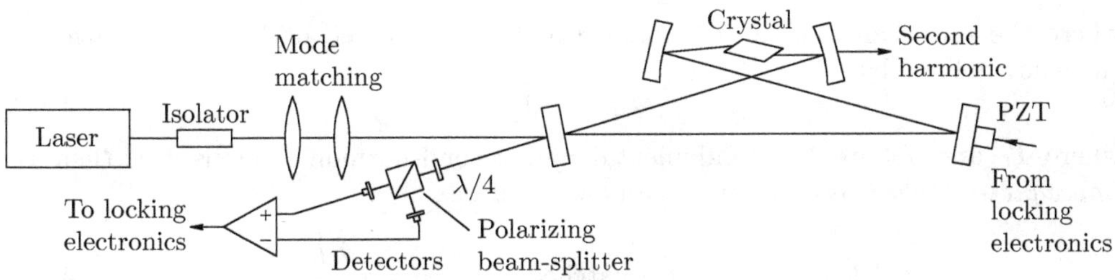


**Fig. 14.20** Schematic of ring cavity frequency doubling system.

shown in the figure (except the isolator) have already been discussed in some detail. The isolator prevents destabilization of the laser due to back reflections from the cavity or other optical components (such as the mode-matching lenses). Single Faraday isolators can provide 33 dB of isolation, which is often enough to avoid feedback problems. The beam is mode matched into the lowest mode of the cavity using the two lenses shown. The cavity is designed to have a small waist size at the crystal, which allows the crystal to perform optimally, as discussed in the previous section.

There are two strategies for minimizing the loss due to the crystal. One can purchase a crystal with *anti-reflection coatings*, which can reduce the loss at each surface to $\approx 0.1\%$. Alternatively, the crystal can be cut so that the circulating beam impinges on it at Brewster's angle. The latter approach requires that the polarization be in the plane of incidence, and the crystal needs to be cut so that this polarization will generate the largest amount of second harmonic. The Brewster cut approach has the advantage of allowing the complete elimination of the astigmatism in the circulating beam, since the astigmatism due to the off-axis incidence on the concave mirrors can be canceled by the astigmatism due to the Brewster-cut crystal. This eases the mode matching and might improve the conversion efficiency, though in practice, a small amount of astigmatism is not a problem.

It has become standard practice to lock the cavity resonant frequency to the laser frequency using the *polarization scheme* (Chapter 4) to generate the discriminant and apply the feedback signal to a piezo-electric actuator attached to one of the mirrors. The Brewster approach facilitates the polarization scheme, since the crystal is a very polarization-sensitive element. Finally, at least one mirror (the output mirror) needs a *dichroic* coating which is highly reflective at the fundamental and transmissive at the second harmonic.

With relatively efficient crystals (such as potassium niobate and KTP), it is possible to convert a large fraction (often more than 50%) of the fundamental to the second harmonic. This modifies somewhat the impedance-matching strategy, since the *depletion of the fundamental* will now be a loss which must be taken into account. One can approach this problem analytically by modifying the expression for the circulating power given in eqn 3.10. Without depletion of the fundamental, the circulating intensity at resonance is

$$\text{No depletion:} \quad I_c = I_0 \frac{t_1^2}{(1 - r_1 r_m)^2}, \tag{14.114}$$

where the terms are defined in Chapter 3. If $\gamma$ is the second-harmonic conversion efficiency defined by

$$I_2 = \gamma I_f^2, \tag{14.115}$$

where $I_f$ and $I_2$ are the fundamental and second-harmonic intensities, then the *impedance-matched* second harmonic power will be

$$\text{No depletion, impedance matched:} \qquad I_2 = \gamma \left( \frac{I_0}{T} \right)^2, \tag{14.116}$$

where $I_0$ is the input intensity to the cavity and $T = t_1^2$. The behavior when the impedance is mismatched is obtained directly from eqn 3.27:

$$\text{No depletion, impedance mismatched:} \qquad I_2 = \gamma \left( \frac{I_0}{T} \right)^2 \left( \frac{4\sigma}{(\sigma + 1)^2} \right)^2, \tag{14.117}$$

where

$$\sigma \approx \frac{T}{L}, \tag{14.118}$$

and $L$ is the sum of all cavity losses exclusive of the transmission through the input coupler.

When depletion is included, the quantity $r_m$ needs to be modified. We will assume that the circulating intensity has already been established at the level, $I_c$. Each pass, of intensity, $I_1$, will generate a second-harmonic *loss* of $\gamma I_1 I_c$. The *fractional loss* in intensity is $\gamma I_1 I_c / I_1 = \gamma I_c$. The *fractional remaining intensity* per pass is therefore $1 - \gamma I_c$. We will modify $r_m$ by multiplying it by the fractional *field* remaining after one round trip due to depletion (this requires a square root). Thus,

$$r_m \Longrightarrow r_m \times \sqrt{1 - \gamma I_c} \approx r_m \left( 1 - \frac{\gamma I_c}{2} \right) = r_m \left( 1 - \tfrac{1}{2} \sqrt{\gamma I_2} \right), \tag{14.119}$$

where we used the expression for the second-harmonic intensity, $I_2 = \gamma I_c^2$, in the last equation. The circulating power with depletion is now

$$\text{Depletion:}\quad I_c = I_0 \frac{t_1^2}{[1 - r_1 r_m (1 - \frac{1}{2}\sqrt{\gamma I_2})]^2}, \tag{14.120}$$

and the second-harmonic intensity is

$$I_2 = \frac{\gamma t_1^4 I_0^2}{[1 - r_1 r_m (1 - \frac{1}{2}\sqrt{\gamma I_2})]^4}. \tag{14.121}$$

This can be simplified somewhat if we assume that $t_1^2 \ll 1$ and $1 - r_m^2 \ll 1$. The round-trip loss (excluding depletion and transmission through the input mirror) is $L$, where

$$L \equiv 1 - r_m^2. \tag{14.122}$$

Substituting this in the equation for $I_2$ and dropping terms of second order in small quantities, the second-harmonic intensity is

$$I_2 = \frac{16\gamma T^2 I_0^2}{[L + T + \sqrt{\gamma I_2}]^4} \qquad T \equiv t_1^2 = 1 - r_1^2. \tag{14.123}$$

The quantity raised to the fourth power in the denominator is just the total loss, including that due to depletion. This is a cubic equation in $\sqrt{I_2}$ and is most easily solved for $I_2$ numerically on a computer. One can, however, obtain several important insights without obtaining an explicit solution.

We recall that a cavity is *impedance matched* when the input coupling is adjusted to obtain the maximum circulating power. A more useful and equivalent definition of impedance matching is that the *reflection vanishes.* If we modify $r_m$ to include depletion, we can define $r'_m$ to be

$$r'_m = r_m \sqrt{1 - \sqrt{\gamma I_2}}. \tag{14.124}$$

When depletion is included, the resonant formulas for $I_c$ and $I_r$ are

$$I_c = I_0 \frac{T}{(1 - r_1 r'_m)^2} \tag{14.125}$$

$$I_r = \left[\frac{r'_m - r_1}{1 - r_1 r'_m}\right]^2. \tag{14.126}$$

From the second equation, we obtain the impedance-matching condition (zero reflection) when $r_1 = r'_m$:

$$r_1 = r'_m \Longrightarrow \sqrt{1 - T} = \sqrt{1 - L}\sqrt{1 - \frac{\gamma I_0}{T}}, \tag{14.127}$$

where, in the last equality, we use the fact that the impedance matched $I_c$ is just $I_0/T$. This is a quadratic equation in $T$ whose solution, when $L \ll 1$, is

$$T_{matched} = L/2 + \sqrt{L^2/4 + \gamma I_0}. \tag{14.128}$$

The impedance-matched second harmonic is just

$$I_2 = \gamma \left( \frac{I_0}{T_{matched}} \right)^2 = \frac{4\gamma I_0^2}{(L + \sqrt{L^2 + 4\gamma I_0})^2}. \tag{14.129}$$

Following Kimble (1993), one can define a *figure of merit*, $\epsilon_{NL} \equiv \gamma/L^2$. Using this, the *conversion efficiency* $(I_2/I_0)$ is

$$\text{Conversion efficiency} = \eta = \frac{4\epsilon_{NL} I_0}{(1 + \sqrt{1 + 4\epsilon_{NL} I_0})^2} \quad \text{(impedance matched)}. \tag{14.130}$$

This reduces to eqn (14.116) when $4\epsilon_{NL} I_0 \ll 1$ (or $4\gamma I_0 \ll L^2$).

From the above, we can draw two conclusions. First, when the depletion loss, $\sqrt{\gamma I_2}$, is comparable to the other losses, $L + T$, the increase of second harmonic with input power (with $T$ fixed) is much slower than the squared dependence which occurs at lower powers. At high input power, the second harmonic increases as $I_0^{2/3}$. If one continually adjusts $T$ for impedance matching as the power is raised, the second harmonic is proportional to $I_0$ at high power. This can be seen from eqn 14.129, which describes the impedance-matched situation.

The second conclusion is that one must change the transmission of the input coupler when using high powers to obtain optimum second harmonic (and impedance matching). As an example, if one has a potassium niobate crystal whose length is 1 cm ($\gamma \approx 0.01$ /W) and a cavity with $T = 0.01$ and $L = 0.01$ (impedance matched without depletion), then depletion becomes important when $I_2 = 40$ mW, which is a modest power (one can obtain ten times this power with some effort). The input power is about 80 mW; thus the conversion efficiency is 50% from a relatively low-power laser. With this power, a 3% input coupler would be needed for impedance matching and optimum conversion. The second-harmonic power using this coupler is 56 mW, a modest improvement over the impedance-mismatched situation. Figure 14.21 plots the second-harmonic power as a function of the input power for the above parameters. The right-hand plot is the same function on a log-log scale, showing the deviation from quadratic dependence on the input at higher powers.

There are a number of problems which can be troublesome when high fundamental and second-harmonic power are present in a doubling crystal. For example, potassium niobate is subject to *blue induced infrared absorption*, which can greatly limit the cavity enhancement due to an increase in the absorption of the fundamental in the presence of significant *second-harmonic* intensity. This phenomenon is discussed in the literature (for example, Shiv (1995)) and can be reduced by heating the crystal. With large fundamental powers, there can be significant absorption-induced heating of the crystal, which can prevent stable phase-matching when the crystal possesses strongly temperature-dependent refractive indices. Active temperature stabilization of the crystal can solve this problem. The thermal effects can also produce considerable *thermal lensing* in the crystal, with consequent degradation of the beam quality. These matters are also discussed in various publications, and the reader is invited to consult the literature to learn more about them (For example, Le Targat (2005) discusses the thermal lensing problem in periodically poled KTP and is able to obtain 75% conversion efficiency by understanding and minimizing the problem.)

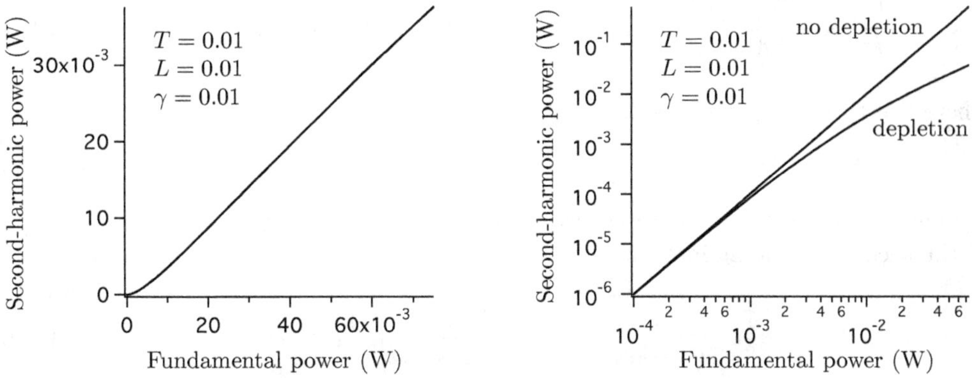


**Fig. 14.21** Second-harmonic power as a function of input power on linear scale (left) and log-log scale (right).

Before leaving this topic, we should mention a less obvious way to increase the second-harmonic power using a cavity. In the foregoing, we have discussed the use of a cavity to increase the circulating power at the *fundamental*, with consequent increase in the power at the second harmonic. It turns out that one can also use a cavity to *increase the nonlinear interaction* by making the cavity *resonant at the second harmonic*. Essentially, the cavity increases the *crystal length* and thereby increases the conversion efficiency by the factor $4/T_2$, where $T_2$ is the transmission of the output coupler of the cavity at the second harmonic. This possibility was pointed out in 1966 by Ashkin (Ashkin (1966)), but was not actually demonstrated until somewhat later (e.g., by Kimble (1993)). The technique is especially useful for low-loss crystals with small nonlinear coefficients. It can be combined with fundamental enhancement to yield good conversion efficiency with only a few milliwatts of input power (Kimble (1993)). When used alone, it is not nearly as efficient as fundamental enhancement (which provides a factor of $\approx 1/T_1^2$) and one might legitimately question whether it is worth the considerable additional equipment complexity.

## 14.8 Sum-frequency generation

The nonlinear interaction responsible for second-harmonic generation can also lead to the generation of sum and difference frequencies when the two input beams are at different frequencies. The three fields are

$$\text{Output wave:} \quad E_j^{(\omega_3)} = \frac{1}{2} E_{3j}(z) e^{i(\omega_3 t - k_3 z)} + \text{cc} \tag{14.131}$$

$$\text{Input wave 1:} \quad E_i^{(\omega_1)} = \frac{1}{2} E_{1i}(z) e^{i(\omega_1 t - k_1 z)} + \text{cc} \tag{14.132}$$

$$\text{Input wave 2:} \quad E_k^{(\omega_2)} = \frac{1}{2} E_{2k}(z) e^{i(\omega_2 t - k_2 z)} + \text{cc.} \tag{14.133}$$

The equation which describes sum-frequency generation is a straightforward variation of eqn 14.45:

$$\frac{dE_{3i}}{dz} = -i\frac{\omega_3}{4}\sqrt{\frac{\mu_0}{\epsilon_3}}\sum_{jk} d_{ijk}E_{1j}E_{2k}e^{i\Delta kz}, \qquad \Delta k = k_3 - k_1 - k_2, \tag{14.134}$$

where the three frequencies satisfy

$$\omega_3 = \omega_1 + \omega_2. \tag{14.135}$$

The result for the *intensity* $I^{(\omega_3)}$ at the sum frequency is obtained in the same way as the second-harmonic intensity:

$$I^{(\omega_3)} = \frac{1}{8}\omega_3^2\sqrt{\frac{\mu_0^3}{\epsilon_1\epsilon_2\epsilon_3}}(d_{ijk})^2 L^2 \frac{\sin^2(\Delta kL/2)}{(\Delta kL/2)^2} I^{(\omega_1)} I^{(\omega_2)}. \tag{14.136}$$

The phase-matching requirement is

$$\text{Phase matching:} \quad \boldsymbol{k}_3 = \boldsymbol{k}_1 + \boldsymbol{k}_2. \tag{14.137}$$

Phase matching can be accomplished either by using the crystal birefringence or by quasi-phase matching, as discussed earlier in this chapter. The procedures are straightforward extensions of those used in second-harmonic generation. One can use a cavity to enhance the power at *each* of the sum frequencies. A single cavity gives a sum-frequency power which is *linear* in the cavity enhancement factor, and one must use two overlapped cavities (e.g., Hemmati (1983)) to obtain a nonlinear enhancement which is comparable to that obtained in cavity-enhanced second-harmonic generation.

## 14.9 Periodically poled optical waveguides

From our treatment of second-harmonic generation, it should be clear that it is highly desirable to have as large an interaction length as possible while keeping the beam diameter small. This suggests the use of an optical waveguide to enhance the second-harmonic generation efficiency. Until recently, it has proven to be very difficult to phase match the second-harmonic generation in a waveguide, since the constraints imposed by the confinement in the waveguide make it very difficult to simultaneously satisfy the phase-matching requirements. With the development of quasi-phase matching, this problem has been largely solved and it has become practical to achieve highly efficient second-harmonic generation in a waveguide. This technique is considerably simpler and more robust than doubling in a build-up cavity, but suffers from a lack of flexibility: a single waveguide can double only a narrow range of wavelengths.

A *periodically poled* (i.e., quasi-phase matched) waveguide is shown in Fig. 14.22. If the fundamental intensity is not depleted, the conversion efficiency is proportional to the square of the length of the waveguide (eqn 14.49). However, in favorable situations, it is likely that there will be at least some *depletion* of the fundamental power as the input wave progresses down the waveguide. Indeed, with efficient nonlinear media, the conversion from the fundamental to the second harmonic can be very nearly 100%. Our treatment so far of second-harmonic generation has assumed that the fundamental power is constant (at least during a single pass of the crystal), and we will now examine what happens when this constraint is removed.

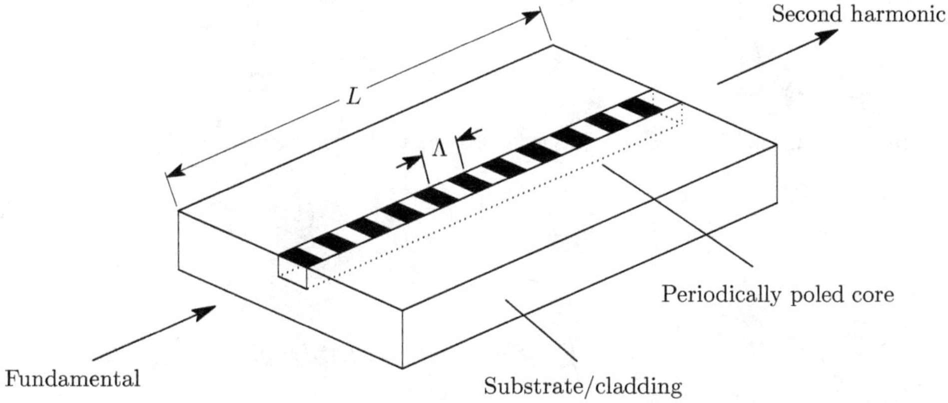


**Fig. 14.22** Second-harmonic generation in a periodically poled waveguide.

We will begin with eqn 14.45, modifying it to take account of the waveguide:

$$\frac{dE_2}{dz} = -i\frac{\omega}{2n_{w,2}}\left(\frac{2}{\pi}\right)\sqrt{\frac{\mu_0}{\epsilon_0}}d_{eff}(E_1)^2 e^{i\Delta\beta_z z} \qquad \Delta\beta = \beta_2 - 2\beta_1 - \frac{2\pi}{\Lambda} \tag{14.138}$$

where the 1, 2 subscripts refer to the fundamental and second-harmonic quantities, respectively, $\Lambda$ is the spatial period of the refractive index modulation and $\beta_i$ are the propagation constants of the two waves in the waveguide. The *waveguide indices*, $n_{w,1}$ and $n_{w,2}$ are the effective indices in the waveguide and are defined by

$$\begin{aligned} n_{w,1} &= \frac{\beta_1 c}{\omega} \\ n_{w,2} &= \frac{\beta_2 c}{2\omega} \end{aligned} \tag{14.139}$$

where $\omega$ is always the frequency of the fundamental wave. The $2/\pi$ factor is due to the symmetric square wave modulation of the refractive index and was discussed in the section on quasi-phase matching. We are assuming that the two fundamental fields have the same polarization (type I SHG) and that the second-harmonic polarization is orthogonal to that of the fundamental. Equation 14.138 was originally derived for plane waves; before proceeding further, we need to take account of the transverse dependence of the fields, which are described by planar waveguide modes. In particular, if the fundamental is the $TE_0$ mode, the second harmonic will be a TM mode, since its polarization is orthogonal to that of the fundamental. The two fields will be written as the product of a function which depends only upon the axial coordinate and one depending upon the transverse coordinates:

$$E_i(x, y, z) = E_i(z)\mathcal{E}_i(x, y) \qquad i = 1, 2 \tag{14.140}$$

where the $\mathcal{E}_i(x, y)$ functions are normalized to unity:

$$\int_{-\infty}^{\infty}\int_{-\infty}^{\infty}(\mathcal{E}_i)^2 dxdy = 1 \qquad i = 1, 2. \tag{14.141}$$

Ultimately, we will want the powers, which are given by

$$P_i = \frac{n_{w,i}}{2}\sqrt{\frac{\epsilon_0}{\mu_0}}\int_{-\infty}^{\infty}\int_{-\infty}^{\infty}|E_i(z)\mathcal{E}_i(x,y)|^2dxdy = \frac{n_{w,i}}{2}\sqrt{\frac{\epsilon_0}{\mu_0}}|E_i|^2 \quad i = 1,2. \quad (14.142)$$

Multiplying eqn 14.138 by $\mathcal{E}_2(x, y)$ and integrating over $x$ and $y$, we obtain

$$\frac{dE_2(z)}{dz} = -i\frac{\omega J}{2n_{w,2}}\left(\frac{2}{\pi}\right)\sqrt{\frac{\mu_0}{\epsilon_0}}d_{eff}(E_1(z))^2e^{i\Delta\beta_z z} \quad (14.143)$$

where $J$ is the *overlap* integral,

$$J = \int_{-\infty}^{\infty}\int_{-\infty}^{\infty}\mathcal{E}_2(x,y)(\mathcal{E}_1(x,y))^2dxdy. \quad (14.144)$$

We define the amplitudes $A_i$:

$$A_i \equiv \left[\frac{\epsilon_0 n_{w,i}^2}{4\mu_0}\right]^{1/4} E_i(z) = \sqrt{P_i}. \quad (14.145)$$

Substituting for the $E_i$, the equation for the growth of the second harmonic is

$$\frac{dA_2}{dz} = -i\kappa A_i^2 e^{i\Delta\beta z} \quad (14.146)$$

where

$$\kappa = \left(\frac{\omega d_{eff}}{\sqrt{2n_2n_1^2}}\right)\left(\frac{2}{\pi}\right)\left(\frac{\mu_0}{\epsilon_0}\right)^{3/4} J. \quad (14.147)$$

We will now save a few steps by asserting *power conservation*:

$$\frac{d}{dz}(P_1 + P_2) = 0 \Longrightarrow P_1 + P_2 = P_1(0) = P_0 \quad (14.148)$$

where $P_0$ is the fundamental power at $z = 0$ (where $P_2 = 0$). We will assume that the fields are perfectly phase-matched ($\Delta\beta = 0$); otherwise the solution involves Jacobi elliptic functions, which adds an unnecessary complication. The harmonic growth equation is now

$$\frac{dA_2}{dz} = -i\kappa(P_0 - |A_2|^2). \quad (14.149)$$

If we assume that $A_1$ is real, then, from the growth equation, $A_2$ is pure imaginary. Writing $A_2 = iA_2'$, where $A_2'$ is real, we obtain:

$$\frac{dA_2'}{dz} = -\kappa A_1^2 = -\kappa(P_0 - A_2'^2) \quad (14.150)$$

This equation can be integrated from $z = 0$ to $z = L$:

$$-\kappa L = \int_0^{A_2'(L)} \frac{dA_2'}{P_0 - A_2'^2}. \quad (14.151)$$

and the result is

$$A_2'(L) = -\sqrt{P_0}\tanh(\sqrt{P_0}\kappa L). \tag{14.152}$$

Squaring this equation, we obtain the power growth,

$$\frac{P_2}{P_0} = \tanh^2(\sqrt{P_0\eta}L) \tag{14.153}$$

where $\eta = \kappa^2$ (in units of $\mathrm{W}^{-1}\mathrm{cm}^{-2}$) is

$$\eta = \frac{\omega^2 d_{eff}^2}{2n_{w,1}^2 n_{w,2}} \left(\frac{2}{\pi}\right)^2 \left(\frac{\mu_0}{\epsilon_0}\right)^{3/2} J^2. \tag{14.154}$$

Summarizing, the conversion efficiencies for the two regimes – significant depletion and no depletion – are:

$$\begin{aligned} \text{No depletion:} \quad & \frac{P_2}{P_0} = \Gamma^2 L^2 \\ \text{Depletion:} \quad & \frac{P_2}{P_0} = \tanh^2(\Gamma L), \end{aligned} \tag{14.155}$$

where

$$\Gamma \equiv \sqrt{P_0\eta}. \tag{14.156}$$

The first equation is, of course, very similar to the one derived much earlier (eqn 14.49). A plot of the conversion efficiency versus $\Gamma L$ appears in Fig. 14.23. When $L$ is somewhat larger than $\Gamma^{-1}$, most of the input power is converted into the second harmonic.

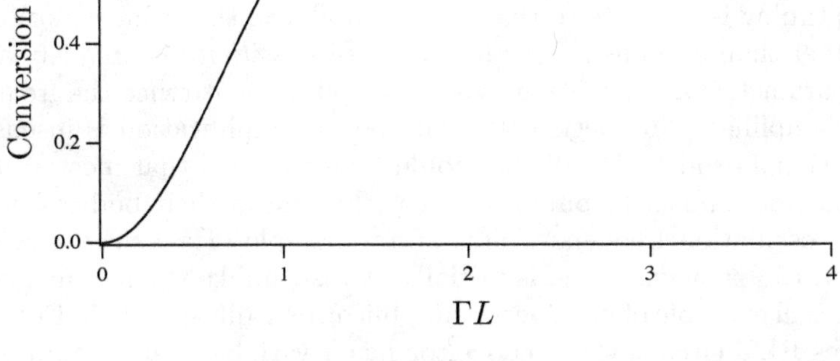


**Fig. 14.23** Plot of the conversion efficiency in a waveguide versus the waveguide length in units of $\Gamma L$).

A very common nonlinear material for periodically poled waveguide doublers is lithium niobate ($LiNbO_3$). The material is transparent from 350 nm to 5200 nm and can, using periodic poling, be phase-matched for second-harmonic generation for any wavelength for which the material is transparent at the fundamental and second harmonic. Waveguides are commercially available for fundamental wavelengths from 780 nm to 1580 nm with values of $\eta$ as high as 0.8/W-cm$^2$. They are available with more than one waveguide per chip and can be "temperature tuned" to extend the wavelength coverage. They are also available with fiber pigtails for use with fiber laser pumps (with a single aligned waveguide between the pigtails). Together with fiber lasers, DFB diode lasers, semiconductor amplifiers, and (possibly) whispering gallery comb generators, periodically poled lithium niobate waveguides (PPLN – pronounced "piplin" – waveguides) should help to hasten the end of the days of cumbersome lasers together with complicated frequency synthesis and measurement devices.

## 14.10 Parametric interactions

The nonlinear susceptibility described above allows the reverse of sum-frequency generation to take place: when a wave at frequency $\omega_3$ propagates in a nonlinear crystal, *amplification* of already existing waves at $\omega_1$ and $\omega_2$ can occur when momentum and energy are conserved ($\boldsymbol{k}_3 = \boldsymbol{k}_1 + \boldsymbol{k}_2$ and $\omega_3 = \omega_1 + \omega_2$). The beam at $\omega_3$ is called the *pump* and the beams at $\omega_1$ and $\omega_2$ are called the *signal* and *idler*. This phenomenon is the optical analogue of the well-known processes of radio-frequency and microwave *parametric amplification*.

Before describing optical parametric processes, we will illustrate the phenomenon with a mechanical and an electrical example. The mechanical example involves the familiar process of "pumping" a swing (it is interesting that the colloquial word *pumping* is exactly applicable to this process). The amplitude of the pendulum oscillation of a child's swing can be increased if the child lowers his center of gravity when the swing is descending, and raises his center of gravity when the swing is ascending. The child is *modulating* the moment of inertia of the swing at *twice the pendulum oscillation frequency* and thereby amplifying the swing's motion. Two aspects of this process are worth noting. First, the child's behavior can only amplify an existing motion and will not start a swing which is at rest; thus, the child is *not driving the swing's oscillation.* Second, the *phase* of the child's input relative to the swing's motion is very important: if the child's timing is opposite to that described above, the swing's motion will be attenuated rather than amplified. This latter *phase sensitivity* is uniquely applicable to *degenerate* parametric amplification, where the pump is at twice the frequency of the signal being amplified. Non-degenerate parametric amplification is insensitive to the phase of the signal or idler. Finally, it should be mentioned that most *seated* swingers use a different mechanism to pump a swing: they move their bodies back and forth at the swing oscillation frequency. This is an example of a *driven oscillator* and is capable of starting a swing which is initially at rest, unlike the parametric approach.

The electrical example of parametric amplification is illustrated in Fig. 14.24, which shows a series RLC circuit whose capacitor has a variable plate separation. The resonant frequency when $R \ll \omega L$ is approximately $\omega_0 = 1/\sqrt{LC}$. We assume that the capacitor is initially *charged.* If the plates are suddenly pulled apart slightly, work is

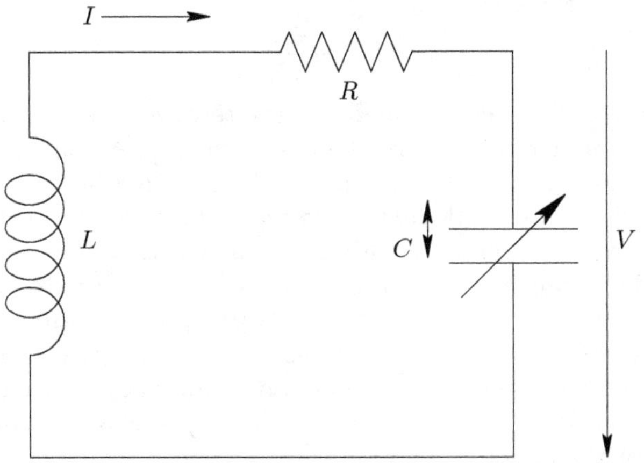


**Fig. 14.24** Circuit illustrating parametric amplification due to a modulated capacitance (via its plate separation).

done on the capacitor (due to the Coulomb attraction between the plates) and the capacitance is slightly reduced while the voltage across the capacitor is increased (since the charge does not change instantaneously and $V = Q/C$). A quarter of a cycle of $\omega_0$ later, the capacitor will be discharged and the plates are quickly restored to their original position (against zero force). After another quarter cycle, the capacitor's charge is opposite to the original charge and the plates are again pulled apart against the Coulomb attraction. If this process is repeated, we see that energy is supplied to the circuit through an external *pumping* process which occurs at twice the circuit's resonant frequency. Again, amplification will only occur for a specific phase of the pumping action. If the circuit were inert (no initial charge on the capacitor nor current through the inductor), nothing would happen when the capacitor's value is modulated. Even though our description involved *sudden* changes in the capacitor plate separation, the same phenomenon would occur if the capacitance were sinusoidally modulated. The equation satisfied by the current is

$$-L\frac{dI}{dt} = RI + \frac{1}{C}\int Idt. \tag{14.157}$$

Differentiating and moving the terms to the left side:

$$\frac{d^2I}{dt^2} + \gamma\frac{dI}{dt} + \omega_0^2 I = 0 \qquad \omega_0 = \frac{1}{\sqrt{LC}}, \quad \gamma = \frac{R}{L}. \tag{14.158}$$

The capacitance modulation is

$$C = C_0 + \Delta C \sin\omega_p t. \tag{14.159}$$

The differential equation for this parametric process is thus

$$\frac{d^2I}{dt^2} + \gamma\frac{dI}{dt} + \omega_0^2(1 + \delta\sin\omega_p t)I = 0, \tag{14.160}$$

where, assuming $\Delta C \ll C_0$,

$$\delta = -\frac{\Delta C}{C_0}. \tag{14.161}$$

From this differential equation, the source of the term *parametric* should be clear: we are modulating a *parameter* of the circuit at frequency $\omega_p$. Actually, we are modulating a rather special kind of parameter: we are modulating a *reactance.* This modulation occurs in actual circuits by employing a *nonlinear reactance*: a reactance whose value depends upon the voltage across it. In our example circuit, we would use a *varicap* diode: a PN diode whose normal voltage-dependent capacitance, when reversed biased, is enhanced relative to ordinary diodes (varicap diodes are also called *varactor* diodes). One can also modulate the *resistance* by exploiting the nonlinear resistance of a diode. Even though *frequency mixing* can occur due to a nonlinear resistance, *parametric amplification cannot take place.* In fact, most common mixers use the nonlinear resistance of their diodes for frequency conversion. However, for an electrical realization of parametric amplification or oscillation, a modulated *nonlinear reactance* is absolutely necessary.

In anticipation of an *exponentially growing* current due to parametric resonance, we will use a trial solution for the current given by

$$I(t) = I_0 \cos(\omega t + \phi) e^{st}, \tag{14.162}$$

where $s$ is real and is the expected *gain constant.* Substituting this into the differential equation and equating terms with the same exponential time dependence, one has

$$\omega_p = 2\omega \tag{14.163}$$

$$\phi = 0 \text{ or } \pi, \tag{14.164}$$

where we have dropped the term at $3\omega$. This result confirms our claim that $\omega_p = 2\omega$ and that the amplification is phase dependent. By equating the real and imaginary parts of the coefficients of the exponential terms to zero, we obtain

$$s = \frac{\omega\delta}{4} - \frac{\gamma}{2} \tag{14.165}$$

$$\omega^2 = \omega_0^2 + s^2 + \gamma s, \tag{14.166}$$

where we assume that $\omega \approx \omega_0$. From the first equation, the condition for amplification ($s > 0$) is

$$\text{Parametric amplification:} \quad \frac{\omega_0 \delta}{2} > \gamma. \tag{14.167}$$

From the second equation, the steady-state solutions ($s = 0$) occurs when $\omega = \omega_0$.

This circuit has only a single solution at $\omega_p = 2\omega$ (degenerate) since it contains a single resonant system. One can construct a circuit with *two resonant systems*, with different resonant frequencies, which are both coupled to a nonlinear reactance. That circuit will display non-degenerate parametric amplification at *both* the signal and idler frequencies. Since our goal is to analyze *optical* parametric processes, we will now turn to the optical case which quite naturally displays the various features of non-degenerate parametric amplification.

In treating optical parametric phenomena, we start with a description of the three fields:

$$\text{Pump wave:} \quad E_j^{(\omega_3)} = \frac{1}{2}E_{3j}(z)e^{i(\omega_3 t - k_3 z)} + \text{cc} \tag{14.168}$$

$$\text{Signal wave:} \quad E_i^{(\omega_1)} = \frac{1}{2}E_{1i}(z)e^{i(\omega_1 t - k_1 z)} + \text{cc} \tag{14.169}$$

$$\text{Idler wave:} \quad E_k^{(\omega_2)} = \frac{1}{2}E_{2k}(z)e^{i(\omega_2 t - k_2 z)} + \text{cc.} \tag{14.170}$$

The pump frequency is necessarily higher than either the signal or the idler frequency for parametric amplification to take place. We will treat the pump field as a parameter and solve the equations for the growth of the signal and idler waves. These two equations are obtained in exactly the same way as the differential equation (eqn 14.134) describing sum-frequency generation. The two equations are

$$\frac{dE_{1i}}{dz} = -\frac{i\omega_1}{4}\sqrt{\frac{\mu_0}{\epsilon_1}}d_{ijk}E_{2j}^* E_{3k}e^{-i\Delta kz} \tag{14.171}$$

$$\frac{dE_{2k}^*}{dz} = \frac{i\omega_2}{4}\sqrt{\frac{\mu_0}{\epsilon_2}}d_{kij}E_{1i}E_{3j}^* e^{i\Delta kz}, \tag{14.172}$$

where

$$\Delta k = k_3 - k_1 - k_2 \tag{14.173}$$

$$\omega_3 = \omega_1 + \omega_2. \tag{14.174}$$

Before solving these equations, we will simplify the notation by using coupling constants $\kappa_1$ and $\kappa_2$ and include absorption terms described by (intensity) absorption coefficients $\alpha_1$ and $\alpha_2$:

$$\begin{aligned} \text{Signal:} \quad \frac{dE_1}{dz} &= -\frac{\alpha_1}{2}E_1 - i\kappa_1 E_2^* E_3 e^{-i\Delta kz} \\ \text{Idler:} \quad \frac{dE_2^*}{dz} &= -\frac{\alpha_2}{2}E_2^* + i\kappa_2 E_1 E_3^* e^{i\Delta kz}, \end{aligned} \tag{14.175}$$

where

$$\begin{aligned} \kappa_1 &= \frac{\omega_1}{4}\sqrt{\frac{\mu_0}{\epsilon_1}}d \\ \kappa_2 &= \frac{\omega_2}{4}\sqrt{\frac{\mu_0}{\epsilon_2}}d, \end{aligned} \tag{14.176}$$

and we have dropped the coordinate subscripts to reduce clutter. In all of the above, we assume that there is little depletion of the *pump*; if this is so, we do not need an equation for $E_3$ and can treat the pump field as a constant.

To simplify matters, we will assume that $\alpha_1 = \alpha_2 = \alpha$. We will look for a solution of the form

$$E_1(z) = E_1(0)e^{Gz-i\Delta kz/2} \tag{14.177}$$
$$E_2(z) = E_2(0)e^{Gz-i\Delta kz/2}. \tag{14.178}$$

These expressions assume a *gain* equal to $G$ and include the $e^{-i\Delta kz/2}$ factor in order to remove the similar factor from the equations. Substituting and dividing out the exponentials, we obtain

$$(G - i\Delta k/2 + \alpha/2)E_1(0) + i\kappa_1 E_3 E_2^*(0) = 0 \tag{14.179}$$
$$-i\kappa_2 E_3^* E_1(0) + (G + i\Delta k/2 + \alpha/2)E_2^*(0) = 0. \tag{14.180}$$

These *homogeneous* equations will only have a non-trivial solution when the determinant of the coefficients vanishes. The solution for $G$ is

$$G = -\frac{\alpha}{2} \pm \sqrt{\kappa_1\kappa_2|E_3|^2 - \left(\frac{\Delta k}{2}\right)^2}. \tag{14.181}$$

We thus see that there is a *parametric gain*, $g$, given by

$$\text{Parametric gain:} \quad g = \sqrt{\kappa_1\kappa_2|E_3|^2 - \left(\frac{\Delta k}{2}\right)^2}, \tag{14.182}$$

where we have taken the positive square root. This gain is maximum when $\Delta k = 0$ (phase matched case). The maximum gain is proportional to the *pump field* and the square root of the product of the signal and idler frequencies. The overall gain, $G$, is reduced by absorption ($\alpha$) and phase mismatching ($\Delta k \neq 0$).

The coupled differential equations can be solved by taking linear combinations of both solutions:

$$E_1(z) = \left(E_{11}e^{gz} + E_{12}e^{-gz}\right) e^{-\alpha/2-i\Delta kz/2} \tag{14.183}$$
$$E_2(z) = \left(E_{21}e^{gz} + E_{22}e^{-gz}\right) e^{-\alpha/2-i\Delta kz/2}, \tag{14.184}$$

and using the boundary conditions that the $z = 0$ fields are $E_{1,2}(0)$. The result is

$$E_1(z) = E_1(0)e^{-i\Delta kz/2-\alpha z/2}\left[\cosh gz + \frac{i\Delta k}{2g}\sinh gz\right] - \frac{i\kappa_1}{g}E_3E_2^*(0)e^{-i\Delta kz/2-\alpha z/2}\sinh gz \tag{14.185}$$

$$E_2(z) = E_2(0)e^{-i\Delta kz/2-\alpha z/2}\left[\cosh gz + \frac{i\Delta k}{2g}\sinh gz\right] - \frac{i\kappa_2}{g}E_3E_1^*(0)e^{-i\Delta kz/2-\alpha z/2}\sinh gz. \tag{14.186}$$

In the following, we will simplify these somewhat cumbersome solutions.

From eqns 14.175, together with a similar one for $E_3$, one can easily show that

$$-\frac{1}{\omega_3}\frac{d}{dz}|E_3|^2 = \frac{1}{\omega_1}\frac{d}{dz}|E_1|^2 = \frac{1}{\omega_2}\frac{d}{dz}|E_2|^2. \qquad (14.187)$$

Since the squared fields are proportional to the photon flux, integrating over the length of the crystal gives the *net flux change* as one traverses the crystal; this is proportional to the power change divided by the frequency. Thus, we obtain the *Manley–Rowe relation*

$$\text{Manley–Rowe relation:} \quad -\Delta\left(\frac{P_3}{\omega_3}\right) = \Delta\left(\frac{P_2}{\omega_2}\right) = \Delta\left(\frac{P_1}{\omega_1}\right), \qquad (14.188)$$

where the deltas are interpreted as the changes in the quantities between the input and output planes. The minus sign is unique to the wave at $\omega_3$, whose frequency is the sum of the other two frequencies. The Manley–Rowe relation suggests (and quantum mechanics confirms) a photon model for optical parametric interactions. Using this model, the absorption of a single photon at the highest frequency is accompanied by the *joint* creation of single photons at two other frequencies which sum to the highest frequency. This simple model can provide a great deal of insight into these processes. For example, it explains why parametric gain cannot occur at the highest of the three frequencies: if the pump is at one of the two lower frequencies, energy conservation would be violated if a photon at the highest frequency were created. Finally, it is worth mentioning that *all* of the nonlinear behaviors discussed in this chapter can be considered to be *parametric interactions*, and this was recognized in the title of one of the early papers on the subject (Boyd (1968)). Thus, all of the behaviors satisfy the Manley–Rowe relationship and can be explained with a simple model involving interactions among three photons in the presence of a nonlinear medium. Of course, only *parametric amplification* can exhibit gain.

Parametric gain is a *coherent* optical gain process which, like laser gain, can lead to oscillation if a pumped medium is placed inside a resonant cavity. A *parametric oscillator* is, in principle, more widely tunable than a laser, since the parametric device is merely limited by the requirement that energy and momentum be conserved. Parametric oscillation generates both a signal and an idler beam, as required by the Manley–Rowe relation. Parametric oscillators can be either *doubly resonant* or *singly resonant*, depending on whether the cavity resonates at both signal and idler frequencies or only at the signal frequency.

A doubly resonant optical parametric oscillator is shown schematically in Fig. 14.25. The oscillation conditions for a doubly resonant oscillator are obtained by requiring that the fields replicate themselves after one round trip in the cavity. The field equations for $\Delta k = 0$ are obtained from eqns 14.185 and 14.186:

$$E_1(z) = e^{-\alpha z/2}\left[E_1(0)\cosh gz - i\sqrt{\frac{\kappa_1}{\kappa_2}}E_2^*(0)\sinh gz\right] \qquad (14.189)$$

$$E_2^*(z) = e^{-\alpha z/2}\left[E_2^*(0)\cosh gz + i\sqrt{\frac{\kappa_2}{\kappa_1}}E_1(0)\sinh gz\right]. \qquad (14.190)$$

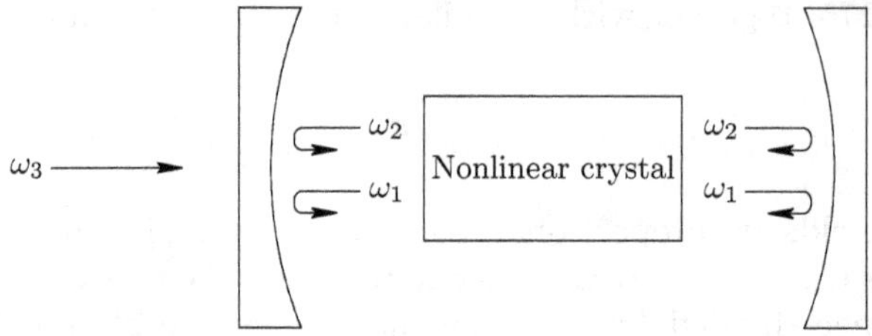


**Fig. 14.25** Schematic of doubly resonant optical parametric oscillator showing pump beam at $\omega_3$ together with signal and idler beams at $\omega_1$ and $\omega_2$.

The reflectivities are assumed to be the same for each mirror, and the *squares* of the field reflectivities at the signal and idler frequencies are

$$\text{Signal:} \quad r_1^2 = R_1 e^{2i\phi_1} \tag{14.191}$$

$$\text{Idler:} \quad r_2^2 = R_2 e^{2i\phi_2}. \tag{14.192}$$

The quantities $R_1$ and $R_2$ are the *power reflectivities* at the signal and idler frequencies. One can now apply the requirement that the field replicates itself after a round trip:

$$E_1(0) = r_1^2 e^{-\alpha L - 2ik_1 L}\left[E_1(0)\cosh gL - i\sqrt{\frac{\kappa_1}{\kappa_2}}E_2^*(0)\sinh gL\right] \tag{14.193}$$

$$E_2^*(0) = (r_2^*)^2 e^{-\alpha L + 2ik_2 L}\left[E_2^*(0)\cosh gL + i\sqrt{\frac{\kappa_2}{\kappa_1}}E_1(0)\sinh gL\right], \tag{14.194}$$

where we include the $2ik_{1,2}z$ terms in the exponents, since we require that the field amplitude *and phase* replicate themselves. Note that the arguments of the hyperbolic functions are $gL$ and not $2gl$ since there is *no parametric gain* in the reverse direction (there is, however, loss in both directions). This is a pair of homogeneous equations in the variables $E_1(0)$ and $E_2^*(0)$ and a non-trivial solution requires that the determinant of the coefficients vanish. From this, we obtain the oscillation condition

$$\begin{aligned}(1 - r_1^2 e^{-\alpha L - 2ik_1 L}\cosh gL)(1 - (r_2^*)^2 e^{-\alpha L + 2ik_2 L}\cosh gL)\\ = (r_1 r_2^*)^2 e^{-2\alpha L + 2i(k_2 - k_1)L}\sinh^2 gL.\end{aligned} \tag{14.195}$$

We will simplify this by making the realistic assumption that the mirror transmissions are greater than the bulk losses and set $\alpha = 0$. We will also assume that the lowest threshold is obtained when *both* the signal and the idler are at axial modes of the cavity and set:

$$2k_1 L - 2\phi_1 = m \times 2\pi \tag{14.196}$$

$$2k_2 L - 2\phi_2 = n \times 2\pi \qquad m, n \;\text{ integers.} \tag{14.197}$$

Then, the oscillation condition becomes

$$\begin{aligned}(1 - R_1\cosh gL)(1 - R_2\cosh gL) = R_1 R_2 \sinh^2 gL\\ \Longrightarrow (R_1 + R_2)\cosh gL - R_1 R_2 = 1.\end{aligned} \tag{14.198}$$

If the mirrors have $R_{1,2} \approx 1$, we can expand the hyperbolic cosine and obtain

$$\text{Doubly resonant oscillation condition:} \quad gL = \sqrt{(1 - R_1)(1 - R_2)}. \tag{14.199}$$

The threshold pumping intensity, $I_{3t}$, at $\omega_3$ can be obtained using eqns 14.176 for the coupling constants and eqn 14.47 for the intensity in terms of the field. The result is

$$\text{Pumping threshold:} \quad I_{3t} = \frac{8}{\omega_1 \omega_2 d^2 L^2} \sqrt{\frac{\epsilon_1 \epsilon_2 \epsilon_3}{\mu_0^3}} (1 - R_1)(1 - R_2). \tag{14.200}$$

An important advantage of doubly resonant parametric amplifiers is the very low threshold compared to the singly resonant oscillator. A possible disadvantage is the difficulty in obtaining two simultaneous cavity resonances at the idler and signal frequencies. If $n_1$ and $n_2$ are the indices of refraction for the signal and idler, it is easy to show that the separation between simultaneous axial resonances is about

$$\text{Separation between resonances} = \frac{\pi c}{(n_1 - n_2)L}. \tag{14.201}$$

For typical refractive indices, this separation might be 100 times the mode separation in the cavity. A small change in the cavity length can cause mode hops with large frequency excursions, which is often undesirable. This situation is not nearly as serious in a singly resonant oscillator.

A singly resonant parametric oscillator simply relaxes the requirement that the cavity be resonant at *both* signal and idler frequencies. We can derive the oscillation condition for a singly resonant oscillator from eqn 14.195 by letting $r_2 = 0$. Setting in addition $\alpha = 0$ and $\Delta k = 0$, we obtain

$$R_1 \cosh gL = 1, \tag{14.202}$$

where we assume the cavity is resonant at $\omega_1$ so that $2k_1 L - 2\phi_1 = m \times 2\pi$. When $R_1 \approx 1$, we obtain

$$\text{Singly resonant oscillation condition:} \quad gL = \sqrt{2(1 - R_1)}. \tag{14.203}$$

Thus, the ratio of thresholds for doubly to singly resonant oscillators is

$$\frac{\text{singly resonant pumping threshold}}{\text{doubly resonant pumping threshold}} = \frac{2}{1 - R_2}, \tag{14.204}$$

where the $R_2$ in the denominator is for the doubly resonant oscillator. From this, we see that the singly resonant threshold is at least ten times greater than the doubly resonant threshold.

A singly resonant oscillator can employ a different and more flexible approach to phase matching. Since the three beams do not need to be *collinear* for momentum conservation (phase matching) to be satisfied, the pump beam is sent in at the appropriate angle to satisfy momentum conservation and ensure that the *signal* beam is along the axis of the cavity. The direction of the idler beam is unimportant, since it is discarded. Fig. 14.26 illustrates the geometry used in singly resonant oscillators. This

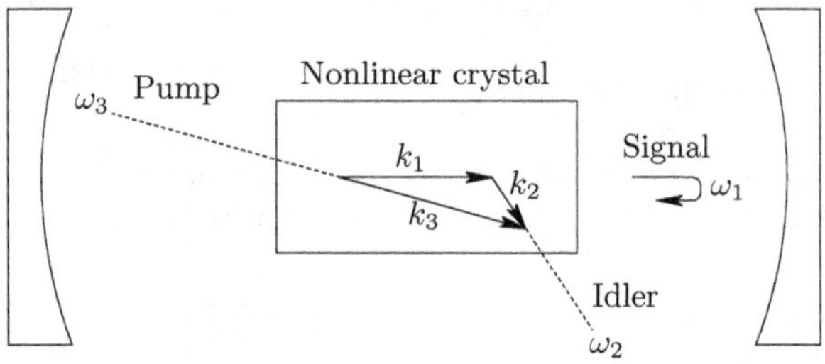


**Fig. 14.26** Schematic of singly resonant optical parametric oscillator illustrating non-collinear phase matching.

phase-matching approach is called *non-collinear phase matching* and is conceptually similar to the generation of a coherent outgoing wave in Cerenkov radiation.

The *gain saturation* mechanism in optical parametric oscillators is somewhat different from saturation in a laser. The situation is actually more straightforward: when the pump intensity is sufficient to produce a threshold gain, further increases in the pumping intensity go entirely to the signal and idler beams, since the gain must be *clamped* at threshold as in a laser. This has the interesting and useful result of *power-limiting* the transmitted pump beam: the transmitted pump power will be independent of the incident power above the oscillation threshold. The *efficiency* can by fairly high in a parametric oscillator, particularly if the idler frequency is somewhat lower than the signal frequency. One can obtain a numerical result for the output power using the Manley–Rowe relationship. If $\Delta P = P_3 - P_{3t}$ is the pumping power excess over threshold, the *number of excess photons* is proportional to $\Delta P/\omega_3$ and these excess photons go into the signal and idler beams, whose photon numbers are proportional to $P_1/\omega_1$ and $P_2/\omega_2$ with the same proportionality constant. Equating these and solving for $P_1$,

$$\text{Power output:} \quad P_1 = \frac{\omega_1(P_3 - P_{3t})}{\omega_3}. \tag{14.205}$$

This approaches $(\omega_1/\omega_3)P_3$ when $P_3 \gg P_{3t}$.

So far, our treatment of parametric phenomena has been based entirely on *classical* principles: given the nonlinear susceptibility, everything follows without resorting to quantum mechanics. There is one parametric process which is fundamentally quantum-mechanical: it is called *spontaneous parametric fluorescence*. This process is roughly analogous to *spontaneous emission* in atoms, which also must be treated quantum-mechanically. Spontaneous emission can be thought of as emission which is *stimulated* by the single quantum per mode of the vacuum, and spontaneous parametric fluorescence can be considered to be parametric amplification of the vacuum field. (A similar analogy exists with spontaneous Raman scattering.)

A diagram of the apparatus demonstrating parametric fluorescence together with a vector diagram illustrating *non-collinear phase matching* appears in Fig. 14.27. Our approach will be to sum all pairs of amplified signal and idler modes assuming a single photon per signal mode (see, for example, Byer (1968) or Yariv (1989)). We begin by determining the number of signal modes whose propagation angle is between $\psi$ and

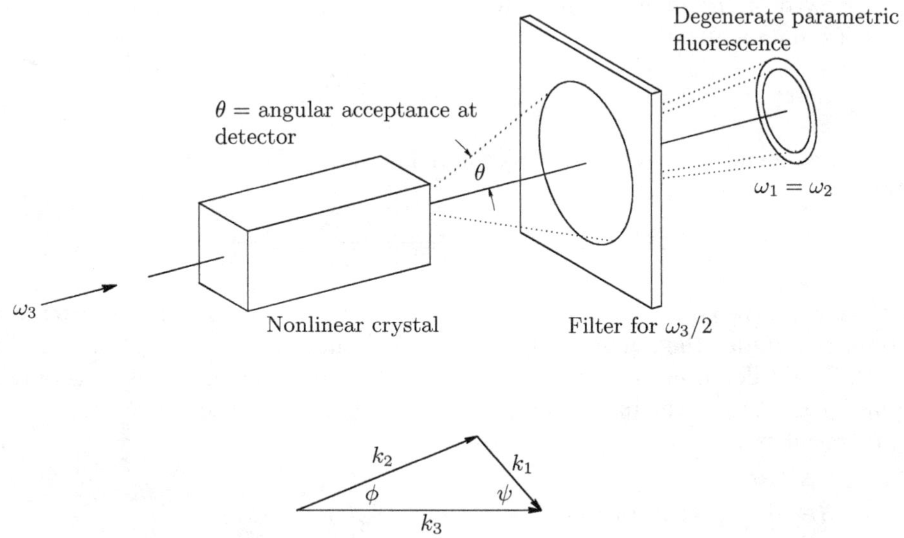


**Fig. 14.27** Schematic of the apparatus demonstrating spontaneous optical parametric fluorescence (top) and the vector diagram illustrating non-collinear phase matching (bottom). The acceptance angle of the detector is $\theta$; all of the angles are exaggerated for clarity. The *degenerate* photons are obtained using a filter whose passband is centered on $\omega_3/2$. For type I phase matching, photons on opposite sides of the conical surface are *entangled.*

$\psi + d\psi$ and whose wave vector magnitude is between $k_1$ and $k_1 + dk_1$. We will assume that the waves satisfy *periodic boundary conditions.* For example, the signal field is a plane wave and is given by

$$E_1(\boldsymbol{r}) = E_0 e^{i\boldsymbol{k}_1\cdot\boldsymbol{r} - i\omega_1 t} \qquad \boldsymbol{k} = \left(\frac{2\pi l}{L_x}, \frac{2\pi m}{L_y}, \frac{2\pi n}{L_z}\right), \tag{14.206}$$

where $l, m, n$ are integers, $L_{x,y,z}$ are the periods along the three directions and the volume of the crystal is $V = L_x L_y L_z$. The volume per mode in $k$-space is $(2\pi)^3/\mathrm{V}$. The number of modes is

$$dN_1 = \frac{2\pi k_1^2 dk_1 \sin\psi d\psi}{(2\pi)^3/V}, \tag{14.207}$$

where the numerator is just the differential volume element in spherical coordinates with $\psi$ taken as the *polar angle* (with the $z$-axis along $\boldsymbol{k}_3$). Due to the cylindrical symmetry, the azimuthal integration yields $2\pi$. The intensity is obtained by multiplying by $\hbar\omega_1$ to obtain the differential energy and then by $c/n_1 V$ to obtain the differential intensity:

$$dI_1 = \frac{k_1^2 c dk_1 \hbar\omega_1 \psi d\psi}{4\pi^2 n_1}, \tag{14.208}$$

where for small $\psi$ we used $\sin\psi \approx \psi$.

The growth of the idler field due to a signal at $z = 0$ (but no idler at $z = 0$) is obtained from eqn 14.186:

$$E_2(z) = -i\kappa_2 z E_3 E_1^*(0) e^{-i\Delta k z/2} f(\Delta k, z), \tag{14.209}$$

where we assume no bulk losses ($\alpha = 0$) and define the *phase-matching factor*, $f(\Delta k, z)$ as

$$f(\Delta k, z) \equiv \frac{\sin\sqrt{(\Delta k/2)^2 - \kappa_1\kappa_2|E_3|^2}z}{\sqrt{(\Delta k/2)^2 - \kappa_1\kappa_2|E_3|^2}z}. \tag{14.210}$$

Note that we have converted the hyperbolic sine to a circular sine by exchanging the two terms under the square root (and we left out the "$i$" since $f(\Delta k, z)$ will be squared). The differential idler power due the differential signal intensity is obtained by squaring $E_2(L)$, converting it to a power and substituting the above expression for $dI_1$. The result is

$$dP_2 = \langle\text{Area}\rangle \times dI_2 = dI_1\,(\kappa_2|E_3|L)^2 \left(\frac{\omega_2}{\omega_1}\right) f^2(\Delta k, L)A, \tag{14.211}$$

where $A$ is the area of the idler beam and the frequency ratio is due to the Manley–Rowe relation. Using the law of sines, one can convert the $\psi$ expression in $dI_1$ to one involving $\phi$:

$$\psi d\psi = \left(\frac{k_2}{k_1}\right)^2 \phi d\phi \qquad (\psi, \phi \ll 1). \tag{14.212}$$

The total idler power is obtained by integrating $dP_2$ over $\phi$ from 0 to $\theta$ and over all $\omega_2$ (using $d\omega_2 = -d\omega_1$):

$$P_2 = \int_{-\infty}^{\infty}\int_0^{\theta} d\omega_2 d\phi (dP_2). \tag{14.213}$$

Since the integral is over $\omega_2$ and $\phi$, we need to convert $\Delta k$ and the $k$ values to functions of these variables. The quantity $\Delta k$ can be obtained using the law of sines when $\phi, \psi \ll 1$ as

$$\Delta k = k_3 - k_2\cos\phi - k_1\cos\psi \approx \frac{k_2 k_3}{k_1}\frac{\phi^2}{2} + k_3 - k_2 - k_1. \tag{14.214}$$

The $k$ values can be expanded around their *collinearly phase-matched* values (where $k_{30} = k_{10} + k_{20}$) as

$$\begin{aligned} k_2 &= k_{20} + \frac{\partial k_2}{\partial\omega_2}(\omega_2 - \omega_{20}) \\ k_1 &= k_{10} + \frac{\partial k_1}{\partial\omega_1}(\omega_1 - \omega_{10}), \end{aligned} \tag{14.215}$$

where the $\omega_{10}$, $\omega_{20}$ frequencies correspond to the collinearly phase-matched $k$ values. The integral can be evaluated with these substitutions and with the assumption that

$\kappa_1\kappa_2|E_3|^2$ can be ignored in comparison to $\Delta k$ in the $f^2(\Delta l, L)$ expression. The result is

$$P_2 = \pi \frac{\beta L P_3}{|b|}\theta^2, \tag{14.216}$$

where

$$b \equiv \left(\frac{\partial k_2}{\partial \omega_2}\right)_{\omega_{20}} - \left(\frac{\partial k_1}{\partial \omega_1}\right)_{\omega_{10}} \tag{14.217}$$

$$\beta \equiv \frac{\hbar \omega_1 \omega_2^4 n_2 d^2}{\pi^2 c^5 n_1 n_3 \epsilon_0^3}. \tag{14.218}$$

The reason for the asymmetry in $\omega_1$ and $\omega_2$ is that we are starting with $\omega_1$ and are summing over all contributions of $\omega_2$ that satisfy the conservation laws and have $k_2$ values which make an angle less than $\theta$ with the $z$-axis. One of the references mentioned above began with $\omega_2$ and obtained the same result with the identity of the idler and signal reversed from ours. Of course, $\omega_1$ and $\omega_2$ are not independent of each other, since they must satisfy $\omega_1 + \omega_2 = \omega_3$.

The *bandwidth* (range of $\omega_2$ values) can be obtained from our above expression for $\Delta k$:

$$\Delta k = \frac{k_2 k_3}{k_1}\frac{\phi^2}{2} + k_3 - k_2 - k_1 = -b(\omega_2 - \omega_{20}) + a\phi^2, \tag{14.219}$$

where

$$a = \frac{k_2 k_3}{2k_1}, \tag{14.220}$$

and $b$ is defined in eqn 14.217 and we used the relation $\omega_2 - \omega_{20} = -(\omega_1 - \omega_{10})$. We will define the bandwidth by the separation between $\Delta k = 0$ and the first node of the sinc function $(\sin(\Delta k L/2)/(\Delta k L/2))$. Thus

$$\Delta k L = \left(a\phi^2 - b(\omega_2 - \omega_{20})\right) L = 2\pi \Longrightarrow \Delta\omega = \omega_2 - \omega_{20} = \frac{a\phi^2 L - 2\pi}{bL}. \tag{14.221}$$

Therefore, the bandwidth is

$$\Delta\omega = \frac{2\pi}{|b|L} \qquad \text{when } \theta = \phi_{max} < \sqrt{\frac{2\pi}{aL}} \tag{14.222}$$

$$\Delta\omega = \frac{a\theta^2}{|b|} \qquad \text{when } \theta > \sqrt{\frac{2\pi}{aL}}. \tag{14.223}$$

In typical cases, the bandwidth is many tens of nm.

The principal use of parametric fluorescence in atomic physics is as a source of *entangled* photons. These can be obtained when the signal and idler frequencies are the same; this situation is called *degenerate spontaneous parametric fluorescence* (or, more commonly, *degenerate parametric down-conversion*). In the degenerate case, the fluorescent beam (at $\omega_1 = \omega_2$) becomes a conical surface shown in Fig. 14.27 for the case when the signal and idler have the same polarization (type I, non-collinear phase matching). If the polarizations of the two beams are orthogonal (type II phase matching), there will be two cones which are displaced from each other and which contain photons of one or the other polarization. The geometries for both type I and type II phase matching are shown in Fig. 14.28.

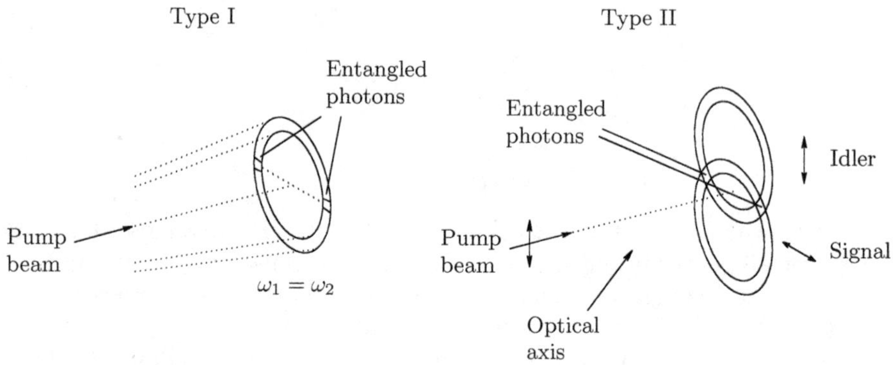


**Fig. 14.28** Detail of beams from degenerate parametric fluorescence for type I and type II phase matching. A negative uniaxial crystal is assumed, so the pump is polarized in the plane of the optical axis (extraordinary wave). The polarizations for the type II case are indicated by the arrows.

Even though non-collinear phase matching offers great tunability, one must still consider the crystal birefringence when designing an experiment using degenerate parametric fluorescence. Since $k_1 = k_2$ and $\omega_3 = 2\omega_1$, the phase matching condition is

$$\boldsymbol{k}_3 = 2\boldsymbol{k}_1 \Longrightarrow \frac{n_3\omega_3}{c} = 2\frac{n_1\omega_1}{c}\cos\phi \Longrightarrow \cos\phi = \frac{n_3}{n_1}. \tag{14.224}$$

This cannot be satisfied unless $n_3 \leq n_1$. Therefore, when using a negative uniaxial crystal (such as BBO), the pump polarization must be along the extraordinary direction (in the plane of the wave vector and the optical axis). For type I phase matching, the signal and idler are ordinary waves. With type II phase matching, both polarizations are generated and each wave of a given polarization will have its own conical surface. The two conical surfaces are displaced from each other by an angle which depends upon the angle that the pump beam makes with the optical axis.

The photons generated by degenerate parametric fluorescence can be *entangled.* The idea of entanglement is best understood using the photon picture of the parametric process. A pump photon spontaneously splits into signal and idler photons of the same frequency. Each individual process must conserve both momentum and energy, so the fluorescent photons have half the energy of the pump photons, and momentum conservation requires that the two outgoing photons be symmetrically displaced about the $k$-vector of the pump beam. The entanglement of the two emitted photons lies in the fact that all of the properties of one photon of the emitted pair can be determined from the measured properties of the other photon. Entangled photons are mutually coherent and therefore can generate interference fringes. The location of the entangled photons on opposite sides of the cone for type I or at the intersections of the cones for type II is a result of the requirement that momentum be conserved. The photons at the intersections of the cones in type II phase matching are, in addition, *polarization-entangled.*

## 14.11 Further reading

The books by Yariv (1989) and Davis (1996) have fairly complete sections on nonlinear optics, while the books by Ghatak and Thyagarajan (1989) and by Milonni and Eberly (1988) have shorter but quite satisfactory discussions of the subject. In addition, the aforementioned books by Yariv and Davis have excellent sections on anisotropic crystals, although the best treatment in the author's opinion is in the classic book by Born and Wolf (1980). A very complete treatment of frequency tuning and frequency and temperature tolerances in quasi-phase matching appears in a paper by Fejer et al. (1992) while an early paper on quasi-phase matching itself is by Somekh and Yariv (1972). The classic paper on harmonic generation using focused beams is by Boyd and Kleinman (1968) and the classic paper on the use of a (standing-wave) cavity to enhance second-harmonic generation is by Ashkin et al. (1966). The latter paper also discusses the use of a cavity resonant at the *second harmonic* for nonlinear enhancement. The paper by Ou and Kimble (1993) also discusses the benefits of resonance on the second harmonic and the related paper by Polzik and Kimble (1991) discusses the effects of depletion and thermal lensing in an enhancement cavity. A good table of Sellmeier coefficients for a number of nonlinear materials appears in the *Tunable Laser Handbook* (Duarte (1995)) and a tabulation of nonlinear coefficients is in the paper by Roberts (1992). A classic treatment of spontaneous parametric fluorescence is in the paper by Byer and Harris (1968) and a discussion of the applications of entangled photons is in the paper by Zeilinger (1999).

## 14.12 Problems

(14.1) Using the Sellmeier coefficients given in the text, determine the phase-matching angle for generating second-harmonic radiation at 231 nm from BBO. (BBO is a negative uniaxial crystal.)

(14.2) For a 1 cm length of BBO used in problem 1, what is the aperture length and the *double-refraction parameter*, $B$?

(14.3) In a negative uniaxial crystal, assume that there exists a phase-matching angle, $\theta_m$, at which the ordinary index at $\omega$ is equal to the extraordinary index at $2\omega$. For small deviations of the propagation angle about $\theta_m$, derive an expression for the corresponding change, $\Delta k(\theta)L$, in the total phase difference between the fundamental and the second harmonic. (This allows one to determine the maximum *angular acceptance* of the fundamental beam.)

(14.4) A bow-tie cavity is used to enhance the second-harmonic generation in a potassium niobate crystal. Assume that the input coupler power transmission is 0.02 and the remaining losses are also 0.02. If the crystal conversion efficiency is 0.01/W and the input power is 100 mW, how much second-harmonic power will be generated? What is the optimum input-coupling transmission, and what will be the second-harmonic power if the optimum input coupling is used?

# 15

# Frequency and amplitude modulation

## 15.1 Introduction

Many of the applications described in this book require the ability to modulate the frequency, phase or amplitude of laser radiation. The devices used for these purposes fall into two broad categories: those that exploit the dependence of the indices of refraction of certain crystals on an applied electric field (the *electro-optic effect*) and those that use an acoustic wave to modulate the index of refraction (via the *acousto-optic* effect). In this chapter we will describe both categories of modulators and the underlying processes that enable them to function.

## 15.2 The linear electro-optic effect

The *linear electro-optic effect* refers to the linear dependence of the refractive indices on an applied electric field in certain crystals. In the absence of an electric field, the crystal is described in some coordinate system by the equation for the *index ellipsoid*,

$$\frac{x^2}{n_x^2} + \frac{y^2}{n_y^2} + \frac{z^2}{n_z^2} = 1, \tag{15.1}$$

where $x, y$ and $z$ are the axes (*principal axes*) along which $\boldsymbol{E}$ is parallel to $\boldsymbol{D}$. (These matters are discussed in some detail in the previous chapter.) When an electric field is present, the index ellipsoid will change to

$$\left(\frac{1}{n^2}\right)_1 x^2 + \left(\frac{1}{n^2}\right)_2 y^2 + \left(\frac{1}{n^2}\right)_3 z^2 + 2\left(\frac{1}{n^2}\right)_4 yz + 2\left(\frac{1}{n^2}\right)_5 xz + 2\left(\frac{1}{n^2}\right)_6 xy = 1. \tag{15.2}$$

The six indexed quantities (of the form $1/n^2$) are parameters which describe the effect of the electric field and reduce to $1/n^2_{x,y,z}$ or zero when the field is not present:

$$\begin{aligned}
\left(\frac{1}{n^2}\right)_1 &= \frac{1}{n_x^2} \qquad \text{(when } E = 0\text{)} \\
\left(\frac{1}{n^2}\right)_2 &= \frac{1}{n_y^2} \\
\left(\frac{1}{n^2}\right)_3 &= \frac{1}{n_z^2} \\
\left(\frac{1}{n^2}\right)_4 &= \left(\frac{1}{n^2}\right)_5 = \left(\frac{1}{n^2}\right)_6 = 0.
\end{aligned} \tag{15.3}$$

Recall that the $x, y$ and $z$ in the above two equations refer to possible *polarization directions* of the optical wave. The electro-optic effect is described by the *changes* in the $(1/n^2)_i$ due to the electric field, $\boldsymbol{E}$:

$$\Delta\left(\frac{1}{n^2}\right)_i = \sum_{j=1}^{3} r_{ij}E_j \qquad i = 1\ldots6. \tag{15.4}$$

Written out, the equation is

$$\begin{pmatrix} \Delta\left(\frac{1}{n^2}\right)_1 \\ \Delta\left(\frac{1}{n^2}\right)_2 \\ \Delta\left(\frac{1}{n^2}\right)_3 \\ \Delta\left(\frac{1}{n^2}\right)_4 \\ \Delta\left(\frac{1}{n^2}\right)_5 \\ \Delta\left(\frac{1}{n^2}\right)_6 \end{pmatrix} = \begin{pmatrix} r_{11} & r_{12} & r_{13} \\ r_{21} & r_{22} & r_{23} \\ r_{31} & r_{32} & r_{33} \\ r_{41} & r_{42} & r_{43} \\ r_{51} & r_{52} & r_{53} \\ r_{61} & r_{62} & r_{63} \end{pmatrix} \begin{pmatrix} E_x \\ E_y \\ E_z \end{pmatrix}. \tag{15.5}$$

The $6 \times 3$ matrix $r_{ij}$ is called the *electro-optic tensor* and is zero if the crystal is symmetric under a coordinate system inversion. To see why this is so, when $\boldsymbol{r} \to -\boldsymbol{r}$, $\boldsymbol{E} \to -\boldsymbol{E}$, but in a symmetric crystal $r_{ij}$ and $\Delta(1/n^2)_i$ are unchanged. Thus we have $\Delta(1/n^2)_i \to -\Delta(1/n^2)_i$ which can only be true if $r_{ij} = 0$ for all $i, j$.

Only a few elements of the electro-optic tensor are independent. Using the crystal symmetry properties, one can list the non-zero values of $r_{ij}$ for the more important crystal classes (as was done for the nonlinear coefficients, $d_{ij}$, in the previous chapter):

- **Tetragonal, class $\bar{4}2m$** – example, ADP, AD*P, KDP, KD*P

$$\begin{pmatrix} 0 & 0 & 0 \\ 0 & 0 & 0 \\ 0 & 0 & 0 \\ r_{41} & 0 & 0 \\ 0 & r_{41} & 0 \\ 0 & 0 & r_{63} \end{pmatrix} \tag{15.6}$$

- **Trigonal, class $3m$** – example lithium niobate, lithium tantalate

$$\begin{pmatrix} r_{11} & 0 & r_{13} \\ -r_{11} & 0 & r_{13} \\ 0 & 0 & r_{33} \\ 0 & r_{42} & 0 \\ r_{42} & 0 & 0 \\ 0 & -r_{11} & 0 \end{pmatrix} \tag{15.7}$$

- **Hexagonal, class** 6 – example, lithium iodate

$$\begin{pmatrix} 0 & 0 & r_{13} \\ 0 & 0 & r_{13} \\ 0 & 0 & r_{33} \\ r_{41} & r_{42} & 0 \\ r_{42} & -r_{41} & 0 \\ 0 & 0 & 0 \end{pmatrix} \tag{15.8}$$

- **Orthorhombic, class** $mm2$ – example, potassium niobate, LBO

$$\begin{pmatrix} 0 & 0 & r_{13} \\ 0 & 0 & r_{23} \\ 0 & 0 & r_{33} \\ 0 & r_{42} & 0 \\ r_{51} & 0 & 0 \\ 0 & 0 & 0 \end{pmatrix} \tag{15.9}$$

Table 15.1 lists the electro-optic tensor elements for the more important modulator materials.

**Table 15.1** Electro-optic tensor elements for a few important crystals (from Yariv (1989).)

| Crystal | Point group symmetry | $r_{ij}$ (in $10^{-12}$ m/V) |
|---|---|---|
| ADP | $\bar{4}2m$ | $r_{41} = 28$ |
| | | $r_{63} = 8.5$ |
| KDP | $\bar{4}2m$ | $r_{41} = 8.6$ |
| | | $r_{63} = 10.6$ |
| $LiNbO_3$ | $3m$ | $r_{13} = 8.6$ |
| | | $r_{22} = 3.4$ |
| | | $r_{33} = 30.8$ |
| | | $r_{42} = 28$ |
| $LiTaO_3$ | $3m$ | $r_{13} = 5.7$ |
| | | $r_{33} = 30.3$ |

## 15.3 Bulk electro-optic modulators

The *formal* approach to determine the behavior of an electro-optic crystal when an electric field is applied is to first *diagonalize* the quadratic form appearing in the equation for the index ellipsoid (eqn 15.2). This is simply a transformation to a coordinate system where the $yz$, $xz$ and $xy$ terms are zero. Then one can obtain the change in the relevant refractive index due to the field (using eqn 15.4) and the consequent change in the propagation behavior of a beam traversing the crystal. Due to the sparse nature of the electro-optic tensor, one can often determine the new axes by inspection. A simple example should illustrate the approach.

The material KDP is a tetragonal, class $\bar{4}2m$ crystal with only two independent electro-optic tensor elements. It is also a uniaxial crystal, whose ellipsoid equation in the presence of an electric field is

$$\frac{x^2+y^2}{n_o^2}+\frac{z^2}{n_e^2}+2r_{41}E_xyz+2r_{41}E_yxz+2r_{63}E_zxy=1, \qquad (15.10)$$

where the factors of 2 are due to their presence in the terms $(1/n^2)_i$ in the index ellipsoid equation for $i = 4, 5, 6$. We will first analyze a *longitudinal* electro-optic modulator, where both the field and wave propagation are along the $z$-axis. The equation for the index ellipsoid is

$$\frac{x^2+y^2}{n_o^2}+\frac{z^2}{n_e^2}+2r_{63}E_zxy=1. \qquad (15.11)$$

From the symmetry of this equation about both the $z$-axis and a plane through the $z$-axis that is tilted at 45° to the $x$-axis, it should be evident that the $xy$ term will be eliminated by a 45° rotation about the $z$-axis. The new coordinates (primed) are related to the original ones by

$$\begin{pmatrix} x' \\ y' \end{pmatrix} = \frac{1}{\sqrt{2}}\begin{pmatrix} 1 & 1 \\ -1 & 1 \end{pmatrix}\begin{pmatrix} x \\ y \end{pmatrix}. \qquad (15.12)$$

Performing the substitution yields

$$\left(\frac{1}{n_0^2}+r_{63}E_z\right)x'^2+\left(\frac{1}{n_o^2}-r_{63}E_z\right)y'^2+\frac{z^2}{n_e^2}=1. \qquad (15.13)$$

The electric field breaks the degeneracy in the $xy$-plane, producing two new and *different* indices whose principal axes are at 45° to the original $x$ and $y$ axes. Assuming that the index change is small, and using

$$\Delta n = -\frac{n^3}{2}\Delta\left(\frac{1}{n^2}\right), \qquad (15.14)$$

one obtains

$$n_{x'} = n_o - \frac{n_o^3}{2}r_{63}E_z \qquad (15.15)$$

$$n_{y'} = n_o + \frac{n_o^3}{2}r_{63}E_z, \qquad (15.16)$$

while $n_z$ is unchanged. If we launch a wave along the $z$-axis that is polarized along the original $x$ direction, the waves whose polarizations are projected along the $x'$- and

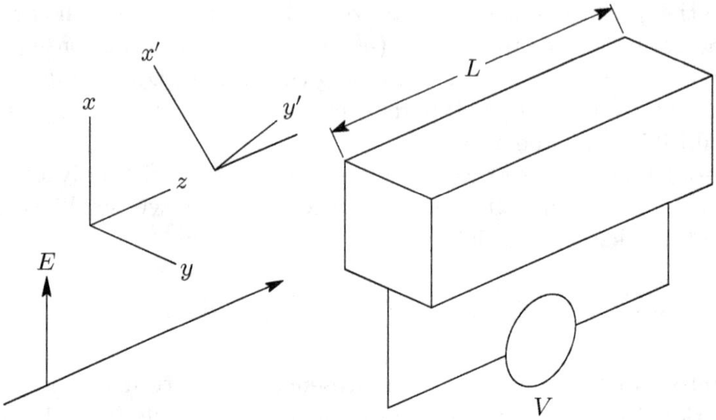


**Fig. 15.1** Geometry of longitudinal electro-optic modulator.

$y'$-axes will propagate at *different* speeds, generating an *elliptically* polarized wave. The two components are

$$E_{x'} = \frac{E_0}{\sqrt{2}} e^{i\omega t - i(\omega/c)n_{x'}z} \tag{15.17}$$

$$E_{y'} = \frac{E_0}{\sqrt{2}} e^{i\omega t - i(\omega/c)n_{y'}z}. \tag{15.18}$$

The *phase difference* between these waves is

$$\Delta\phi = \frac{\omega}{c}(n_{y'} - n_{x'})z = \frac{\omega n_o^3 r_{63} E_z}{c} z, \tag{15.19}$$

where the wave enters the crystal at $z = 0$. If the crystal length is $L$, the longitudinal field is approximately $V/L$ for an applied voltage $V$ and the phase difference is $\pi$ radians when

$$\text{Longitudinal:} \quad V_\pi = \frac{\lambda}{2n_o^3 r_{63}}. \tag{15.20}$$

This quantity is independent of the crystal length and is called the *half-wave voltage* of the crystal for operation as a longitudinal modulator. For KDP at the He–Ne emission wavelength (633 nm), $V_\pi \approx 9$ kV, a very large value. The geometry is illustrated in Fig. 15.1. If a polarizer is placed after the crystal and aligned along the $y$-axis, the transmission of a vertically polarized beam will be zero when $V = 0$ and at its maximum value when $V = V_\pi$, since at the half-wave voltage the modulator behaves like a half-wave plate and rotates the polarization by twice the angle (90°) that the fast axis (the $x'$-axis) makes with the electric field vector. Thus, the modulator/polarizer combination provides *amplitude modulation.* The modulation will not be a linear function of the voltage for large voltage excursions. If the polarization of the laser beam is along the $x'$- or $y'$-axis, the modulator will produce pure phase modulation with a phase deviation of

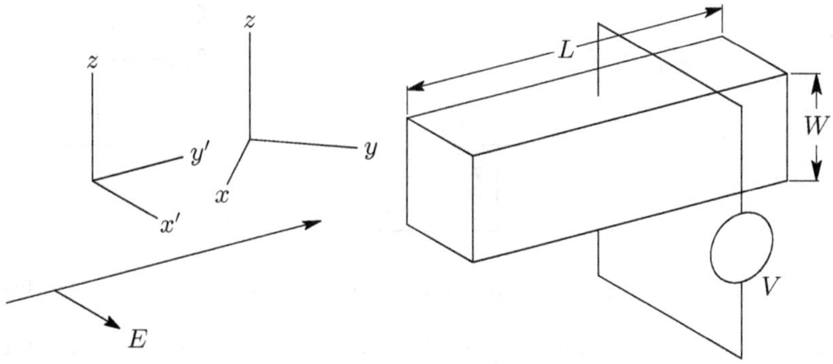


**Fig. 15.2** Geometry of transverse electro-optic modulator set up for pure phase modulation.

$$\text{Longitudinal:} \quad \phi_{deviation} = \frac{\pi n_o^3 r_{63} L}{\lambda} E_z = \frac{\pi n_o^3 r_{63}}{\lambda} V. \qquad (15.21)$$

Longitudinal modulators are rarely used because of the large and impractical half-wave voltage. A significant reduction in $V_\pi$ can be obtained using the *transverse* configuration, where the electric field is applied perpendicular to the wave propagation direction. If this is done, the phase shift can be increased by increasing the *aspect ratio* (length/width) of the modulator until the aperture is too small to allow easy passage of the laser beam. A possible geometry in KDP is for the field to be applied along the $z$-axis and for the beam to propagate along either the $x'$- or $y'$-axis. The crystal will need to be *cut* so that the faces which are normal to the beam propagation direction are at 45° to the original $x$- or $y$-axis. The geometry is shown in Fig. 15.2. When used as a pure phase modulator (polarization perpendicular to applied field), the change in index is

$$\Delta n = \frac{n_o^3}{2} r_{63} E_z, \qquad (15.22)$$

and the phase deviation is

$$\text{Transverse:} \quad \phi_{deviation} = \frac{\pi n_o^3 r_{63}}{\lambda} \left(\frac{L}{W}\right) V, \qquad (15.23)$$

where $W$ is the width of the crystal along the $z$-axis. Unlike for a longitudinal modulator, the phase shift depends upon the aspect ratio. The half-wave voltage is

$$\text{Transverse:} \quad V_\pi = \frac{\lambda}{n_o^3 r_{63}} \frac{W}{L}. \qquad (15.24)$$

The half-wave voltage can be reduced to a practical value by using a sufficiently large aspect ratio.

Bulk modulators (those whose dimensions are much larger than the wavelength of light) typically do not have aspect ratios of greater than 20 and therefore have half-wave voltages of 100 volts or greater. For fixed-frequency applications, a simple resonant step-up transformer is usually sufficient to drive such a modulator. A variant

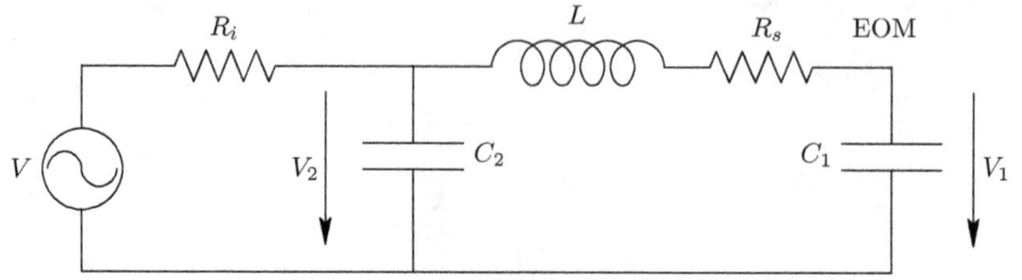


**Fig. 15.3** A simple circuit for driving an electro-optic modulator at a fixed frequency.

of the transformer which might be simpler to construct is a coil with a capacitive divider. It is actually exactly analogous to a step-up transformer, where the turns ratio is equal to the ratio of the capacitor values. Geometrically, it is a capacitively loaded pi-network and is shown in Fig. 15.3. The electro-optic modulator capacitance is $C_1$. The coil is chosen to resonate with $C_1$ at approximately the operating frequency. The input capacitor ($C_2$) is quite a bit larger than $C_1$ and is chosen to impedance match the network to the generator whose internal impedance is $R_i$. Matching will take place when

$$R_i = Q^2 \left(\frac{C_1}{C_2}\right)^2 R_s = R_p \left(\frac{C_1}{C_2}\right)^2, \tag{15.25}$$

where $R_p$ is the parallel resistance of the coil, which is $Q^2$ times the series resistance, $R_s$. The $Q$ of the coil is equal to its reactance divided by its series resistance. When the network is matched, the step-up ratio ($V_1/V_2$) is simply $C_2/C_1$. The criterion for the validity of this approach is

$$Q \gg \frac{C_2}{C_1}. \tag{15.26}$$

This relation also establishes the maximum step-up ratio ($\approx Q$). These relations are easily derived from the rules governing the combination of impedances. However, there is a simple explanation which avoids algebra if we accept the analogy to a step-up transformer with turns ratio $C_2/C_1$. The usual transformation law from series to parallel resistance of a circuit element is $R_p = Q^2 R_s$ (when $Q \gg 1$). We assume that most of the voltage is dropped across $C_1$ (compared to that across $C_2$), so the parallel resistance can be considered to appear across $C_1$. The rule for transforming impedances across a transformer is to multiply by the square of the turns ratio, yielding the above expression for the matching to the generator. If we assume that the circulating current in the series resonant circuit consisting or $L$, $C_1$ and $C_2$ is much greater than the current injected by the generator (this is equivalent to the condition $Q \gg C_2/C_1$), then the two capacitors behave like a voltage divider giving us the step-up ratio of $C_2/C_1$.

This circuit is perfectly adequate for driving an electro-optic modulator at a fixed frequency whose associated wavelength is much greater than the length of the modulator. However, for very high frequencies or a large bandwidth, one must use a *traveling-wave* modulator, described in the next section.

## 15.4 Traveling-wave electro-optic modulators

At very high frequencies, the applied field can change significantly during the transit of the modulator by the laser beam. This *transit-time* effect will reduce the size of the phase deviation and occurs when the length of the modulator approaches the wavelength of the applied radio-frequency field. The solution is a *traveling-wave modulator*: the modulating signal is launched onto a transmission line which is in close contact with the electro-optic material and travels in the same direction as the optical beam. Clearly, if the speed of the radio-frequency wave is the same as the optical wave, there will be no transit time effects. This is closely analogous to *phase matching* in sum-frequency generation.

In analyzing a traveling-wave modulator, we assume that the *optical phase change*, $d\phi$, due to a *radio-frequency* field, $E_{rf}(z)$, over a distance $dz$ is

$$d\phi = \alpha E_{rf}(z)dz, \tag{15.27}$$

where $\alpha$ is a constant. The *total* phase change, $\Delta\phi$, is obtained by integrating

$$\Delta\phi = \alpha \int_0^L E_{rf}(z)dz, \tag{15.28}$$

where $L$ is the crystal length and the entrance to the crystal is at $z = 0$. The radio-frequency traveling wave is a function of both $t$ and $z$ and is given by

$$E_{rf}(z,t) = E_{rf,0}e^{i(\omega_m n_m/c)z - i\omega_m t}, \tag{15.29}$$

where $n_m$ and $\omega_m$ are the refractive index and frequency of the traveling radio-frequency wave. One must eliminate the time dependence in performing the integral. This is done by recognizing that the *optical* wave has progressed a distance $z$ in time $t$ given by

$$z = \frac{c}{n}t \Longrightarrow t = \frac{n}{c}z, \tag{15.30}$$

where we assume that it enters the crystal at $t = 0$ and where $n$ is the index for the *optical* wave. If we substitute the $t$ expression into the field equation, we are assuming that we are traveling at the same speed as an optical wavefront. The result of the substitution is

$$E_{rf}(z) = E_{rf,0}e^{i\omega_m z(n_m/c - n/c)} = E_{rf,0}e^{i\omega_m z\Delta n/c}, \tag{15.31}$$

where $\Delta n = n_m - n$. Integrating, we obtain

$$\begin{aligned}\Delta\phi &= \frac{\alpha E_{rf,0}c}{i\omega_m \Delta n}(e^{i\omega_m \Delta n L/c} - 1) \\ &= \frac{2\alpha E_{rf,0}c}{\omega_m \Delta n}e^{i\omega_m \Delta n L/2c}\sin(\omega_m \Delta n L/2c) \\ &= \alpha E_{rf,0}L e^{i\omega_m \Delta n L/2c}\operatorname{sinc}(\omega_m \Delta n L/2c).\end{aligned} \tag{15.32}$$

Note that $\Delta\phi = \alpha E_{rf,0}L$ when either $\omega_m = 0$ (DC case) or $\Delta n = 0$ (perfectly phase-matched case). One can assume that the maximum *useful* bandwidth is from $\omega_m = 0$ to the frequency, $\omega_c$, where the argument of the sinc is $\pi/2$ ($\operatorname{sinc}(\pi/2) = 0.64$).

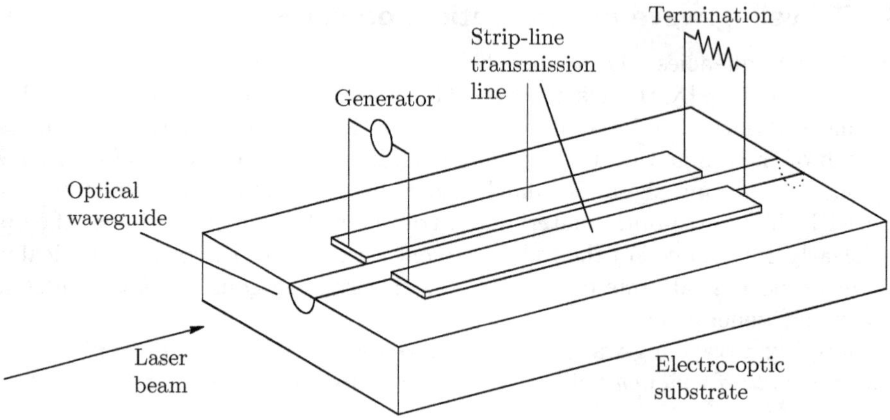


**Fig. 15.4** A traveling-wave electro-optic modulator using an optical waveguide. It combines a bandwidth greater than 1 GHz with a half-wave voltage of about 10 V.

$$\text{Bandwidth:} \quad \omega_c = \frac{\pi c}{\Delta n L}. \tag{15.33}$$

The transit-time limitation without a traveling wave is obtained by setting $n_m = 0$ (infinite RF speed; i.e., no traveling wave). Then the maximum frequency is

$$\text{Transit time limitation:} \quad \omega_c = \frac{\pi c}{n L}. \tag{15.34}$$

Observe that at the latter value of $\omega_c$, $L = \lambda/2$.

A typical traveling-wave phase modulator is shown in Fig. 15.4. It is manufactured with a *strip line* transmission line for the radio-frequency wave and an optical waveguide for the optical field. The substrate is often lithium niobate, which works well at about 1.5 $\mu$m. In addition to the large bandwidth (at least several GHz), the small width of the strip line produces a very large electric field and allows a half-wave voltage as low as 10 V. One application of this device is in sweeping a frequency-locked laser over a wide frequency range. The laser can be locked by making one of the phase-modulation sidebands resonant with one of the reference cavity's modes; changing the radio-frequency will sweep the laser without harming its frequency lock.

## 15.5 Acousto-optic modulators

In the previous section we described phase modulation due to an electric field which causes a first-order change in the indices of refraction of a non-centrosymmetric crystal. A high-frequency sound wave can also modulate the indices of refraction of a medium via the *acousto-optic effect*, and a number of devices, called *acousto-optic modulators*, exploit this phenomenon to modulate the frequency, amplitude or propagation direction of a laser beam.

The amplitude of a sound wave in an elastic medium is described by the *strain tensor*, $S_{ij}$, a second-order tensor defined by

$$S_{ij} = \begin{cases} \dfrac{\partial \xi_i}{\partial x_j} + \dfrac{\partial \xi_j}{\partial x_i} & i \neq j \\ \\ \dfrac{\partial \xi_i}{\partial x_i} & i = j, \end{cases} \tag{15.35}$$

where $\xi_i$ is the displacement of the medium in the direction of $x_i$ due to the sound wave and $x_i$ is a Cartesian coordinate. This is a *symmetric tensor* and therefore has only six independent elements. When describing the acousto-optic effect, it is the usual practice to redefine the strains in analogy with the six coefficients of the general equation for an index ellipsoid:

$$\begin{aligned} S_1 &= S_{11} \\ S_2 &= S_{22} \\ S_3 &= S_{33} \\ S_4 &= S_{23} \\ S_5 &= S_{31} \\ S_6 &= S_{12}. \end{aligned} \tag{15.36}$$

Hopefully, the distinction between the singly subscripted quantities and doubly subscripted quantities will not be confusing. Off-diagonal strains $(S_{ij}, i \neq j)$ describe a *shear wave*, and diagonal strains describe a *compression wave*. The type of wave propagating in the medium will depend upon the characteristics of the *transducer* which launches the wave. We describe the acousto-optic effect using an equation similar to that for the electro-optic effect:

$$\Delta \left( \frac{1}{n^2} \right)_i = \sum_j p_{ij} S_j, \tag{15.37}$$

where the material constants, $p_{ij}$, are called the *strain-optic coefficients* and the $1/n^2$ quantities are defined in the same way as in eqn 15.2. In an isotropic solid, most of these constants vanish and one can readily show that the non-vanishing coefficients satisfy

$$p_{11} = p_{22} = p_{33} \tag{15.38}$$

$$p_{12} = p_{21} = p_{13} = p_{31} = p_{23} = p_{32} \tag{15.39}$$

$$p_{44} = p_{55} = p_{66} = (p_{11} - p_{12})/2. \tag{15.40}$$

All of the remaining coefficients are zero. Thus, there are only two independent coefficients ($p_{11}$ and $p_{12}$) in this case.

The geometry of a simple acousto-optic device is shown in Fig. 15.5. The sound wave, whose frequency, wave vector and wavelength are $\Omega, K$ and $\Lambda$, is launched by a transducer in the positive $x$-direction. An absorber at the top of the modulator ensures that the wave is a *traveling wave*. The laser beam is normally incident from the left along the $z$-direction. We will consider light polarizations along the $y$-axis (perpendicular to the plane of the figure) or along the $x$-axis. The spatial refractive index

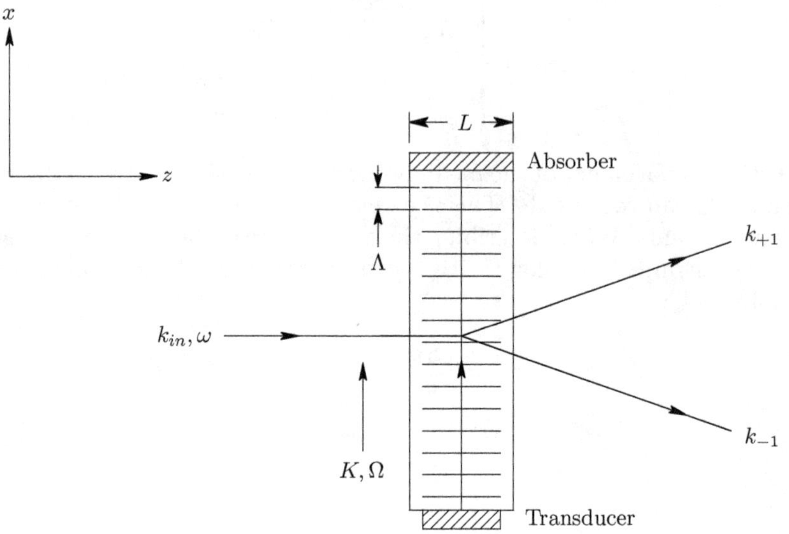


**Fig. 15.5** Geometry of acousto-optic modulator in *Raman–Nath* regime.

modulation described by eqn 15.37 will form a moving, *phase grating* and initially the grating will be assumed to be very thin, which is the requirement for so-called *Raman–Nath* (also called Debye–Sears) diffraction. We will shortly quantify the meaning of a "thin" grating.

We will assume that the sound field is a *compressive* wave traveling in an isotropic medium, so that the only non-zero strain component is $S_1$. From eqn 15.37, the changes to the refractive index for the two possible light polarizations are

$$x\text{-polarization:} \quad \Delta n_1 = -\tfrac{1}{2} n_1^3 p_{11} S_1 \tag{15.41}$$

$$y\text{-polarization:} \quad \Delta n_2 = -\tfrac{1}{2} n_2^3 p_{12} S_1, \tag{15.42}$$

where we have used $\Delta(1/n^2) = -2\Delta n/n^3$. To allow both cases, we will lump the constants into a constant $C$:

$$\Delta n(x, z, t) = CS(x, z, t), \tag{15.43}$$

where $S$ is the sound wave amplitude and we include explicit time and space dependences in $\Delta n$ and $S$. For a sound wave traveling in the $x$-direction, $S$ is

$$S(x, z, t) = S_0 e^{iKx - i\Omega t}. \tag{15.44}$$

The electric field, $E(x, z, t)$, of a light wave propagating in the $z$-direction in the absence of a sound wave is

$$\text{No sound wave:} \quad E(x, z, t) = E_0 e^{ink_0 z - i\omega t}, \tag{15.45}$$

where $n$ is the refractive index of the medium and $k_0$ is the vacuum wave vector ($\omega/c$) of the light. In the presence of a sound wave,

$$\text{With sound wave:} \quad E(x,z,t) = E_0 e^{i(n+\Delta n)k_0 z - i\omega t}. \tag{15.46}$$

The explicit time and spatial dependence of $\Delta n$ is

$$\Delta n = CS_0 e^{iKx - i\Omega t} = \Delta n_0 e^{iKx - i\Omega t}. \tag{15.47}$$

If we let $\phi(x, L, t)$ be the total phase shift experienced by the light as it traverses the crystal,

$$\phi(x, L, t) = nk_0 L + k_0 \int_0^L \Delta n(x, z, t) dz = nk_0 L + k_0 L \Delta n_0 e^{iKx - i\Omega t}. \tag{15.48}$$

By taking the real part of this expression, we see that there is a sinusoidal "corrugation" of the phase shift in the $x$-direction; this structure acts like a *phase grating*. The electric field at the exit plane is

$$E(x, L, t) = E_0 e^{i\phi(x,L,t) - i\omega t} = E_0 e^{ink_0 L - i\omega t} e^{ik_0 L \Delta n_0 \cos(Kx - \Omega t)}, \tag{15.49}$$

where we have replaced the exponential expression for the phase shift by its real part. The second exponential can be expanded in Bessel functions using the well-known relation,

$$e^{i\beta \cos\theta} = \sum_{m=-\infty}^{\infty} i^m J_m(\beta) e^{im\theta}. \tag{15.50}$$

The result is

$$E(x, L, t) = E_0 e^{-i\omega t} \sum_{m=-\infty}^{\infty} i^m J_m(k_0 L \Delta n_0) e^{i(nk_0 L + mKx - m\Omega t)}. \tag{15.51}$$

The result of the modulation will be the generation of a number of beams at different frequencies and angles relative to the input beam. At the exit face of the medium, the $k$-vector of the $m$th beam has an $x$-component given by $mK$ and a $z$-component given by $nk_0$. Converting these to angles relative to the input beam, we have the following association between the beam frequencies and exit angles, $\theta_m$:

$$\omega + m\Omega \Longleftrightarrow \theta_m = \tan^{-1} \frac{mK}{nk_0} \approx \frac{m\lambda}{\Lambda}, \tag{15.52}$$

where $\lambda = 2\pi/nk_0$ and we assume that $\theta_m \ll 1$. The intensities of the various orders are determined by the argument to the Bessel functions; this argument is called the *Raman–Nath parameter*, $\nu$:

$$\text{Raman–Nath parameter:} \quad \nu = k_0 L \Delta n_0. \tag{15.53}$$

The electric field, $E_m$, of the $m$th order at $z = L$ is

$$E_m = E_0 (i)^m J_m(\nu) e^{ink_0 L + imKx - i(\omega + m\Omega)t}. \tag{15.54}$$

Two assumptions were made in the above calculation. First, it was assumed that the width of the sound field ($L$ in the figure) is small enough to ignore optical diffraction.

Quantitatively, a "piece" of the wave of width $\delta x$ will suffer an angular spread of $\lambda/\delta x$ and will have a width of $\delta x + L\lambda/\delta x$ at the exit face. We choose $\delta x$ to minimize the total width at $z = L$ and find that $\delta x = \sqrt{L\lambda}$ and the total width at the exit face is $2\sqrt{L\lambda}$. We can ignore optical diffraction when the spread is less than the wavelength $\Lambda$ of the sound wave:

$$\text{Criterion for Raman–Nath diffraction:} \quad L \ll \frac{\Lambda^2}{4\lambda}. \tag{15.55}$$

It is customary to use the *Klein–Cook parameter*, $Q$, to distinguish Raman–Nath diffraction from the alternative (called *Bragg* diffraction):

$$\text{Klein–Cook parameter:} \quad Q \equiv \frac{2\pi L\lambda}{\Lambda^2} = L\frac{K^2}{k}. \tag{15.56}$$

Using $Q$, Raman–Nath diffraction occurs when

$$\text{Raman–Nath diffraction:} \quad Q \ll 1. \tag{15.57}$$

The second assumption is that $\Delta n_0$ is small enough that there is no *additional bending* of the optical wave due to the transverse refractive index gradient. We will not belabor this point, since our focus is on Bragg diffraction where this phenomenon is largely absent. In the Raman–Nath regime, we can ignore index gradients when

$$\text{Criterion for ignoring ray bending:} \quad \nu \ll 2/Q. \tag{15.58}$$

Since $Q \ll 1$ in the Raman–Nath regime, the Raman–Nath parameter can be somewhat greater than unity and many diffraction orders can be present in the output beam (only two are shown in the figure). As an example of Raman–Nath diffraction, consider $TeO_2$, a common material in acousto-optic modulators. The sound velocity, $v_m$, in this material is 4.2 $\times 10^3$ m/s and the refractive index is about 2.26. The maximum modulation frequency, $f_m$, at 633 nm for $L = 1$ cm is

$$\text{Raman–Nath diffraction in 1 cm TeO}_2\text{:} \quad f_m < v_m\sqrt{\frac{n}{2\pi L\lambda_0}} = 31.7 \text{ MHz}. \tag{15.59}$$

We thus see that Raman–Nath diffraction is a relatively low-frequency phenomenon compared to Bragg diffraction, which occurs at frequencies greater than $\approx 100$ MHz.

Before leaving the topic of Raman–Nath diffraction, we will discuss a *figure of merit* which will help in comparing the diffraction efficiencies of various modulator materials. From eqn 15.42, we see that $\Delta n \propto n^3 pS$. For *weak* diffraction ($\nu < 1$), only first-order diffraction will occur, and $J_1(\nu) \propto \nu$. Thus, the *power*, $P$, in the diffractive wave is proportional to

$$P \propto (\Delta n)^2 \propto n^6 p^2 S^2, \tag{15.60}$$

where the first proportionality is due to $E \propto J_1(\nu) \propto \nu \propto \Delta n$ when $\nu < 1$. From acoustics, the intensity, $I_s$, of the acoustical wave is given by

$$I_s = \tfrac{1}{2}\rho v_m^3 S^2, \tag{15.61}$$

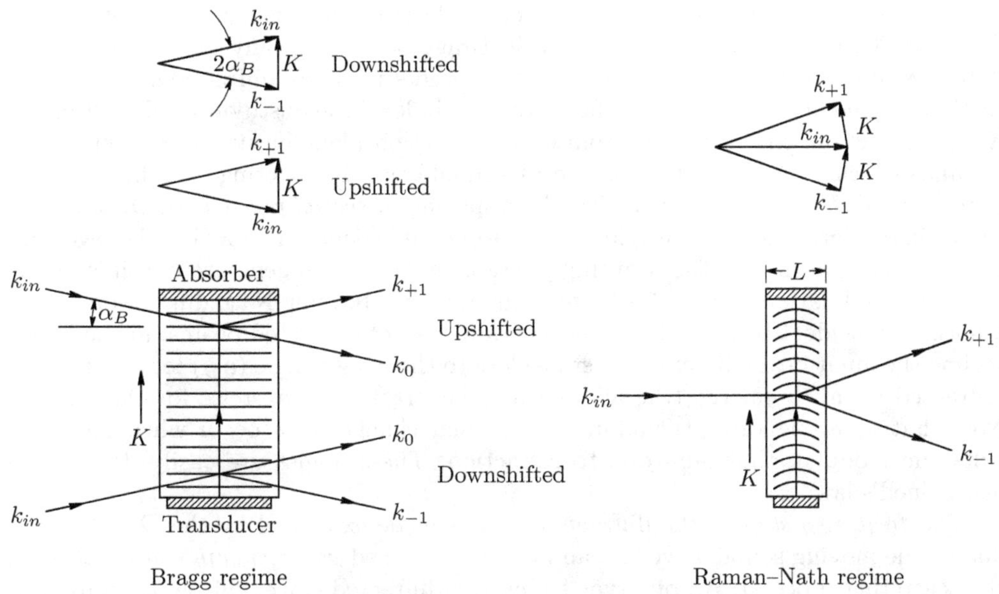


**Fig. 15.6** Geometry of acousto-optic modulator in Bragg regime (left) and Raman–Nath regime (right).

where $\rho$ is the density of the material. The ratio of the diffracted beam power to the acoustical intensity is proportional to the figure of merit, $M_2$:

$$\text{Figure of merit:} \quad M_2 = \frac{n^6 p^2}{\rho v_m^3}. \tag{15.62}$$

The strong dependence on the refractive index favors optically dense materials for acousto-optic devices.

At higher acoustical frequencies, the above approximations break down and we are in the so-called Bragg regime. Before addressing the theory of Bragg diffraction analytically, we will obtain the basic results using the model of a *photon–phonon collision.* This model is illustrated in Fig. 15.6 for both the Raman–Nath and Bragg regimes. We require that momentum be conserved in a collision between the incoming photon and the phonon due to the acoustical wave:

$$\hbar \boldsymbol{k'}_{photon} = \hbar \boldsymbol{k}_{photon} + \hbar \boldsymbol{K}_{phonon}, \tag{15.63}$$

where the prime refers to the photon after the collision. Since the photon momenta ($\hbar k$, $\hbar k'$) are much greater than the phonon momentum ($\hbar K$), the *magnitude* of the photon momentum is essentially unchanged in the collision while the angular deflection is $2\alpha_B$. From this, we obtain the *Bragg condition*

$$\text{Bragg condition:} \quad \sin\alpha_B = \frac{K}{2k} = \frac{\lambda}{2\Lambda}, \tag{15.64}$$

where we have dropped the subscripts and distinguish the photon quantities from the phonon quantities by case. This is the familiar expression for Bragg diffraction of X-rays in crystals where the $\Lambda$ is one of the lattice plane spacings. One difference from the X-ray case is that acousto-optic Bragg scattering only generates a *single* diffracted order while there are many higher orders in X-ray diffraction. This is due to the *sinusoidal* modulation of the refractive index in acousto-optic diffraction; in X-ray diffraction, the scattering from each diffraction plane occurs over a very small distance which, from Fourier theory, results in additional scattering from higher order harmonics of the inverse of the lattice-plane spacing (a *spatial frequency*). Of course, in the Raman–Nath regime, multiple orders are possible, since the medium behaves like a thin phase grating and the resulting *phase modulation* can generate multiple orders when the modulation index ($\nu$) is large enough. Unlike Raman–Nath diffraction, where a normally incident beam will *always* generate a set of diffracted beams, in the Bragg regime the input beam direction must be close to the Bragg angle ($\alpha_B$) for a significant diffracted beam to emerge (the allowed angular error is an exercise for the reader). Note that we are ignoring the additional angular changes that occur when the beam exits the modulator medium due to refraction. These angles are easily determined using Snell's law.

The *frequency shift* of the diffracted beam can be explained by the *Doppler shift* due to the moving sound wave (it can also be explained using *energy conservation* in the scattering process). An observer facing the diffracted wave will see a component of the sound velocity given by $v_m \sin\alpha_B$. The Doppler shift of the light due to this is

$$\text{Doppler shift} = \frac{c}{\lambda}\frac{v_m \sin\alpha_B}{c} = \frac{c}{\lambda}v_m\frac{\lambda}{2\Lambda c} = \frac{\Omega}{2}. \tag{15.65}$$

The incoming wave will see the *negative* of this same Doppler shift, with the result that the net Doppler shift will be $\Omega$ for the upshifted beam (and $-\Omega$ for the downshifted one).

The right side of Fig. 15.6 illustrates Raman–Nath diffraction using the same collision model. It is interesting to note that an alternative (but equivalent) explanation for the criterion for Raman–Nath diffraction can be obtained from this model. Instead of considering diffraction of the *light beam*, we examine the diffraction of the *sound wave*. For small $L$ or low acoustical frequencies, the sound wavefronts will no longer be plane but will have some curvature, as shown in the figure. From the vector diagram, one can see that we will obtain two simultaneous diffraction orders when the angle between $k_{+1}$ and $k_{-1}$ is equal to the angle between the two $K$ vectors which are normal to the sound wavefronts on each side of the curved wavefronts. From simple diffraction theory, the latter is about $\Lambda/L = 2\pi/KL$ and, from geometry, the former is about $2K/k$. Equating these, we obtain

$$\text{Raman–Nath:}\quad \frac{2\pi}{LK} \approx \frac{2K}{k} \Longrightarrow L \approx \frac{\pi k}{K^2}, \tag{15.66}$$

which is essentially the same upper limit on the length as was obtained earlier. Thus, Bragg diffraction will occur when the sound wavefronts are planes, and Raman–Nath diffraction will occur when there is enough wavefront curvature to allow two or more simultaneous diffracted orders. The criterion for Bragg diffraction can be expressed using the Klein–Cook parameter:

$$\text{Bragg regime:} \quad Q \gg 1. \tag{15.67}$$

A theoretical analysis of Bragg diffraction is somewhat more complicated than that of Raman–Nath diffraction. We will use the *coupled-wave* method, closely following the approach in Ghatak (1989). We will make the *small angle approximation*, which assumes that the incoming wave makes an angle of much less than 1 radian with the $z$-axis. We start with the scalar wave equation for each electric field component and substitute into it the sum of the three possible wave solutions (incoming wave and two possible diffracted waves) together with a spatially modulated dielectric permittivity. The latter is given by

$$\epsilon = \epsilon_u + \Delta\epsilon \sin(\Omega t - Kx), \tag{15.68}$$

where $\epsilon_u$ is the permittivity in the absence of the sound wave. In the absence of sources, each electric field component satisfies the *scalar wave equation*

$$\nabla^2 E - \mu_0 \epsilon_u \frac{\partial^2 E}{\partial t^2} \approx \mu_0 \Delta\epsilon \sin(\Omega t - Kx) \frac{\partial^2 E}{\partial t^2}, \tag{15.69}$$

where $E$ is any electric field component. The use of the scalar wave equation obscures polarization issues, but it is simple and will provide the desired result. We can consider the right-hand side to be a *source* term.

Using our knowledge of what to expect from Bragg diffraction, we assume there are only two possible outgoing waves: one at frequency $\omega + \Omega$ and a second at frequency $\omega - \Omega$. The total electric field will therefore be of the form

$$E = E_0 + E_+ + E_-, \tag{15.70}$$

where

$$E_0 = A_0(x, z) e^{i\omega t - i\boldsymbol{k}\cdot\boldsymbol{r}} \tag{15.71}$$
$$E_+ = A_+(x, z) e^{i(\omega+\Omega)t - i\boldsymbol{k}^+\cdot\boldsymbol{r}} \tag{15.72}$$
$$E_- = A_-(x, z) e^{i(\omega-\Omega)t - i\boldsymbol{k}^-\cdot\boldsymbol{r}}, \tag{15.73}$$

where $A_0$ and $A_\pm$ are slowly varying functions of the coordinates. We will leave out some of the details in the following, but will undertake the following operations:

- We will eliminate the $y$-dependence in the differential equation.
- We will substitute $E$ into the differential equation and neglect terms proportional to $\partial^2 A/\partial x^2$ and $\partial^2 A/\partial z^2$, since the amplitudes are slowly varying over the distance of an optical wavelength.
- We will move everything to the left-hand side and equate the coefficients of $e^{i\omega t}$ and $e^{i(\omega\pm\Omega)t}$ to zero, since these exponentials are linearly independent.
- Finally, we will assume that the amplitudes depend only on $z$ and therefore $\partial A/\partial x \approx 0$

It turns out (as expected) that it is not possible to obtain a solution which includes *both* $E_+$ and $E_-$. In the following we will treat the case of a non-zero $E_+$ ($E_- = 0$).

After a bit of algebra, two coupled differential equations emerge:

$$\frac{d\tilde{A}_0}{dz} = \kappa \tilde{A}_+ e^{i\Delta k_z z} \tag{15.74}$$

$$\frac{d\tilde{A}_+}{dz} = -\kappa \tilde{A}_0 e^{-i\Delta k_z z}, \tag{15.75}$$

where

$$\kappa = \frac{\omega^2 \mu_0 \Delta\epsilon}{4\sqrt{k_z k_z^+}} \tag{15.76}$$

$$\Delta k_z = k_z - k_z^+ \tag{15.77}$$

$$\tilde{A}_0 = \sqrt{\frac{k_z}{2\omega\mu_0}} A_0 \tag{15.78}$$

$$\tilde{A}_+ = \sqrt{\frac{k_z^+}{2\omega\mu_0}} A_+. \tag{15.79}$$

The quantity $\kappa$ is the *coupling constant* between the incident and diffracted waves due to the sound wave, and the square modulii of the tilde amplitudes are equal to the *powers* in the incident and diffracted waves.

The two coupled equations are solved by differentiating one and substituting the other to obtain a second-order differential equation. The latter is solved in the usual way subject to the initial conditions:

$$\tilde{A}_0(z = 0) = 1 \tag{15.80}$$

$$\tilde{A}_+(z = 0) = 0, \tag{15.81}$$

where we assume that the incident power is unity. The solutions are

$$P_0(z) = |\tilde{A}_0|^2 = \cos^2 \delta z + \left(\frac{\Delta k_z}{2\delta}\right)^2 \sin^2 \delta z \tag{15.82}$$

$$P_+(z) = |\tilde{A}_+|^2 = \left(\frac{\kappa}{\delta}\right)^2 \sin^2 \delta z, \tag{15.83}$$

where

$$\delta = \sqrt{\kappa^2 + (\Delta k_z)^2/4}. \tag{15.84}$$

When the diffracted wave makes the same angle with the $z$-axis as the incident wave (i.e., satisfies the Bragg condition), $\Delta k_z = 0$ and the powers are

$$P_0(z) = \cos^2 \kappa z \tag{15.85}$$

$$P_+(z) = \sin^2 \kappa z. \tag{15.86}$$

From eqn 15.42 and the fact that $\epsilon = \epsilon_0 n^2$:

$$|\Delta\epsilon| = \epsilon_0 n^4 p S. \tag{15.87}$$

Substituting this into eqn 15.76, we obtain

$$\kappa = \frac{\omega n^3 p S}{4c \cos \alpha_B}, \tag{15.88}$$

where $k_z = k_z^+$ and we are assuming incidence at the Bragg angle, so $k_z = k_z^+ = (\omega/c) n \cos \alpha_B$. Using the expressions for the sound intensity (eqn 15.61) and the figure of merit (eqn 15.62), we obtain

$$\kappa = \frac{\pi}{\sqrt{2}\lambda_0 \cos \alpha_B} \sqrt{M_2 I_s}. \tag{15.89}$$

One will obtain unity diffraction efficiency (theoretically) when $\kappa L = \pi/2$. The corresponding sound intensity is

$$\text{Maximum Bragg diffraction:} \quad I_s = \frac{\lambda_0^2 \cos^2 \alpha_B}{2 M_2 L^2}. \tag{15.90}$$

Finally, the acoustical power is obtained by multiplying this by $LH$ where $H$ is the width of the modulator (its extent in the $y$-direction):

$$\text{Maximum Bragg diffraction:} \quad P_s = \frac{\lambda_0^2 \cos^2 \alpha_B}{2 M_2} \frac{H}{L}. \tag{15.91}$$

The acoustical power requirements are proportional to $\lambda_0^2$; hence, it is more difficult to modulate infrared beams than visible beams. Since $P_s$ is inversely proportional to $M_2$, one would like a medium with a large index of refraction ($M_2 \propto n^6$) and a low sound speed (though this will harm the modulation bandwidth).

Acousto-optic modulators have two *bandwidths*. One is the *modulation bandwidth*, which is the maximum modulation frequency of an amplitude modulated drive signal for which the modulation will be transferred to the diffracted beam. The second bandwidth is the *range of drive frequencies* that will produce a diffracted beam. The first is approximately the inverse of the transit time of the sound wave across the (focused) laser beam. If one focuses to a waist $\omega_0$, the modulation bandwidth is approximately

$$\text{Modulation bandwidth} \approx \frac{v_m}{2\omega_0}. \tag{15.92}$$

Increasing the modulation bandwidth by tightening the focus can have the undesired side effect of reducing the diffraction efficiency, since the angular spread of the input beam will be large enough that a portion of the input beam will not satisfy the Bragg condition. The other bandwidth (range of drive frequencies) is due to the violation of the Bragg condition when the drive frequency is changed. Although the spread of the sound wave is not as extreme as in a Raman–Nath modulator, there is some angular spread, given by $\approx \Lambda/L$. The change of the Bragg angle due to a change in drive frequency of $\Delta f_m$ is $(\lambda/2v_m)\Delta f_m$, where $f_m$ is the average drive frequency. Equating these and solving for $\Delta f_m$

$$\text{Tuning bandwidth:} \quad \Delta f_m \approx \frac{2\Lambda v_m}{\lambda L} = \frac{2\Lambda^2}{\lambda L} f_m. \tag{15.93}$$

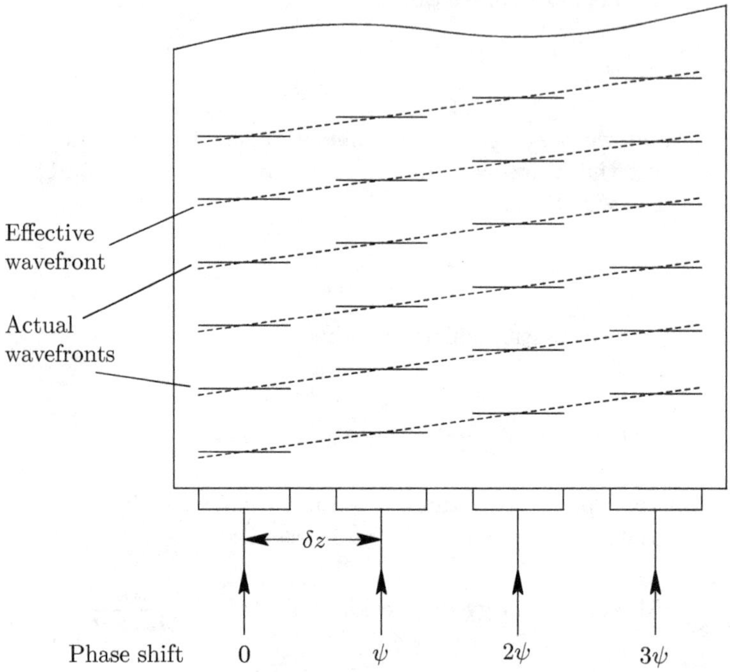


**Fig. 15.7** Phased array of transducers used to increase frequency range of Bragg regime acousto-optic modulator. The dashed line is the *effective* tilted wavefront generated by the several phase-shifted transducers.

One unfortunate consequence of this relationship is that $L$ must be decreased to obtain a large $\Delta f_m$, and this reduces the diffraction efficiency. A solution is to use a *phased-array transducer*, shown schematically in Fig. 15.7. Four transducers are shown in the figure; a larger or smaller number can be used. Phase shifters are provided for each transducer so that adjacent transducers differ in phase by $\psi$ radians. A phase shift of $\psi$ corresponds to a *wavefront* displacement of $\Lambda(\psi/2\pi)$ and an angular *tilt* of the composite wavefront by angle $(\Lambda/\delta z)(\psi/2\pi) = \psi v_m/(2\pi\Omega\delta z)$, where $\delta z$ is the separation between the transducers. This angle is inversely proportional to the sound frequency, $\Omega$. Since the Bragg angle is an arcsine of the frequency, the tracking will not be exact. A tolerable strategy is to match the angles at the end of the range, $\Delta f_m$. The transducer separation needed for this is

$$\text{Tracking at endpoints of range:} \quad \delta z = \frac{n v_m^2}{(f_0^2 - \Delta f_m^2/4)\lambda}, \tag{15.94}$$

where $f_0 = \Omega_0/2\pi$ is the center frequency in Hz.

We will end this section with a discussion of the applications of acousto-optic modulators. The most obvious application is as an *amplitude modulator*. The radio-frequency drive signal is modulated (perhaps using an inexpensive double-balanced mixer) and the power in the diffracted beam will be a monotonic function of the signal

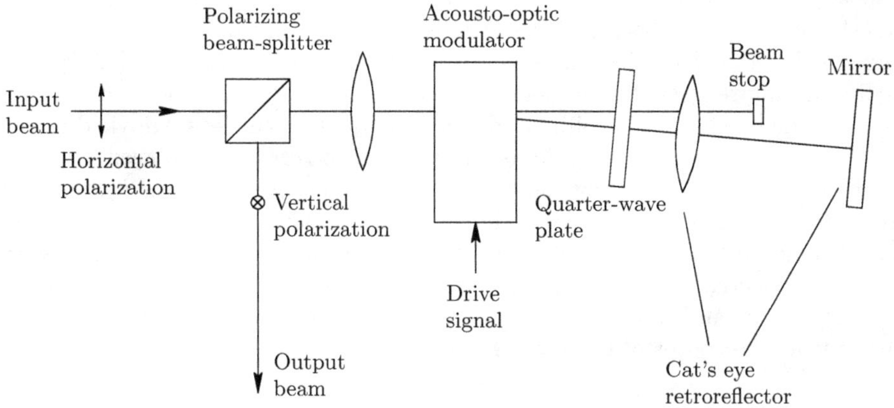


**Fig. 15.8** Double-pass scheme for doubling frequency shift and eliminating angular shift in acousto-optic modulator used as a frequency shifter.

sent to the radio-frequency modulator. Unfortunately, the modulation is not *linear*, since the power in the diffracted beam is proportional to $\cos^2 \kappa L$ and $\kappa$ is proportional to the *amplitude* of the transducer drive. For small signal modulation, one should operate the modulator at the most linear point in its diffraction efficiency function: this occurs when $\kappa L \approx \pi/4$. Acousto-optic modulators have been very successfully used in this way to stabilize the intensity of a laser beam.

Bragg modulators are also used to deflect laser beams by a controlled angle. The input (and diffracted) angle is set to the middle of the acoustical frequency range and the input beam angle is held constant as the sound frequency is swept. The *change* in the diffracted beam angle, $\delta\alpha$, for a change in the sound frequency, $\delta\Omega$, is simply

$$\delta\alpha = \frac{\lambda}{v_m}\delta\Omega. \tag{15.95}$$

The maximum angular sweep range and the modulation bandwidth are determined by the bandwidth considerations discussed previously.

Perhaps the most common application of an acousto-optic modulator in an atomic physics laboratory is as a *frequency shifter*. The diffracted beam will be frequency shifted by exactly $\pm\Omega$, as discussed previously. One of the undesirable side effects of sweeping the frequency of a laser beam in this way is the accompanying *angular change* in the diffracted beam. The usual solution is to use the modulator in the *double-pass configuration* shown in Fig. 15.8. The basic idea is to reflect the diffracted beam back along its path and into the modulator for a second pass. Since the second pass makes exactly the same angle with the acoustical wave as the first pass, the beam will receive the same magnitude and sign of frequency shift as it did in the first pass. The beam, as it emerges the second time, will be collinear with the incident beam for all frequency shifts. Thus, the net frequency shift will be twice the acoustical frequency and the angular shift will be zero. The beam can be separated from the incident beam using the combination of polarizing beam-splitter and quarter-wave plate shown in the figure; its operation is explained in Chapter 4.

The key to the operation of the device is a means of ensuring that the beam be exactly retro-reflected back into the modulator for any diffraction angle. This is accomplished using a modified *cat's-eye retro-reflector* consisting of the lens and mirror shown in the figure. Usually, cat's-eye retro-reflectors position the mirror one focal length from the lens so that a collimated beam incident on the lens will yield a reflected collimated beam at the same angle as the incident beam but displaced from it by a small distance. A better arrangement is to place the lens *one focal length from the center of the modulator.* Since the incident beam is focused into the modulator, the beam will emerge collimated from the second lens and can be sent normal to the mirror. The reflected beam will then enter the modulator at the ideal angle for the second pass. One can replace the mirror/lens with a concave mirror whose center of curvature is at the center of the modulator.

## 15.6 Further reading

The electro-optic effect and associated devices are well covered in the books by Yariv (1989) and Davis (1996). In particular, the latter has a good, though short, discussion of traveling-wave electro-optic modulators. Acousto-optic devices are adequately described in the two above texts. An excellent discussion of the distinction between the Raman–Nath and Bragg regimes together with a treatment of the Bragg regime using a coupled wave analysis appear in the book by Ghatak and Thyagarajan (Ghatak (1989)). A useful book devoted entirely to acousto-optics is by Korpel (1997). There are numerous papers on acousto-optics; two which were useful are Korpel (1981) and Young and Yao (1981). A description of a double-pass acousto-optic modulator appears in the paper by Donley et al. (2005).

## 15.7 Problems

(15.1) Lithium niobate is a negative uniaxial crystal which is often used as an electro-optic phase modulator. Derive an expression for the phase shift in a crystal of length $l$ and width $d$ in terms of the electro-optic coefficients $r_{13}$ and $r_{33}$ and the indices of refraction ($n_e$ and $n_o$) for a wave propagating in the $y$-direction and whose polarization makes a 45° angle with the $z$-axis in the $xz$-plane. The electric field is applied in the $z$-direction, which is also the optical axis. Using this expression, find the half-wave voltage for a He–Ne laser beam ($\lambda$ =633 nm) in a crystal where $l$ = 10 mm and $d$ =0.5 mm (use the Sellmeier equation to determine the refractive indices).

(15.2) For the modulator in problem 1, the light is now polarized along the $z$-axis, which will result in pure phase modulation (unlike in the previous problem). What is the half-wave voltage for this case?

(15.3) Assume that we have a 6 MHz acoustical wave propagating in water at a speed of 1500 m/s. What is the maximum width of the modulator "cell" for Raman–Nath diffraction of a He–Ne laser beam ($\lambda = 633$ nm$, n = 1.33$)? If the cell is 1 cm wide, show that the diffraction is Raman–Nath and calculate the angle between the direct beam (perpendicular to the sound wave) and the first diffracted wave.

(15.4) An acousto-optic modulator is constructed from dense flint glass, which has the following parameters: $n = 1.92$, $p = 0.25$, $v_m = 3.1 \times 10^3$ m/s and $\rho = 6.3 \times 10^3$ kg/m$^3$. Assume that the modulator is 50 mm long and 2 mm wide and the acoustic frequency is 40 MHz. Calculate the figure of merit, $M_2$, and the Bragg diffraction angle for a He–Ne beam (633 nm). What is the acoustical power for maximum diffraction?

(15.5) Determine the allowed *angular* deviation from the Bragg condition of an acousto-optic modulator where the diffraction efficiency is reduced by a factor of 2 from its maximum value.

# References


Agger, S., Povlsen, J. H. and Varming, P., "Single-frequency thulium-doped distributed-feedback fiber laser", Opt. Lett. **29**, 1503 (2004).

Ashcroft, N. W. and Mermin, N. D., *Solid State Physics* (Brooks Cole, 1976).

Ashkin, A., Boyd, G. D. and Dziedzic, J. M., "Resonant Optical Second Harmonic Generation and Mixing", IEEE Journ. Quant. Elect. **2**, 109 (1966).

Born, M. and Wolf, E., *Principles of Optics* (6th Ed, Pergamon Press, 1980).

Botez, D., "Analytical Approximation of the Radiation Confinement Factor for the $TE_0$ Mode of a Double Heterojunction Laser", IEEE Journ. Quant. Elect. **14**, 230 (1978).

Boyd, G. D. and Kleinman, D. A., "Parametric Interaction of Focused Gaussian Light Beams", Journ. Appl. Phys. **39**, 3597 (1968).

Byer, R. L. and Harris, S. E., "Power and Bandwidth of Spontaneous Parametric Emission", Phys. Rev. **168**, 1064 (1968).

Byer, R. L., "Diode-pumped Solid-State Lasers", Science **239**, 742 (1988).

Carmon, T., Yang, L. and Vahala, K. J., "Dynamical thermal behavior and thermal self-stability of microcavities", Optics Express **12**, 4742 (2004).

Condon, E. U. and Shortley, G. H., *The Theory of Atomic Spectra* (Cambridge University Press, 1964).

Davis, C. C., *Lasers and Electro-Optics* (Cambridge University Press, 1996).

Day, T., Gustafson, E. K. and Byer, R. L., "Sub-Hertz Relative Frequency Stabilization of Two-Diode Laser-Pumped Nd:YAG Lasers Locked to a Fabry–Pérot Interferometer", IEEE Journ. Quant. Elect. **28**, 1106 (1992).

Del'Haye, P., Schliesser, A., Arcizet, O., Wilken, T., Holzwarth, R. and Klippenberg, T. J., "Optical frequency comb generation from a monolithic microresonator", Nature **450**, 1214 (2007).

Del'Haye, P., Arcizer, O., Schliesser, A., Holzwarth, R. and Kippenberg, T. J., "Full Stabilization of a Microresonator-Based Optical Frequency Comb", Phys. Rev. Lett. **101**, 053903 (2008).

Del'Haye, Herr, T., Gavartin, E., Holzwarth, R. and Kippenberg, T. J., "Octave-Spanning Tunable Frequency Combs on a Chip", from CLEO 2010 proceedings.

Digonnet, M. F. and Shaw, H., "Analysis of a Tunable Single Mode Optical Fiber Coupler", IEEE Journ. Quant. Elect. **QE-18**, 746 (1982).

Digonnet, M.F., ed., *Rare Earth Doped Fiber Lasers and Amplifiers* (2nd ed., Marcel Dekker, Inc., 2001).

Dong, L., Loh, W. H., Caplen, J. E., Minelly, J. D., Hsu, K. and Reekie, L., "Efficient single-frequency fiber lasers with novel photosensitive Er/Yb optical fibers", Opt. Lett. **22**, 694 (1997).

Donley, E. A., Heavner, F. L., Tataw, M. O. and Jefferts, S. R., "Double-pass acousto-optic modulator system", Rev. Sci. Instrum. **76** 063112 (2005).

Dorf, R. C., *Modern Control Systems*, (Addison-Wesley, 1967).

Drever, R. W. P., Hall, J. L., Kowalski, F. V., Hough, J., Ford, G. M. and Ward, H., "Laser Phase and Frequency Stabilization Using an Optical Resonator", Appl. Phys. B **31**, 97 (1983).

Duarte, F. J. ed., *Tunable Lasers Handbook*, (Academic Press, 1995).

Elliot, D. S., Roy, R. and Smith, S. J., "Extracavity laser band-shape and bandwidth modification", Phys. Rev. A **26**, 12 (1982).

Fejer, M. M., Magel, G. A., Jundt, D. H. and Byer, R. L., "Quasi-Phase-Matched Second Harmonic Generation: Tuning and Tolerances", IEEE Journ. Quant. Elect. **28**, 2631 (1992).

Foot, C. J., *Atomic Physics* (Oxford University Press, 2005).

Fork, R. L., Martinez, O. E. and Gordon, J. P., "Negative dispersion using pairs of prisms", Opt. Lett. **9** 150 (1984).

Freed, C. and Haus, H. A., "Photoelectron Statistics Produced by a Laser Operating Below and Above the Threshold of Oscillation", IEEE Journ. Quant. Elect. **2**, 190 (1966).

Gardner, F. M., *Phaselock Techniques* (3rd ed., Wiley-Interscience, 2004).

Ghatak, A. K. and Thyagarajan, K., *Optical Electronics* (Cambridge University Press, 1989).

Ghatak, A. K. and Thyagarajan, K., *Introduction to Fiber Optics* (Cambridge University Press, 2000).

Gloge, D., "Weakly Guiding Fibers", Appl. Opt. **10**, 2252 (1971).

Gruber, J. B. and Conway, J. G., "Electronic Energy Levels and Quantum States of Tm (IV)", Journ. Chem. Phys. **32**, 1178 (1960).

Guan, W. and Marciante, J. R., "Single-polarization, single-frequency, 2 cm ytterbium-doped fibre laser", Electr. Lett. **43** (2007).

Geng, G., Wang, Q., Lou, T., Jiang, S. and Amzajerdian, F., "Single-frequency narrow-linewidth Tm-doped fiber laser using silicate glass fiber", Opt. Lett. **34**, 3493 (2009).

Hall, J. L., Hollberg, L., Baer, T. and Robinson, H. G., "Optical heterodyne saturation spectroscopy", Appl. Phys. Lett. **39**, 680 (1981).

Hall, J. L, Taubman, M. S. and Ye, J., "Laser Stabilization", in Optical Society of America, *Handbook of Optics, Volume IV* (McGraw Hill, 2000).

Hansch, T. W. and Couillaud, B., "Laser Frequency Stabilization by Polarization Spectroscopy of a Reflecting Reference Cavity", Optics. Comm. **35**, 441 (1980).

Harris, M. L., Cornish, S. L., Tripathi, A. and Hughes, I. G., "Optimization of sub-Doppler DAVLL on the rubidium D2 line", J. Phys. B At. Mol. Phys. **41**, 085401 (2008).

Hecht, E., *Optics* (4th Ed, Addison Wesley, 2002).

Helmcke, J., Lee, S. A. and Hall, J. L., "Dye laser spectrometer for ultrahigh spectral resolution: design and performance", Appl. Opt. **21**, 1686 (1982).

Hemmati, H., Bergquist, J. C. and Itano, W. M., "Generation of continuous-wave 194 nm radiation by sum-frequency mixing in an external ring cavity", Opt. Lett. **8**, 73 (1983).

Henry, C. H., "Theory of the Lindwidth of Semiconductor Lasers", IEEE Journ. Quant. Elect. **18**, 259 (1982).

Henry, C. H. and Kazarinov, R. F., "Instability of Semiconductor Lasers Due to Optical Feedback from Distant reflectors", IEEE Journ. Quant. Elect. **22**, 294 (1986).

Herzberg, G., *Atomic Spectra and Atomic Structure* (Dover Publications, NY, 1944).

Jackson, J. D., *Classical Electrodynamics* (3rd Ed, John Wiley & Sons, 1999).

Jauncey, I. M., Reekie, L., Townsend, J. E. and Payne, D., N., "Single-Longitudingal-Mode Operation on an $Nd^{3+}$-Doped Fibre Laser", Electron. Lett. **24**, 24 (1988).

Jenkins, F. A. and White, H. E., *Fundamentals of Optics* (3rd Ed, McGraw-Hill, 1957).

Joyce, W. B and Dixon, R. W., "Analytic approximation for the Fermi energy of an ideal Fermi gas", Appl. Phys. Lett. **31**, 354 (1977).

Kapon, E. ed., *Semiconductor Lasers I* (Academic Press, 1999).

Ou, Z. Y. and Kimble, H. J., "Enhanced conversion efficiency for harmonic generation with double resonance", Opt. Lett. **18**, 1053 (1993).

Kittel, C., *Introduction to Solid State Physics* (8th Ed, Wiley, 2004).

Kogelnik, H. and Li, T., "Laser Beams and Resonators", Appl. Opt. **5**, 1550 (1966).

Kogelnik, H. and Shank, C. V., "Coupled-Wave Theory of Distributed Feedback Lasers", J. Appl. Phys. **43**, 2327 (1972).

Korpel, A., "Acousto-Optics – A Review of Fundamentals", Proc. of IEEE **69**, 48 (1981).

Korpel, A., *Acousto-Optics* (Marcel Dekker, Inc., 1997).

Labachelerie, M. de and Passedat, G., "Mode-hop suppression of Littrow grating-tuned lasers", Appl. Opt. **32**, 269 (1993).

Lang, R. L., Park, R., Mehuys, D., O'Brien, S., Major, J. and Welch, D., "Numerical Analysis of Flared Semiconductor Laser Amplifiers", IEEE Journ. Quant. Elect. **29** 2044 (1993).

Lathi, B. P., *Signals, Systems and Communication* (John Wiley and Sons, 1965).

Liu, Y., Lin, J., Huang, G., Guo, Y. and Duan, C., "Simple empirical analytic approximation to the Voigt profile", J. Opt. Soc. Am. B **18**, 666 (2001).

Loh, W. H., Samson, B. N., Dong, L., Cowle, G. J. and Hsu, K., "High Performance Single Frequency Fiber Grating-Based Erbium:Ytterbium-Codoped Fiber Lasers", Journ. Lightwave Tech. **16** 114 (1998).

Loudon, R., *The Quantum Theory of Light* (2nd Ed, Clarendon Press, Oxford, 1983).

Ma, K.-S., Jungner, P., Ye, J. and Hall, J. L., "Delivering the same optical frequency at two places: accurate cancellation of phase noise indroduced by an optical fiber or other time-varying path", Opt. Lett. **19**, 1777 (1994).

Maguire, L. P., van Bijnen, R. M. W., Mese, E. and Scholten, R. E., "Theoretical calculation of saturated absorption for multi-level atoms", J. Phys. B **39**, 2709 (2006).

Marcuse, D., "The Coupling of Degenerate Modes in Two Parallel Dielectric Waveguides", B. S. T. J., **50**, 1791 (1971).

Marcuse, D., "Coupled-Mode Theory for Anisotropic Optical Waveguides", B. S. T. J. **54**, 985 (1975).
Marcuse, D., "Loss Analysis of Single-Mode Fiber Splices", B. S. T. J. **56**, 703 (1977).
Mears, R. J., Reekie, L., Poole, S. B. and Payne, D. N., "Neodymium-doped silica single-mode fibre lasers", Electron. Lett. **21**, 738 (1985).
Merzbacher, E., *Quantum Mechanics* (John Wiley and Sons, Inc., 1961).
Milonni, P. W. and Eberly, J. H., *Lasers*, (John Wiley and Sons, 1988).
Mogensen, F., Olesen, H. and Jacobsen, G., "Locking Conditions and Stability Properties for a Semiconductor Laser with External Light Injection", IEEE Journ. Quant. Elect. **21**, 184 (1985).
Mollenauer, L. F. and Stolen, R. H., "The soliton laser", Opt. Lett. **9**, 13 (1984).
Mor, O. and Arie, A., "Performance Analysis of Drever-Hall Laser Frequency Stabilization Using a Proportional + Integral Servo", IEEE Journ. Quant. Elect. **33**, 532 (1997).
Nilsson, A. C., Gustafson, E. K. and Byer, R. L., "Eigenpolarization Theory of Monolithic Nonplanar Ring Oscillators", IEEE Journ. Quant. Elect. **25**, 767 (1989).
Paschotta, R., Spühler, G. J., Sutter, D. H., Matuschek, N. and Keller, U., "Double-chirped semiconductor mirror for dispersion compensation in femtosecond lasers", Appl. Phys. Lett. **75** 2166 (1999).
Pearman, C. P., Adams, C. S., Cox, S. G., Griffin, P. F., Smith, D. A. and Hughes, I. G., "Polarization spectroscopy of a closed atomic transition: applications to laser frequency locking", J. Phys. B. **35**, 5141 (2002).
Petermann, K., *Laser Diode Modulation and Noise* (Springer, 1991).
Petitbon, I., Gallion, P., Debarge, G. and Chabran, C., "Locking Bandwidth and Relaxation Oscillations of on Injection-Locked Semiconductor Laser", IEEE Journ. Quant. Elect. **24**, 148 (1988).
Polzik, E. S. and Kimble, H. J., "Frequency doubling with $KNbO_3$ in an external cavity", Opt. Lett. **16**, 1400 (1991).
Richter, L. E., Mandelberg, H. I., Kruger, M. S. and McGrath, P. A., "Linewidth Determination from Self-Heterodyne Measurements with Subcoherence Delay times", IEEE Journ. Quant. Elect. **22**, 2070 (1986).
Roberts, D. A., "Simplified Characterization of Uniaxial and Biaxial Nonlinear Optical Crystals: A Plea for Standardization of Nomenclature and Conventions", IEEE Journ. Quant. Elect. **28**, 2057 (1992).
Savchenko, A. A., Matsko, A. B., Strekalov, D., Mohageg, M., Ilchenko, V. and Maleki, L., "Low Threshold Optical Oscillations in a Whispering Gallery Mode $CaF_2$ Resonator", Phys. Rev. Lett. **93**, 243905 (2004).
Schülzgen, A., Li, L., Nguyen, D., Spiegelberg, Ch., Rogojan, R. M., Laronche, A., Albert, J. and Peyghambarian, N., "Distributed feedback fiber laser pumped by multimode laser diodes", Opt. Lett. **33**, 614 (2008).
Shiv, L., Sorensen, J. L., Polzik, E. S. and Mizell, G., "Inhibited light-induced absorption in $KNbO_3$", Opt. Lett. **20**, 2270 (1995).
Siegman, A. E., *Lasers* (University Science Books, 1986).
Slater, J. C., *Quantum Theory of Atomic Structure, Vol. I* (McGraw-Hill Book Co., NY, 1960).

Snitzer, E., "Cylindrical Dielectric Waveguide Modes", Journ. Opt. Soc. Am. **51**, 491 (1961).

Snitzer, E., "Optical Maser Action of Nd+3 in a Barium Crown Glass", Phys. Rev. Lett. **7**, 444 (1961).

Snyder, A. W., "Coupled-Mode Theory for Optical Fibers", Journ. Opt. Soc. Am. **62**, 1267 (1972).

Snyder, A. W. and Love, J. D., *Optical Waveguide Theory*, (Chapman and Hall, 1983).

Somekh, S and Yariv, A., "Phase Matching by Periodic Modulation of the Nonlinear Optical Properties", Opt. Comm. **6**, 301 (1972).

Spiegelberg, Ch., Geng, J., Hu, Y., Kaneda, Y., Jiang, S. and Peyghambarian, N., "Low-Noise Narrow Linewidth Fiber Laser at 1550 nm", Journ. Lightwave Tech. **22**, 57 (2004).

Suhara, T., *Semiconductor Laser Fundamentals* (Marcel Dekker, Inc., 2004).

Svelto, O., *Principles of Lasers* (4th ed, Springer, 2004).

Tamura, K., Haus, H. A. and Ippen, E. P., "Self-starting Additive Pulse Mode-locked Erbium Fibre Ring Laser", Elect. Lett. **28**, 2226 (1992).

Le Targat, R., Zondy, J. J. and Lemonde, P., "75%-Efficiency blue generation from an intracavity PPKTP frequency doubler", Opt. Comm. **247**, 471 (2005).

Tkach, R. W. and Chraplyvy, "Regimes of Feedback Effects in 1.5-$\mu$m Distributed Feedback Lasers", IEEE Journ. Lightwave Tech. **LT-4**, 1675 (1986).

Torgerson, J. and Nagourney, W., "Tunable cavity coupling scheme using a wedged plate", Optics Comm. **161**, 264 (1999).

Ulrich, R., Rashleigh, S. C. and Eickhoff, W., "Bending-induced birefringence in single-mode fibers", Opt. Lett. **5**, 273 (1980).

Verdeyen, J. T., *Laser Electronics* (3rd Ed, Prentice Hall, 1995).

Woodgate, G. K., *Elementary Atomic Structure* (Oxford University Press, 2000).

Yariv, A., *Quantum Electronics* (3rd ed., John Wiley and Sons, 1989).

Ye, J. and Cundiff, S. T (ed.), *Femtosecond Optical Frequency Comb Technology* (Springer, 2005).

Young, E. H. Jr. and Yao, S.-K., "Design Considerations for Acousto-Optic Devices", Proc. of IEEE **69**, 54 (1981).

Yue, C., Pend, J. D., Liao, Y. B. and Zhou, B. K., "Fiber ring resonator with finesse of 1260", Electron. Lett. **24**, 622 (1988).

Zeilinger, A., "Experiment on the foundations of quantum physics", Rev. Mod. Phys. **71**, S288 (1999).

Zhu, X., Shi, W., Zong, J., Nguyen, D., Norwood, R. A., Chavez-Pirson, A. and Peyghambarian, N., "976 nm single-frequency distributed Bragg reflector fiber laser", Opt. Lett. **37**, 4167 (2012).

Zimmermann, M., Gohle, C., Holzwarth, R., Udem, T. and Hänsch, T. W., "Optical clockwork with an offset-free difference-frequency comb: accuracy of sum- and difference-frequency generation", Opt. Lett. **29**, 310 (2004).

# Index

The manufacturer's authorised representative in the EU for product safety is Oxford University Press España S.A. of el Parque Empresarial San Fernando de Henares, Avenida de Castilla, 2 – 28830 Madrid (www.oup.es/en or product.safety@oup.com). OUP España S.A. also acts as importer into Spain of products made by the manufacturer.

www.ingramcontent.com/pod-product-compliance
Lightning Source LLC
Chambersburg PA
CBHW060101300126
39022CB00005B/78

* 9 7 8 0 1 9 9 6 6 5 4 8 8 *